# Essentials of Fluid Dynamics

# Essentials of Fluid Dynamics

Edited by **Haley Adison**

Clanrye International
New Jersey

Published by Clanrye International,
55 Van Reypen Street,
Jersey City, NJ 07306, USA
www.clanryeinternational.com

**Essentials of Fluid Dynamics**
Edited by Haley Adison



International Standard Book Number: 978-1-63240-568-5 (Hardback)



Printed in China.

# Contents

# Preface

Fluid dynamics is the branch of fluid mechanics that studies the flow of fluids. It is further divided into hydrodynamics which studies the liquids in motion and aerodynamics which studies the gases in motion. It is an upcoming field of science that has undergone rapid development over the past few decades. This discipline is applicable to a wide range of areas such as modeling fission weapon detonation, predicting weather patterns, calculating moments and forces on aircraft, traffic engineering, etc. This book elucidates new techniques and their applications in a multidisciplinary approach. Some of the diverse topics covered in this text address the varied branches that fall under this category. Through this book, we attempt to further enlighten the readers about the new concepts in this field. It will serve as a reference to a broad spectrum of readers including engineers and students.

All of the data presented henceforth, was collaborated in the wake of recent advancements in the field. The aim of this book is to present the diversified developments from across the globe in a comprehensible manner. The opinions expressed in each chapter belong solely to the contributing authors. Their interpretations of the topics are the integral part of this book, which I have carefully compiled for a better understanding of the readers.

At the end, I would like to thank all those who dedicated their time and efforts for the successful completion of this book. I also wish to convey my gratitude towards my friends and family who supported me at every step.

**Editor**

# The Influence of Boundaries on the Stability of Compositional Plumes

**Khaled S. Al-Mashrafi, Ibrahim A. Eltayeb**

Department of Mathematics and Statistics, College of Science, Sultan Qaboos University, Muscat, Oman
Email: P001175@student.squ.edu.om

## Abstract

**The influence of boundaries on the dynamics of a compositional plume is studied using a simple model in which a column of buoyant fluid rises in a less buoyant fluid bounded by two vertical walls with a finite distance apart. The problem is governed by four dimensionless parameters: The Grashoff number, *R*, which is a measure of the difference in concentration of light material of the plume to its surrounding fluid, the Prandtl number, $\sigma$, which is the ratio of viscosity, $\nu$, to thermal diffusivity, $\kappa$, the thickness of the plume, $2x_0$, and the distance, *d*, between the two vertical walls relative to the salt-finger length scale. The influence of the boundary on the fluxes of material, heat, and buoyancy is examined to find that the buoyancy flux possesses a local maximum for moderate to small thicknesses of the plume when they lie close to the wall. This has the effect of introducing a region of instability for thin plumes near the wall with an asymptotically larger growth rate. In addition, the presence of the boundary suppresses the three-dimensional instabilities present in the unbounded domain and allows only two-dimensional instabilities for moderate to small distances between the bounding walls.**

## Keywords

**Compositional Plumes; Flux; Stability; Growth Rate; Bounded Domain**

## 1. Introduction

Studies on the dynamics of fluid alloys are relevant to industrial (e.g., Rees and Worster [1] and references therein), environmental (e.g., Wells *et al.* [2] and references therein) and geophysical (e.g., Loper [3], Moffatt [4], Al-Lawatia*et al.* [5]), applications. Consequently, there has been considerable interest in studying the various aspects of the dynamics of fluid alloys.

In industrial applications, one of the problems the iron casting industry faces is the appearance of freckles in iron bars causing their weakness. When iron ore is poured into molds or designs, air trapped at the bottom of the design rises in the form of thin filaments into the liquid iron. When the iron solidifies, these filaments form trapped air pockets that appear as very thin black strips along the outer surface of the iron bar and lead to a weakness in the iron bar. The experimental work of Copley *et al.* [6] showed the appearance of plumes rising as thin filaments from a mushy layer. This work was extended by a number of authors (see, e.g. Huppert [7], Chen and Chen [8], Tait and Jaupart [9], Jellinek *et al.* [10], Classen *et al.* [11], Aussillous *et al.* [12], Pol *et al.* [13]). It has been observed that the behaviour of plumes emanating from mushy layers depends on whether they are near the walls of the container or not (see, e.g., Hellawell *et al.* [14]).

Theoretical studies of a compositional plume rising in a fluid of infinite extent have shown that the plume is unstable (see, e.g., Eltayeb and Loper [15]) even for small Grashoff numbers. Moreover, this is found to be true even if the plume is subject to rotation or in the presence of a magnetic field even if another plume is also present [5] [16]-[18].

The main purpose of this study is to examine the influence of boundaries on the dynamics of compositional plumes. For this purpose, we introduce boundaries to the model discussed by Eltayeb and Loper [16]. This is a simple model that neglects material diffusion, which is known to be small, compared to viscous and thermal diffusion [15]. A finite column of compositionally buoyant fluid contained between two vertical interfaces, referred to as a Cartesian plume, is rising in a fluid bounded by two parallel walls enclosing the plume (see **Figure 1**). The neglect of material diffusion allows us to adopt a function of concentration of light material that is simple with the consequence that an analytical solution is obtained. This allows us to examine the dynamics of the plume in a bounded region in the whole parameter space and thus get some insight into the influence of the boundaries on the dynamics of a plume.

In Section 2, we formulate the problem, which involves four dimensionless parameters: the Grashoff number, $R$, which measures the ratio of the buoyancy force to the viscous force, the Prandtl number, $\sigma$, which measures the ratio of viscosity force to the thermal diffusivity, defined by

$$R = \frac{UL}{\nu}, \quad \sigma = \frac{\nu}{\kappa}, \tag{1}$$

where $U$ and $L$ are characteristic velocity and length-scale, respectively (see Equations (9) and (10) below), and $\nu$ and $\kappa$ are kinematic viscosity and thermal diffusivity, respectively, and the thickness of the plume, $2x_0$, and the distance between the two bounding walls, $d$, made dimensionless using the length scale $L$.

In Section 3, we use a top-hat profile of the concentration of light material to obtain a solution representing a plume of thickness $2x_0$, rising between two rigid sidewalls a distance, $d$, apart. We discuss the influence of the presence of the sidewalls on the basic state flow and temperature as well as the associated fluxes of material, heat, and buoyancy. It is found that the presence of the boundaries increases the amplitude of the basic state vertical velocity of the plume flow. The fluxes of material, heat and buoyancy are presented as contours in the $(x_0, a_2)$ plane, where $a_2$ is a measure of the distance between plume and the nearest sidewall.

In Section 4, we examine the stability of the plume. This poses an eigenvalue problem for the growth rate $\Omega$ of the perturbations. In the absence of the walls, the plume is always unstable for small values of the Grashoff number, $R$, and the instability takes the form of one of two uncoupled modes, depending on the values of the dimensionless parameters: a varicose (V) mode, in which the two interfaces of the plume are out-of-phase or a sinuous (S) mode, in which case the two interfaces are in-phase. When the walls are introduced, the same two modes persist but are modified by the presence of the walls. We refer to them below as the modified varicose (MV) and modified sinuous (MS) modes. The stability results are discussed in Section 5. In particular, it will be shown that the influence of the sidewalls is quite complicated. While it tends to stabilise the plume if it is equidistant from the two sidewalls, it can destabilise the plume if it is nearer to one sidewall than to the other. Some concluding remarks are made in Section 6.

## 2. Formulation of the Problem

We consider a two-component incompressible fluid in which the concentration of the solvent component (light material) is $C$ and the temperature is $T$. The two fluids have the same kinematic viscosity, $\nu$, and thermal diffusivity, $\kappa$. The system is governed by the equations of motion, mass, heat, concentration of the light material, and state. These equations are

$$\rho_r \left[ \frac{\partial \boldsymbol{u}}{\partial t} + (\boldsymbol{u} \cdot \nabla) \boldsymbol{u} \right] = -\nabla p + \rho_r \nu \nabla^2 \boldsymbol{u} - \rho g \hat{z}, \tag{2}$$

$$\nabla \cdot \boldsymbol{u} = 0, \tag{3}$$

$$\frac{\partial T}{\partial t} + \boldsymbol{u} \cdot \nabla T = \kappa \nabla^2 T, \tag{4}$$

$$\frac{\partial C}{\partial t} + \boldsymbol{u} \cdot \nabla C = 0, \tag{5}$$

$$\frac{\rho}{\rho_r} = 1 - \alpha (T - T_r) - \beta (C - C_r), \tag{6}$$

where $\boldsymbol{u}$ is the velocity vector, $p$ the pressure, $g$ the uniform acceleration of gravity, $\hat{z}$ is the upward unit vector, $t$ the time, $\alpha$ the coefficient of thermal expansion, $\beta$ the coefficient of compositional expansion, $\rho$ the density, $(\rho_r, T_r, C_r)$ reference values, and we have assumed that the fluid is Boussinesq. The Equations (2)-(6) allow a hydrostatic balance governed by

$$\frac{\mathrm{d}p_h}{\mathrm{d}z} + \rho g = 0, \frac{\mathrm{d}^2 T_h}{\mathrm{d}z^2} = 0, \boldsymbol{u}_h = 0, C_h = C_r. \tag{7}$$

Motivated by the experimental work on plumes rising from mushy layers, we take a temperature profile

$$T_h = \gamma z + T_r, \tag{8}$$

where $\gamma$ is a positive constant and $z$ is the vertical coordinate measured vertically upwards, so that the temperature increases with height making the fluid stably stratified thermally and any instabilities will be due to transport of material.

We now cast the Equations (2)-(6) into dimensionless form. It is found that in order to maintain the effects of temperature variations and compositional variations, we use the salt-finger length scale defined by

$$L = \left( \frac{\nu \kappa}{\alpha \gamma g} \right)^{\frac{1}{4}}, \tag{9}$$

and a velocity unit with the definition

$$U = \beta \tilde{C} \left( \frac{g \kappa}{\alpha \nu \gamma} \right)^{\frac{1}{2}}, \tag{10}$$

so that the ensuing motions are driven by the plume flow transporting the light material, $C$, upwards. Here $\tilde{C}$ is the maximum amplitude of the concentration of light material. We further choose $\beta \tilde{C} / \alpha$, $L/U$ and $\rho_r \beta \tilde{C} \left( \nu g^3 \kappa / \alpha \gamma \right)^{1/4}$ as units of temperature, time and pressure, respectively, and express the equations in dimensionless form as

$$R \left[ \frac{\partial \boldsymbol{u}}{\partial t} + (\boldsymbol{u} \cdot \nabla) \boldsymbol{u} \right] = -\nabla \left( p + \frac{z}{\beta \tilde{C}} \right) + \nabla^2 \boldsymbol{u} + (T - T_r + C - C_r) \hat{z}, \tag{11}$$

$$\nabla \cdot \boldsymbol{u} = 0, \tag{12}$$

$$R\sigma \left[ \frac{\partial T}{\partial t} + \boldsymbol{u} \cdot \nabla T \right] = \nabla^2 T, \tag{13}$$

$$\frac{\partial C}{\partial t} + \boldsymbol{u} \cdot \nabla C = 0. \tag{14}$$

Here the dimensionless parameters $R$ and $\sigma$ are the Grashoff and Prandtl numbers defined in Equation (1) above.

We define a Cartesian coordinate system $O(x, y, z)$ in which $Oz$ is vertically upwards and $Ox$, $Oy$ are horizontal with the $x$-axis normal to the bounding walls (see **Figure 1**). A column of fluid of finite thickness, $2x_0$, rising vertically upwards in a fluid of different concentration and bounded on either side by vertical walls,

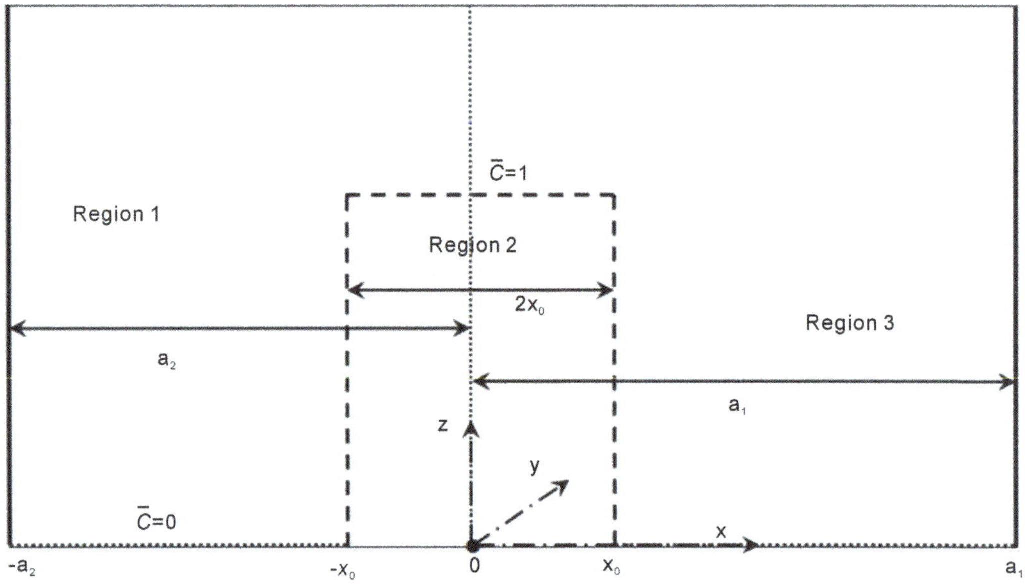


**Figure 1.** The geometry of the problem showing the profile of the basic state concentration of light material representing a plume of width, $2x_0$, and concentration, 1, rising vertically in a fluid of width, $d$, and concentration, 0. Two vertical planes bound the plume on either side such that the center of the plume is a distance $a_1$ from the wall on the right and $a_2$ from the wall on the left.

a distance $d$ apart. We choose the origin such that the plume interfaces are situated at $x = \pm x_0$ and the walls at $x = -a_2$ and $x = a_1$. The region is unbounded in the $y$ and $z$ directions. In comparison with the plumes observed in experiments on mushy layers (see, e.g., Huppert [7]), our model is different in that it is unbounded in the $y$ and $z$ directions. We feel that both assumptions can be adopted for the following reasons: First, the studies in [16] [17] showed good agreement between the stability results of the circular cylindrical plume and the Cartesian plume. Secondly, experimental work on mushy layers and the formation of plumes shows that fully developed plumes rise to heights 200 times their thickness [14], and we can approximate the situation for a fully developed plume by considering it infinite in the vertical direction.

We can now take the flow variables to have the form

$$\boldsymbol{u}(x,y,z,t) = \mathbf{0} + \bar{w}(x)\hat{z} + \varepsilon \boldsymbol{u}^{\dagger}(x,y,z,t), \tag{15}$$

$$C(x,y,z,t) = C_r + \bar{C}(x) + \varepsilon C^{\dagger}(x,y,z,t), \tag{16}$$

$$p(x,y,z,t) = p_h + \bar{p}(x) + \varepsilon p^{\dagger}(x,y,z,t), \tag{17}$$

$$T(x,y,z,t) = T_h + \bar{T}(x) + \varepsilon T^{\dagger}(x,y,z,t), \tag{18}$$

such that the variables with subscript $h$ represent hydrostatic balance and given (in dimensionless form) by

$$T_h = T_r + \frac{(z - z_r)}{\sigma R}, \tag{19}$$

$$p_h = p_r - \frac{(z - z_r)}{\beta \tilde{C}} + \frac{(z - z_r)^2}{2\sigma R}. \tag{20}$$

The variables with an "overbar" are basic state variables dependent only on the horizontal coordinate $x$, because the horizontal variations of the vertical plume flow caused by the difference in composition between the plume and the surrounding fluid imposes a horizontal variation of temperature.The variables with a "dagger" indicate a perturbation of small amplitude $\varepsilon(\ll 1)$.

Substituting the expressions (15)-(18) into the system (11) - (14), the terms independent of $\varepsilon$ give the basic state equations, which depend on $x$ only

$$-\frac{\mathrm{d}\overline{p}}{\mathrm{d}x}\hat{\boldsymbol{x}}+\left(\frac{\mathrm{d}^2\overline{w}}{\mathrm{d}x^2}+\overline{C}+\overline{T}\right)\hat{z}=0, \tag{21}$$

$$\frac{\mathrm{d}^2\overline{T}}{\mathrm{d}x^2}=\overline{w}(x). \tag{22}$$

These equations are discussed in section 3 below.

The order $\varepsilon$ terms in the equations provide the linearised perturbation equations as follows

$$R\left[\frac{\partial \boldsymbol{u}^\dagger}{\partial t}+\overline{w}\hat{z}\cdot\nabla\boldsymbol{u}^\dagger+\left(\boldsymbol{u}^\dagger\cdot\nabla\overline{w}\right)\hat{z}\right]=-\nabla p^\dagger+\nabla^2\ \boldsymbol{u}^\dagger\ +\left(T^\dagger+C^\dagger\right)\hat{z} \tag{23}$$

$$\nabla\cdot\boldsymbol{u}^\dagger=0, \tag{24}$$

$$\sigma R\left[\frac{\partial T^\dagger}{\partial t}+\overline{w}\frac{\partial T^\dagger}{\partial z}+\boldsymbol{u}^\dagger.\nabla\overline{T}\right]+\boldsymbol{u}^\dagger\cdot\hat{z}=\nabla^2T^?, \tag{25}$$

$$\frac{\partial C^\dagger}{\partial t}+\overline{w}\frac{\partial C^\dagger}{\partial z}+\boldsymbol{u}^\dagger\cdot\nabla\overline{C}=0. \tag{26}$$

The perturbation equations are solved in Section 4 below.

## 3. The Basic State

Equation (14) is automatically satisfied for the basic state and we are free to choose a concentration function $\overline{C}(x)$. Since we are extending the study by Eltayeb and Loper [16], we will adopt their choice of $\overline{C}(x)$. Thus

$$\overline{C}(x)=\begin{cases}1, & |x|\le x_0\\ 0, & -a_2\le x<-x_0, x_0<x\le a_1\end{cases}. \tag{27}$$

Consider the basic state Equations (21) and (22). Define

$$F(x)=\overline{T}(x)-\mathrm{i}\overline{w}(x). \tag{28}$$

Then

$$\overline{p}=0, \frac{\mathrm{d}^2F}{\mathrm{d}x^2}-\mathrm{i}F=\mathrm{i}\overline{C}. \tag{29}$$

The Equation (29) is subject to the boundary conditions

$$\left.\begin{array}{l}\text{(i) } F,\dfrac{\mathrm{d}F}{\mathrm{d}x}\quad \text{continuous across } x=\pm x_0\\ \text{(ii) } F=0 \text{ at } x=a_1,-a_2\end{array}\right\}. \tag{30}$$

The solution is

$$F(x)=\begin{cases}-A\sinh(ka_1)\sinh\left[k(x+a_2)\right]; & -a_2\le x<-x_0\\ -A\sinh(ka_1)\sinh\left[k(x+a_2)\right]+\cosh\left[k(x+x_0)\right]-1; & -x_0\le x\le x_0\\ A\sinh(ka_2)\sinh\left[k(x-a_1)\right]; & x_0<x\le a_1\end{cases}, \tag{31}$$

where $A$ and $k$ are defined by

$$A=\frac{2\sinh(kx_0)}{\sinh(kd)},\ k=\frac{1}{\sqrt{2}}(1+\mathrm{i}),\ d=a_1+a_2. \tag{32}$$

A sample of the profiles of the solutions $\overline{T}(x)$ and $\overline{w}(x)$ is plotted for different values of the plume thickness $2x_0$, the distance $a_2$ and $d=10$ in **Figure 2**. The profiles are symmetric when the plume is situated half-way between the sidewalls. The oscillatory nature of the velocity profile introduces negative flow

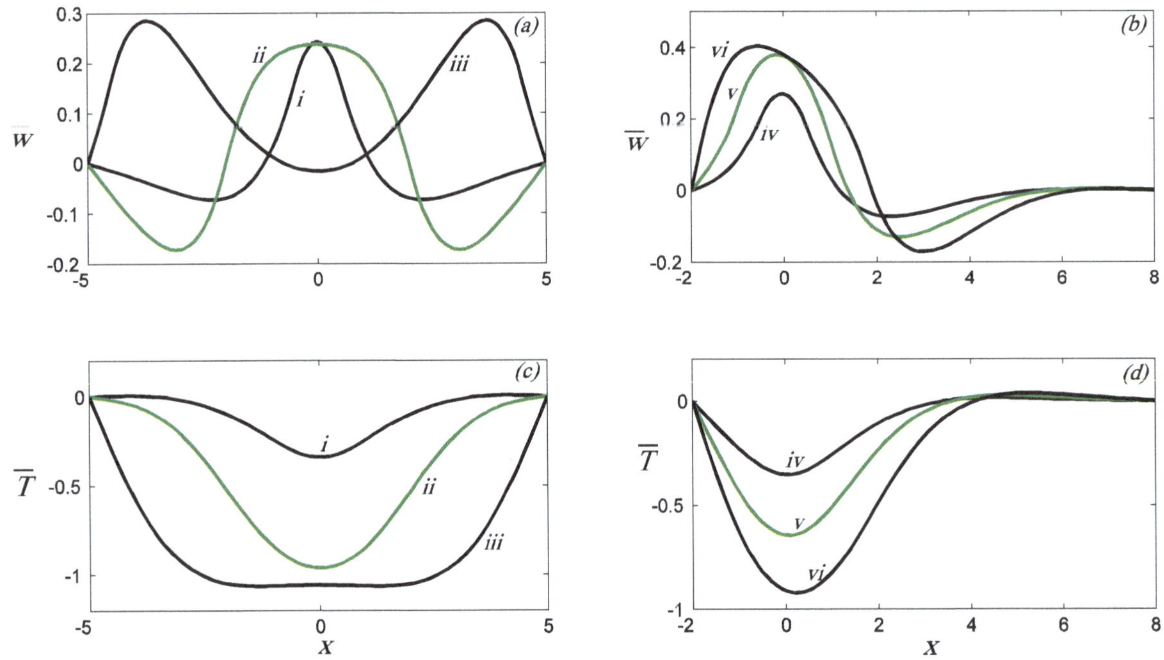


**Figure 2.** The profiles of the basic state velocity, $\overline{w}(x)$, and temperature, $\overline{T}(x)$, for different values of plume thickness, $2x_0$, and distance, $a_2$, from the wall on the left when $d = 10$. (a) and (c) refer to $\overline{w}$ and $\overline{T}$, respectively, when the plume is positioned half-way between the two sidewalls and the labels i, ii, iii correspond to $x_0 = 0.5, 2, 4.5$, respectively. (b) and (d) refer to $\overline{w}$ and $\overline{T}$ when $a_2 = 2$ and the labels iv, v, vi correspond to $x_0 = 0.5, 1, 1.8$, respectively. Note that when the plume is wide, the flow is oscillatory within the plume and it slows down in the middle of the plume, while the flow of the plume is enhanced in the center of the plume when the plume approaches the wall.

(*i.e.*, downwards flow) within the plume when it is wide, and this has an effect on the net transport of material by the plume. The wide plume is also associated with a temperature profile that is almost uniform in the main body of the plume. If the position of the plume moves towards a sidewall, symmetry is broken. Here the downward flow outside the plume is partially suppressed in the narrow region between the plume and the nearest wall and strengthened on the far side. Such behavior will lead to the modification of the modes of instability in the absence of the sidewalls.

The basic state solution is associated with fluxes of heat, $F_H$, material, $F_m$, and buoyancy, $F_B$, which are non-dimensionalised using the units $\beta^2\tilde{C}^2\left(g\kappa^3/\nu\alpha^7\gamma^3\right)^{1/4}$, $\beta\tilde{C}^2\left(g\kappa^3/\nu\alpha^3\gamma^3\right)^{1/4}$ and $\beta^2\tilde{C}^2\left(g\alpha\kappa^3/\nu\gamma^3\right)^{1/4}$, respectively. They are given by

$$F_H = \frac{1}{2}\int_{-a_2}^{a_1} \overline{w}(x)\overline{T}(x)\mathrm{d}x,\ F_m = \frac{1}{2}\int_{-a_2}^{a_1} \overline{w}(x)\overline{C}(x)\mathrm{d}x,\ F_B = F_H + F_m. \tag{33}$$

(cf. [15]). The integration is straightforward and leads to

$$F_m = \frac{1}{2}\mathrm{Im}\left\{\frac{1}{k}\left[2A\sinh(kx_0)\sinh(ka_1)\sinh(ka_2) - \sinh(2kx_0)\right]\right\}, \tag{34}$$

$$F_H = -\frac{1}{4}\mathrm{Im}\left\{\frac{1}{k}\left[\frac{\sinh^2(kx_0)}{\sinh^2(kd)}F_{H1} + A\sinh(ka_1)F_{H2} + F_{H3}\right]\right\}, \tag{35}$$

where $F_{H1}$, $F_{H2}$ and $F_{H3}$ are given by

$$F_{H1} = \sinh^2(ka_1)\left\{\sinh\left[2k(x_0+a_2)\right] - 2k(x_0+a_2)\right\} - \sinh^2(ka_2)\left\{\sinh\left[2k(x_0-a_1)\right] - 2k(x_0-a_1)\right\} \tag{36}$$

$$F_{H2} = 2\cosh\left[k\left(x_0 + a_2\right)\right] - \frac{1}{2}\cosh\left[k\left(3x_0 + a_2\right)\right] - \frac{3}{2}\cosh\left[k\left(a_2 - x_0\right)\right] - 2kx_0\sinh\left[k\left(a_2 - x_0\right)\right], \tag{37}$$

$$F_{H3} = \frac{\sinh\left(4kx_0\right)}{4} - 2\sinh\left(2kx_0\right). \tag{38}$$

The fluxes are presented in the $(a_2, x_0)$ plane in **Figure 3**. The presence of the sidewalls has complicated the behavior of the fluxes as compared to the case of infinite surrounding fluid. For a fixed position of the plume (*i.e.*, fixed $a_2$) relative to the wall, gradual increase in the thickness of the plume is associated with an increase in the downward heat flux. For plumes of thickness less than about 2, the heat flux is almost a constant as the plume moves towards a sidewall. For plumes with larger thickness, the heat flux increases as the wall is approached. The upward material flux behaves similarly if the distance from the wall is less than about 4.5. For larger distances from the sidewalls, the material flux increases as $x_0$ increases from zero reaching a maximum before it decreases to a minimum and starts to increase again to a larger value as $x_0$ approaches $a_2$ and the plume interface approaches a sidewall. The buoyancy flux, which is the net system flux, possesses two local maxima and a minimum. The local maximum with the largest value is situated on the boundary at $x_0 = 1.3$, while the other one is situated half-way between the two sidewalls and about the same value of $x_0$. The minimum occurs for $x_0 = 3.7$ and lies half-way between the sidewalls. The buoyancy flux per unit area, illustrated in (*d*), has the same general behavior as the buoyancy flux but the positions of the two local maxima and minimum are different.

## 4. Solution of the Eigenvalue Problem

In this section, we solve the eigenvalue problem posed by the perturbation Equations (23)-(26) and the relevant boundary conditions to obtain expressions for the growth rate. Our interest lies in the instability produced by the buoyant fluid in the plume. We assume that the interface at the plane $x = x_0$ is given a small harmonic disturbance of the form

$$x = x_0 + \varepsilon\exp\left(\Omega t + \mathrm{i}\left(my - nz\right)\right) + c.c., \tag{39}$$

where $m$ and $n$ are the horizontal and vertical wavenumbers, $c.c.$ refers to the complex conjugate, and $\Omega$ is a complex constant, which can be expressed as

$$\Omega = \Omega_r + \mathrm{i}\Omega_i. \tag{40}$$

$\Omega_r$ and $\Omega_i$ will be referred to as the real and imaginary parts of $\Omega$. The stability of the plume is determined by the sign of $\Omega_r$. If it is negative for all possible values of the wavenumbers $m$ and $n$, then the plume is stable, but the system is rendered unstable if any pair $(m, n)$ of wavenumbers gives a positive value of $\Omega_r$. If the preferred mode occurs for $m$, $n$ both non-zero, it is referred to as a 3-dimensional mode but if any one of them vanishes it is 2-dimensional. If the maximum value of $\Omega_r$ vanishes, the plume is neutrally stable.

The disturbance (39) will propagate into the system, and affect the second interface and the variables of the system to produce the perturbations. The disturbance at the interface $x = -x_0$ can be written in the form

$$x = -x_0 + \varepsilon\eta_1\exp\left(\Omega t + \mathrm{i}\left(my - nz\right)\right) + c.c., \tag{41}$$

where $\eta_1$ is the amplitude of the displacement of the interface at $x = -x_0$, and will be determined by the solution.

The perturbation variables produced by the disturbance (39) can be expressed in the form

$$\left\{\boldsymbol{u}^\dagger, C^\dagger, T^\dagger, p^\dagger\right\} = \left\{-\mathrm{i}nu, nmv, w, C, T, -\mathrm{i}np\right\}\exp\left(\Omega t + \mathrm{i}\left(my - nz\right)\right) + c.c., \tag{42}$$

where the factors $-\mathrm{i}n$, $nm$ and $-\mathrm{i}n$ are introduced in the variables $u$, $v$ and $p$, respectively, for convenience.

Substituting the variables (42) into (23)-(26), we obtain the following ordinary differential equations in $x$

$$Du - m^2v + w = 0, \tag{43}$$

$$\Delta u - Dp = R\bar{\Omega}u, \tag{44}$$

$$\Delta v - p = R\bar{\Omega}v, \tag{45}$$

$$\Delta w + T + C + n^2p = R\left(\bar{\Omega}w - \mathrm{i}nuD\bar{w}\right), \tag{46}$$

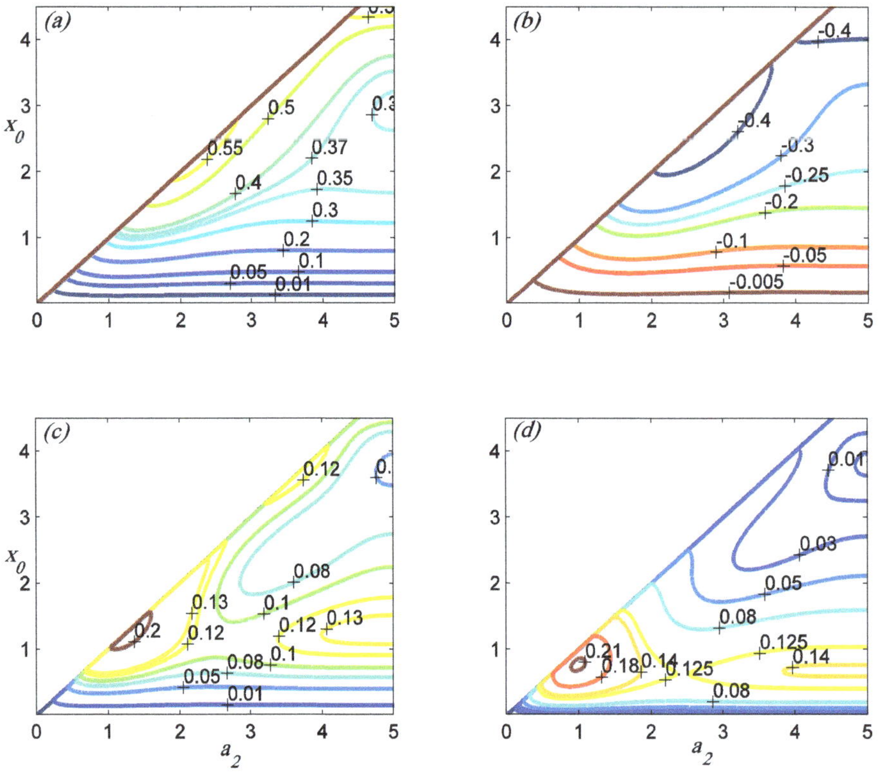


**Figure 3.** The contours of the basic state fluxes of the top-hat bounded plume in the $(a_2, x_0)$ plane; (a) material flux, $F_m$, (b) heat flux, $F_H$, (c) buoyancy flux, $F_B$ and (d) buoyancy flux per unit area $(F_B/x_0)$ when $d = 10$. Note that the heat and material fluxes are larger at the sidewall, the buoyancy flux has a maximum value of 0.215, when $(a_2, x_0) \approx (1.22, 1.22)$ and the buoyancy flux per unit area has a maximum 0.212 at $(a_2, x_0) \approx (1,1)$, and both maxima lie on the sidewall. The fluxes are shown for half the interval because they are symmetric about the middle plane between the sidewalls.

$$\Delta T - w = \sigma R\left(\bar{\Omega} T - \mathrm{i} n u D\bar{T}\right), \tag{47}$$

$$\bar{\Omega} C = 0. \tag{48}$$

Here we have used

$$b^2 = m^2 + n^2,\ D \equiv \frac{\mathrm{d}}{\mathrm{d}x},\quad \Delta \equiv D^2 - b^2,\quad \bar{\Omega} = \Omega - \mathrm{i} n \bar{w}(x). \tag{49}$$

The boundary conditions across the interfaces are (see, Eltayeb and Loper [16])

$$u, v, w, T, p, C, Dv, DT \quad \text{are continuous across} \quad x = \pm x_0, \tag{50}$$

$$Dw\left(x_0^+\right) - Dw\left(x_0^-\right) = -1,\quad Dw\left(-x_0^+\right) - Dw\left(-x_0^-\right) = \eta_1, \tag{51}$$

$$-\mathrm{i} n u\left(x_0\right) = \Omega - \mathrm{i} n \bar{w}\left(x_0\right),\quad -\mathrm{i} n u\left(-x_0\right) = \left[\Omega - \mathrm{i} n \bar{w}\left(-x_0\right)\right]\eta_1. \tag{52}$$

$$u = v = w = 0 \quad \text{at} \quad x = a_1, x = -a_2. \tag{53}$$

In addition, the sidewalls are maintained at the hydrostatic temperature so that

$$T = 0 \quad \text{at} \quad x = a_1, x = -a_2. \tag{54}$$

Equation (48) gives

$$C\left(x\right) = 0. \tag{55}$$

It was found that it is useful to derive the following three equations. First, differentiate (45) once and subtract (44) to get

$$\Delta\varsigma = R\left[-invD\overline{w}+\overline{\Omega}\varsigma\right],\quad \varsigma = Dv-u, \tag{56}$$

Where $\varsigma$ is related to the vertical component of vorticity. Secondly, differentiate (43) once and subtract (44) to obtain

$$n^2u = m^2\varsigma - D\left(w+p\right) - R\overline{\Omega}u\,. \tag{57}$$

Thirdly, apply the operator $\Delta$ to (43) and use (44)-(46) to find

$$\Delta p - T = 2\mathrm{i}RnuD\overline{w}. \tag{58}$$

The previous studies on a compositional plume showed that the plume flow is unstable for small value of Grashoff number [15]-[17]. This dimensionless number measures the strength of the plume, resulting from the maximum amplitude of the basic concentration. It transpires that instability is also present for small values of $R$ here too. We then write

$$f\left(x,y,z,t\right)=\sum_{r=0}^{\infty} f_r\left(x,y,z,t\right)R^r,\quad \Omega=\sum_{r=1}^{\infty}\Omega_r R^{r-1}, \tag{59}$$

where $f\left(x,y,z,t\right)$ indicates any of the perturbation variables $u$, $v$, $w$, $p$ and $T$.

Substituting the expressions (59) into the system (43)-(47), (56)-(58) and the associated boundary conditions (50)-(54), and equating the coefficients of $R^r\left(r=0,1,2,\cdots\right)$ to zero, we get systems of ordinary differential equations which can be solved successively to find an expression for the growth rate. The two systems obtained for $R^0$ (referred to as Problem 0) and $R^1$ (referred to as problem 1) are sufficient to determine the stability of the interfaces, to leading order. It is found that the instability is present only in part of the parameter space, and it is necessary to consider the next order of the growth rate governed by Problem 1.

### 4.1. Problem 0

The coefficients of $R^0$ in the system (45)-(47), (57)-(58) then consist of the equations

$$\Delta v_0 - p_0 = 0, \tag{60}$$

$$\Delta w_0 + T_0 + n^2 p_0 = 0, \tag{61}$$

$$\Delta T_0 - w_0 = 0, \tag{62}$$

$$\Delta p_0 - T_0 = 0 \tag{63}$$

$$n^2 u_0 = -D\left(w_0 + p_0\right), \tag{64}$$

noting that (56) and the appropriate conditions imply that $\varsigma_0 = 0$ everywhere. Taking note of (64), the boundary conditions can then be expressed as

$$D\left(w_0+p_0\right)=v_0=w_0=T_0=0 \text{ at } x=a_1, x=-a_2, \tag{65}$$

$$v_0, w_0, T_0, p_0, Dv_0, DT_0, D\left(w_0+p_0\right) \text{ are continuous across } x=\pm x_0, \tag{66}$$

$$-inu\left(x_0\right)=\Omega-in\overline{w}\left(x_0\right),\quad -inu\left(-x_0\right)=\left[\Omega-in\overline{w}\left(-x_0\right)\right]\eta_1. \tag{67}$$

$$-inu_0\left(x_0\right)=\Omega_1-in\overline{w}\left(x_0\right),\quad -inu_0\left(-x_0\right)=\left[\Omega_1-in\overline{w}\left(-x_0\right)\right]\eta_1. \tag{68}$$

We operate on Equation (61) with $\Delta$, and use equations (62) and (63) to get

$$\Delta^3 w_0 + \Delta w_0 + n^2 w_0 = 0. \tag{69}$$

The solution of the system (60)-(64) subject to the boundary conditions (65)-(67) is given by

$$\left\{v_0^{(i)}, w_0^{(i)}, T_0^{(i)}, p_0^{(i)}\right\}\left(x\right)=\sum_{j=1}^{3}\left\{1,\mu_j^3,\mu_j^2,\mu_j\right\}\left[A_j^{(i)}\cosh\left(\lambda_j x\right)+B_j^{(i)}\sinh\left(\lambda_j x\right)\right], \tag{70}$$

$$u_0^{(i)}(x)=\sum_{j=1}^{3}\lambda_j\left[A_j^{(i)}\sinh(\lambda_j x)+B_j^{(i)}\cosh(\lambda_j x)\right], \tag{71}$$

where the superscript "$i$" in the solution refers to the region of the problem defined by

$$i=\begin{cases}1; & -a_2\le x<-x_0\\ 2; & -x_0\le x\le x_0\\ 3; & x_0<x\le a_1\end{cases}, \tag{72}$$

(see **Figure 1**) and $\mu_j\,(j=1,2,3)$ are the roots of the cubic equation

$$\mu_j^3+\mu_j+n^2=0, \tag{73}$$

with $\lambda_j$ given by

$$\lambda_j=\sqrt{\mu_j+b^2}. \tag{74}$$

The constants $A_j^{(i)}$ and $B_j^{(i)}$ for $(i,j=1,2,3)$ are given by

$$A_j^{(1)}=\frac{-F_j\sinh(\lambda_j a_2)}{\sinh(\lambda_j d)}\left[\sinh\{\lambda_j(x_0-a_1)\}+\eta_1\sinh\{\lambda_j(x_0+a_1)\}\right], \tag{75}$$

$$A_j^{(2)}=\frac{-F_j}{\sinh(\lambda_j d)}\left[\sinh(\lambda_j a_2)\sinh\{\lambda_j(x_0-a_1)\}-\eta_1\sinh(\lambda_j a_1)\sinh\{\lambda_j(x_0-a_2)\}\right], \tag{76}$$

$$A_j^{(3)}=\frac{F_j\sinh(\lambda_j a_1)}{\sinh(\lambda_j d)}\left[\sinh\{\lambda_j(x_0+a_2)\}+\eta_1\sinh\{\lambda_j(x_0-a_2)\}\right], \tag{77}$$

$$B_j^{(1)}=\frac{-F_j\cosh(\lambda_j a_2)}{\sinh(\lambda_j d)}\left[\sinh\{\lambda_j(x_0-a_1)\}+\eta_1\sinh\{\lambda_j(x_0+a_1)\}\right], \tag{78}$$

$$B_j^{(2)}=\frac{-F_j}{\sinh(\lambda_j d)}\left[\cosh(\lambda_j a_2)\sinh\{\lambda_j(x_0-a_1)\}+\eta_1\cosh(\lambda_j a_1)\sinh\{\lambda_j(x_0-a_2)\}\right], \tag{79}$$

$$B_j^{(3)}=\frac{-F_j\cosh(\lambda_j a_1)}{\sinh(\lambda_j d)}\left[\sinh\{\lambda_j(x_0+a_2)\}+\eta_1\sinh\{\lambda_j(x_0-a_2)\}\right], \tag{80}$$

with

$$F_j=\frac{\mu_j^2}{\lambda_j(2\mu_j+3n^2)}. \tag{81}$$

The application of the boundary conditions (68) gives an expression for the growth rate $\Omega_1$ and the displacement of the interface $\eta_1$. This leads to

$$(\Omega_1/in)^2+S_1(\Omega_1/in)+S_2=0, \tag{82}$$

$$\eta_1=\frac{-N_{j-}}{(\Omega_1/in)-\overline{w}(-x_0)+M_{j+}}, \tag{83}$$

in which

$$S_1=N_{j+}+M_{j+}-\overline{w}(x_0)-\overline{w}(-x_0), \tag{84}$$

$$S_2=\{N_{j+}-\overline{w}(x_0)\}\{M_{j+}-\overline{w}(-x_0)\}-N_{j-}M_{j-}, \tag{85}$$

$$N_{j\pm}=\sum_{j=1}^{3}\frac{-\lambda_j F_j}{\sinh(\lambda_j d)}\sinh\{\lambda_j(x_0-a_1)\}\cosh\{\lambda_j(x_0\pm a_2)\}, \tag{86}$$

$$M_{j\pm}=\sum_{j=1}^{3}\frac{-\lambda_j F_j}{\sinh(\lambda_j d)}\sinh\{\lambda_j(x_0-a_2)\}\cosh\{\lambda_j(x_0\pm a_1)\}. \tag{87}$$

Thus

$$\Omega_1=\frac{in}{2}\left(-S_1\pm\sqrt{D_p}\right),\ D_p=S_1^2-4S_2. \tag{88}$$

The properties of the roots of the cubic Equation (73) render $N_{j\pm}$, $M_{j\pm}$ real and hence the discriminant $D_p$ is real. It follows that $\Omega_1$ is imaginary if $D_p\geq 0$ and complex when $D_p<0$. In the absence of the sidewalls, the two modes are such that the two interfaces of the plume are either in-phase giving a sinuous (S) solution or out-of-phase giving a varicose (V) solution. In both cases, $\Omega_1$ is imaginary and the disturbances are neutral at this level of approximation of the growth rate. The introduction of the boundaries has destroyed the symmetry unless the plumes are situated halfway between the sidewalls.

It is informative to establish the relationship between the modes of the bounded plume defined by (88) and those of the unbounded one particularly that we expect the modes of the bounded plume to reduce to sinuous and varicose when the plume is positioned half-way between the two sidewalls. We take the limit $a_1,a_2\to\infty$, and find that

$$N_{j+}\to 0,\ N_{j-}\to\frac{1}{2}\sum_{j=1}^{3}\frac{\mu_j^2E_j}{\left(2\mu_j+3n^2\right)},\ \text{as}\ a_1,a_2\to\infty, \tag{89}$$

$$M_{j+}\to 0,\ M_{j-}\to\frac{1}{2}\sum_{j=1}^{3}\frac{\mu_j^2E_j}{\left(2\mu_j+3n^2\right)},\ \text{as}\ a_1,a_2\to\infty, \tag{90}$$

$$S_1\to -2\bar{w}(x_0),\ S_2\to\left(\bar{w}(x_0)\right)^2-\left(\frac{1}{2}\sum_{j=1}^{3}\frac{\mu_j^2E_j}{\left(2\mu_j+3n^2\right)}\right)^2,\ \text{as}\ a_1,a_2\to\infty, \tag{91}$$

and

$$D_p\to\left(\sum_{j=1}^{3}\frac{\mu_j^2E_j}{\left(2\mu_j+3n^2\right)}\right)^2,\ \text{as}\ a_1,a_2\to\infty, \tag{92}$$

where $E_j$ is defined by

$$E_j=\exp\left(-2\lambda_j x_0\right). \tag{93}$$

Substituting these expressions into the Equation (88) for $\Omega_1$, we get

$$\Omega_1\to in\left\{\bar{w}(x_0)\pm\frac{1}{2}\sum_{j=1}^{3}\frac{\mu_j^2E_j}{\left(2\mu_j+3n^2\right)}\right\},\ \eta_1\to\mp 1,\ \text{as}\ a_1,a_2\to\infty. \tag{94}$$

The growth rate (94) is the same as the growth rate of the Cartesian plume obtained in Eltayeb and Loper [16] and the values of the displacement $\eta_1$ shows that the phase of the interface at $x=-x_0$ is either out-of-phase (varicose mode) with $\eta=-1$ or in-phase (sinuous mode) with $\eta=1$. It thus follows that the upper sign in (88) refers to a modification of the varicose mode, which we shall refer to as the modified varicose mode (MV) while the other will be denoted by the modified sinuous (MS) mode. The growth rate will be denoted by $\Omega^{(k)}$, where $k=MV,MS$ for the modified varicose and sinuous modes, respectively.

## 4.2. Problem 1

The coefficients of $R^1$ in the perturbation Equations (44)-(47), (56)-(58) give the set

$$\Delta u_1 = M_u, \tag{95}$$

$$\Delta v_1 - p_1 = M_V, \tag{96}$$

$$\Delta w_1 + T_1 + n^2 p_1 = M_w, \tag{97}$$

$$\Delta T_1 - w_1 = M_T, \tag{98}$$

$$\Delta p_1 - T_1 = M_p, \tag{99}$$

$$n^2 u_1 = m^2 \varsigma_1 - D(w_1 + p_1) - \bar{\Omega}_1 u_0, \tag{100}$$

$$\Delta \varsigma_1 = -\mathrm{i}m(w_0 - n^2 v_0) D\bar{w}, \tag{101}$$

in which

$$\begin{aligned} M_u &= Dp_1 + (\Omega_1 - \mathrm{i}n\bar{w})u_0, \quad M_v = (\Omega_1 - \mathrm{i}n\bar{w})v_0, \\ M_w &= (\Omega_1 - \mathrm{i}n\bar{w})w_0 - \mathrm{i}nu_0 D\bar{w}, \quad M_p = 2\mathrm{i}nu_0 D\bar{w}, \\ M_T &= \sigma(\Omega_1 - \mathrm{i}n\bar{w})T_0 - \mathrm{i}n\sigma u_0 D\bar{T}. \end{aligned} \tag{102}$$

The associated boundary conditions are

$$v_1 = w_1 = T_1 = D(p_1 + w_1) = 0 \ \text{ at } \ x = a_1, x = -a_2, \tag{103}$$

$$v_1, w_1, T_1, p_1, Dv_1, DT_1, Dw_1, D(p_1 + w_1) \ \text{ are continuous across } \ x = \pm x_0, \tag{104}$$

$$\Omega_2 = -\mathrm{i}nu_1(x_0), \ \ \Omega_2 \eta_1 = -\mathrm{i}nu_1(-x_0). \tag{105}$$

The equations and boundary conditions (95)-(105) are solved in the Appendix A. They lead to the growth rate $\Omega_2$ given by

$$\Omega_2^{(k)} = -f_\eta^{(k)} \left\{ \Omega_{21}^{(k)} + \Omega_{22}^{(k)} + \Omega_{23}^{(k)} + \Omega_{24}^{(k)} + \frac{\mathrm{i}}{n}\hat{g}^{(k)} \right\}, \tag{106}$$

where $f_\eta^{(k)}$ is given by

$$f_\eta^{(k)} = \frac{1}{1 + \left(\eta_1^{(k)}\right)^2}, \tag{107}$$

and $k$ takes the symbols $MV$ or $MS$. The expressions $\Omega_{21}^{(k)}$, $\Omega_{22}^{(k)}$, $\Omega_{23}^{(k)}$, $\Omega_{24}^{(k)}$ and $\hat{g}^{(k)}$ are given by

$$\Omega_{21}^{(k)} = -\mathrm{i}\Omega_1^{(k)} \int_{-a_2}^{a_1} \left\{ nu_0^{(k)} G^{(k)} - \frac{1}{n} w_0^{(k)} H^{(k)} \right\} \mathrm{d}x, \tag{108}$$

$$\Omega_{22}^{(k)} = \frac{\mathrm{i}\Omega_1^{(k)}}{n} \sum_{j=1}^{3} C_j \int_{-a_2}^{a_1} \left\{ w_0^{(k)} + \sigma\mu_j T_0^{(k)} \right\} H_j^{(k)} \mathrm{d}x, \tag{109}$$

$$\Omega_{23}^{(k)} = \int_{-a_2}^{a_1} \left( -u_0^{(k)} \left\{ n^2 \bar{w} G^{(k)} + H^{(k)} D\bar{w} \right\} + \bar{w} w_0^{(k)} H^{(k)} \right) \mathrm{d}x, \tag{110}$$

$$\Omega_{24}^{(k)} = \sum_{j=1}^{3} -C_j \int_{-a_2}^{a_1} H_j^{(k)} \left[ -\bar{w}\left\{ w_0^{(k)} + \sigma\mu_j T_0^{(k)} \right\} + \left\{ (2\mu_j^2 + 1) D\bar{w} - \sigma\mu_j D\bar{T} \right\} u_0^{(k)} \right] \mathrm{d}x, \tag{111}$$

$$\hat{g}^{(k)} = g^{(k)}(-a_2) + g^{(k)}(a_1), \tag{112}$$

in which $G^{(k)}$, $H^{(k)}$, $H_j^{(k)}$, $C_j$, $g^{(k)}(-a_2)$ and $g^{(k)}(a_1)$ are given in the Appendix.

It is noteworthy that because of the properties of the cubic Equation (73) for $\mu$, the zeroth order variables are all real. For wavenumbers for which the zeroth order solution is neutrally stable, $\Omega_1$ is purely imaginary and the non-homogeneity of the equations of problem 1 are all imaginary. It then follows that the variables with subscript 1 are all imaginary. When we employ this result into the expression for $\Omega_2$, we find that $\Omega_2$ is real, and consequently it will determine the stability of the plume outside the unstable regions of the zeroth order.

## 5. Discussions of the Results

The growth rates given by the expressions (88) and (106) were computed in the parameter space $(x_0, a_2, \sigma, m, n)$. For a given set of the parameters $x_0, a_2, \sigma$, the growth rate is maximized over $m$ and $n$.

The maximum value, $\Omega_c$, of $\mathrm{Re}(\Omega)$ and the corresponding wavenumbers $m_c, n_c$ and the vertical wave speed $U_c\left(=\mathrm{Im}(\Omega_c)/n_c\right)$ define the preferred mode of instability for that set of parameters.

First we consider Equation (88). This growth rate at this level of approximation is independent of $\sigma$. As we mentioned previously, the stability of the plume at the leading order of approximation depends on $D_p$. In **Figure 4** we show the isolines of $D_p$ in the wavenumber plane for some representative values of $x_0$ and $a_2$. It is found that $D_p$ is negative for small values of $x_0$ and $a_2 - x_0$ indicating that instability at zeroth order is possible only if the plume is very thin and is close to the wall. Indeed, the maximisation of the growth rate (88) when $D_p < 0$ shows that instability is possible only for values of $a_2 - x_0$ not exceeding 0.25 and the unstable modes are two-dimensional and propagate vertically upwards (**Figure 5**). In the calculations, the discrimnant and the growth rate are scaled up by $10^2$ as adopted by previous authors, in order to facilitate comparison with the results in the absence of boundaries.

Computations of the growth rate (106) showed that the plume is always unstable at a growth rate of $O(R)$. The maximum growth rate at any particular point in the parameter space $(x_0, a_2, \sigma)$ can belong to the MS or the MV mode depending in a complicated way on the relative magnitudes of the parameters. As any one parameter is varied keeping the other two fixed, the preferred mode of one type can change to the other mode when the parameter reaches a certain value. Moreover, variations of a parameter can also lead to a mode of particular type (*i.e.*, MS or MV) changing from two-dimensional to three-dimensional, or the reverse, when the parameter increases through a certain value. This is due to the fact that the expression (106) can possess more than one local maximum and as the parameter is increased, the larger of the two maxima decreases and the smaller increases until a value is reached when the smaller one overtakes the originally larger one and becomes preferred. **Figure 6** illustrates such behaviour for a sample of the parameters.

In **Figure 7** we illustrate the dependence of the preferred mode of instability on the Prandtl number, $\sigma$, in a way that allows comparison with the limiting case of no sidewalls. For small values of the Prandtl number the MS mode is preferred while the MV mode is preferred for large Prandtl numbers. This agrees well with the case of no sidewalls [16]. The value, $\sigma_0$ of the Prandtl number at which the mode changes from MS to MV depends on the distance between the plume and the nearest wall. As the sidewall gets closer, $\sigma_0$ increases indicating that the presence of the boundaries tends to suppress the MV mode. The presence of the boundaries also tends to stabilise the plume as the growth rate is reduced in magnitude with the decrease in $d$. It is noteworthy that whatever the values of $d$ or $\sigma$, the MS mode is three-dimensional and the MV is two-dimensional when the plume is equidistant from the sidewalls.

**Figure 8** illustrates the dependence of the preferred mode parameters on the thickness of the plume when it takes different positions relative to the sidewalls. We can observe that 1) when the plume is close to a sidewall, the preferred mode is two-dimensional (with $m_c = 0$), 2) for moderate to large values of $a_2$, the preferred mode is of the MS type when the plume is thin but changes to MV and then back to MS as it approaches the wall, 3) in all cases the growth rate increases from its value for small thickness to a maximum before it decreases to a small value as the plume increases and approaches the sidewall. We should point out here that when the plume is very close to a sidewall, the region enclosed between the plume and wall may be so thin that diffusion may not be negligible. The inclusion of diffusion in this particular case was examined in detail both on the modification to the profile (27) of the basic concentration and on the equation (26) of the perturbations. While the basic state variables $\bar{w}(x), \bar{T}(x)$ are almost identical, the stability is slightly influenced by the presence of diffusion.

In contrast with the unbounded plume where instability is $O(R)$ everywhere in the parameter space, the instability of the bounded Cartesian plume has instabilities with growth rates of $O(1)$ and $O(R)$. The region in the parameter space where there is instability with the larger growth rate (*i.e.*, $O(1)$), is small and depends on the distance between the walls. In **Figure 9**, the regime diagram for the two instabilities is shown when $d = 10$.

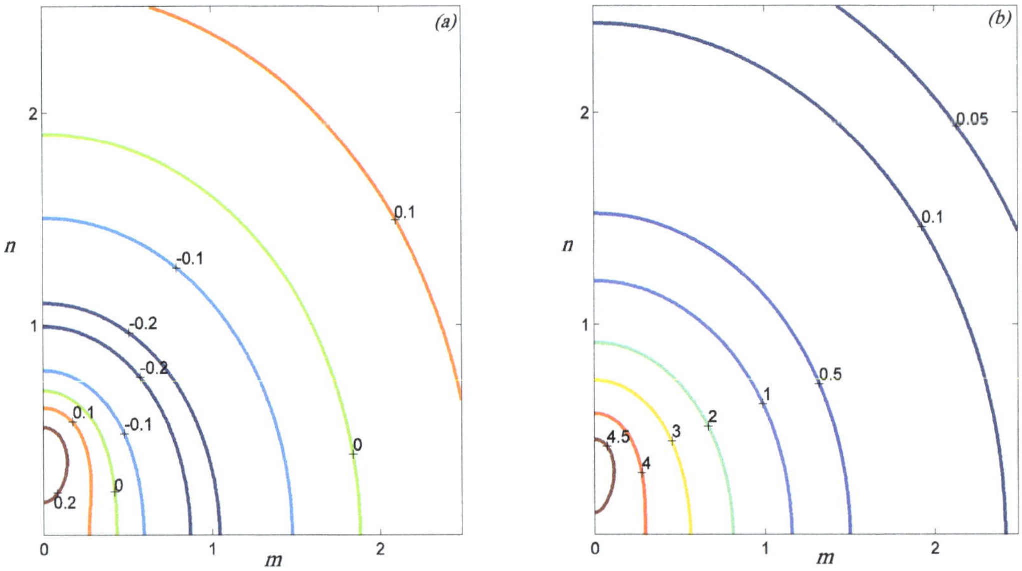


**Figure 4.** The contours of the discriminant, $D_p$, in the $(m,n)$ plane when $x_0 = 0.2, d = 10$ for two different values of $a_2$: (a) $a_2 = 0.3$ and (b) $a_2 = 2$. $D_p$ here is scaled by $10^2$ for ease of presentation. Note that a region of negative values of the discriminant appears when the plume is close to the wall as in (a), which indicates instability of the plume at zero order.

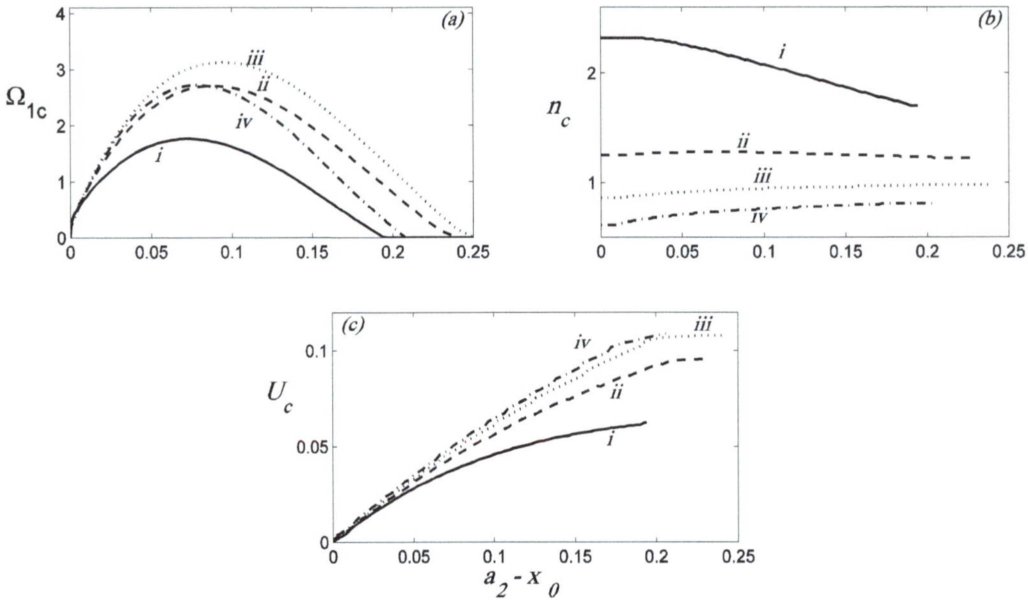


**Figure 5.** The preferred mode of instability with growth rate of order $O(1)$ as a function of the distance from the sidewall, $a_2 - x_0$, for four different values of $x_0$; (*i*) = 0.1, (*ii*) = 0.2, (*iii*) = 0.3 and (*iv*) = 0.4, when $d = 10$. Note the dependence of the magnitude of the growth rate on the thickness of the plume and its distance from the wall.

This instability with growth rate $O(1)$ occurs only if the plume is relatively thin (of thickness not more than about half the salt-finger length scale) and its distance from the sidewall does not exceed about 0.25. We also note that when the plume is very close to the wall the growth rate becomes smaller.

The preferred mode is associated with plume interfaces that are determined by (39) and (41). The amplitude at

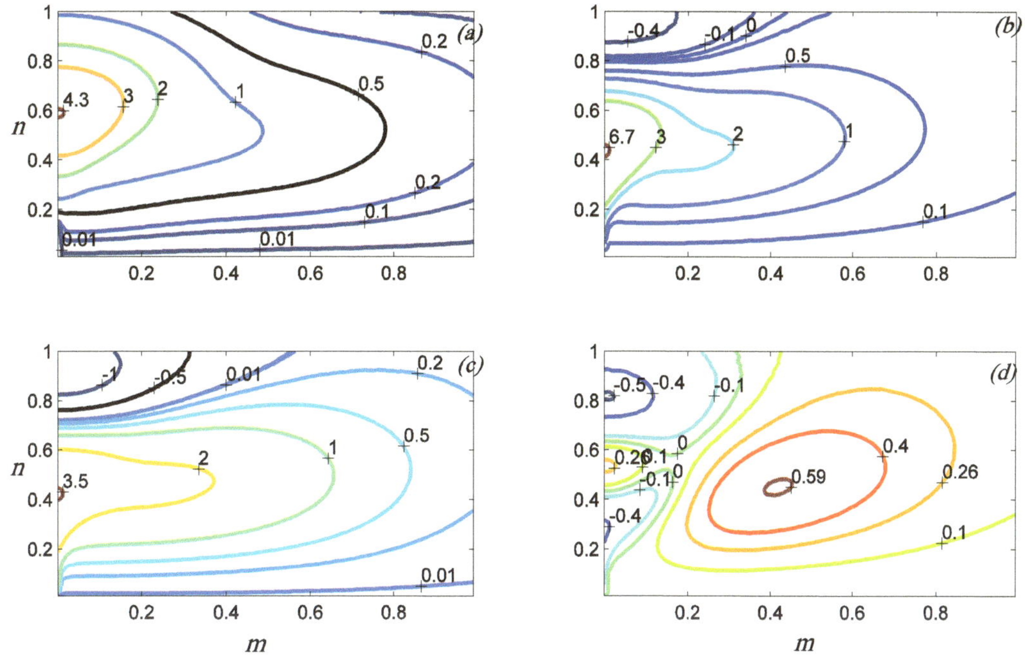


**Figure 6.** Contours of the growth rate of the modes MV, as in (a) and (c), and MS, as in (b) and (d), in situations when the preferred mode changes from one type to another. Here $x_0 = 2$, $\sigma = 10$, and $d = 10$, and $a_2 = 3$ for (a), (b) and $a_2 = 5$ for (c), (d). (a), (c) refer to the MV mode and (b) , (d) refer to the MS mode. Note that the MS mode is preferred for $a_2 = 3$ and the MV mode is preferred when $a_2 = 5$.

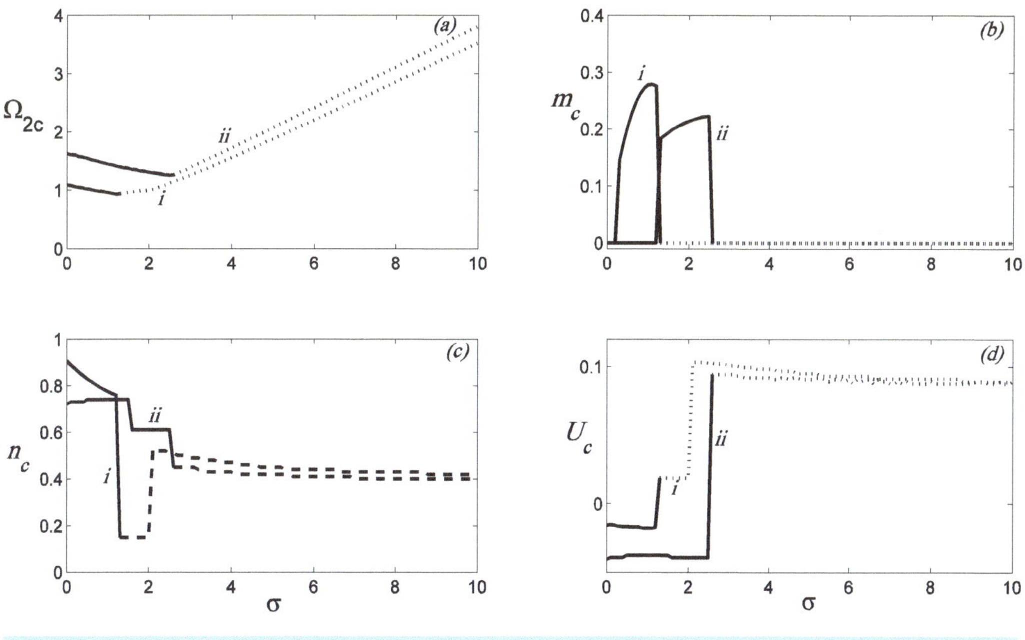


**Figure 7.** Illustration of the influence of the sidewalls on the stability of the Cartesian plume. The preferred mode as a function of Prandtl number, $\sigma$, when $x_0 = 2$ and the plume is situated halfway between the sidewalls (*i.e.*, $a_1 = a_2$). The curves i and ii refer to two different distances between the sidewalls: i) $d = 10$, and ii) $d = 20$. The solid curve refers to the MS mode while the broken one refers to the MV mode. Note the decrease in the growth rate as the distance $d$ between the sidewalls is reduced. The presence of the boundaries also decreases the range of $\sigma$ for which the MS mode is preferred.

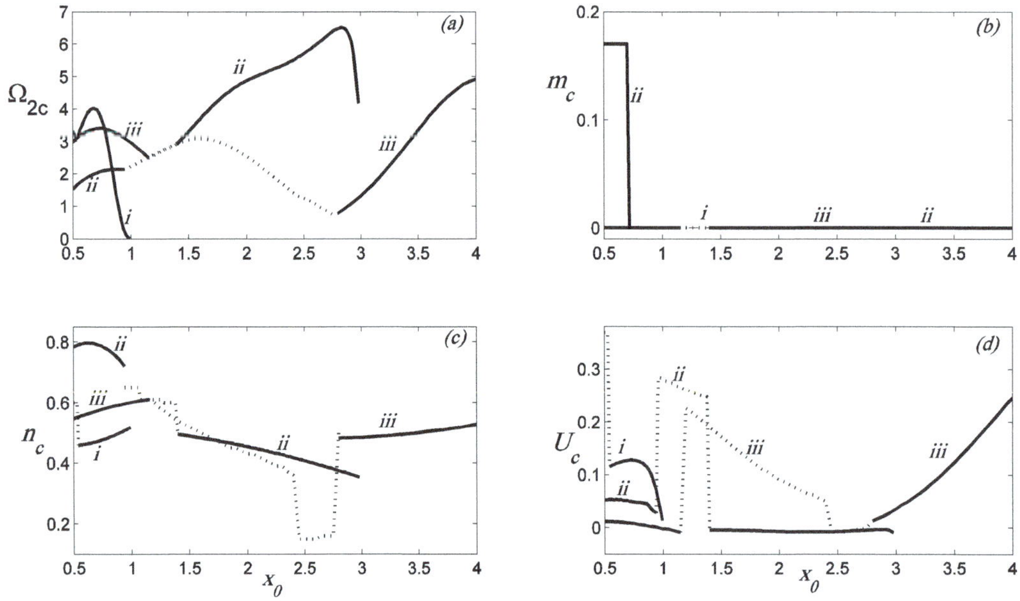


**Figure 8.** The preferred mode parameters as a function of $x_0$ for $\sigma = 7$ and $d = 10$ for three different values of $a_2$: (i) 1, (ii) 3, (iii) 5. The solid curve refers to the MS mode while the broken one refers to the MV mode. The modes are two-dimensional except in the case of thin plumes when $a_2 = 3$. Note the complicated behaviour of the preferred mode as the thickness of the plume changes at different positions relative to the boundaries.

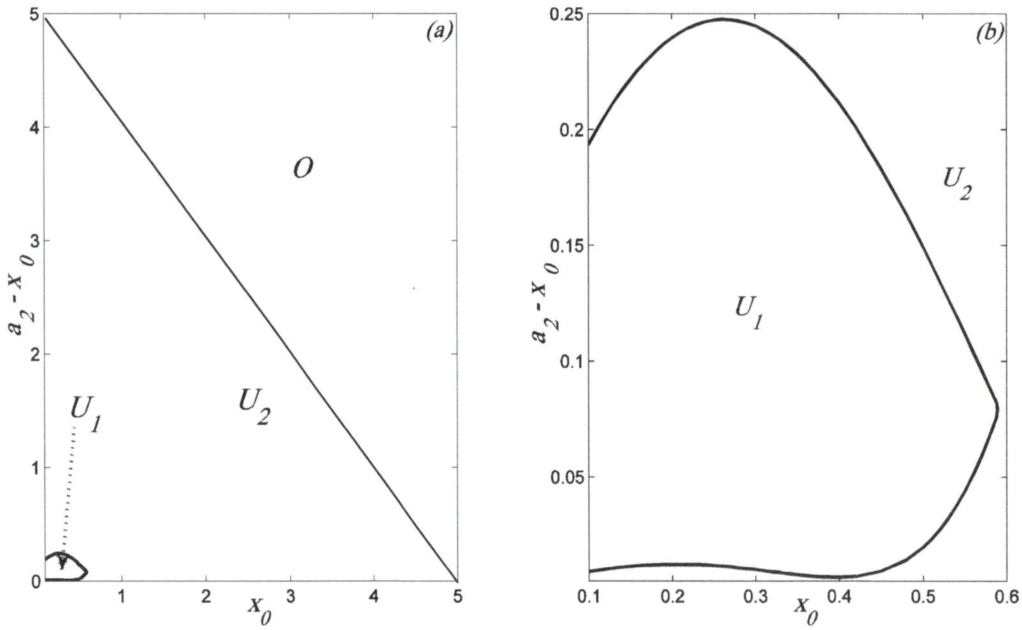


**Figure 9.** The regime diagram of the bounded Cartesian plume in the plane $x_0, a_2 - x_0$ for fixed $d = 10$. In (a), the regions labelled $U_1$ and $U_2$ refer to instabilities with growth rate $O(1)$ and $O(R)$, respectively. The area $O$ is outside the domain since cannot exceeds. Subfigure (b) shows a magnification of thearea for $a_2 - x_0 \le 0.25$ and $x_0 \le 0.6$ of figure (a).

$x = x_0$ is fixed at the value 1 while the amplitude at the interface at $x = x_0$ is determined by $\eta_1$, which is

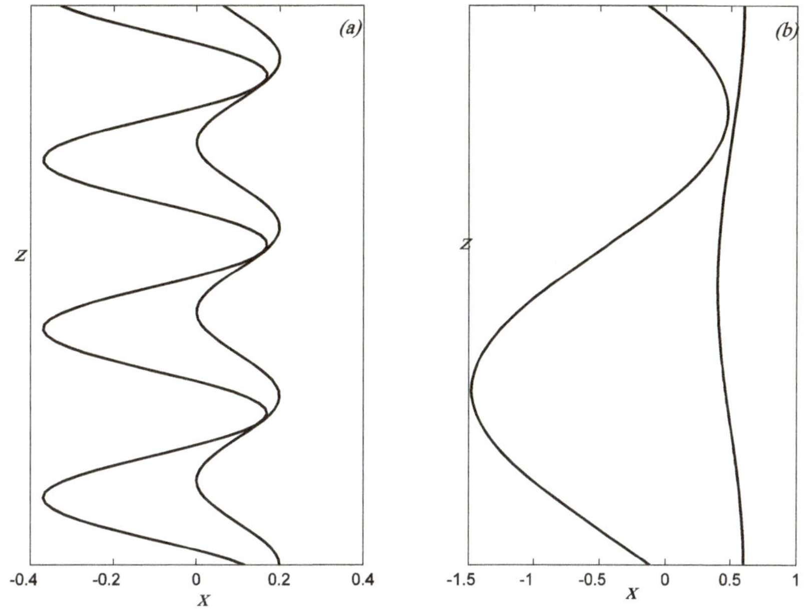


**Figure 10.** A sample of the profiles of the interfaces of the unstable mode for two values of the pair $(x_0, a_2)$ when $d = 10$. The profiles are magnified for clarity by the same factor $\varepsilon(=0.1)$. (a) $x_0 = 0.1$, $a_2 = 0.2$, $n_c = 2.08$, $m_c = 0$ and (b) $x_0 = 0.5$, $a_2 = 0.6$, $n_c = 0.63$, $m_c = 0$. Note that the interface pro- files are very close at regular intervals.

determined by the parameters of the preferred mode for any prescribed values of $x_0, a_2, \sigma, d$. In **Figure 10** we give samples of the profiles of the interfaces relating to some preferred modes. It is noteworthy that the interfaces are very close at regular points across the length of the plume and this may indicate a tendency to break into blobs.

## 6. Conclusions

The dynamics of a plume of buoyant fluid, in the form of a channel of finite width, rising in a less buoyant fluid contained between two parallel sidewalls, a distance $d$ apart, has been investigated. It is found that:

1) The plume is associated with a vertical flow that is balanced by a down flow on either side of the plume, and the flow inside the plume can develop a reverse (downward) flow around the center of the plume if the plume is wide enough.

2) The flow and concomitant temperature transport material upwards and heat downwards in such a way that the net upward buoyancy flux is positive, and possesses two local maxima and a minimum.

3) The instability of the interfaces has the following main properties:

a) The instability can take one of two modes, which are modifications of the sinuous and varicose modes of the plume in the absence of sidewalls but here modified by the lack of symmetry due to the different positions of the plume relative to the sidewalls,

b) If the plume is close to a sidewall, the instability has a growth rate $O(1)$ on the convective time scale, provided $x_0$ does not exceed a certain value,

c) For plumes away from the sidewalls, the instability has a growth rate of $O(R)$,

d) The presence of the boundaries tend to stabilise the plume when it is equidistant from the sidewalls and the growth rate of the unstable mode is reduced as the sidewalls approach the plume,

e) When the Prandtl number is small, the modified sinuous (MS) mode is preferred while the modified varicose (MV) mode is preferred for large values of the Prandtl number,

f) The preferred MS mode is generally 3-dimensional while the MV mode is generally 2-diemnsional,

g) The profiles of the unstable plume indicate that the instability might lead to the break-up of the plume into blobs that rise to the top.

4) The relatively large growth rates of the instability when the plume is close to a sidewall may be due to heat flux emitted by the boundary.

The role of diffusion has been neglected in the present study because it is generally very small. However, it can be expected that it maybe potent when the plume is close to a sidewall. This has been analysed (but not included here) and found to provide small correction. Diffusion may also be potent in thin boundary layers at the interfaces of the plume at $x = \pm x_0$, where the concentration profile experiences a jump. This is expected to be analysed in a future study.

An attempt was made to compare the present results with experimental observations but we have not been able to identify a detailed experimental study on the influence of the boundaries on the plumes rising from mushy layers. However, the results obtained here agree with the general observations of Hell a well *et al.* [14].

## References

[1] Rees Hones, D.W. and Worster, M.G. (2013) Fluxes through Steady Chimneys in a Mushy Layer during Binary Alloy Solidification. *Journal of Fluid Mechanics*, **714**, 127-151. http://dx.doi.org/10.1017/jfm.2012.462

[2] Wells, A.J., Wettlaufer, J.S. and Orszag, S.A. (2011) Brine Fluxes from Growing Sea Ice. *Geophysical Research Letters*, **38**, Article ID: L04501. http://dx.doi.org/10.1029/2010GL046288

[3] Loper, D.E. (1978) The Gravitationally Powered Dynamo. *Geophysical Journal International*, **54**, 389-404. http://dx.doi.org/10.1111/j.1365-246X.1978.tb04265.x

[4] Moffatt, H.K. (1989) Liquid Metal MHD and the Geodynamo. In: Lielpeteris, J. and Moreau, R., Eds., *Liquid Metal Magnetohydrodynamics*, Kluwer Academic Publishers, Dordrecht, 403-412. http://dx.doi.org/10.1007/978-94-009-0999-1_49

[5] Al-Lawatia, M.A., Elbashir, T.B.A., Eltayeb, I.A., Rahman, M.M. and Balakrishnan, E. (2011) The Dynamics of Two Interacting Compositional Plumes in the Presence of a Magnetic Field. *Geophysical & Astrophysical Fluid Dynamics*, **105**, 586-615. http://dx.doi.org/10.1080/03091929.2010.518316

[6] Copley, S.M., Giamel, A.F., Johnson, S.M. and Hornbecker, M.F. (1970) The Origin of Freckles in Unidirectionally Solidified Casting. *Metallurgical Transactions*, **1**, 2193-2204. http://dx.doi.org/10.1007/BF02643435

[7] Huppert, H.E. (1990) The Fluid Mechanics of Solidification. *Journal of Fluid Mechanics*, **212**, 209-240. http://dx.doi.org/10.1017/S0022112090001938

[8] Chen, C.F. and Chen, F. (1991) Experimental Study of Directional Solidification of Aqueous Ammonium Chloride Solution. *Journal of Fluid Mechanics*, **227**, 567-586. http://dx.doi.org/10.1017/S0022112091000253

[9] Tait, S. and Jaupart, C. (1992) Compositional Convection in a Reactive Crystalline Mush and Melt Differentiation. *Journal of Geophysical Research*, **97**, 6735-6756. http://dx.doi.org/10.1029/92JB00016

[10] Jellinek, A.M., Kerr, R.C. and Griffiths, R.W. (1999) Mixing and Compositional Stratification Produced by Natural Convection: 1. Experiments and Their Application to Earth's Core and Mantle. *Journal of Geophysical Research*, **104**, 7183-7201. http://dx.doi.org/10.1029/1998JB900116

[11] Classen, S., Heimpel, M. and Christensen, U. (1999) Blob Instability in Rotating Compositional Convection. *Geophysical Research Letters*, **26**, 135-138. http://dx.doi.org/10.1029/1998GL900227

[12] Aussillous, P., Sederman, A.J., Gladden, L.F., Huppert, H.E. and Worster, M.G. (2006) Magnetic Resonance Imaging of Structure and Convection in Solidifying Mushy Layers. *Journal of Fluid Mechanics*, **522**, 99-125. http://dx.doi.org/10.1017/S0022112005008451

[13] Pol, S., Fernando, H.J.S. and Webb, S. (2010) Evolution of Double Diffusive Convection in Low-Aspect Ratio Containers. *American Physical Society*, **55**, 81.

[14] Hellawell, A., Sarazin, J.R. and Steube, R.S. (1993) Channel Convection in Partly Solidified Systems. *Philosophical Transactions of the Royal Society A*, **345**, 507-544. http://dx.doi.org/10.1098/rsta.1993.0143

[15] Eltayeb, I.A. and Loper, D.E. (1991) On the Stability of Vertical Double-Diffusive Interfaces. Part 1. A Single Plane Interface. *Journal of Fluid Mechanics*, **228**, 149-181.

[16] Eltayeb, I.A. and Loper, D.E. (1994) On the Stability of Vertical Double-Diffusive Interfaces. Part 2. Two Parallel Interfaces. *Journal of Fluid Mechanics*, **267**, 251-271. http://dx.doi.org/10.1017/S0022112094001175

[17] Eltayeb, I.A. and Loper, D.E. (1997) On the Stability of Vertical Double-Diffusive Interfaces. Part 3. Cylindrical Interfaces. *Journal of Fluid Mechanics*, **353**, 45-66. http://dx.doi.org/10.1017/S0022112097007374

[18] Eltayeb, I.A. (2006) The Stability of a Compositional Plume Rotating in the Presence of a Magnetic Fluid. *Geophysical & Astrophysical Fluid Dynamics*, **100**, 429-455. http://dx.doi.org/10.1080/03091920600799541

## Appendix A: Derivation of the Expression for the Growth Rate of the Cartesian Plume

Here we derive the solvability condition for the first order system (*i.e.*, problem 1) in order to obtain an expression for the growth rate $\Omega_2$. Elimination of all variables $w_1$ and $T_1$ from Equations (97)-(99) in favor of $p_1$ gives

$$\Delta^3 p_1 + \Delta p_1 + n^2 p_1 = M_{tot}, \tag{A.1}$$

in which $M_{tot}$ is defined by

$$M_{tot} = \left(\Delta^2 + 1\right) M_p + \Delta M_T + M_w. \tag{A.2}$$

Next, we consider a function $G(x)$ satisfying the homogeneous form of (95)

$$\Delta G = 0, x \neq \pm x_0, \tag{A.3}$$

and satisfies the following conditions

$$G(a_1) = 0, \quad G(-a_2) = 0, \tag{A.4}$$

$$G \text{ is continuous across } x = \pm x_0\ , \tag{A.5}$$

$$DG\left(-x_0^+\right) - DG\left(-x_0^-\right) = -\eta_1, \quad DG\left(x_0^+\right) - DG\left(x_0^-\right) = -1. \tag{A.6}$$

It then follows that

$$G^{(k)}(x) = \frac{1}{b}\begin{cases} -F_{1b}^{(k)} \sinh\left[b\left(x + a_2\right)\right]; & -a_2 \le x < -x_0 \\ \left(-F_{1b}^{(k)} \sinh\left[b\left(x + a_2\right)\right] - \eta_1^{(k)} \sinh\left[b\left(x + x_0\right)\right]\right); & -x_0 \le x \le x_0, \\ F_{2b}^{(k)} \sinh\left[b\left(x - a_1\right)\right]; & x_0 < x \le a_1 \end{cases} \tag{A.7}$$

where $F_{1b}^{(k)}$ and $F_{2b}^{(k)}$ are given by

$$F_{1b}^{(k)} = \frac{-1}{\sinh(bd)}\left\{\eta_1^{(k)} \sinh\left(b\left(a_1 + x_0\right)\right) + \sinh\left(b\left(a_1 - x_0\right)\right)\right\}, \tag{A.8}$$

$$F_{2b}^{(k)} = \frac{-1}{\sinh(bd)}\left\{\eta_1^{(k)} \sinh\left(b\left(a_2 - x_0\right)\right) + \sinh\left(b\left(a_2 + x_0\right)\right)\right\}, \tag{A.9}$$

and $k$ refers to the modes MS and MV. We now multiply Equation (95) by $G^{(k)}(x)$, integrate by parts from $x = -a_2$ to $x = a_1$, and use the conditions for $u_1$, $G$ and $p_1$ at the interfaces and the boundaries to obtain

$$\Omega_2^{(k)} = i n f_\eta^{(k)} \int_{-a_2}^{a_1} p_1^{(k)} DG^{(k)} dx + M_0^{(k)}, \tag{A.10}$$

where $f_\eta^{(k)}$ and $M_0^{(k)}$ are defined by

$$f_\eta^{(k)} = \frac{1}{1 + \left(\eta_1^{(k)}\right)^2}, M_0^{(k)} = i n f_\eta^{(k)} \int_{-a_2}^{a_1} \left(\Omega_1^{(k)} - i n \overline{w}\right) u_0^{(k)} G^{(k)} dx. \tag{A.11}$$

Now we define a function $H^{(k)}(x)$ by

$$H^{(k)}(x) = DG^{(k)}(x), \tag{A.12}$$

and introduce the function

$$S^{(k)}(x) = H^{(k)}(x) + \sum_{j=1}^{3} C_j H_j^{(k)}(x) \tag{A.13}$$

and note that

$$H^{(k)}(x)=\begin{cases}-F_{1b}^{(k)}\cosh\left[b(x+a_2)\right]; & -a_2\le x<-x_0\\ -F_{1b}^{(k)}\cosh\left[b(x+a_2)\right]-\eta_1^{(k)}\cosh\left[b(x+x_0)\right]; & -x_0\le x\le x_0\,,\\ F_{2b}^{(k)}\cosh\left[b(x-a_1)\right]; & x_0<x\le a_1\end{cases}\tag{A.14}$$

and

$$H_j^{(k)}(x)=\begin{cases}-F_{1j}^{(k)}\cosh\left[\lambda_j(x+a_2)\right]; & -a_2\le x<-x_0\\ -F_{1j}^{(k)}\cosh\left[\lambda_j(x+a_2)\right]-\eta_1^{(k)}\cosh\left[\lambda_j(x+x_0)\right]; & -x_0\le x\le x_0\,,\\ F_{2j}^{(k)}\cosh\left[\lambda_j(x-a_1)\right]; & x_0<x\le a_1\end{cases}\tag{A.15}$$

$$C_j=\frac{-n^2}{3n^2+2\mu_j},\tag{A.16}$$

where $F_{1b}^{(k)}$ and $F_{2b}^{(k)}$ are defined by (A.8) and (A.9), while $F_{1j}^{(k)}$ and $F_{2j}^{(k)}$ are similarly defined by

$$F_{1j}^{(k)}=\frac{-1}{\sinh(\lambda_j d)}\left\{\eta_1^{(k)}\sinh\left(\lambda_j(a_1+x_0)\right)+\sinh\left(\lambda_j(a_1-x_0)\right)\right\},\tag{A.17}$$

$$F_{2j}^{(k)}=\frac{-1}{\sinh(\lambda_j d)}\left\{\eta_1^{(k)}\sinh\left(\lambda_j(a_2-x_0)\right)+\sinh\left(\lambda_j(a_2+x_0)\right)\right\}.\tag{A.18}$$

We multiply equation (A.1) by $S^{(k)}(x)$ and integrate from $x=-a_2$ to $x=a_1$, noting that $H^{(k)}(x)$ and $H_j^{(k)}(x)$ satisfy the equations

$$\Delta H^{(k)}=0,\Delta H_j^{(k)}=\mu_j H_j^{(k)},\tag{A.19}$$

and have the proprieties

$$DH^{(k)}(a_1)=DH^{(k)}(-a_2)=0,\quad DH_j^{(k)}(a_1)=DH_j^{(k)}(-a_2)=0,\tag{A.20}$$

$$H^{(k)}(x_0^+)-H^{(k)}(x_0^-)=-1,\quad H_j^{(k)}(x_0^+)-H_j^{(k)}(x_0^-)=-1,\tag{A.21}$$

$$H^{(k)}(-x_0^+)-H^{(k)}(-x_0^-)=-\eta_1^{(k)},\quad H_j^{(k)}(-x_0^+)-H_j^{(k)}(-x_0^-)=-\eta_1^{(k)},\tag{A.22}$$

$$DH^{(k)}\text{ and }DH_j^{(k)}\text{ are continuous across }x=\pm x_0,\tag{A.23}$$

to obtain the relation

$$n^2\int_{-a_2}^{a_1}H^{(k)}p_1^{(k)}\mathrm{d}x=\int_{-a_2}^{a_1}H^{(k)}\left(M_p^{(k)}+M_w^{(k)}\right)\mathrm{d}x+\sum_{j=1}^{3}C_j\int_{-a_2}^{a_1}\left(\left(\mu_j^2+1\right)M_p^{(k)}+M_w^{(k)}+\mu_jM_T^{(k)}\right)H_j^{(k)}\mathrm{d}x+\hat{g}^{(k)},\tag{A.24}$$

where $\hat{g}^{(k)}$ is defined in (112) and $g^{(k)}(a_1)$ and $g^{(k)}(-a_2)$ are given by

$$g^{(k)}(a_1)=\sum_{j=1}^{3}-C_j\mu_jF_{2j}^{(k)}\left(DT_1^{(k)}(a_1)+\mu_jp_1^{(k)}(a_1)\right),\tag{A.25}$$

$$g^{(k)}(-a_2)=\sum_{j=1}^{3}-C_j\mu_jF_{1j}^{(k)}\left(DT_1^{(k)}(-a_2)+\mu_jp_1^{(k)}(-a_2)\right).\tag{A.26}$$

We now eliminate the integral involving $p_1$ between (A.10) and (A.24) to obtain an expression for $\Omega_2^{(k)}$

$$\Omega_2^{(k)}=\frac{\mathrm{i}}{n}f_\eta^{(k)}\left[\int_{-a_2}^{a_1}H^{(k)}\left(M_p^{(k)}+M_w^{(k)}\right)\mathrm{d}x+\sum_{j=1}^{3}C_j\int_{-a_2}^{a_1}\left\{H_j^{(k)}\left(\left(\mu_j^2+1\right)M_p^{(k)}+M_w^{(k)}+\mu_jM_T^{(k)}\right)\right\}\mathrm{d}x\right]+\frac{\mathrm{i}}{n}f_\eta^{(k)}\hat{g}^{(k)}+M_0^{(k)},\tag{A.27}$$

which is the expression for the growth rate in terms of the zeroth order variables and the basic state only.

2

# Computational Fluid Dynamics Study on Water Flow in a Hollow Helical Pipe

**Ebrahim Ahmadloo[1*], Najmeh Sobhanifar[2], Fatemeh Sadat Hosseini[3]**

[1]Young Researchers and Elite Club, Darab Branch, Islamic Azad University, Darab, Iran
[2]Young Researchers and Elite Club, Shoushtar Branch, Islamic Azad University, Shoushtar, Iran
[3]Young Researchers and Elite Club, Shiraz Branch, Islamic Azad University, Shiraz, Iran
Email: *Ebrahimahmadloo@yahoo.com, sobhanifar.najme@yahoo.com, polytak2000@gmail.com

---

## Abstract

**Although curved pipes are used in a wide range of applications, flow in curved pipes is relatively less well known than that in straight ducts. This paper presents a computational fluid dynamics study of isothermal laminar single-phase flow of water in a hollow helical pipe at various Reynolds numbers. The ranging of Reynolds numbers of fluid was from 703.2 to 1687.7. The three dimensional governing equations for mass and momentum have been solved. It was found that with increasing Reynolds number and creation of centrifugal forces, a high velocity and pressure region occurs between two tubes, at the outer side of the hollow helical pipe walls. Friction factor decreases as the tendency for turbulence increases.**

## Keywords

**Computational Fluid Dynamics, Water Flow, Hollow Helical Pipe, Pressure Gradient, Velocity**

---

## 1. Introduction

Although curved pipes are used in a wide range of applications, flow in curved pipes is relatively less well known than that in straight ducts. In recent years, Helical coiled tubes have been more attention because of their practical applications such as compactness, easy manufacture and high efficiency in heat and mass transfer. They are extensively used in compact heat-exchangers, heat-exchanger network, heating or cooling coils in the piping systems, intake in air-crafts, fluid amplifiers, coil steam generators, refrigerators, nuclear reactors, thermo siphons, and other heat transfer equipment involving phase change, chemical plants as well as in the food and drug industries [1].

*Corresponding author.

In particular, helical coils are also used as steam generators in some "generation IV" nuclear reactors like IRIS [2].

During the designing and operating processes of helical coiled tubing steam generators, the prediction of pressure drop of single-phase flow and two-phase mixtures is an essential step for the adjustment of pump pressure head and even calculation of heat transfer conditions. Despite considerable progress that has been made in the past several decades this, and many different correlations (such as that of Berger and Talbot [3], Akagawa *et al.* [4], Chen and Zhou [5], Unal *et al.* [6], Nariai *et al.* [7], Watanabe *et al.* [8], Guo *et al.* [9]), were proposed.

Naphon and Wongwises [10], in particular, recently reviewed more than 100 studies addressing both single and two-phase flow through curved tubes. Two-phase flow pressure drop in helical coils, in particular, received considerable attention, both in adiabatic and diabatic flow conditions.

M. Moawed [11] investigated the forced convection from helical coiled tubes with different parameters. Their results showed that, for the same $P/d_o$ (Pitch of coil tube/Outer diameter of coil tube), the higher values of Nusselt number can be obtained with a high value of $D/d_o$ (Coil diameter/Outer diameter of coil tube) while the small value of Nusselt number can be obtained with a small value of $D/d_o$.

Austen and Soliman [12] studied laminar flow and heat transfer in helically coiled tubes with substantial pitch. The influence of pitch on the pressure drop and heat transfer characteristics of helical coils was explored for the condition of uniform input heat flux. Water was used as the test fluid. Significant pitch effects were noted in the friction factor and Nusselt number results at low Reynolds number.

Choi *et al.* numerically studied the steady laminar flows in coiled hollow ducts and observed the evolution of secondary flow and the effect of radius ratio on the flow development. It was concluded that the flow in a curved hollow duct was not necessarily fully developed earlier when the radius ratio was larger owing to the complicated interaction between the viscous and centrifugal forces [13]. Hollow helical pipes could not only provide more compactness in volume, but also controlled main flow and secondary flow due to the inner-wall boundary effects, making the friction factor and heat transfer characteristics different from straight circular tubes [14]. For hollow helical pipe, the experimental results of Xin *et al.* indicated that the single-phase and two-phase flow pressure drops in hollow helical pipe differ from that in helical pipe with a circular cross-section [15].

## 2. Analysis and Modeling

### 2.1. Experiment

**Figure 1** shows the geometry of test section. The setup consisted of a hollow helical pipe with 9 turns, inner coil of 9.53 mm diameter (I.D.), and outer coil of 6.22 mm. The coil has stretched lengths of 2.286 m. The diameter and the pitch of the outer coil are 196.85 mm and 25.4 mm, respectively [6].

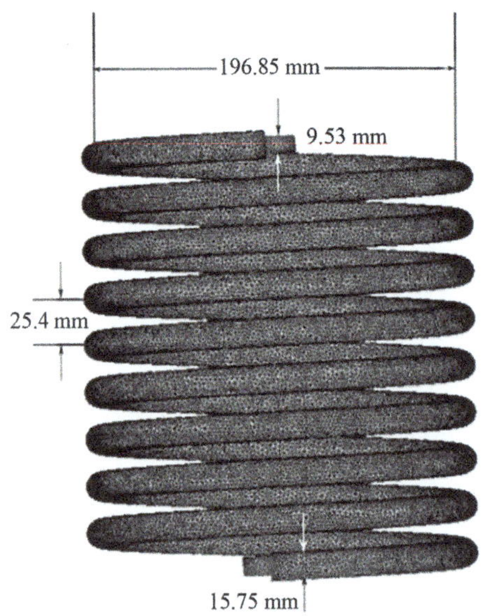


**Figure 1.** The schematic of hollow helical pipe.

## 2.2. Governing Equations

### 2.2.1. Continuity Equation

$$\rho\frac{\partial u}{\partial x}+\rho\frac{\partial v}{\partial y}+\rho\frac{\partial w}{\partial z}=0 \tag{1}$$

### 2.2.2. Momentum Equations

$$\rho\frac{\partial u}{\partial t}+\rho u\frac{\partial u}{\partial x}+\rho v\frac{\partial u}{\partial y}+\rho w\frac{\partial u}{\partial z}=\mu\left(\frac{\partial^2 u}{\partial x^2}+\frac{\partial^2 u}{\partial y^2}+\frac{\partial^2 u}{\partial z^2}\right)-\frac{\partial p}{\partial x} \tag{2-a}$$

$$\rho\frac{\partial v}{\partial t}+\rho u\frac{\partial v}{\partial x}+\rho v\frac{\partial v}{\partial y}+\rho w\frac{\partial v}{\partial z}=\mu\left(\frac{\partial^2 v}{\partial x^2}+\frac{\partial^2 v}{\partial y^2}+\frac{\partial^2 v}{\partial z^2}\right)-\frac{\partial p}{\partial y} \tag{2-b}$$

$$\rho\frac{\partial w}{\partial t}+\rho u\frac{\partial w}{\partial x}+\rho v\frac{\partial w}{\partial y}+\rho w\frac{\partial w}{\partial z}=\mu\left(\frac{\partial^2 w}{\partial x^2}+\frac{\partial^2 w}{\partial y^2}+\frac{\partial^2 w}{\partial z^2}\right)-\frac{\partial p}{\partial z} \tag{2-c}$$

## 2.3. Boundary Conditions

At the inlet according to different Reynolds numbers for laminar flow, corresponding normal speeds were used. No turbulence option was chosen for analysis (laminar flow). At outlet, pressure boundary condition was applied. No slip boundary condition was used at walls.

## 2.4. Solution Method

Convergence criterion of 1.0E−5 for RMS of all of the equations was used. Hybrid scheme was used for discretization of all equations and SIMPLEC algorithm was used for pressure-velocity coupling. The effect of gravitational force is applied in this analysis. In this analysis, a superficial velocity of 0.101, 0.113, 0.162 and 0.243 $ms^{-1}$ for Reynolds numbers of 703.2, 787.6, 1125.3 and 1687.7 at the inlet are specified, respectively.

## 3. Results and Discussion

**Figure 2** shows the contour plots of velocity at four positions; 1 × 180°, 6 × 180°, 9 × 180° and outlet. Four Reynolds number of 703.2, 787.6, 1125.3 and 1687.7 for laminar flow were also applied. As can be seen in the **Figure 2**, we have more uniformed high speed between two tubes, at the outer side of the hollow helical pipe walls, as the Reynolds number enhances. This is due to the effect of centrifugal force, but slowest speed is related to near regions to the inner tube walls. When is creating centrifugal forces, high velocity region is observed at the outer side of the hollow helical pipe, whereas in the case of flow in a straight pipe, high velocity region is observed at the centre of the pipe. The location of high velocity region shifted slightly down in successive turns of pipe, and the band of very high water velocity (dark red band) is slightly wider, too. **Figure 3** shows the contour plots of pressure gradient at four positions; 1 × 180°, 6 × 180°, 9 × 180° and outlet of hollow helical pipe for water flow with 703.2, 787.6, 1125.3 and 1687.7 Reynolds numbers for laminar flow. The curvature of the pipe causes centrifugal force to act on the fluid. This leads to a higher velocity in the outer side of the pipe, as compared to the flow in a straight pipe, where high velocity region is observed at the centre of the pipe. The acceleration forces acting on the fluid flow in the pipe create high pressure region at outer side of the pipe wall. In other words, raising the pressure on the outer side of the wall caused the buoyancy forces. It was equal with 522.163, 591.548, 905.118 and 1481.64 Pascal for Reynolds numbers of 703.2, 787.6, 1125.3 and 1687.7 as calculated and depicted in Equation (3). Then friction factor was calculated from Equation (4) as its results are presented in **Table 1** and **Figure 4**. The results were too close to the experimental data.

$$Re=\frac{\rho u\left(d_o-d_i\right)}{\mu} \tag{3}$$

The friction factor for single phase flow is determined using:

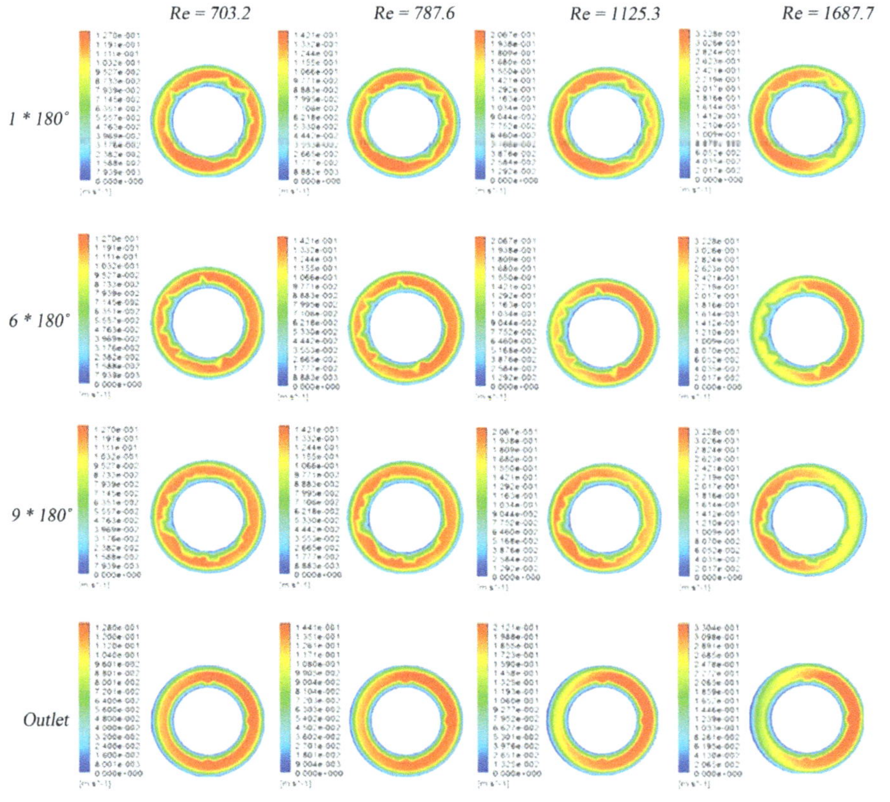


**Figure 2.** The contour plots of velocity at different positions.

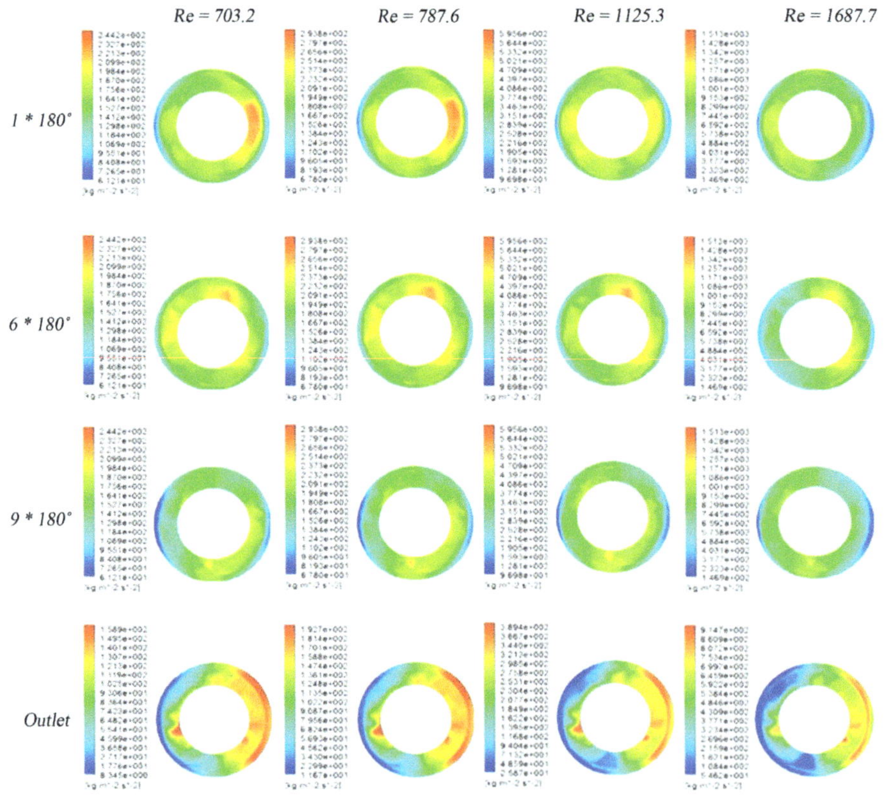


**Figure 3.** The contour plots of pressure gradient at different positions.

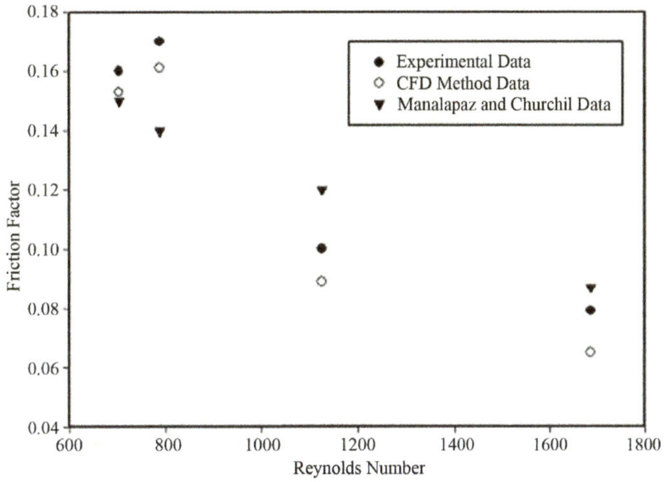


**Figure 4.** Friction factor of laminar single liquid phase flow of water in hollow helical pipe versus various Reynolds number.

**Table 1.** Values of friction factor versus Reynolds number with computed average and mean average errors.

| Reynolds No. | $f$ (CFD Method) | $f$ (Experimental) | AE[a] |
|---|---|---|---|
| 703.2 | 0.153 | 0.16 | 0.04375 |
| 787.6 | 0.161 | 0.17 | 0.05294 |
| 1125.3 | 0.089 | 0.1 | 0.11 |
| 1687.7 | 0.065 | 0.079 | 0.1772 |
| | | MAE[b] | 0.09597 |

[a] $AE = \left|\frac{f^{CFD} - f^{exp}}{f^{exp}}\right|$; [b] $MAE = \sum_{i=1}^{NP} \frac{\left|\frac{f^{CFD} - f^{exp}}{f^{exp}}\right|}{NP}$; NP, number of data points.

$$f = \frac{(\Delta p/L)(d_o - d_i)}{\frac{1}{2}\rho u^2} \quad (4)$$

It seems that the friction factor results can be predicted with reasonable accuracy in the complete laminar flow regions. The critical Reynolds numbers predicted by Srinivasan's equation (Srinivasan *et al.* 1970) is 1139 for this test section.

**Figure 4** shows the values of friction factor of laminar single liquid phase flow of water in hollow helical pipe versus various Reynolds numbers, graphically. As well depicted in this figure, we computed the values of experimental data of friction factor with CFD method with a high accuracy compared with Manalapaz and Churchil method [16].

The calculated and experimental values of friction factor with computed average and mean average errors can be observed in **Table 1**. The results were very satisfactory.

Therefore, we would like to present and suggest a new technique namely CFD method, for simulation and prediction of various processes in industrial applications. In this approach, we are able to design the best geometry of equipments in different industries with the highest accuracy and maximum of efficiency.

## 4. Conclusion

Laminar single-phase of water flow through a hollow helical pipe is simulated using computational fluid dynamics (CFD) model. It was found that centrifugal forces created a high velocity region at the outer side of the hollow helical pipe walls. The acceleration forces acting on the fluid flow in the pipe create high pressure region at outer side of the pipe wall. Friction factor decreases with tend to increasing turbulency, although the results of MAE of correlations—CFD model showed a very close values compared with experimental data.

## References


[1] Ali, S. (2001) Pressure Drop Correlations for Flow through Regular Helical Coil Tubes. *Fluid Dynamics Research*, **28**, 295-310. http://dx.doi.org/10.1016/S0169-5983(00)00034-4

[2] Carelli, M.D., Conway, L.E., Oriani, L., Petrović, B., Lombardi, C.V., Ricotti, M.E., Barroso, A.C.O., Collado, J.M., Cinotti, L., Todreas, N.E., Grgić, D., Moraes, M.M., Boroughs, R.D., Ninokata, H., Ingersoll, D.T. and Oriolo, F. (2004) The Design and Safety Features of the IRIS Reactor. *Nuclear Engineering and Design*, **230**, 151-167. http://dx.doi.org/10.1016/j.nucengdes.2003.11.022

[3] Berger, S.A., Talbot, L. and Yao, L.S. (1983) Flow in Curved Pipes. *Annual Review of Fluid Mechanics*, **15**, 461-512. http://dx.doi.org/10.1146/annurev.fl.15.010183.002333

[4] Akagawa, K., Sakaguchi, T. and Ueda, M. (1971) Study on Gas-Liquid Two-Phase Flow in Helically Coiled Tubes. *Bulletin of JSME*, **14**, 564-571. http://dx.doi.org/10.1299/jsme1958.14.564

[5] Chen, X.J. and Zhou, F.D. (1981) An Investigation of Flow Pattern and Frictional Pressure Drop Characteristics of Air-Water Two-Phase Flow in Helical Coils. *Procceedings of the Fourth Miami International Conference on Alternate Energy Sources*, 120-129.

[6] Unal, H.C., Van Gasself, M.L.G. and Von't Versalt, P.M. (1981) Dryout and Two-Phase Flow Pressure Drop in Sodium Heated Helically Coiled Steam Generators Tubes at Elevated Pressure. *International Journal of Heat and Mass Transfer*, **24**, 285-298. http://dx.doi.org/10.1016/0017-9310(81)90036-3

[7] Nariai, H., Kobayashi, M. and Matsuoka, T. (1982) Friction Pressure Drop and Heat Transfer Coefficient of Two-Phase Flow in Helically Coiled Tube Once-Through Steam Generator for Integrated Type Marine Water Reactor. *Journal of Nuclear Science and Technology*, **109**, 936-947. http://dx.doi.org/10.1080/18811248.1982.9734239

[8] Watanabe, O., Tajima, O. and Shimoya, O. (1986) Flow and Heat Transfer of Gas and Liquid Two-Phase Flow in Helical Coils. *Transactions on JSME*, **52**, 1857-1864. http://dx.doi.org/10.1299/kikaib.52.1857

[9] Guo, L.J., Chen, X.J., Zhang, S.K. and Feng, Z.P. (1994) Correlation for Predicting Pressure Drop of Single and Two-Phase Flow through Horizontal Helically Coiled Tubes. *Proceedings of the Third International Symposium on Multiphase Flow and Heat Transfer*, Xi'an, 514-521.

[10] Naphon, P. and Wongwises, S. (2006) A Review of Flow and Heat Transfer Characteristics in Curved Tubes. *Renewable and Sustainable Energy Reviews*, **10**, 463-490. http://dx.doi.org/10.1016/j.rser.2004.09.014

[11] Moawed, M. (2011) Experimental Study of Forced Convection from Helical Coiled Tubes with Different Parameters. *Energy Conversion and Management*, **2**, 1150-1156. http://dx.doi.org/10.1016/j.enconman.2010.09.009

[12] Austen, D.S. and Soliman, H.M. (1988) Laminar Flow and Heat Transfer in Helically Coiled Tubes with Substantial Pitch. *Journal of Experimental Thermal and Fluid Science*, **1**, 183-194. http://dx.doi.org/10.1016/0894-1777(88)90035-0

[13] Choi, H.K. and Park, S.O. (1992) Laminar Entrance Flow in Curved Annular Ducts. *International Journal of Heat and Fluid Flow*, **13**, 41-49. http://dx.doi.org/10.1016/0142-727X(92)90058-H

[14] Karahalios, G.T. (1990) Mixed-Convection Flow in a Heated Curved Pipe with Core. *Physics of Fluids*, **2**, 2164-2175. http://dx.doi.org/10.1063/1.857803

[15] Xin, R.C., Awwad, A., Dong, Z.F. and Ebadian, M.A. (1997) An Experimental Study of Single-Phase and Two-Phase Flow Pressure Drop in Annular Helicoidal Pipes. *International Journal of Heat and Fluid Flow*, **18**, 482-488. http://dx.doi.org/10.1016/S0142-727X(97)80006-9

[16] Manalapaz, R.L. and Churchill, S.W. (1980) Fully Developed Laminar Flow in a Helically Coiled Tube of Finite Pitch. *Chemical Engineering Communications*, **7**, 57-78. http://dx.doi.org/10.1080/00986448008912549

## Nomenclature

$d$: Diameter, m
$D$: Coil diameter, m
$L$: Pipe length, m
$P$: Pressure, Pa = N/m$^2$
$\Delta p$: Pressure difference, Pa
$r$: Tube radius, $d/2$
$R$: Coil mean radius, $D/2$
$X$: Spatial position, m
$u$: Velocity, m/s

## Greek Symbols

$\rho$ : Density of fluid, kg/m$^3$
$\mu$: Dynamic viscosity, Pa·s = N·s/m$^2$

## Subscripts

$i$: Inlet
$o$: Outlet

## Dimensionless Groups

$f$: Friction factor, $(\Delta p/L)(d_o - d_i)\Big/\left(\frac{1}{2}\rho u^2\right)$

$Re$: Reynolds number, $\rho u L/\mu$

3

# A Note on Hydromagnetic Flow of an Oldroyd-B Fluid near an Infinite Plate Induced by Half Rectified Sine Pulses

**Arun Kumar Ghosh[1], Sanjib Kumar Datta[2], Pulakesh Sen[3]**

[1]Jadavpur University, Kolkata, India
[2]Department of Mathematics, University of Kalyani, Kalyani, India
[3]Department of Mathematics, D. N. C. College, Aurangabad, India
Email: akg10143@yahoo.co.in, sanjib_kr_datta@yahoo.co.in, psendnc2011@gmail.com

## Abstract

**An initial value problem concerning the motion of an incompressible, electrically conducting, viscoelastic Oldroyd-B fluid bounded by an infinite rigid non-conducting plate is solved. The unsteady motion is generated impulsively from rest in the fluid due to half rectified sine pulses subjected on the plate in its own plane in presence of an external magnetic field. It is assumed that no external electric field is acting on the system and the magnetic Reynolds number is very small. The operational method is used to obtain exact solutions for the fluid velocity and the shear stress on the wall. Quantitative analysis of the results is presented with a view to disclose the simultaneous effects of the external magnetic field and the fluid elasticity on the flow and the wall shear stress for different periods of pulsation of the plate. It is also shown that the classical and hydromagnetic Rayleigh solutions appear as the limiting cases of the present analysis.**

## Keywords

**Hydromagnetic, Pulsatile Flow, Oldroyd-B Fluid, Contour Integral**

## 1. Introduction

The fluid flow generated by pulsatile motion of the boundary is found to have important applications in aerospace science, nuclear fusion, astrophysics, atmospheric science, cosmical gas dynamics, seismology and physiological fluid dynamics. Ghosh [1] examined the motion of an incompressible viscous fluid in a channel bounded by two coaxial cylinders when the inner cylinder is set in motion by the pulses of longitudinal impulses. Subsequently, Chakraborty and Ray [2] studied the unsteady magnetohydrodynamic coquette flow between two parallel plates when the upper plate is set in motion impulsively from rest by random pulses subjected on it. Makar [3] presented the solution of magnetohydrodynamic flow between two parallel plates when the velocity tooth pulses are imposed on the upper plate to initiate the fluid motion impulsively from rest and the induced magnetic field is neglected. Bestman and Njoku [4] constructed the solution of the same problem as that of author [3] without ignoring the effect of induced magnetic field and using the methodology different from that of author [3] to arrive at the solution of the problem. Ghosh and Sarkar [5] considered the hydromagnetic channel flow of an electrically conducting dusty fluid induced by tooth pulses. These authors have shown that the methodology followed by the authors [3] [4] leads to the same exact solution of the problem. Ghosh and Ghosh [6] solved the same problem as that of authors [5] replacing the boundary condition at the upper plate of the channel by rectified sine pulses instead of tooth pulses as encountered by authors [5]. In a subsequent paper Ghosh and Ghosh [7] also examined the hydromagnetic flow of a two-phase fluid near a pulsating plate with a view to its application in the analysis of suspension boundary layers. On the other hand, Datta & Datta [8] [9] discussed the heat transfer to pulsatile flow of a dusty fluid in pipes and channels with a view to their application in the analysis of blood flow. In spite of the above works, the study of pulsatile flows in Oldroyd-B fluids is equally important because these fluids encounter both elasticity and memory effects exhibited by most polymeric and biological liquids. Tanner [10] solved the classical Rayleigh problem associated with an Oldroyd-B fluid while Ghosh and Mitra [11] constructed solutions of Stokes problem for a linear viscoelastic fluid of Oldroyd-B type. Ghosh and Debnath [12] analysed the heat transfer to pulsatile flow of a viscoelastic Oldroyd-B fluid between two parallel plates. Exact solutions for some simple flows of an Oldroyd-B fluid were given by Rajagopal and Bhatnagar [13]. Some simple flows of an Oldroyd-B fluid were discussed by Hayat, Siddiqui & Asghar [14]. Exact solutions of MHD flows of an Oldroyd-B fluid over an infinite oscillatory plate were obtained by Hayat *et al.* [15]. Stokes' first problem for an Oldroyd-B fluid in a porous half space was solved by Tan and Masuoka [16]. Javadzadegan *et al.* [17] discussed the pulsatile flow of viscous and viscoelastic fluids in constricted tubes with a view to analyse the viscoelastic behaviour of blood using the Oldroyd-B fluid model. Most recently, Ghosh and Sana [18] presented solutions of hydromagnetic rotating flow of an Oldroyd-B fluid near an oscillating plate. In the present paper, we intend to construct solution of hydromagnetic flow of an Oldroyd-B fluid near an infinite plate induced by half rectified sine pulses with a view to its application in more generalised physical situations. It appears that besides various applications of pulsatile flow mentioned above, the present investigation seems to be useful in the analysis of suspension boundary layers, the blood flow during irregular heartbeats, flows in the liquid core of the earth during earthquakes, spreading of protective coating on electronic and optical devices and in the detection of magnetic effect on electrically conducting flows of biological fluids.

The problem is concerned with the unsteady hydromagnetic flow of a semi-infinite expanse of an incompressible, electrically conducting viscoelastic Oldroyd-B fluid bounded by an infinite rigid non-conducting plate. The motion is supposed to generate from rest in the fluid due to unidirectional motion of the plate in its own plane caused by half rectified sine pulses applied periodically on it. The operational method is used to obtain exact solutions for the fluid velocity and the shear stress exerted by the fluid on the pulsating plate. The quantitative analysis is made to examine the influence of the elasticity, the magnetic field and the time-period of pulsation on the fluid velocity and also on the wall shear stress. It is shown that hydromagnetic and hydrodynamic Rayleigh solutions appear as special cases of the present analysis.

## 2. Basic Equations

The constitutive equations for an Oldroyd-B fluid [14] [15] are

$$T = -p \cdot I + S \tag{2.1}$$

$$S + \lambda_1 \frac{Ds}{Dt} = \mu \left[ 1 + \lambda_2 \frac{D}{Dt} \right] A_1 \tag{2.2}$$

where $T$ = Cauchy stress tensor, $p$ = fluid pressure, $I$ = identity tensor, $S$ = extra stress tensor and $\mu$ = viscosity coefficient, $\lambda_1$ = relaxation time, $\lambda_2$ = retardation time (assumed constants).

The tensor $A_1$ is defined as

$$A_1 = \nabla V + (\nabla V)^{\mathrm{T}} \tag{2.3}$$

In a Cartesian system, $\frac{D}{Dt}$ (upper convected time derivative) operating on any tensor $B_1$ is

$$\frac{DB_1}{Dt} = \frac{\partial B_1}{\partial t} + (V \cdot \nabla) B_1 - (\nabla V) B_1 - B_1 (\nabla V)^{\mathrm{T}} \tag{2.4}$$

It is to be mentioned here that this model includes the viscous fluid as a particular case for $\lambda_1 = \lambda_2$, the Maxwell fluid when $\lambda_2 = 0$ and an Oldroyd-B fluid when $0 < \lambda_2 < \lambda_1 < 1$.

The stress equations of motion for an incompressible electrically conducting Oldroyd-B fluid in presence of an external magnetic field are

$$(\nabla V) = 0 \tag{2.5}$$

$$\rho \left[ \frac{\partial V}{\partial t} + (V \cdot \nabla) V \right] = \nabla \cdot T + J \times B \tag{2.6}$$

$$(\nabla B) = 0 \tag{2.7a}$$

$$(\nabla \times B) = \mu_o J \tag{2.7b}$$

$$(\nabla \times E) = -\frac{\partial B}{\partial t} \tag{2.8a}$$

$$J = \sigma [E + V \times B] \tag{2.8b}$$

where $V(u, v, w)$ = fluid velocity, $\rho$ = fluid density, $J$ = current density, $B$ = magnetic flux density, $E$ = electric field, $\mu_o$ = magnetic permeability (assumed constant), and $\sigma$ = electrical conductivity (assumed finite).

It is to be noted here that the electromagnetic effect entering in Equations (2.6) to (2.8b) are approximation under the assumption of the small magnetic Reynolds number. More precisely, the above equations hold good if the condition $V \ll \sigma L$, in which $V$ is the characteristic velocity and $L$ is the characteristic scale of motion, is satisfied. It is also noteworthy that the inequality given above holds in many cases of interest.

## 3. Formulation of the Problem

In this problem, we consider the motion of a viscoelastic electrically conducting Oldroyd-B fluid where the $x$-axis is taken in the direction of flow with the origin at the plate and y-axis perpendicular to it. The initial motion is generated in the fluid due to the half rectified sine pulses applied on the plate. A uniform magnetic field of strength $B_0$ is acting parallel to $y$-axis. We assume that no external electric field is acting on the fluid and the magnetic Reynolds number is very small. This implies that the current in the fluid is mainly due to the induced electric field and the applied magnetic field remains essentially unaltered by the electric current flowing in the fluid. We also assume that the induced magnetic field produced by the motion of the fluid is negligible compared to the applied magnetic field so that the Lorentz force term in Equation (2.6) becomes $-\sigma B_0^2 V$. Since the motion is a plain one and the plate is infinitely long, we assume that all the physical variables are independent of $x$ and $z$. Then from the equation of continuity (2.5) and from the physical conditions of the problem, we take

$$V = \{u(y,t), 0, 0\} \tag{3.1a}$$

and

$$S = S(y,t) \tag{3.1b}$$

The equations of motion in (2.6) then reduces to

$$\rho \frac{\partial u}{\partial t} = -\frac{\partial p}{\partial x} + \frac{\partial S_{xy}}{\partial y} - \sigma B_0^2 u \tag{3.2a}$$

$$\frac{\partial p}{\partial y} = \frac{\partial S_{yy}}{\partial y} \tag{3.2b}$$

$$\frac{\partial p}{\partial z}=0 \tag{3.2c}$$

It follows from Equations (2.2), (3.1a) and (3.1b) that

$$S_{xx}+\lambda_1\left[\frac{\partial S_{xx}}{\partial t}-2S_{xy}\frac{\partial u}{\partial y}\right]=-2\mu\lambda_2\left(\frac{\partial u}{\partial y}\right)^2 \tag{3.3a}$$

$$S_{xy}+\lambda_1\left[\frac{\partial S_{xy}}{\partial t}-S_{yy}\frac{\partial u}{\partial y}\right]=\mu\left(\frac{\partial u}{\partial y}\right)+\lambda_2\mu\frac{\partial^2 u}{\partial y\partial t} \tag{3.3b}$$

$$S_{yy}+\lambda_1\frac{\partial S_{yy}}{\partial t}=0 \tag{3.3c}$$

The Equation (3.3c) gives

$$S_{yy}=A(y)\mathrm{e}^{-t/\lambda_1} \tag{3.4}$$

where $A(y)$ is an arbitrary function of $y$. But $S_{yy}$ is known to be zero for $t < 0$. This implies that $A(y)$ must be zero. Hence $S_{yy}$ is zero always. Consequently, from (3.2a) and (3.3b) and in the absence of the pressure gradient along $x$ direction, we get

$$\left(1+\lambda_1\frac{\partial}{\partial t}\right)\frac{\partial u}{\partial t}=\nu\left(1+\lambda_2\frac{\partial}{\partial t}\right)\frac{\partial^2 u}{\partial y^2}-\frac{\sigma B_0^2}{\rho}\left(1+\lambda_1\frac{\partial}{\partial t}\right)u \tag{3.5}$$

which on introducing the dimensionless quantities given by $\bar{u}=\frac{u}{U_0},\bar{y}=\frac{y}{\sqrt{\nu\lambda_1}},\bar{t}=\frac{t}{\lambda_1},k=\frac{\lambda_2}{\lambda_1}(\le 1)$ and $n=\frac{\sigma B_0^2\lambda_1}{\rho}$ and on dropping the bars, we get

$$\left(1+\frac{\partial}{\partial t}\right)\frac{\partial u}{\partial t}=\left(1+k\frac{\partial}{\partial t}\right)\frac{\partial^2 u}{\partial y^2}-n\left(1+\frac{\partial}{\partial t}\right)u \tag{3.6}$$

The problem now reduces to solving Equation (3.6) subject to initial and boundary conditions:

$$u(y,t)=f(t)\text{ on }y=0\text{ when }t>0 \tag{3.7a}$$

$$u(y,t)\to 0\text{ as }y\to\infty\text{ when }t>0 \tag{3.7b}$$

and

$$u(y,0)=\frac{\partial u(y,0)}{\partial t}=0\text{ for }0\le y<\infty \tag{3.7c}$$

where $f(t)$ representing the half rectified sine pulses, as shown in **Figure 1**, is a periodic function of time with period $T$.

## 4. Solution of the Problem

In view of the nature of $f(t)$ (as shown in **Figure 1**) mentioned above the mathematical form of $u(0, t)$ may be written as

$$u(0,t)=E_0\left[\sin\frac{2\pi t}{T}H(t)+\sum_{m=1}^{\infty}\sin\frac{2\pi(t-mT/2)}{T}H(t-mT/2)\right] \tag{4.1}$$

where $H(t)$ is the Heaviside unit step function defined as

$$H(t-T)=0, t<T\text{ and }H(t-T)=1, t\ge T.$$

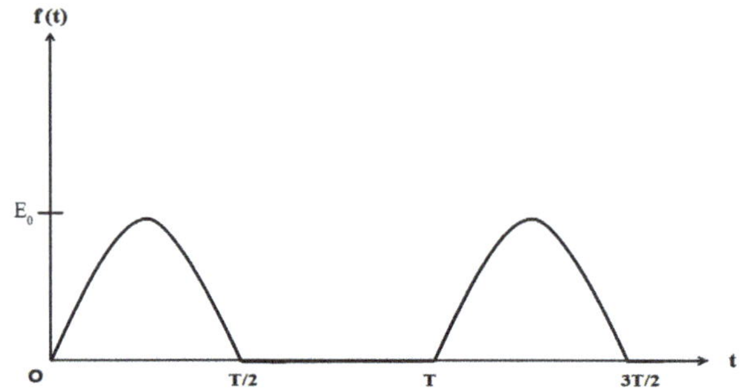


**Figure 1.** Half-rectified sine pulses with period *T*.

The use of Laplace transform method for the solution of the Equation (3.6) with initial conditions in Equation (3.7c), provides the transformed equation for the fluid velocity in the form

$$\frac{d^2\bar{u}}{dy^2}-\frac{(s+n)(1+s)}{1+ks}\bar{u}=0 \tag{4.2}$$

with the boundary conditions

$$\bar{u}\to 0 \text{ as } y\to\infty \tag{4.3}$$

and

$$\bar{u}(s)=E_0\frac{\pi\alpha}{(\alpha^2s^2+\pi^2)(1-e^{-\alpha s})} \quad \text{at } y=0 \tag{4.4}$$

where $s$ is the Laplace Transform variable and $\alpha$ = T/2.

The transformed solution for the fluid velocity $\bar{u}(y,s)$ becomes

$$\bar{u}(y,s)=E_0\frac{\pi\alpha}{(\alpha^2s^2+\pi^2)(1-e^{-\alpha s})}\exp\left\{-y\sqrt{\frac{(s+n)(1+s)}{1+ks}}\right\} \tag{4.5}$$

The inversion of Equation (4.5) gives

$$u(y,t)=\frac{E_0}{2\pi i}\int_{\gamma-i\infty}^{\gamma+i\infty}\frac{\pi\alpha e^{st}}{(\alpha^2s^2+\pi^2)(1-e^{-\alpha s})}\exp\left\{-y\sqrt{\frac{(s+n)(1+s)}{1+ks}}\right\}ds \tag{4.6}$$

The inversion integral has poles at $s=0,\pm\frac{i\pi}{\alpha}$ and a series of poles at $s=\pm i\beta_m,\beta_m=\frac{2m\pi}{\alpha}$, $m=1,2,\cdots$ and branch points at $s=-\frac{1}{k},-1$ and $-n$ as shown in the contour drawn in **Figure 2** in the complex $s$-plane.

Evaluating Equation (4.6) with the help of Cauchy's Residue theorem applied to the contour in **Figure 2**, we get

$$\begin{aligned}\frac{u(y,t)}{E_0}=&\frac{1}{\pi}e^{-y\sqrt{n}}+\frac{e^{-Ay}}{2}\sin\left(\frac{\pi t}{\alpha}-By\right)+\sum_{m=1}^{\infty}\frac{2\pi e^{-Cy}}{\pi^2-\alpha^2\beta_m^2}\cos(\beta_m t-Dy)\\&+\frac{1}{\pi}\int_{1/k}^{\infty}\frac{\pi\alpha e^{-xt}}{(\alpha^2x^2+\pi^2)(1-e^{\alpha x})}\sin\left\{\frac{y}{\sqrt{k}}\sqrt{\frac{(x-1)(x-n)}{x-\frac{1}{k}}}\right\}dx\\&+\frac{1}{\pi}\int_{n}^{1}\frac{\pi\alpha e^{-xt}}{(\alpha^2x^2+\pi^2)(1-e^{\alpha x})}\sin\left\{\frac{y}{\sqrt{k}}\sqrt{\frac{(1-x)(x-n)}{\frac{1}{k}-x}}\right\}dx\end{aligned} \tag{4.7}$$

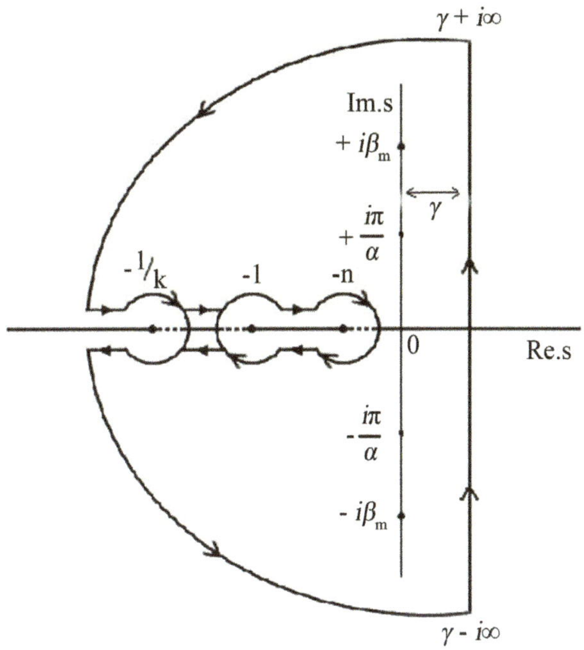


**Figure 2.** The Contour integral path for the Equation (4.7) showing the points of singularity and branch points.

where $A,B=\frac{1}{\sqrt{2\left(1+\frac{k^2\pi^2}{\alpha^2}\right)}}\sqrt{\sqrt{a^2+b^2}\pm a}$, $C,D=\frac{1}{\sqrt{2\left(1+k^2\beta_m^2\right)}}\sqrt{\sqrt{c^2+d^2}\pm c}$

$$a=n\left(1+\frac{k\pi^2}{\alpha^2}\right)-(1-k)\frac{\pi^2}{\alpha^2},\quad b=\frac{\pi}{\alpha}\left[n(1-k)+1+k\frac{\pi^2}{\alpha^2}\right],\quad c=n\left(1+k\beta_m^2\right)-\beta_m^2(1-k)$$

$$d=\beta_m\left[\left(1+k\beta_m^2\right)+n(1-k)\right],\quad \alpha=\frac{T}{2}\text{ and }\beta_m=\frac{2m\pi}{\alpha}.$$

On the other hand, the solution corresponding to the classical viscous fluid can be obtained from Equation (4.7) in the limit $k\to 1$. This solution is given by

$$\frac{u(y,t)}{E_0}=\frac{1}{\pi}e^{-y\sqrt{n}}+\frac{e^{-A_1y}}{2}\sin\left(\frac{\pi t}{\alpha}-B_1y\right)+\sum_{m=1}^{\infty}\frac{2\pi e^{-C_1y}}{\pi^2-\alpha^2\beta_m^2}\cos\left(\beta_m t-D_1y\right)$$
$$+\frac{1}{\pi}\int_n^{\infty}\frac{\pi\alpha e^{-xt}}{\left(\alpha^2x^2+\pi^2\right)\left(1-e^{\alpha x}\right)}\sin\left\{y\sqrt{x-n}\right\}dx \tag{4.8}$$

where $A_1,B_1=\frac{1}{\sqrt{2}}\sqrt{\sqrt{n^2+\frac{\pi^2}{\alpha^2}}\pm n}$ and $C_1,D_1=\frac{1}{\sqrt{2}}\sqrt{\sqrt{n^2+\beta_m^2}\pm n}$

Further, when $E_0=\pi$ and $T\to 0$, the Equation (4.7) reduces to

$$u(y,t)=e^{-y\sqrt{n}}-\frac{1}{\pi}\int_{1/k}^{\infty}\frac{e^{-xt}}{x}\sin\left\{\frac{y}{\sqrt{k}}\sqrt{\frac{(x-1)(x-n)}{x-\frac{1}{k}}}\right\}dx-\frac{1}{\pi}\int_n^{1}\frac{e^{-xt}}{x}\sin\left\{\frac{y}{\sqrt{k}}\sqrt{\frac{(1-x)(x-n)}{\frac{1}{k}-x}}\right\}dx \tag{4.9}$$

which describes the Hydromagnetic Rayleigh solution for an Oldroyd-B fluid and coincides with the findings of authors Ghosh and Mitra [11].

On by putting $k=1$ in the Equation (4.9), we get

$$u(y,t) = e^{-y\sqrt{n}} - \frac{1}{\pi}\int_n^\infty \frac{e^{-xt}}{x}\sin\left\{y\sqrt{x-n}\right\}dx = \frac{1}{2}\left[e^{y\sqrt{n}} erfc\left(\eta+\sqrt{nt}\right) + e^{-y\sqrt{n}} erfc\left(\eta-\sqrt{nt}\right)\right],\quad \eta = \frac{y}{2\sqrt{t}} \tag{4.10}$$

which is the well-known solution of the hydromagnetic Rayleigh problem (cf. [19]).

On the other hand, if $n \to 0$ we get from the Equation (4.10) the classical Rayleigh solution as

$$u(y,t) = erfc(\eta) \tag{4.11}$$

The fluid velocity given by the Equation (4.7) attains the steady-state in the limit $t \to \infty$ and the ultimate flow becomes

$$\frac{u(y,t)}{E_0} = \frac{1}{\pi}e^{-y\sqrt{n}} + \frac{e^{-Ay}}{2}\sin\left(\frac{\pi t}{\alpha} - By\right) + \sum_{m=1}^{\infty}\frac{2\pi e^{-Cy}}{\pi^2 - \alpha^2\beta_m^2}\cos(\beta_m t - Dy) \tag{4.12}$$

The ultimate steady flow given by Equation (4.12) contains the effect of both the elasticity and the magnetic field and consists of several distinct boundary layers in the vicinity of the plate due to the effect of pulsation. One is a Hartman layer of thickness of the order $\sqrt{\frac{\nu\lambda_1}{n}}$, the other is a Stokes-Hartman layer of thickness of order $\sqrt{\frac{\nu\lambda_1}{A}}$ and the rests are multiple Stokes-Hartman boundary layers of thicknesses of order $\sqrt{\frac{\nu\lambda_1}{C}}$.

Since $C > A > n$, the thickness of the Hartman layer is greater than the Stokes-Hartman layers which decrease with the increase of the elastic parameter ($k$) and the magnetic field ($n$). However, in the limit $T \to 0$ ($\alpha \to 0$, $\beta_m \to \infty$) *i.e.* in absence of pulsation, all the boundary layers coalesce into a single classical Hartman layer of thickness of order $\sqrt{\frac{\nu\lambda_1}{n}}$ which appears in the vicinity of the plate. This shows that the elasticity of the fluid can manifest its effect only in presence of pulsation.

The exact solution for the stress on the plate $y = 0$, in dimensionless form, is given by

$$\begin{aligned}\frac{\tau_0}{E_0} &= \frac{\sqrt{n}}{\pi} + \frac{1}{2}\left\{A^*\sin\left(\frac{\pi t}{\alpha}\right) + B^*\cos\left(\frac{\pi t}{\alpha}\right)\right\} \\ &+ \sum_{m=1}^{\infty}\frac{2\pi}{\pi^2-\alpha^2\beta_m^2}\left\{C^*\cos(\beta_m t) - D^*\sin(\beta_m t)\right\} \\ &- \frac{\sqrt{k}}{\pi}\int_{1/k}^{\infty}\frac{\pi\alpha e^{-xt}}{(\alpha^2x^2+\pi^2)(1-e^{\alpha x})}\sqrt{\frac{\left(x-\frac{1}{k}\right)(x-n)}{x-1}}dx \\ &- \frac{\sqrt{k}}{\pi}\int_{n}^{1}\frac{\pi\alpha e^{-xt}}{(\alpha^2x^2+\pi^2)(1-e^{\alpha x})}\sqrt{\frac{\left(\frac{1}{k}-x\right)(x-n)}{1-x}}dx\end{aligned} \tag{4.13}$$

where $A^*, B^* = \frac{1}{\sqrt{2\left(1+\frac{\pi^2}{\alpha^2}\right)}}\sqrt{\sqrt{a_1^2+b_1^2} \pm a_1}$, $C^*, D^* = \frac{1}{\sqrt{2(1+\beta_m^2)}}\sqrt{\sqrt{c_1^2+d_1^2} \pm c_1}$

$a_1 = n\left(1+\frac{k\pi^2}{\alpha^2}\right) + (1-k)\frac{\pi^2}{\alpha^2}$, $b_1 = \frac{\pi}{\alpha}\left[1+k\frac{\pi^2}{\alpha^2} - n(1-k)\right]$, $c_1 = n(1+k\beta_m^2) + \beta_m^2(1-k)$

$d_1 = \beta_m\left[(1+k\beta_m^2) - n(1-k)\right]$, $\alpha = \frac{T}{2}$ and $\beta_m = \frac{2m\pi}{\alpha}$.

which when $k \to 1$ yields

$$\begin{aligned}\frac{\tau_0}{E_0} &= \frac{\sqrt{n}}{\pi} + \frac{1}{2}\left\{A_1^* \sin\left(\frac{\pi t}{\alpha}\right) + B_1^* \cos\left(\frac{\pi t}{\alpha}\right)\right\} \\ &+ \sum_{m=1}^{\infty} \frac{2\pi}{\pi^2 - \alpha^2 \beta_m^2}\left\{C_1^* \cos(\beta_m t) - D_1^* \sin(\beta_m t)\right\} \\ &- \frac{1}{\pi}\int_n^{\infty} \frac{\pi \alpha e^{-xt}}{(\alpha^2 x^2 + \pi^2)(1 - e^{\alpha x})}\sqrt{x - n}\,dx\end{aligned} \tag{4.14}$$

where $A_1^*, B_1^* = \frac{1}{\sqrt{2}}\sqrt{\sqrt{n^2 + \frac{\pi^2}{\alpha^2}} \pm n}$, $C_1^*, D_1^* = \frac{1}{\sqrt{2}}\sqrt{\sqrt{n^2 + \beta_m^2} \pm n}$

On the other hand, if $T \to 0$ and $E_0 = \pi$, we have from Equation (4.13), the shear stress corresponding to Oldroyd-B fluid on the non-pulsating wall becomes

$$\tau_0 = \sqrt{n} + \frac{\sqrt{k}}{\pi}\int_{1/k}^{\infty} \frac{e^{-xt}}{x}\sqrt{\frac{\left(x - \frac{1}{k}\right)(x - n)}{x - 1}}dx + \frac{\sqrt{k}}{\pi}\int_n^1 \frac{e^{-xt}}{x}\sqrt{\frac{\left(\frac{1}{k} - x\right)(x - n)}{1 - x}}dx \tag{4.15}$$

which when $k \to 1$ provides the shear stress on the wall corresponding to hydromagnetic Rayleigh problem as

$$\tau_0 = \sqrt{n} + \frac{1}{\pi}\int_n^{\infty} \frac{e^{-xt}}{x}\sqrt{x - n}\,dx \tag{4.16}$$

and in the limit $n \to 0$, yields the familiar result

$$\tau_0 = \frac{1}{\pi}\int_0^{\infty} \frac{e^{-xt}}{\sqrt{x}}dx = \frac{1}{\sqrt{\pi t}} \tag{4.17}$$

## 5. Numerical Results

Unlike the generating flow in Rayleigh problem and the oscillatory result given by Stokes solution the nature of the pulses imparted on the plate produces both developing and retarding motions propagating unidirectionally in the fluid during the duration of each pulse. The quantitative analysis of the result (4.7) representing the fluid velocity is shown in **Figure 3(a)** for the increasing motions corresponding to $t = 0.1$ and 0.25 and the retarding motions for $t = 0.45$ and 0.7 when $T = 1.0$, $n = 0.1$ and the elastic parameter $k$ assuming the values 0.3, 0.7 and 1.0 in each case. It appears that the fluid velocity near the plate increases with $k$ when the motion is developing and decreases with $k$ when the motion is retarding. For example, at $y = 0.2$, $n = 0.1$, $T = 1.0$, and $k$ changes from 0.3 to 0.7, the fluid velocity increases by 46.4% and 17.3% respectively when $t = 0.1$, 0.25. Similarly, at $y = 0.2$, $n = 0.1$, $T = 1.0$ and $k$ changes from 0.3 to 0.7, the fluid velocity decreases by 13.95% and 38.75% respectively when $t = 0.45$, 0.7. Such a result is expected in the presence of pulses although it is not true in the non-pulsatile case.

We also observe that the above increasing and decreasing effects produced by the elastic parameter $k$ is further enhanced with the increase of magnetic field ($n$). It is to be noted here that the value of the elastic parameter $k = 1$, represents the Newtonian case and the fluid will be more and more viscoelastic for lesser and lesser values of $k$. On the other hand, the magnetic field ($n$) produces damping effect on the flow irrespective of its nature. This fact is presented in **Figure 3(b)**. Similarly, the effects of the magnetic field and the elasticity of the fluid associated with both the increasing and decreasing motions for $T = 0.5$ and $T = 2.0$ are shown respectively in **Figure 4(a)** and **Figure 4(b)**. A comparative study of the developing flow produced by the impulsively moved plate and the pulsatile plate is shown in **Figure 4(c)** for $T = 0$ and $T = 1.0$ respectively. It is seen that the velocity of developing flow near the plate in the pulsatile case ($T = 1.0$) is higher than that produced by the impulsively moved plate ($T = 0$). Moreover, contrary to pulsatile case the fluid velocity near the plate diminishes with the increase of the fluid elasticity ($k$) in the non-pulsatile case although magnetic field has a damping effect on both types of flow. The solution for the hydromagnetic and the classical Rayleigh solutions corresponding to $k = 1$, $n = 0.9$ and $k = 1$, $n = 0$ respectively are represented graphically in **Figure 4(c)**. The effects of k and non the developing and retarding flows with the increase of time period of pulsation ($T$) are presented in **Figure 5(a)** and

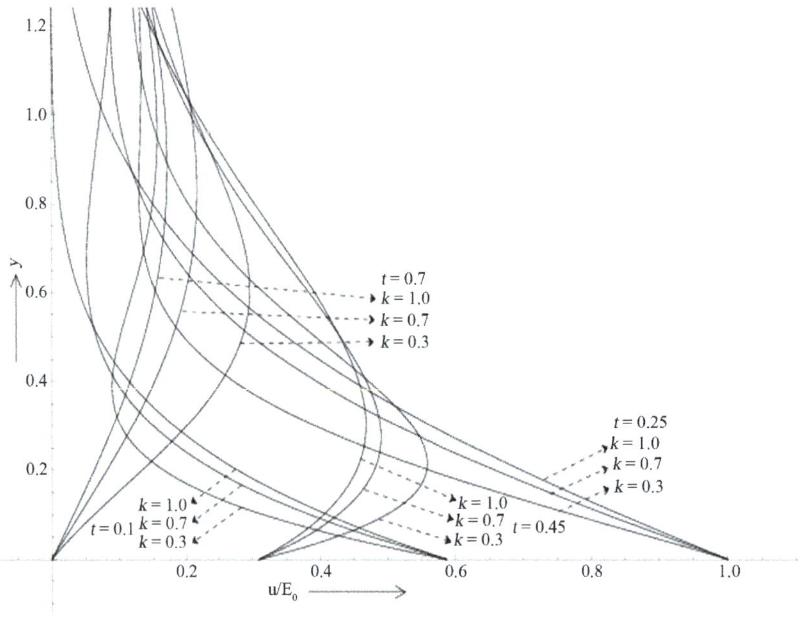


(a)

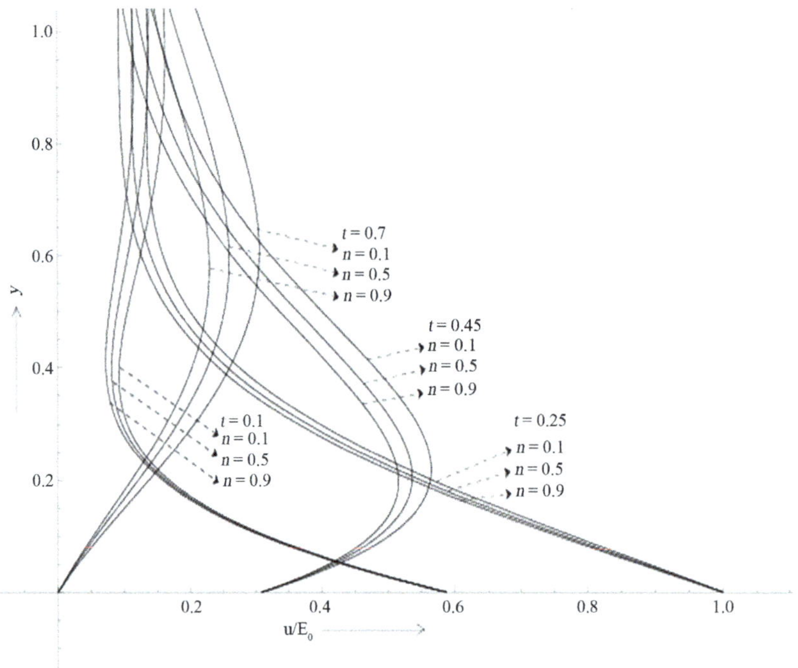


(b)

**Figure 3.** (a) Effect of elasticity ($k$) on the fluid velocity for the developing and retarding motions when $n = 0.1$ and $T = 1.0$; (b) Effect of the magnetic field ($n$) on the fluid velocity for the developing and retarding motions when $k = 0.3$ and $T = 1.0$.

**Figure 5(b)**. It is found that the developing flows gradually diminishes with the increase of time period $T$ while a reverse effect is found in the case of retarding motions. The steady-state solutions for fixed values of $n$ and $k$ and for increasing values of $T$ are shown in **Figure 6**. It is clearly reflected from the nature of the velocity profiles shown in **Figure 6** that the structure of the hydromagnetic boundary layer is significantly different from those of pulsatile boundary layers. The thicknesses of the pulsatile boundary layers near the plate decrease with the increase of time period of pulsation which is expected from steady-state solution.

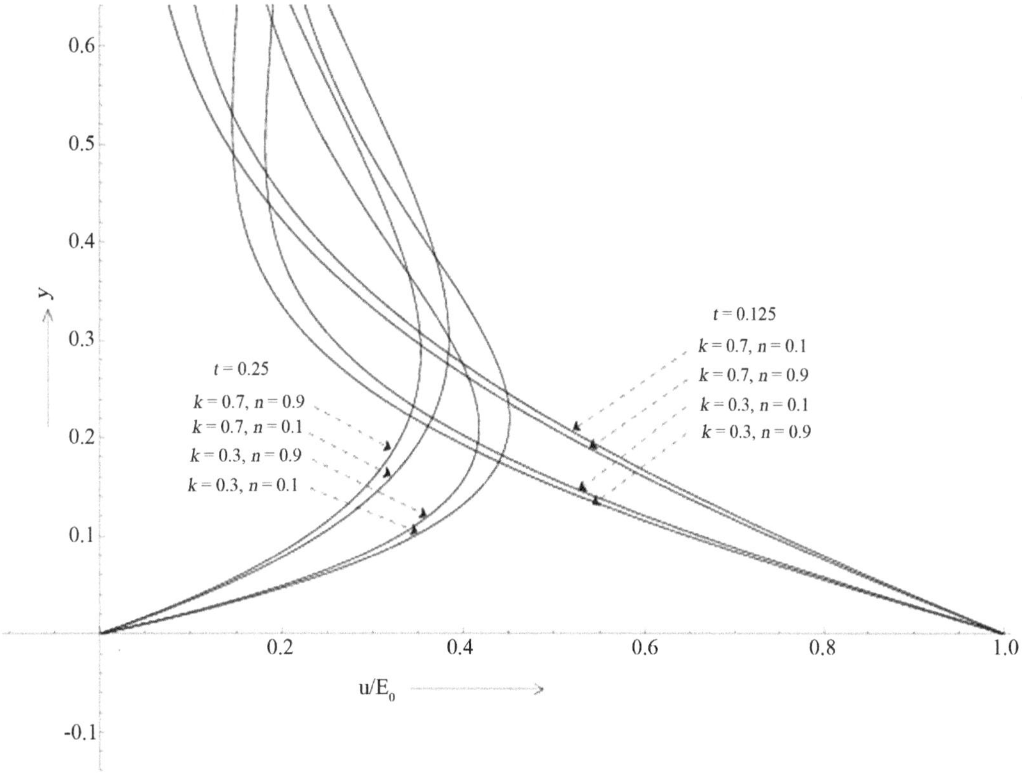
t = 0.125
k = 0.7, n = 0.1
k = 0.7, n = 0.9
k = 0.3, n = 0.1
k = 0.3, n = 0.9
t = 0.25
k = 0.7, n = 0.9
k = 0.7, n = 0.1
k = 0.3, n = 0.9
k = 0.3, n = 0.1
y
u/E0

(a)

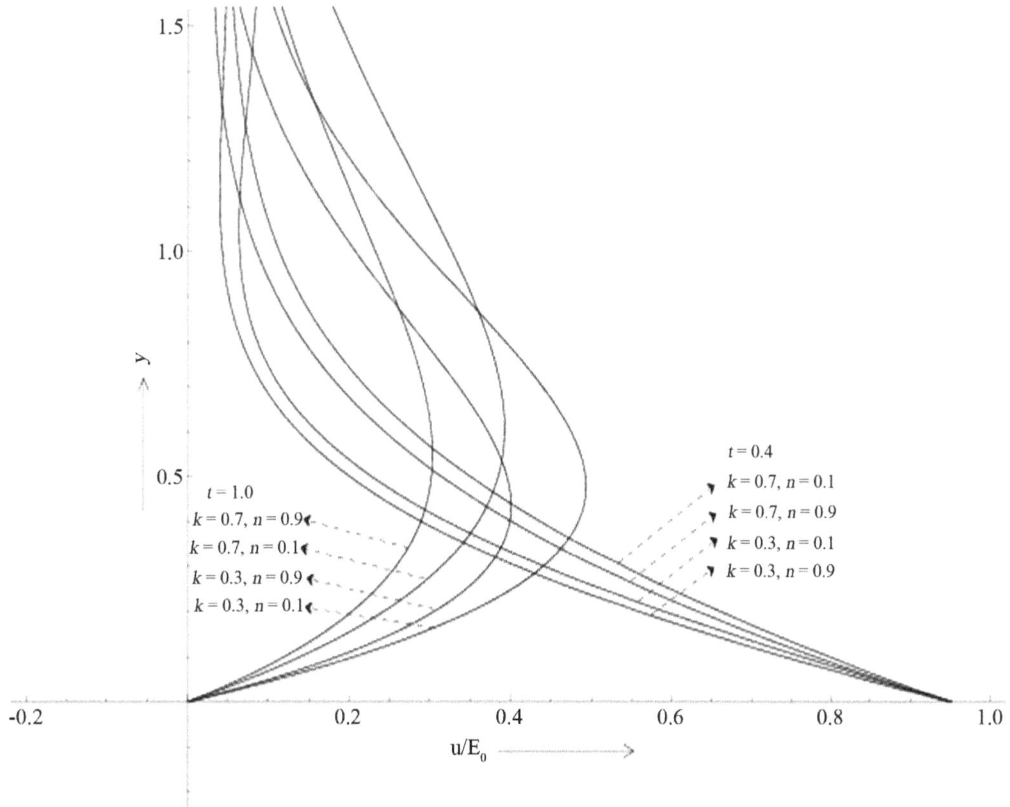
t = 0.4
k = 0.7, n = 0.1
k = 0.7, n = 0.9
k = 0.3, n = 0.1
k = 0.3, n = 0.9
t = 1.0
k = 0.7, n = 0.9
k = 0.7, n = 0.1
k = 0.3, n = 0.9
k = 0.3, n = 0.1
y
u/E0

(b)

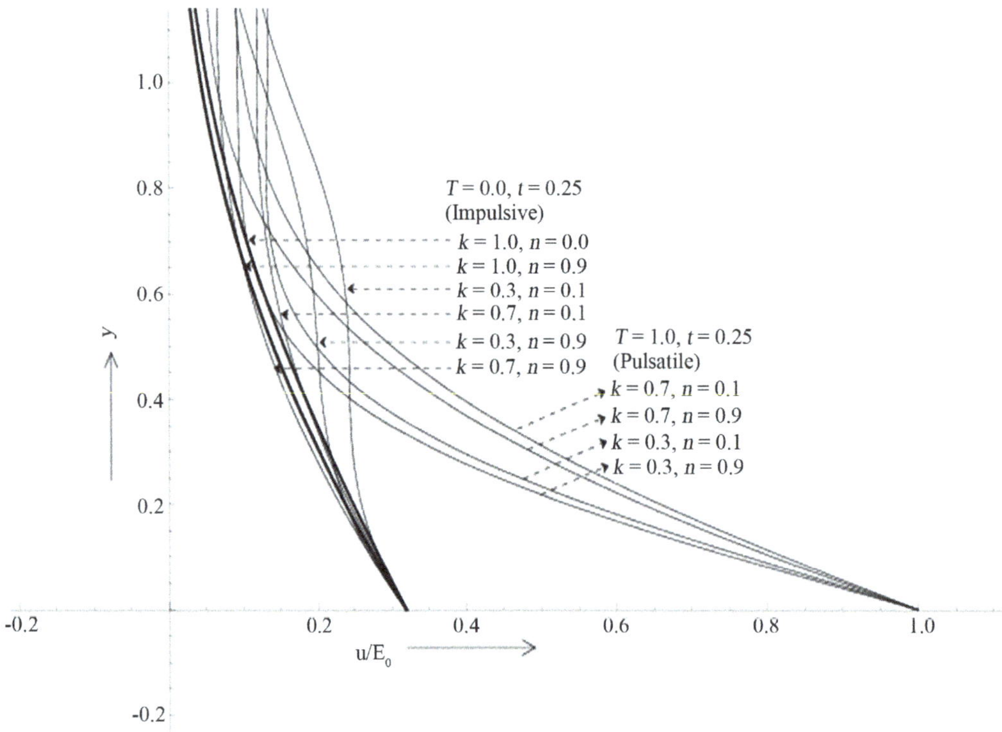


(c)

**Figure 4.** (a) Effects of the magnetic field ($n$) and the elasticity ($k$) on the fluid velocity for the developing and retarding flows when $T = 0.5$; (b) Effects of the magnetic field ($n$) and the elasticity ($k$) on the fluid velocity for the developing and retarding flows when $T = 2.0$; (c) Effects of the magnetic field ($n$) and the elasticity ($k$) on the fluid velocity for both the Pulsatile and Impulsive motion of the plate at $t = 0.25$.

The distribution of shear stress on the plate is plotted in the **Figures 7(a)-(c)** with the advancement of time, respectively, for the cases $T = 0.5$, 1.0 and 2.0. The limiting result associated with the Rayleigh stress at $T = 0.0$ is presented in **Figure 7(d)**. The effects of the elasticity ($k$) and the magnetic field ($n$) have also considered in each case. It is seen that the wall shear stress fluctuates with time in a manner similar to that of the pulses on the plate for all non-zero values of the time period ($T$). Moreover, unlike Rayleigh-stress, the shear stress on the plate in each case becomes negative during the period of rectification. This is expected because at the beginning of rectification, the decelerating motion of the plate starts which drives the fluid towards its destination by exerting shear stress in opposite direction (negative shear stress). Such a situation prevail still the plate acquires sufficient momentum to overcome the negative shear stress during its next accelerated motion. Besides, the maximum value of the shear stress is found to occur at about $t = 0.075\lambda_1$, $0.15\lambda_1$ and $0.3\lambda_1$ respectively when $T =$ 0.5, 1.0 and 2.0. Similar values can also be obtained for minimum values of shear stress on the plate which differs with $T$. In general, the stress on the plate decreases with the increase of the time period of pulsation ($T$) irrespective of the nature of the flow. On the other hand, as in the case of impulsively started motion, increase of the elastic parameter ($k$) increases the stress which is further enhanced with the increase of the magnetic field ($n$) when the flow is developing. However, a reverse effect is found when the motion is retarding.

## 6. Conclusion

The present problem is concerned with the investigation of unsteady motion generated from rest in a semi-infinite expanse of an Oldroyd-B fluid due to multiple frequency unidirectional motion of the plate bounding the fluid instead of that generated impulsively from rest by the movement of the plate with constant velocity (Stokes first problem) orby oscillation of the plate in its own plane with a given frequency (Stokes second problem). As a result, the present problem differs widely from those mentioned above as it provides a unidirectional positive flow for all values of time and Stokes first problem both in hydrodynamic and hydromagnetic situations emerges

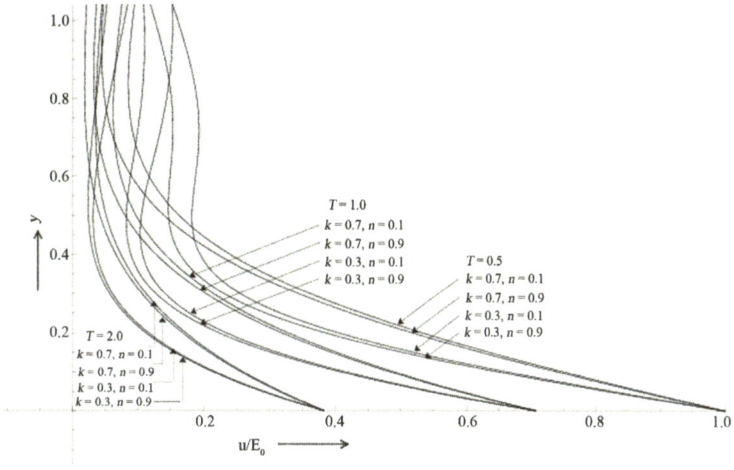


(a)

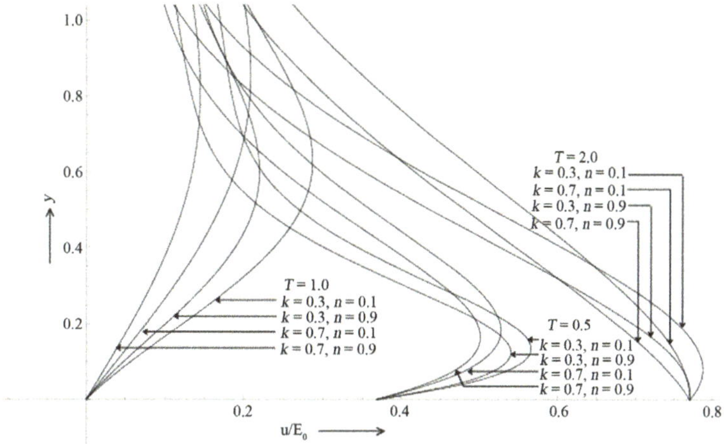


(b)

**Figure 5.** (a) Effects of the magnetic field ($n$) and the elasticity ($k$) on the fluid velocity for the developing motion when $T = 0.5$, 1.0, 2.0 at $t = 0.125$; (b) Effects of the magnetic field ($n$) and the elasticity ($k$) on the fluid velocity for the retarding motion when $T = 0.5$, 1.0, 2.0 at $t = 0.72$.

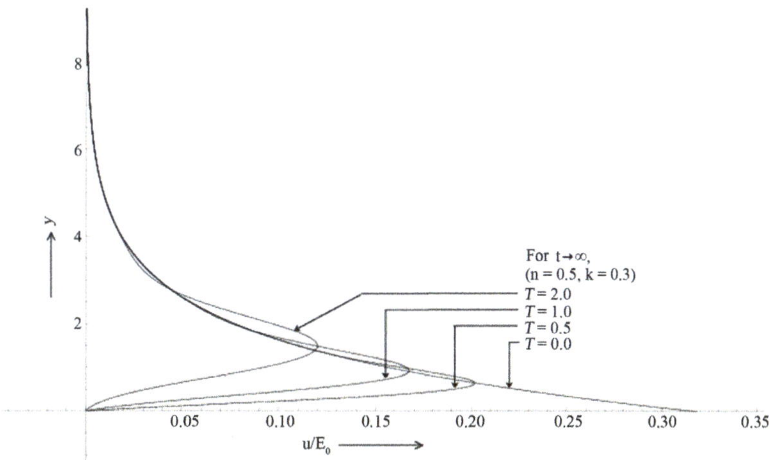


**Figure 6.** The steady-state solution for the fluid motion at $n = 0.5$, $k = 0.3$ when $T = 0.5$, 1.0, 2.0 as $t \to \infty$.

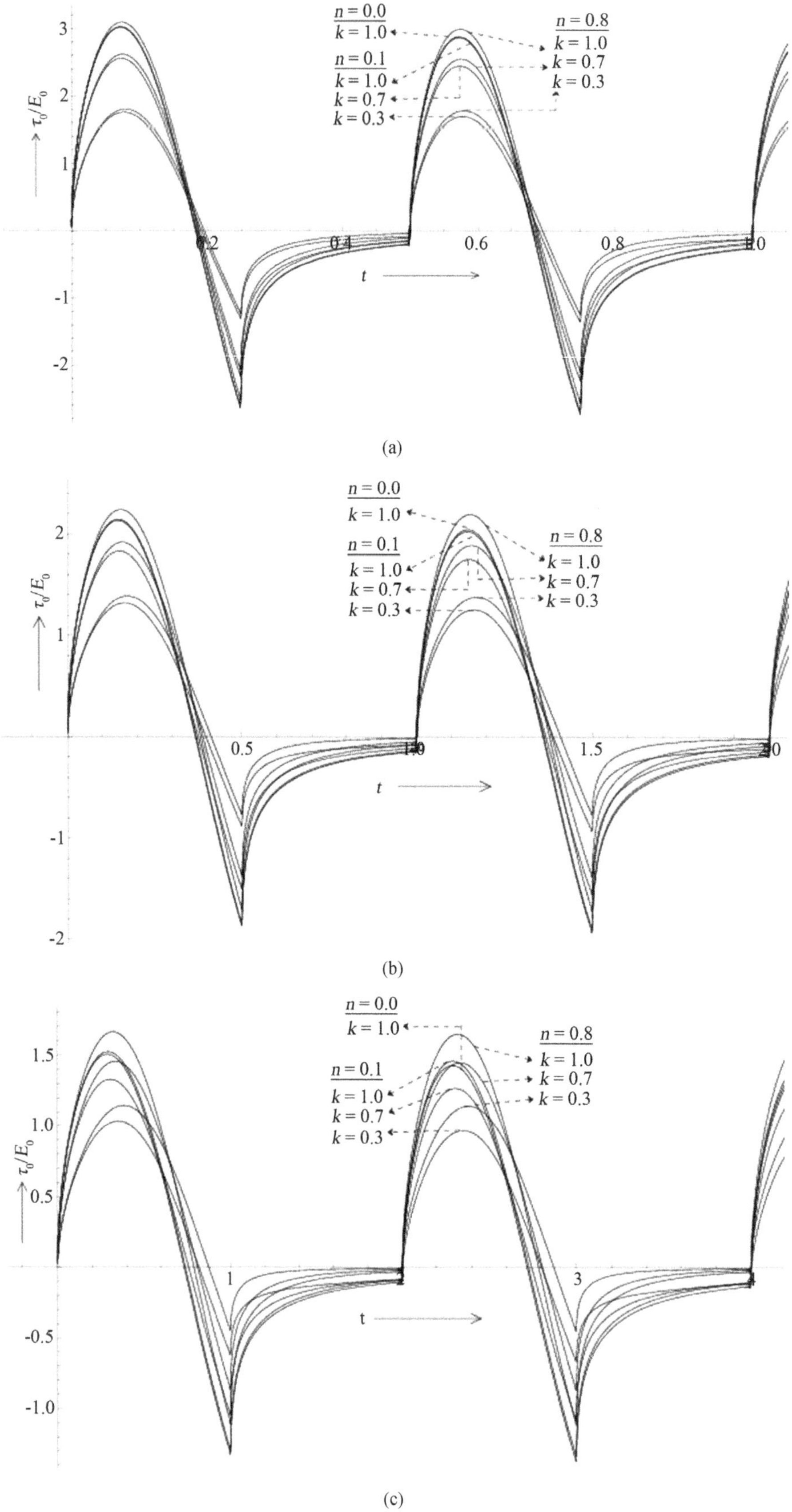

τ0/E0
t
n = 0.0
k = 1.0
n = 0.1
k = 1.0
k = 0.7
k = 0.3
n = 0.8
k = 1.0
k = 0.7
k = 0.3
(a)
(b)
(c)

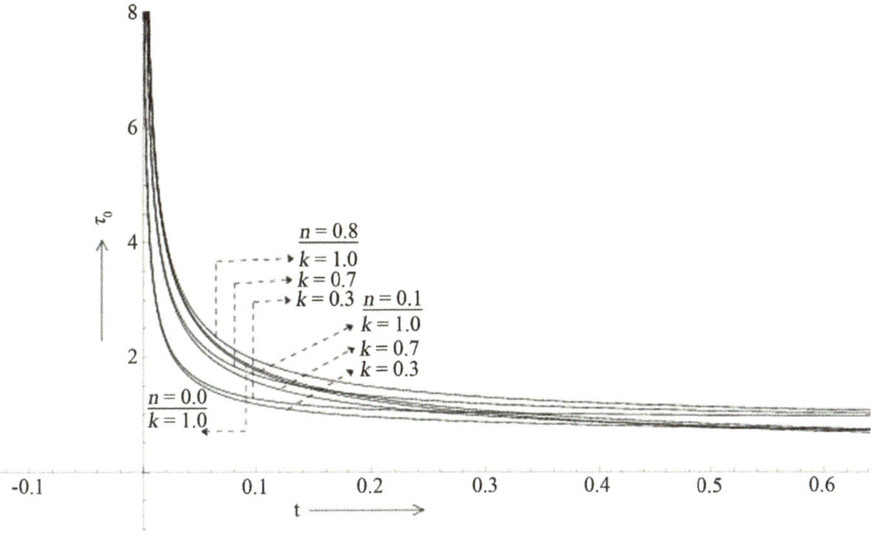


(d)

**Figure 7.** (a) Effects of the magnetic field ($n$) and the elasticity ($k$) on the shear stress at the plate $y = 0$ when $T = 0.5$; (b) Effects of the magnetic field ($n$) and the elasticity ($k$) on the shear stress at the plate $y = 0$ when $T = 1.0$; (c) Effects of the magnetic field ($n$) and the elasticity ($k$) on the shear stress at the plate $y = 0$ when $T = 2.0$; (d) Effects of the magnetic field ($n$) and the elasticity ($k$) on the shear stress at the plate $y = 0$ when $T = 0.0$.

as special cases of it. The present problem although idealized bears a good resemblance with the blood flow in large arteries during irregular heartbeats, some atmospheric and oceanic flows driven by pressure pulses, air flows during traffic movements, movement of aircraft in some strategic positions and geophysical flows during earthquakes. Consequently the present study will be helpful to discuss the structure of the boundary layers and the effects of flow parameters associated with the pulsatile motion of viscoelastic fluids. Additionally, as the major findings of the present analysis we notice that in the presence of pulsation, the elastic parameter increases the developing flow near the plate, whereas, in the absence of pulsation such a motion decreases with the increase of the elasticity ($k$). We further notice that the retarding motion near the plate caused by pulsation decreases with the increase of the elastic parameter $k$. In all cases, the effect of ($k$) on the flow is enhanced with the increase of $n$, although the magnetic field ($n$) has an overall effect of damping the motion irrespective of its nature. The intensity of the pulsatile flow near the plate is always greater than the impulsively generated flow. The magnetic field and the fluid elasticity require some time to exert their influence on the wall shear stress for all values of time period ($T$). Finally, the elastic parameter ($k$) and the magnetic field ($n$) increase the stress value for all values of $T$ which is much pronounced near the ends of the accelerated and retarded motions for the pulsatile case and near the beginning of the developing motion when the flow started impulsively from rest. However, for fixed values of $k$ and $n$ the stress value of the pulsatile flow decreases with the increase of the time period ($T$).

## References


[1] Ghosh, A.K. (1965) Flow of Viscous Liquid Set up between Two Co-Axial Cylinders Due to Pulses of Longitudinal Impulses Applied on the Inner Cylinder. *Journal of Science and Engineering Research*, **9**, 222.

[2] Chakraborty, A. and Ray, J. (1980) Unsteady Magnetohydrodynamic Couetteflow between Two Plates When One of the Plates Is Subjected to Random Pulses. *Journal of the Physical Society of Japan*, **48**, 1361-1364. http://dx.doi.org/10.1143/JPSJ.48.1361

[3] Makar, M.N. (1987) Magnetohydrodynamic Flow between Two Plates When One of the Platesis Subjected to Tooth Pulses. *Acta Physica Polonica*, **A71**, 995.

[4] Bestman, A.R. and Njoku, F.I. (1988) On Hydromagnetic Channel Flow Induced by Toothpulses. Miramare-Trieste.

[5] Ghosh, A.K. and Sarkar, K. (1995) On Hydromagnetic Channel Flow of a Dusty Fluid Induced by Tooth Pulses. *Journal of the Physical Society of Japan*, **64**, 1489-1500. http://dx.doi.org/10.1143/JPSJ.64.1489

[6] Ghosh, S. and Ghosh, A.K. (2004) On Hydromagnetic Channel Flow of a Particulate Suspension Induced by Rectified Sine Pulses. *Journal of the Physical Society of Japan*, **73**, 1506-1513. http://dx.doi.org/10.1143/JPSJ.73.1506

[7] Ghosh, S. and Ghosh, A.K. (2004) On Hydromagnetic Flow of a Two-Phase Fluid near a Pulsating Plate. *Indian Journal of Pure and Applied Mathematics*, **36**, 529.

[8] Datta, N., Dalal, D.C. and Mishra, S.K. (1993) Unsteady Heat Transfer to Pulsatile Flow of a Dustyviscous Incompressible Fluid in a Channel. *International Journal of Heat and Mass Transfer*, **36**, 1783-1788. http://dx.doi.org/10.1016/S0017-9310(05)80164-4

[9] Datta, N. and Dalal, D.C. (1995) Pulsatile Flow and Heat Transfer of a Dusty Fluid through an Infinitely Long Annular Pipe. *International Journal of Multiphase Flow*, **21**, 515-528. http://dx.doi.org/10.1016/0301-9322(94)00064-Q

[10] Tanner, R.I. (1962) Note on the Rayleigh Problem for a Viscoelastic Fluid. *Zeitschrift fur Angewandte Mathematik und Physik*, **8**, 573.

[11] Ghosh, A.K. and Mitra, S. (1977) On Stokes' Problems for Linear Viscoelastic Fluids. *Bulletinde L'Academie Polonaise des Sciences*, **25**, 55.

[12] Ghosh, A.K. and Debnath, L. (1992) On Heat Transfer to Pulsatile Flow of a Viscoelastic Fluid. *Acta Mechanica*, **93**, 169-177. http://dx.doi.org/10.1007/BF01182582

[13] Rajagopal, K.R. and Bhatnagar, R.K. (1995) Exact Solutions for Some Simple Flows of an Oldroyd-B Fluid. *Acta Mechanica*, **113**, 233-239. http://dx.doi.org/10.1007/BF01212645

[14] Hayat, T., Siddiqui, A.M. and Asghar, S. (2001) Some Simple Flows of an Oldroyd-B Fluid. *International Journal of Engineering Science*, **39**, 135-147. http://dx.doi.org/10.1016/S0020-7225(00)00026-4

[15] Hayat, T., Hutter, K., Asghar, S. and Siddiqui, A.M. (2002) MHD Flows of an Oldroyd-B Fluid. *Mathematical and Computer Modelling*, **36**, 987-995. http://dx.doi.org/10.1016/S0895-7177(02)00252-2

[16] Tan, W.C. and Masuoka, T. (2005) Stokes' First Problem for an Oldroyd-B Fluid in a Porous Half Space. *Physics of Fluids*, **17**, Article ID: 023101. http://dx.doi.org/10.1063/1.1850409

[17] Javadjzadegan, A., Esmaeili, M., Majidi, S. and Fakhimghanbarzadeh, B. (2009) Pulsatile Flow of Viscous and Viscoelastic Fluids in Constricted Tubes. *Journal of Mechanical Science and Technology*, **23**, 2456-2467. http://dx.doi.org/10.1007/s12206-009-0713-9

[18] Ghosh, A.K. and Sana, P. (2009) On Hydromagnetic Rotating Flow of an Oldroyd-B Fluid near an Oscillating Plate. *Zeitschrift für angewandte Mathematik und Physik*, **60**, 1135-1155. http://dx.doi.org/10.1007/s00033-009-8060-3

[19] Yang, H.T. and Healy, J.V. (1973) The Stokes Problem for a Conducting Fluid with Suspension of Particles. *Applied Scientific Research*, **21**, 378.

# 4

# Weis-Fogh Mechanism Mathematic Model of Wave Power Generation Device

**Xin Chen, Quan Kuang, Yang Li, Songbo Wang, Yunling Ye, Shesheng Zhang***

Wuhan University of Technology, Wuhan, China
Email: *sheshengz@qq.com

## Abstract

**A Weis-Fogh mechanism wave power generation system is designed, its physical model and mathematical model are discussed, and the component expressions of fluid dynamic expression are derived. Adopting numerical integral algorithm, the work done by fluid force acting on wing is calculated.**

## Keywords

**Wave, Weis-Fogh, Power Generation, Mathematical Model**

## 1. Introduction

Saving energy and reducing contaminant emission are important researching projects for economics and human living environment. As we know, using wave energy to generate electricity can reduce oil and coal consumption, and decease dust and emissions of harmful substances emission. So it attracts scientists to carry out power generation of scientific research by using wave energy [1]. According to the statistics data, there is huge amount of wave energy stored in the ocean, which can be used to transfer to electricity energy [2]. There are many devices to absorb wave energy, such as floating rope round of absorbing wave energy device [3], matrix of relative movement absorbing wave energy device [4], buoy type absorption wave energy device [5] and floating type absorption wave energy device [6], according to the fluctuation characteristics of waves. All of those devices didn't use the theory of Weis-Fogh [7]. According to the fluid mechanics theory of Weis-Fogh mechanism [8], Weis-Fogh mechanism is a simulation of the wasp flying V vibration device, which is different from conventional produce wing lift mechanism [9]. Rigid body test measures the lift coefficient as high as 7 - 8 when the wing opens [10] [11]. Thus, Weis-Fogh mechanism is possible to be used in the case of wave energy power generation. But it needs a lot of scientific research.

*Corresponding author.

The paper will consider Weis-Fogh wave power generation device, calculating its fluid doing work acted on the wing.

## 2. The Working Principle of the Generation Set

Weis-Fogh wave power energy generation device makes of two parts mainly: one is Weis-Fogh absorbing wave energy mechanism; another is the power generation mechanism. As **Figure 1** shows, Weis-Fogh absorbing wave energy mechanism is below water surface, others are power generation mechanism. Those two parts are connected with push-rods which make the power generation device and Weis-Fogh absorbing wave energy mechanism as an integral device. One endpoint of the push-rod is connected with wing AB, the other point is connected with power generation. In **Figure 1**, wave makes the wing AB moving up and down, and wing AB drives the push-rod moving up and down at same times. As this way, the power generation device transfers kinetic energy into electrical energy.

### 2.1. Weis-Fogh Wave Power Device

The wing AB's movements of Weis-Fogh mechanism wave power device will be described as below paper. According to **Figure 2**, at the beginning, the wing attaches the low wall tightly. Acted by the wave force, the endpoint of wing B leaves the wall, the wing opens, around the A point rotation. Such movement is called as opening stage, see **Figure 2(a)**. After the angle is opened to the $\beta$, the point A leaved away from bottom wall, and all point of wing AB moves up in same speed until the point B meets the top wall. In this case, wing AB moves up without angle turning, see **Figure 2(b)**. After point B attaches top wall, the wing makes close motion until the point A meets the top wall, such movements is called the closing stage, see **Figure 2(c)**. As the same way, acted by wave force, the point A leaves the wall, the wing opens, around the B point rotation. Such movement is also called as opening stage. After the angle is opened to the $\beta$, the point B leaved away from bottom wall, and all point of wing AB moves down in same speed until the point A meets the low wall. In this case, wing AB moves up without angle turning. After point A attaches low wall, the wing makes close motion until the point B meets the low wall, Such movements is also called the closing stage. In such way, the wing finishes one cycle movement.

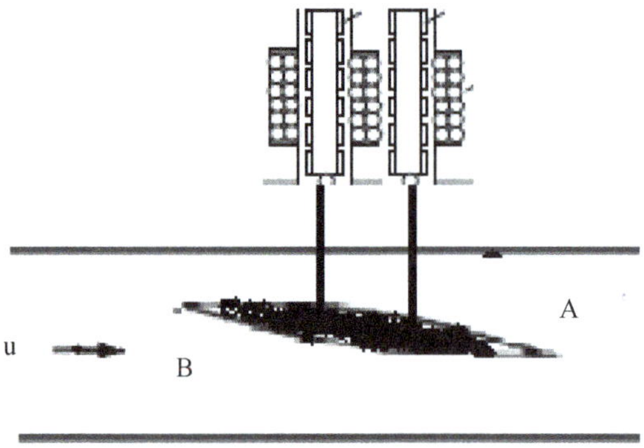


**Figure 1.** Weis-Fogh wave power generation device.

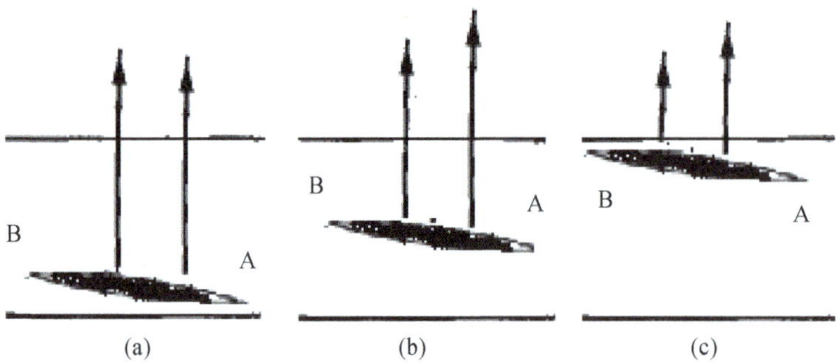


**Figure 2.** The movements of the Weis-Fogh wave power generation device.

## 2.2. Power Generation Device

The power generation device is shown as **Figure 1**, it connected with Weis-Fogh absorbing wave energy mechanism by push-rod, and transports wave energy into electrical energy. There are two push-rods by which can control the wing rotation angle.

## 2.3. Force Model

According to the Weis-Fogh wave energy device, the wing is acted by fluid force. During the opening stage, the endpoint of wing A doesn't leave the wall, the fluid force acting on wing can be calculated by using the theory of Weis-Fogh mechanism. Between the opening stage and the closed stage, the fluid force acting on wing can be calculated by fluid-solid coupling method. At the opening stage and at the closing stage, the fluid is supposed as ideal.

## 2.4. The Equations of Motion of the Wing

During the opening stage, the low wall supports wing at point A, the push-rod thrusts wing, and fluid force acts on the wing. As the influence of above forces, the wing rotates around the endpoint of wing A, the motion equation is:

$$J\pi\ddot{\alpha} = M_f - M_L \tag{1}$$

where $J$ is the moment of inertia, $M_f$ is the torque acted by fluid force on the wing, $M_L$ is the torque which the push-rod acts on the wing. Let opening angle is $(\pi\alpha_0)$ and the velocity is $\Omega_0$ when $T=0$.

Above equation is also used to the closing stage.

Between the opening stage and the closing stage, the wing just moves up and down, the corresponding equation of motion is:

$$m\ddot{y} = F_y - F_L \tag{2}$$

Here $m$ is mass of wing, $F_y$ is the force of the fluid to the wing projected on the $y$ direction, $F_L$ is the push from push-rod. When $T=0$, let $y$ equal to $y_0$, velocity is $V_{y_0}$.

This paper will consider the fluid power during opening stage and closing stage.

# 3. Fluid Power at Opening and Closing Stage

During this period, the wing's thickness is considered very thin compared with its length. For using theory of Weis-Fogh mechanism, the wing can be supposed as a tablet. As **Figure 3(a)** shows, the fluid domain surrounded by boundary $GEFA_1BA_2FG$ is mapped into up $\zeta$ plane. The boundary point $G$ on physic plane mappings to the real axis point $\xi=(-\infty)$ on mapping plane. Point $E$ mappings to point $\xi=b$, point $A_1$ mappings to point $\xi=-1$, point $B$ mappings to point $\xi=a$, point $A_2$ mappings to point $\xi=1$, point $F$ mappings to point $\xi=(+\infty)$.

By conformal mapping of complex function theory, the physical to the mapping transformation formula for plane area is [7]:

$$\begin{gathered} z = Kh(\xi)e^{\alpha\pi i} \\ h(\xi) = \int_{-1}^{\xi} h'(t)\,dt \end{gathered} \tag{3}$$

where $K$ is the transform constant, function $h'$ is expressed as:

$$h'(t) = (t-b)^{-1}(t+1)^{-\alpha}(a-t)(1-t)^{\alpha-1} \tag{4}$$

where $K$, $a$, $b$ are constant and related to these expressions below:

$$\begin{gathered} C = Kh(a), \quad d = K\pi, \\ a = b-(b-1)^{1-\alpha}(1+b)^{\alpha}. \end{gathered} \tag{5}$$

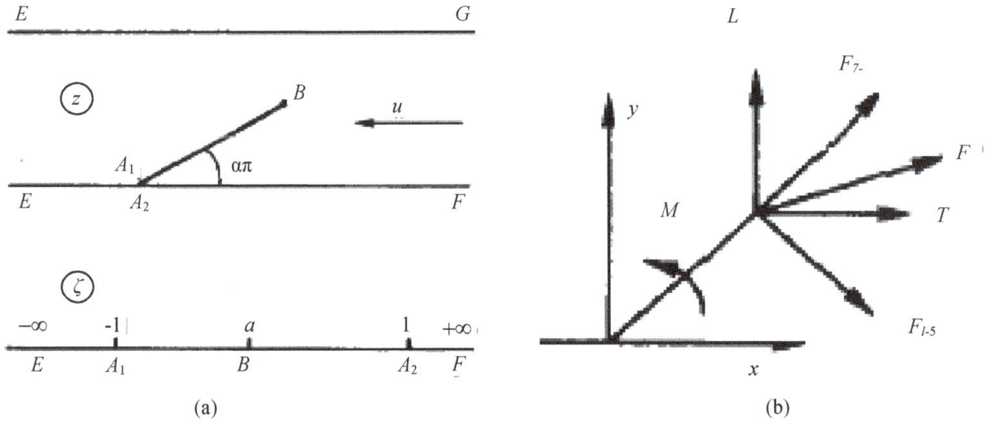


**Figure 3.** (a) The coorespondings between physic plane and mapping plane; (b) Force direction.

Here $C$ is the length of wing, $d$ is grid spacing. The steps of calculation is: getting the constant $K$ firstly, then solve the constant $a$, finally calculating the constant $b$.

The complex potential is:

$$W = \Omega G - Ku\ln(\zeta - b),$$
$$G = \frac{K^2}{2\pi}\int_{-1}^{1}\frac{h^2}{\zeta - t}\mathrm{d}t. \tag{6}$$

Here $\Omega$ is rotating angular velocity, $u$ is the flow speed. According Bernoulli equation, we have:

$$p = p - \rho(\mathrm{Re}W)'_t - 0.5\rho\left|W_\zeta h'(\zeta')\right|^2$$

Integral the pressure along the wing surface, we get the force acting on the wing:

$$i\frac{F}{\rho}\mathrm{e}^{-\alpha\pi i} = \Omega' F_1 - u' F_3 + \Omega^2 F_2 - \Omega u F_4 + u^2 F_5 + i\left(\Omega^2 F_7 - \Omega u F_8 + u^2 F_9\right) \tag{7}$$

where:

$$F_1 = -\mathrm{Re}\int_{-1}^{1} GKh'\mathrm{d}t, \quad F_3 = -\mathrm{Re}K\int_{-1}^{1}\ln(t-b)Kh'\mathrm{d}t,$$
$$F_{22} = -0.5\int_{-1}^{1}G_\zeta^2\left|z'_\zeta\right|^{-2}Kh'\mathrm{d}t, \quad F_{42} = -K\int_{-1}^{1}\frac{G_\zeta}{t-b}\left|z'_\zeta\right|^{-2}Kh'\mathrm{d}t,$$
$$F_{52} = -0.5K\int_{-1}^{1}\left|(t-b)z'_\zeta\right|^{-2}Kh'\mathrm{d}t.$$

$$F_2 = \mathrm{Re}F_{42} + \frac{F_{1a}}{\pi}, \quad F_7 = \mathrm{Im}F_{22},$$
$$F_4 = \mathrm{Re}F_{42} + \frac{F_{3a}}{\pi}, \quad F_8 = \mathrm{Im}F_{42},$$
$$F_4 = \mathrm{Re}F_{52}, \quad F_9 = \mathrm{Im}F_{52}.$$

Here $(-F_7)$, $(-F_8)$, $(-F_9)$ are frontal appeal, leading from the wing tips to the root, shown as **Figure 3(b)**.

The force mapping on $x$ axis calls push.

$$T = \Omega' T_1 - u' T_3 + \Omega^2 T_2 - \Omega u T_4 + u^2 T_5 \tag{8}$$

where

$$T_1 = F_1\sin(\alpha\pi), \quad T_3 = F_3\sin(\alpha\pi), \quad T_5 = 0,$$
$$T_2 = F_2\sin(\alpha\pi) + F_7\cos(\alpha\pi),$$
$$T_4 = F_4\sin(\alpha\pi) + F_8\cos(\alpha\pi).$$

The force mapping on $y$ axis calls lift

$$L = \Omega' L_1 - u' L_3 + \Omega^2 L_2 - \Omega u L_4 + u^2 L_5 \tag{9}$$

where

$$L_1 = -F_1\cos(\alpha\pi), \quad L_3 = -F_3\cos(\alpha\pi),$$
$$L_2 = -F_2\cos(\alpha\pi) + F_7\sin(\alpha\pi),$$
$$L_4 = -F_4\cos(\alpha\pi) + F_8\sin(\alpha\pi),$$
$$L_5 = -F_5\cos(\alpha\pi) + F_9\sin(\alpha\pi).$$

correspondingly, the moment is

$$\frac{M}{\rho} = \Omega' M_1 - u' M_3 + \Omega^2 M_2 - \Omega u M_4 + u^2 M_5 \tag{10}$$

where

$$M_1 = \mathrm{Re}\int_{-1}^{1} GKhh'\mathrm{d}t, \quad M_3 = \mathrm{Re}\int_{-1}^{1} \ln(t-b) K^3 Hh'\mathrm{d}t,$$
$$M_{22} = 0.5K\int_{-1}^{1} G_R^2 \frac{h}{h'}\mathrm{d}t, \quad M_{42} = 0.5K\int_{-1}^{1} \frac{G_\alpha}{t-b}\frac{h}{h'}\mathrm{d}t,$$
$$M_5 = 0.5K\int_{-1}^{1} (t-b)^{-2}\frac{h}{h'}\mathrm{d}t,$$
$$M_2 = M_{22} + \frac{\partial M_1}{\pi\partial\alpha}, \quad M_4 = M_{42} + \frac{\partial M_3}{\pi\partial\alpha}.$$

Integration of the moment along the angle is the work of the wing

$$\frac{W}{\rho} = \Omega' W_1 - u' W_3 + \Omega^2 W_2 - \Omega u W_4 + \mathrm{u}^2 W_5,$$
$$W_J = \int_0^\beta M_J \mathrm{d}(\alpha\pi), \quad j = 1,2,3,4,5. \tag{11}$$

## 4. The Calculation of Fluid Power

According the above equations, the code is written to calculate the fluid power. **Figure 4(a)** shows that the components of force acting on wing varied with the angles when the grid spacing and the wing chord length ratio is 0.5. The curves' widths vary large with the component of the subscript. When the dimensionless angle $\alpha$ approaches to 0.5, $F_1$ and $F_2$ tend to infinity, $F_4$ and $F_5$ tend to 0, $F_{7\text{-}9}$ tend to constants.

When the grid spacing and the wing chord length ratio is 0.5, **Figure 4(b)** shows the component's moment varied with the angles. The curves' widths vary large with the component of the subscript. The horizontal axis is dimensionless angle $\alpha = \beta/\pi$. When the dimensionless angle $\alpha$ approaches to 0.5, the moment $W_2$ tends to infinity, and the moment $W_{1\text{-}3}$ and $F_3$ tend to constants.

## 5. Conclusion

The fluid force acting on the wing is the most important for Weis-Fogh wave power generation device. It is

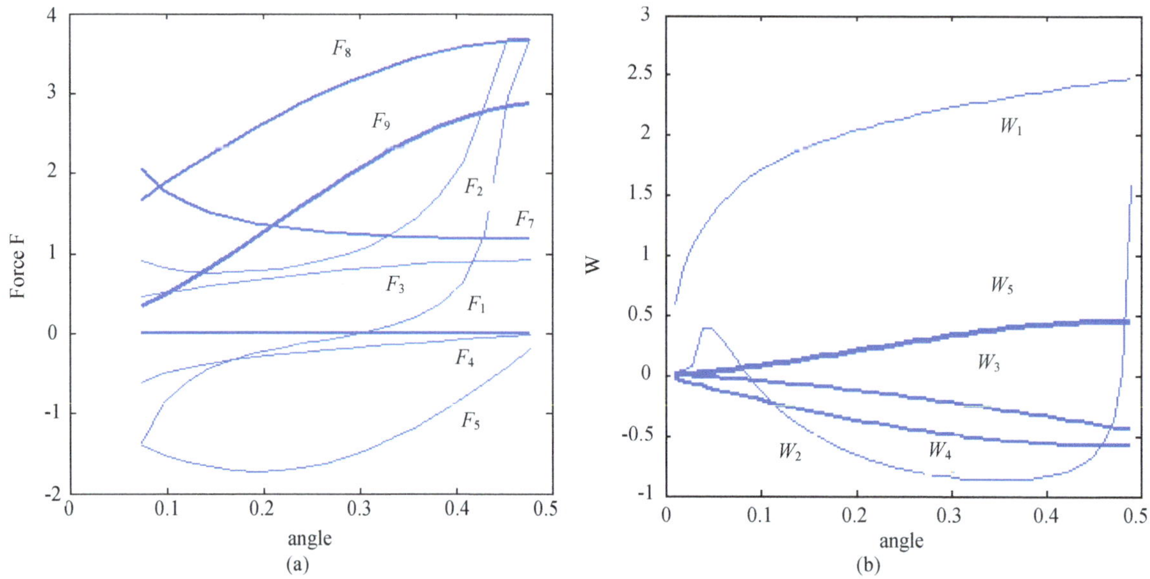


**Figure 4.** (a) Component varied with the angle; (b) The component moments varied with the angle.

always calculated before device designed. According to the Weis-Fogh mechanism theory and conformal transformation theory, the fluid dynamics model of double push-rod cascade wave power generation device is built, and at opening stage and close stage, the expression formula of work done by fluid force acting on wing is derived by using integral method. Such formula is dependent with coming velocity $u$ and wing turning velocity, and can be calculated and saved in the disk. It can increase the speed of calculating fluid force and work.

## Acknowledgements

The paper is financially supported by China National Science & Technology Innovation Fund (No. 20141049705002).

## References

[1] Kang, Q., Xiao, X., Lie, Z.X. and Huang, L.P. (2013) Control Strategy Optimization of Direct Drive Power Output Type Generation System. *Automation of Electric Power Systems*, **3**, 24-29.

[2] Jiao, Y.F. and Liu, Y.L. (2010) The Current Status and Prospect of Wave Power Generation. *Chinese High-Tech Enterprises*, **12**, 89-90.

[3] Qu, Y.M. (2011) Floating Sheave Wave Power Generation Technology. *Rural Electrification*, **6**, 56-57.

[4] Wang, K., Shen, R.H. and Wen, H.A. (2014) A Wave Power Generation—Wave Energy Matrix of Relative Motion Unit Mechanism. *Modern Machinery*, **2**, 40-42.

[5] Wang, S.C., Che, R.W., Hu, C.S., Zhang, Y.H. and You, W. (2007) Research on a Float Type Wave Power Generation Mechanism. *Journal of Solar Energy*, **11**, 1266-1271.

[6] Li, D., Bai, B.D., Yu, Q. and Zhu, B.F. (2011) Research on Floating Type of Wave Power Generation System. *Journal of Solar Energy*, **10**, 1566-1569.

[7] Zhang, S.S., *et al.* (1996) Hydrodynamic Model of Weis-Fogh Mechanism Hydrofoil. *Journal of Hydrodynamics*, **4**, 35-39.

[8] Zhang, S.S., *et al.* (1999) Research and Progress of Weis-Fogh Mechanism Hydrodynamics. *Journal of Hydrodynamics*, **3**, 55-60.

[9] Tsutahara, M. and Kimura, T. (1987) An Application of the Weis-Fogh Mechanism to Ship Propulsion. *Journal of Fluid Engineering*, **109**, 107-113. http://dx.doi.org/10.1115/1.3242629

[10] Weis-Fogh, T. (1973) Quick Estimates of Flight Fitness in Hovering Animals Including Novel Mechanism for Lift Production. *The Journal of Experimental Biology*, **59**, 169-230.

[11] Maxworthy, T. (1979) Experiments on the Weis-Fogh Mechanism of Lift Generation by Insects in Hovering Flight. Part 1. Dynamics of the "Fling". *Journal of Fluid Mechanics*, **93**, 47-63. http://dx.doi.org/10.1017/S0022112079001774

# 5

# The Stability of a Rotating Cartesian Plume in the Presence of Vertical Boundaries

**Khaled S. Al-Mashrafi[1], Ibrahim A. Eltayeb[2]**

[1]Section of Applied Sciences, Department of Human Resources Development, General Directorate of Education in Eastern Region, Ministry of Education, Sur, Oman
[2]Department of Mathematics and Statistics, College of Science, Sultan Qaboos University, Muscat, Oman
Email: P001175@student.squ.edu.om

## Abstract

**The effect of two fixed vertical boundaries, a finite distance apart, on the dynamics of a column of buoyant fluid rising in a less buoyant fluid is investigated in the presence of vertical rotation. It is shown that the presence of the boundaries introduces two main effects on a rotating plume. They tend to stabilise the plume but succeed only reducing the value of the growth rate and the plume remains unstable for all finite values of the distance between the boundaries and the plume. In the absence of the sidewalls, two modes of the instability were found known as the sinuous mode and the varicose mode. The influence of the boundaries is such that it reduces the growth rate of the varicose mode more than that of the sinuous mode and consequently the modified sinuous mode is always preferred in the presence of the boundaries.**

## Keywords

**Compositional Plumes, Stability, Growth Rate, Bounded Domain, Rotation**

## 1. Introduction

The study of the dynamics of compositional plumes is important for many real life applications in industry ([1]-[6]), geophysics ([7]-[25]) and environment ([26]-[34]). While the presence of the compositional plumes can be harmful (e.g., in iron bars), it is useful in geophysics (e.g., the hot compositional plumes, that rise from the inner core boundary of the Earth into the outer core interact with the rotation and magnetic field of the Earth and may contribute to the Geodynamo). Such a wide range of applications has motivated many studies on various aspects of the dynamics of compositional plumes. These studies are experimental and theoretical. The experimental works on the dynamics of compositional plumes observed that the plume flow seems to be stable (Sample and

Hellawell [35]; Chen and Chen [36]; Hellawell *et al.* [37]). The laboratory studies by Hellawell *et al.* [37] find that the plumes are thin and long, and its top part tends to break up and disappears. Classen *et al.* [38] experimentally studied the dynamics of compositional plumes under the influence of vertical rotation to find that the plumes are unstable and break into blobs. On the other hand, the theoretical works on the stability of the plumes showed that the Cartesian plume is always unstable in the absence of rotation (Eltayeb and Loper [16]) and in the presence of rotation (Eltayeb and Hamza [18]). These studies assumed that the plume rises vertically in a fluid of unbounded domains. While the experimental studies were conducted in bounded regions to show that the plume was stable, the theoretical models were conducted in unbounded domains to find that the plume was unstable. Thus it is of interest to examine the influence of the vertical boundaries on the dynamics of the plumes. The mathematical model by Al Mashrafi and Eltayeb [6] investigated the influence of the two fixed vertical boundaries on the dynamics of the plumes. They tested the stability of non-rotating Cartesian plumes in a bounded domain to find that the presence of two vertical boundaries affects the stability, but the plumes remain unstable. Moreover, they found that the plume was stable when it was close to the boundary but had a large thickness and the material diffusion is potent in the thin layer between the plume and the nearest boundary.

Motivated by real life applications and laboratory results, we study here the influence of vertical rotation on the dynamics of bounded Cartesian plume. In general, the purpose of this study is to extend the theoretical model by Al Mashrafi and Eltayeb [6] on the dynamics of a Cartesian compositional plume in bounded regions to include the action of vertical rotation. The model by Al Mashrafi and Eltayeb [6] consisted of a column of buoyant fluid of finite thickness, $2x_0$, rising vertically in another less buoyant fluid bounded by two fixed vertical walls located at $x = a_1$ and $x = -a_2$. The system was infinite in the $y$ and $z$ directions. In the current study, we consider that the whole system rotates about the vertical with a uniform angular speed, $\omega$ (see **Figure 1**).

In Section 2, we formulate the model mathematically and state the boundary conditions of the system. The presence of rotation introduces an additional parameter, $\tau$, which is a measure of the Coriolis force relative to the viscous force, and this parameter referred to hereinafter as the rotation parameter, defined by

$$\tau\left(=\sqrt{Ta}\right)=\frac{\omega L^2}{\nu}, \tag{1}$$

where $Ta$ is the Taylor number, $\nu$ is the kinematic viscosity and $L$ is the unit of length (see Equation (6) below). In Section 3, we investigate the influence of the vertical boundaries on the linear stability of a rotating

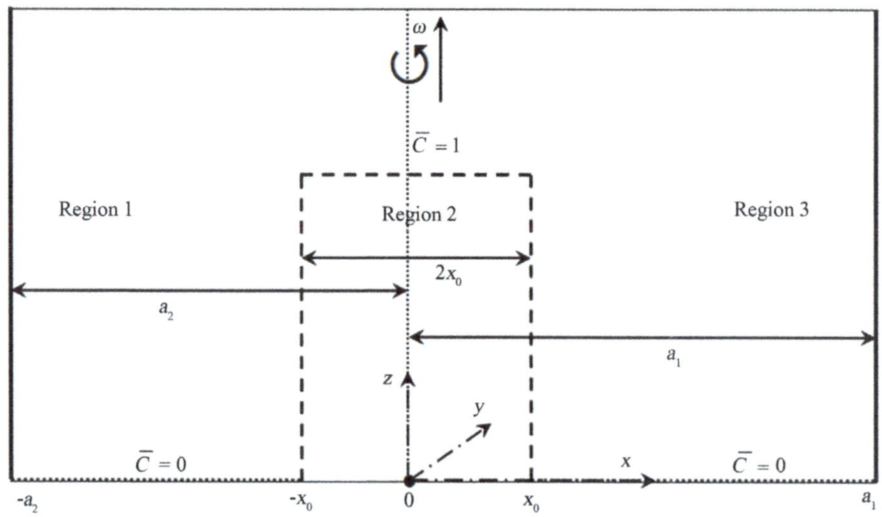


**Figure 1.** The geometry of the problem showing the profile of the basic state concentration of light material, $\bar{C}(x)$, representing a plume of width, $2x_0$, and concentration, 1, rising vertically in a finite fluid of width, $d(=a_1+a_2)$, and concentration, 0. The system rotates uniformly about the vertical with angular speed, $\omega$. The plume divided the system into three regions: region 2 represents the plume whereas regions 1 and 3 represent the surrounding fluid.

bounded Cartesian plume. The problem of the rotating plume was studied by Eltayeb and Hamza [18] in the absence of boundaries. In Section 4, we discuss the effect of the boundaries on the stability of a rotating plume. The growth rate is maximised over the wave numbers plane $(m,n)$ in the parameter space. In Section 5, we make some concluding remarks.

## 2. Formulation of the Problem

We consider a two-component fluid, in which the concentration of the solvent component (light material) is $C$ and the temperature is $T$, rotating uniformly about the vertical with angular velocity $\omega$. The fluid has kinematic viscosity, $\nu$, and thermal diffusivity, $\kappa$, and material diffusion is negligible. The dimensionless equations of the system have been derived by Eltayeb and Hamza [18]. They are

$$R\left[\frac{\partial \boldsymbol{u}}{\partial t}+(\boldsymbol{u}\cdot\nabla)\boldsymbol{u}\right]+\tau\hat{z}\times\boldsymbol{u}=-\nabla\left(p+\frac{z}{\beta\tilde{C}}\right)+\nabla^2\boldsymbol{u}+(T-T_r+C-C_r)\hat{z}, \tag{2}$$

$$\nabla\cdot\boldsymbol{u}=0, \tag{3}$$

$$R\sigma\left[\frac{\partial T}{\partial t}+\boldsymbol{u}\cdot\nabla T\right]=\nabla^2 T, \tag{4}$$

$$\frac{\partial C}{\partial t}+\boldsymbol{u}\cdot\nabla C=0. \tag{5}$$

Here $R$, $\sigma$ and $\tau$ are the Grashoff number, the Prandtl number and the rotation parameter, respectively, defined by

$$R=\beta\tilde{C}\left(\frac{g\kappa^3}{\alpha^3\gamma^3\nu^5}\right)^{\frac{1}{4}}\left(=\frac{UL}{\nu}\right),\quad \sigma=\frac{\nu}{\kappa},\quad \tau=\frac{\omega L^2}{\nu}\left(=\sqrt{Ta}\right), \tag{6}$$

where $Ta$ is the Taylor number and $U$, $L$ and $\tilde{C}$ are characteristic units of velocity, length and concentration, respectively (Al Mashrafi and Eltayeb [6]), and $\boldsymbol{u}$, $p$, $g$, $\hat{z}$, $t$, $\alpha$, $\beta$, $\rho$, $\gamma$ are the velocity vector, the pressure, the constant gravitational acceleration, the upward unit vector, the time, the thermal expansion coefficient, the compositional expansion coefficient, the density, the uniform temperature gradient and the subscript "$r$" in the Equation (2) refers to reference values.

We consider a basic concentration profile

$$\bar{C}(x)=\begin{cases}0, & -a_2\le x<-x_0,\ x_0<x\le a_1\\ 1, & -x_0\le x\le x_0\end{cases}, \tag{7}$$

which defines a plume of thickness, $2x_0$, rising with velocity $\bar{w}(x)$ in the presence of mean temperature $\bar{T}(x)$ and mean pressure $\bar{p}(x)$ such that

$$-\frac{\mathrm{d}\bar{p}}{\mathrm{d}x}\hat{x}+\left(\frac{\mathrm{d}^2\bar{w}}{\mathrm{d}x^2}+\bar{C}+\bar{T}\right)\hat{z}=0, \tag{8}$$

$$\frac{\mathrm{d}^2\bar{T}}{\mathrm{d}x^2}=\bar{w}(x), \tag{9}$$

which are the same equations obtained in the absence of rotation. The presence of rotation does not affect the basic state, and the solution of (8) and (9) is the same as in the absence of rotation. We include it here for easy reference:

$$F(x)=\begin{cases}-A\sinh(ka_1)\sinh[k(x+a_2)]; & -a_2\le x<-x_0\\ -A\sinh(ka_1)\sinh[k(x+a_2)]+\cosh[k(x+x_0)]-1; & -x_0\le x\le x_0\\ A\sinh(ka_2)\sinh[k(x-a_1)]; & x_0<x\le a_1\end{cases}, \tag{10}$$

where $F(x)$, $A$ and $k$ are defined by

$$F(x)=\overline{T}(x)-\mathrm{i}\overline{w}(x),\quad A=\frac{2\sinh(kx_0)}{\sinh(kd)},\quad k=\frac{1}{\sqrt{2}}(1+\mathrm{i}),\quad \mathrm{i}^2=-1. \tag{11}$$

## 3. The Stability Analysis

In this section, we use the perturbation Equations (14)-(17) to investigate the linear stability of the basic state solution given by (10). We assume that the interface at the plane $x=x_0$ is given a small harmonic disturbance of the form

$$x=x_0+\varepsilon E+c.c.,\qquad E=\exp\{\Omega t+\mathrm{i}(my-nz)\}, \tag{12}$$

where $m$ and $n$ are the wavenumber components in the $(y,z)$ plane, $\Omega$ is the growth rate and $c.c.$ refers to the complex conjugate.

The disturbance (12) will propagate into the fluid, and affect the second interface and the variables of the system to produce the perturbations. Consequently, the interface at $x=-x_0$ can be written in the form

$$x=-x_0+\varepsilon\eta_1 E+c.c., \tag{13}$$

where $\eta_1$ is a measure of the amplitude of the interface and it is evaluated as a part of the solution. The perturbations introduced into the system are governed by the dimensionless equations (see Al Mashrafi and Eltayeb [6])

$$R\left[\frac{\partial \boldsymbol{u}^\dagger}{\partial t}+\overline{w}\hat{z}\cdot\nabla\boldsymbol{u}^\dagger+\left(\boldsymbol{u}^\dagger\cdot\nabla\overline{w}\right)\hat{z}\right]+\tau\hat{z}\times\boldsymbol{u}^\dagger=-\nabla p^\dagger+\nabla^2\boldsymbol{u}^\dagger+\left(T^\dagger+C^\dagger\right)\hat{z}, \tag{14}$$

$$\nabla\cdot\boldsymbol{u}^\dagger=0, \tag{15}$$

$$\sigma R\left[\frac{\partial T^\dagger}{\partial t}+\overline{w}\frac{\partial T^\dagger}{\partial z}+\boldsymbol{u}^\dagger\cdot\nabla\overline{T}\right]+\boldsymbol{u}^\dagger\cdot\hat{z}=\nabla^2 T^\dagger, \tag{16}$$

$$\frac{\partial C^\dagger}{\partial t}+\overline{w}\frac{\partial C^\dagger}{\partial z}+\boldsymbol{u}^\dagger\cdot\nabla\overline{C}=0, \tag{17}$$

where $\boldsymbol{u}^\dagger$, $p^\dagger$, $T^\dagger$ and $C^\dagger$ are the perturbations in velocity, pressure, temperature and concentration, respectively.

The perturbation variables take the form

$$\left\{\boldsymbol{u}^\dagger,C^\dagger,T^\dagger,p^\dagger\right\}=\left\{-\mathrm{i}nu(x),nmv(x),w(x),C(x),T(x),-\mathrm{i}np(x)\right\}E+c.c., \tag{18}$$

in which the factors $-\mathrm{i}n$, $nm$, and $-\mathrm{i}n$ are introduced in the variables $u(x)$, $v(x)$ and $p(x)$, respectively, for convenience.

Substituting the variables (18) into the Equations (14)-(17) and after some arrangements we get the following equations

$$\Delta w+T+n^2p=R\left(\overline{\Omega}w-\mathrm{i}nuD\overline{w}\right), \tag{19}$$

$$\Delta T-w=\sigma R\left(\overline{\Omega}T-\mathrm{i}nuD\overline{T}\right), \tag{20}$$

$$\Delta p-T=2\mathrm{i}nRuD\overline{w}+\mathrm{i}m\tau\varsigma, \tag{21}$$

$$\Delta\varsigma=R\left(-\mathrm{i}nvD\overline{w}+\overline{\Omega}\varsigma\right)+\frac{\mathrm{i}}{m}\tau w, \tag{22}$$

$$\left(D^2-m^2\right)v=D\varsigma-w, \tag{23}$$

$$n^2u=-D(p+w)+m^2\varsigma-\mathrm{i}\tau mv-R\overline{\Omega}u, \tag{24}$$

$$C=0, \tag{25}$$

subject to the boundary conditions

$$D(w+p)=v=w=T=\varsigma=0 \quad \text{at } x=a_1,\ x=-a_2, \tag{26}$$

$$v,\ w,\ T,\ p,\ \varsigma,\ Dv,\ DT,\ D(w+p) \ \text{are continuous across } x=\pm x_0, \tag{27}$$

$$\langle Dw(x_0)\rangle=\langle\bar{C}(x_0)\rangle,\quad \langle Dw(-x_0)\rangle=\eta_1\langle\bar{C}(-x_0)\rangle, \tag{28}$$

$$-inu(x_0)=\Omega-in\bar{w}(x_0),\ -inu(-x_0)=\left[\Omega-in\bar{w}(-x_0)\right]\eta_1, \tag{29}$$

where the variable $\varsigma$ is related to the vertical component of vorticity and is defined as

$$\varsigma=Dv-u, \tag{30}$$

and we have introduced the notation

$$D\equiv\frac{\mathrm{d}}{\mathrm{d}x},\quad \Delta\equiv D^2-m^2-n^2,\quad \langle f(\alpha)\rangle=f(\alpha^-)-f(\alpha^+),\quad \bar{\Omega}=\Omega-in\bar{w}(x). \tag{31}$$

(cf. Al Mashrafi and Eltayeb [6]).

We use the same method adopted in Al Mashrafi and Eltayeb [6] and expand the perturbation variables and the growth rate in the small parameter $R$, thus

$$f(x,y,z,t)=\sum_{s=0}^{\infty}f_s(x,y,z,t)R^s,\quad \Omega=\sum_{s=1}^{\infty}\Omega_s R^{s-1},\quad R\ll 1, \tag{32}$$

where $f(x,y,z,t)$ indicates any of the perturbation variables $u$, $v$, $w$, $p$, $\varsigma$ and $T$.

It turned out that the leading order terms in the equations determines the stability of the system. The relevant equations and the boundary conditions are then obtained from (19)-(24) and (26)-(29) by neglecting the terms with $R$. In order to facilitate comparison with the results of the non-rotating case, we shall maintain the subscript 0. The equations are given by

$$\Delta w_0+T_0+n^2p_0=0, \tag{33}$$

$$\Delta T_0-w_0=0, \tag{34}$$

$$\Delta p_0-T_0-im\tau\varsigma_0=0, \tag{35}$$

$$\Delta\varsigma_0-\frac{\mathrm{i}}{m}\tau w_0=0, \tag{36}$$

$$\left(D^2-m^2\right)v_0-D\varsigma_0+w_0=0, \tag{37}$$

$$n^2u_0=-D(p_0+w_0)+m^2\varsigma_0-\mathrm{i}\tau m v_0. \tag{38}$$

The associated boundary conditions are obtained from (26)-(29) by introducing the subscript 0 to all the variables and the subscript 1 to $\Omega$.

The system (33)-(38) together with the boundary conditions poses an eigenvalue problem for the growth rate, $\Omega$, which determines the stability of the system. The real part $\Omega_r$ governs the variations of the amplitude of the disturbance with time, and hence it determines the stability of the disturbance. If it is negative for all possible values of the wavenumbers $m$ and $n$, then the plume is stable, while if at least one pair of $m$ and $n$ gives a positive value, then the plume is unstable. If $\Omega_r$ vanishes for all values of the wavenumbers, the plume is neutrally stable. If the preferred mode occurs for $m$, $n$ both non-zero, it is referred to as a 3-dimensional mode (oblique), and if $m=0$, it is called 2-dimensional (vertical). The case $n=0$ and $m\neq 0$ is found not to occur. The imaginary part $\Omega_i$ determines the phase speed of the disturbance. The vertical phase speed $u_z$ and the horizontal phase speed $u_h$ are defined as

$$u_z=\frac{\Omega_i}{n},\qquad u_h=\frac{\Omega_i}{m}=\frac{nu_z}{m}. \tag{39}$$

We note that $u_h$ is defined only if $m\neq 0$.

We operate on Equation (33) with $\Delta^2$, and use Equations (34)-(36) to get

$$\Delta^3 w_0 + \Delta w_0 + n^2\left(1-\tau^2\right)w_0 = 0. \tag{40}$$

The general solution of the differential Equation (40) can be written in the form

$$w_0^{(i)}(x) = \sum_{j=1}^{3}\mu_j^3\left[A_j^{(i)}\cosh\left(\lambda_j x\right) + B_j^{(i)}\sinh\left(\lambda_j x\right)\right], \tag{41}$$

where the superscript $(i)$ refers to the three regions of the system (see **Figure 1**), $A_j^{(i)}$, $B_j^{(i)}$ $(i,j=1,2,3)$ are constants, and $\mu_j$ are the roots of the cubic equation

$$\mu_j^3 + \mu_j + n^2\left(1-\tau^2\right) = 0, \tag{42}$$

with $\lambda_j$ is given by

$$\lambda_j = \left(\mu_j + b^2\right)^{1/2}. \tag{43}$$

We use the Equations (34)-(36) and the solution (41) to obtain

$$\left\{T_0^{(i)}, p_0^{(i)}, \varsigma_0^{(i)}\right\}(x) = \sum_{j=1}^{3}\left\{\mu_j^2, \mu_j\left(1-\tau^2\right), \frac{\mathrm{i}}{m}\tau\mu_j^2\right\}\left[A_j^{(i)}\cosh\left(\lambda_j x\right) + B_j^{(i)}\sinh\left(\lambda_j x\right)\right]. \tag{44}$$

The Equation (37) can be solved in the form a complementary function and particular solution to find

$$v_0^{(i)}(x) = \frac{1}{m}\sum_{j=1}^{3}\frac{\mu_j^2}{\mu_j + n^2}\left\{-m\mu_j C_j^{(i)} + \mathrm{i}\tau\lambda_j D_j^{(i)}\right\} + \frac{1}{m}\left\{C^{(i)}\cosh(mx) + D^{(i)}\sinh(mx)\right\}, \tag{45}$$

where $C^{(i)}$, $D^{(i)}$ are constants, and $C_j^{(i)}$, $D_j^{(i)}$ are given by

$$\left.\begin{aligned} C_j^{(i)} &= A_j^{(i)}\cosh\left(\lambda_j x\right) + B_j^{(i)}\sinh\left(\lambda_j x\right) \\ D_j^{(i)} &= B_j^{(i)}\cosh\left(\lambda_j x\right) + A_j^{(i)}\sinh\left(\lambda_j x\right) \end{aligned}\right\}. \tag{46}$$

Now we apply the boundary conditions at $x = \pm x_0$, $x = a_1$, $x = -a_2$ and solve the resulting algebraic equations for the constants $A_j^{(i)}$, $B_j^{(i)}$, $C^{(i)}$, $D^{(i)}$ for $i,j = 1,2,3$. The details of the solution are given in the **Appendix A**.

The growth rate is given by the quadratic equation

$$\left(\Omega_1/in\right)^2 + S_1\left(\Omega_1/in\right) + S_2 = 0, \tag{47}$$

where $S_1$ and $S_2$ are given in the **Appendix A**.

Solving Equation (47) yields

$$\Omega_1 = \frac{in}{2}\left(-S_1 \pm \sqrt{S_1^2 - 4S_2}\right), \tag{48}$$

and the displacement $\eta_1$ is given by

$$\eta_1 = \frac{F_{1-}}{\Omega_1 - in\bar{w}\left(-x_0\right) - F_{e-}}, \tag{49}$$

in which $F_{1-}$ and $F_{e-}$ are defined in the **Appendix A**.

We note that, as in the absence of rotation, the system has two modes. The upper sign in the expression (48) corresponds to the modified varicose mode (MV) and the lower sign refers to the modified sinuous mode (MS).

## 4. Discussions of the Results

The growth rate (48) is evaluated numerically in the $(m,n)$ plane as a function of the parameters $x_0$, $a_2$, $d$ and $Ta$ for both modes. The contours of $\Omega_1$ in the $(m,n)$ plane are plotted for sample values of the parameters $x_0$, $a_2$, $Ta$ and $d = 10$ in the **Figures 2-4**.

In **Figure 2**, we present a comparison between the contours of the two modes MV and MS for different values of the Taylor number, $Ta$, when $x_0 = 0.5$, $a_2 = 1$ and $d = 10$. We note that the MV mode always possesses a minimum with negative growth rate and one or two maxima with positive growth rates while the MS mode has

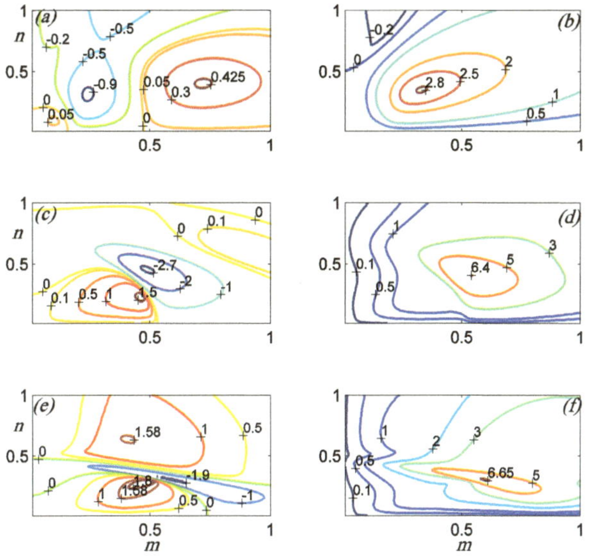


**Figure 2.** Contours of the growth rate, $\Omega_1$, of the modes MV, as in (a), (c), (e) and MS, as in (b), (d), (f) for the rotating bounded Cartesian plume where $x_0 = 0.5$, $a_2 = 1$, $d = 10$. Here $Ta = 0.5$ for (a), (b); $Ta = 2$ for (c), (d); and $Ta = 5$ for (e), (f). Note that the preferred mode is MS.

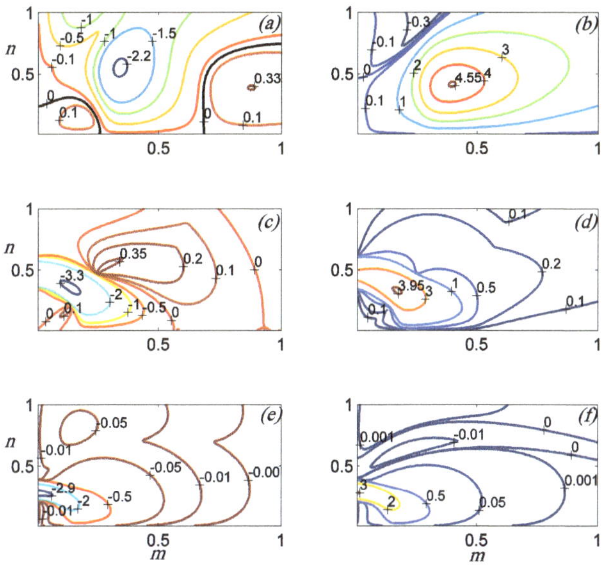


**Figure 3.** Contours of the growth rate, $\Omega_1$, of the modes MV, as in (a), (c), (e) and MS, as in (b), (d), (f) for $x_0 = 0.5$, $Ta = 0.8$, $d = 10$. Here $a_2 = 1$ for (a), (b); $a_2 = 3$ for (c), (d); and $a_2 = 5$ for (e), (f). Note that the preferred mode of the instability is 2-dimensional when $a_2 = 5$.

a minimum with negative growth rate only when $Ta$ is small and one maximum with positive growth rate which is larger than that of the corresponding MV mode. This indicates that the MS mode is preferred for this set of parameters. The local maxima correspond to 3-dimensional modes and the largest maximum always increases with the increase in the rotation parameter $Ta$.

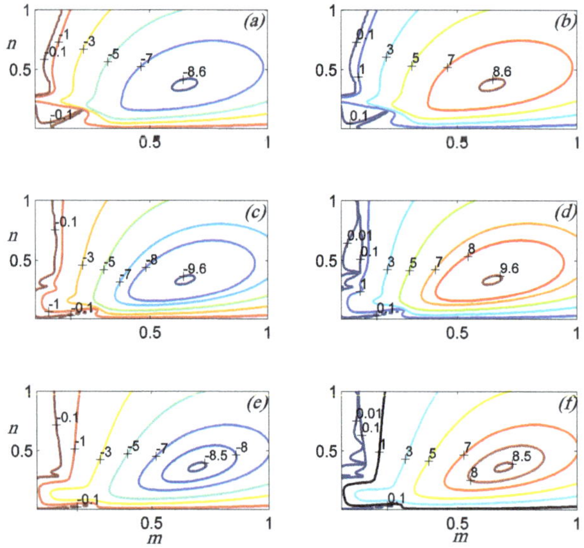


**Figure 4.** Isolines of the growth rate, $\Omega_1^{MV}$, as in (a), (c), (e) and $\Omega_1^{MS}$ as in (b), (d), (f) for $a_2 = 5$, $Ta = 7$, $d = 10$. Here $x_0 = 1$ for (a), (b); $x_0 = 2$ for (c), (d); and $x_0 = 3$ for (e), (f). Note that at fixed point $(m, n)$, the values of $\Omega_1^{MV}$ and $\Omega_1^{MS}$ are nearly same but there different in the sign.

**Figure 3** presents the influence of the distance $a_2$ between the plume and the nearest sidewall on the stability of the plume when the thickness of the plume and the rotation parameter are held fixed at $x_0 = 0.5$ and $Ta = 0.8$. The figure is plotted for three different values of $a_2$: $a_2 = 1$ for subfigures (a), (b); $a_2 = 3$ for subfigures (c), (d); and $a_2 = 5$ for subfigures (e), (f). When the plume is situated half-way between the sidewalls $(a_2 = 5)$, the MV mode has a negative growth rate everywhere and hence stable while the MS mode has a growth rate that is positive everywhere and hence is preferred. The local maximum possesses a vanishing horizontal wave number and hence propagates vertically. As the plume moves towards a wall, the growth rates of both modes increase and the MV mode develops local maxima with positive growth rates but they are not preferred because the MS mode local maximum increases as well and becomes 3-dimenaional.

**Figure 4** illustrates the influence of the plume thickness on the contours of the growth rate for fixed values of the rotation parameter and the distance between the two sidewalls when the plume is situated half-way between the two sidewalls. As $x_0$ increases from $x_0 = 1$, the growth rate for MS mode, which is positive everywhere, increases and that of the MV mode, which is negative everywhere, decreases until $x_0$ reaches a critical value, $x_{0c}$, when the growth rate of the MS mode decreases and that of the MV mode increases.

The contours of **Figures 2-4** indicate that the preferred mode of instability is the modified sinuous (MS) mode. This indication is quantified by calculating the maximum growth rate and the associated wave numbers and phase speeds for different values of the parameters. For fixed values of $x_0$, $a_2$, $d$ and $Ta$, we maximize over the wave numbers $m$ and $n$ by demanding that

$$\frac{\partial \Omega_1^{(k)}}{\partial n} = \frac{\partial \Omega_1^{(k)}}{\partial m} = 0, \quad k = \{MV, MS\}. \tag{50}$$

The solution of (50) gives the values $m_c$ and $n_c$ at maximum growth rate, $\Omega_{1c}$, calculated from (48) for these values of $m$ and $n$ to give, together with the corresponding value $U_c$ of the phase speed, the parameters $(\Omega_{1c}, m_c, n_c, U_c)$ of the preferred mode. For fixed parameters $x_0$, $a_2$, $d$ and $Ta$, and particular mode, all possible local maxima of $\Omega_1^{(k)}$ are identified and the largest value taken together with the corresponding wavenumbers and phase speeds as defining the preferred mode for that set of parameters for that mode. This is carried out for both modified varicose (MV) and modified sinuous (MS) modes, and the largest is chosen as the preferred mode. A sample of the results is given in **Figures 5-9**.

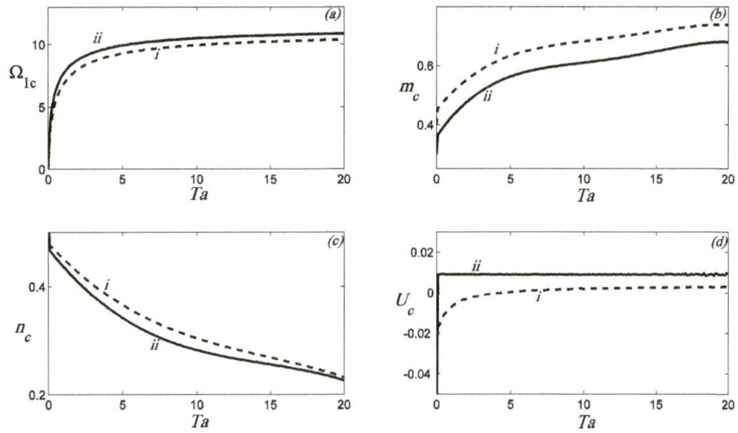


**Figure 5.** The preferred mode parameters $\Omega_{1c}, m_c, n_c, U_c$ are plotted in (a), (b), (c), (d) as a function of $Ta$, for $x_0 = 2$, and two different values of $d$ and the plume half-way between the sidewalls. The roman numbers $i$ and $ii$ refer to $d = 10, a_2 = 5$ and $d = 20, a_2 = 10$, respectively.

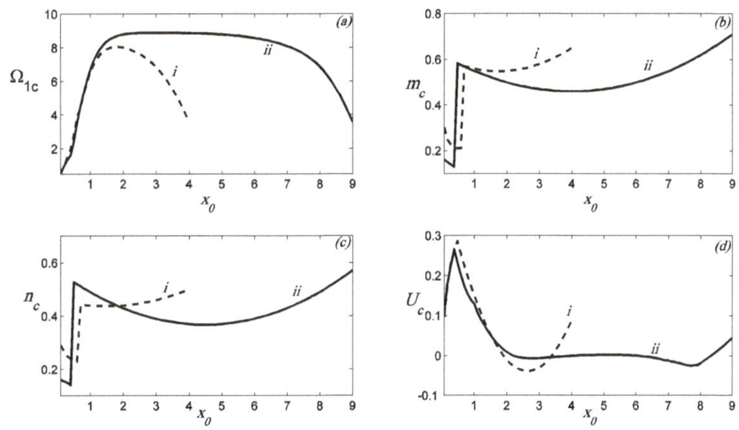


**Figure 6.** The preferred mode parameters $\Omega_{1c}, m_c, n_c, U_c$ are plotted in (a), (b), (c), (d) as a function of $x_0$, for $T_a = 2$, and two different values of $d$ and the plume half-way between the sidewalls. The roman numbers $i$ and $ii$ refer to $d = 10, a_2 = 5$ and $d = 20, a_2 = 10$, respectively.

**Figure 5** and **Figure 6** illustrate the influence of the distance between the boundaries on the preferred mode of instability of the rotating plume. In **Figure 5**, the preferred mode parameters are plotted as a function of $Ta$ for fixed $x_0$ and two values of the distance between the sidewalls, $d$, for a plume situated half-way between the sidewalls. It is found that the preferred mode is the MS mode and the influence of the boundaries tends to stabilise the plume but only reduces the growth rate slightly. The horizontal and vertical wavenumbers are increased while the phase speeds are reduced as the distance, $d$, between the two walls is reduced.

**Figure 6** presents the preferred mode of instability as a function of $x_0$ and fixed $Ta$ where the plume is equidistant from the two walls and $d$ takes the values 10 and 20. The growth rate increases rapidly as $x_0$ increases from zero until it reaches a maximum when the plume occupies the middle half of the region between the sidewalls. As the thickness of the plume increases further, the growth rate decreases if $d$ is small but stays at the maximum value for increasing $x_0$ if the distance $d$ is large. In both case, the growth rate drops to zero as the wall is approached and the plume nearly fill the whole region between the sidewalls. The wavenumbers decrease as $x_0$ increases from zero but they soon jump to larger values and increase with $x_0$. In both cases the MS mode is preferred.

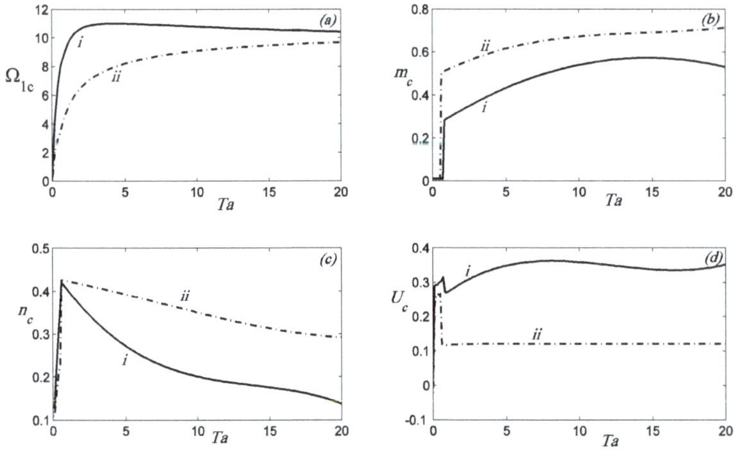


**Figure 7.** The preferred mode parameters $\Omega_{1c}, m_c, n_c, U_c$ are plotted in (a), (b), (c), (d), respectively, as a function of $Ta$, for $x_0 = 1$, $d = 10$ and two different values of $a_2$. The solid curves refer to $a_2 = 2$ and the broken curves refer to $a_2 = 5$. Note that preferred mode is MS in all cases.

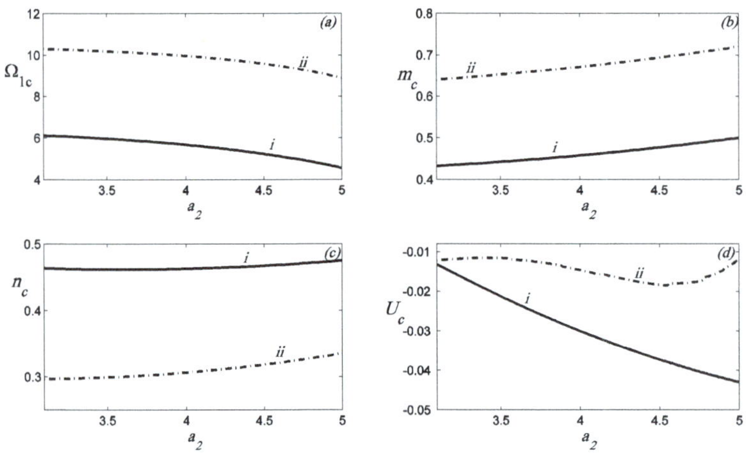


**Figure 8.** The preferred mode parameters $\Omega_{1c}, m_c, n_c, U_c$ are plotted in (a), (b), (c), (d), respectively, as a function of $a_2$, for $x_0 = 3$, $d = 10$ and for two different values of $Ta$. The roman numbers $i$ and $ii$ refer to $Ta = 0.5$ and $Ta = 10$, respectively. Note that the maximum growth decreases as the plume moves away from the wall to the centre.

**Figure 7** illustrates the dependence of the preferred mode on $Ta$ for $x_0 = 1$, $d = 10$ and $a_2$ takes two values: $a_2 = 2$ and $a_2 = 5$. As the Taylor number increases from $Ta = 0$, the maximum growth rate $\Omega_{1c}$ increases rapidly until $Ta$ reaches a certain value that increases as the distance between plume and the nearest sidewall increases, after which the growth rate varies much more slowly. The wavenumbers of the preferred mode show sudden changes at a small value of $Ta$ indicating a change of local maximum as $Ta$, increases through a value, $Ta_e$. The horizontal wavenumber, $m_c$, is zero for small $Ta$ corresponding to 2-dimensional motions, but as $Ta$ reaches $Ta_e$, $m_c$ jumps to a nonzero value and increases thereafter. The vertical wavenumber increases rapidly as $Ta$ increase to $Ta_e$ and then decreases as $Ta$ increases further. The vertical phase speed of the preferred mode also suffers a change at $Ta = Ta_e$ and the jump depends on how far the plume is from the nearest sidewall.

In **Figure 8**, we present the dependence of the preferred mode on the distance $a_2$ between the plume and the nearest sidewall. We note that the growth rate decreases when the plume moves towards the location half-way

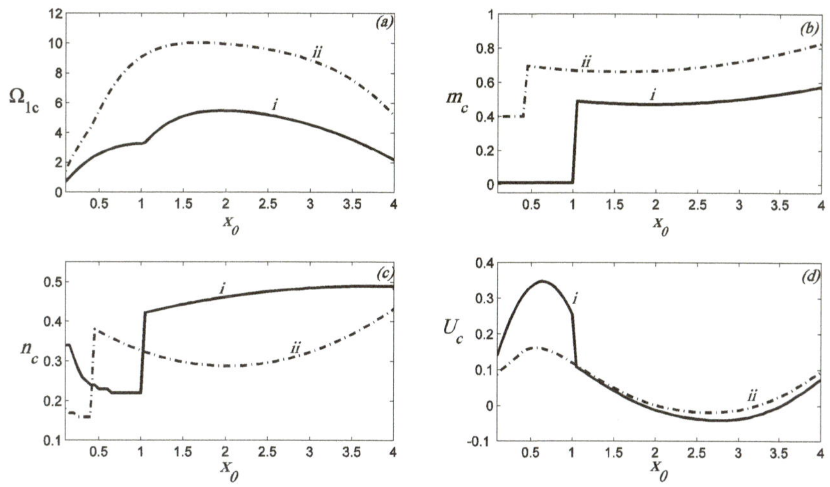


**Figure 9.** The preferred mode parameters $\Omega_{1c}, m_c, n_c, U_c$ are depicted in (a), (b), (c), (d) as a function of $x_0$, for $a_2 = 5$ and $d = 10$. The solid (broken) curves refer to $Ta = 0.5(10)$. Note that the preferred mode of instability is modified sinuous mode in all cases. When the plume is wide, the growth rate decreases whenever the plume moves to the wall.

between the two sidewalls. Both wavenumber components increase steadily as the distance between plume and the nearest sidewall increases. The vertical phase speed however behaves differently as $a_2$ increases; it decreases steadily with increasing $a_2$ if $Ta$ is small but when $Ta$ is large, it decreases slowly reaching a minimum before it increases at a moderate rate. In all cases, it is the MS mode that provides the preferred mode.

**Figure 9** shows the dependence of the critical modes $(\Omega_{1c}, m_c, n_c, U_c)$ on $x_0$. The growth rate increases gradually until $x_0$ reaches a critical thickness, $x_{0c}$, and then it decreases. This indicates that there is a critical thickness representing the most unstable plume. The behaviour of the wavenumber components and phase speed with $x_0$ is quite complicated. The horizontal wavenumber vanishes for small $x_0$ when $Ta$ is small. For large, $m_c$, behaves in a similar way when $x_0$ is close to $x_{0c}$. The vertical wavenumber, on the other hand decreases as $x_0$ increases from zero and jumps to a larger value when the local maximum changes, and then varies very slowly until $x_0$ is almost 4, when $d = 10$.

## 5. Conclusions

The dynamics of a fully developed plume of buoyant fluid, in the form of a channel of finite width, $2x_0$, rising in a less buoyant fluid contained between two parallel vertical walls, a distance $d$ apart, and two fluids rotate uniformly about the vertical have been investigated.

In the absence of boundaries ([18]), it was found that the stability problem depended on the parameters: the Grashoff number, $R$, the Taylor number, $Ta$, and the thickness of the plume, $2x_0$. The magnitude of the growth rate was of the order $O(1)$ for $R \ll 1$ and the instability took one of the two types: sinuous mode or varicose mode.

The presence of the vertical boundaries here introduces two dimensionless parameters: the distance between the plume and the nearest wall, $a_2$, and the distance between the two vertical sidewalls, $d$. The introduction of the boundaries modifies the two modes to be the modified sinuous mode (MS) and the modified varicose mode (MV). It is shown here that the boundaries introduce two main effects to the rotating Cartesian plume studied by Eltayeb and Hamza [18]. First, the sidewalls tends to stabilise the plume but succeed only reducing the growth rate and the plume remains unstable for all values of the Taylor number and the distance from the nearest sidewall. Second, the presence of the sidewalls suppresses the modified varicose mode and allows the modified sinuous only to be unstable. The preferred mode can be 3-dimensional or 2-dimensional depending on the values of the parameters of the system. When the preferred mode is 2-dimensional, propagation can be upwards or downwards.

## References


[1] McDonald, R.J. and Hunt, J.D. (1970) Convection Fluid Motion within the Interdendritic Liquid of a Casting. *Metallurgical Transactions*, **1**, 1787-1788. http://dx.doi.org/10.1007/BF02642039

[2] Copley, S.M., Giamel, A.F., Johnson, S.M. and Hornbecker, M.F. (1970) The Origin of Freckles Inunidirectionally Solidified Castings. *Metallurgical Transactions*, **1**, 2193-2204. http://dx.doi.org/10.1007/BF02643435

[3] Emms, P.W. and Fowler, A.C. (1994) Compositional Convection in the Solidification of Binary Alloys. *Journal of Fluid Mechanics*, **262**, 111-139. http://dx.doi.org/10.1017/S0022112094000443

[4] Worster, M.G. (1997) Convection in Mushy Layers. *Annual Review of Fluid Mechanics*, **29**, 91-122. http://dx.doi.org/10.1146/annurev.fluid.29.1.91

[5] Aussillous, P., Sederman, A.J., Gladden, L.F., Huppert, H.E. and Worster, M.G. (2006) Magnetic Resonance Imaging of Structure and Convection in Solidifying Mushy Layers. *Journal of Fluid Mechanics*, **522**, 99-125. http://dx.doi.org/10.1017/S0022112005008451

[6] Al Mashrafi, K.S. and Eltayeb, I.A. (2014) The Influence of Boundaries on the Stability of Compositional Plumes. *Open Journal of Fluid Dynamics*, **4**, 83-102. http://dx.doi.org/10.4236/ojfd.2014.41007

[7] Verhoogen, J. (1961) Heat Balance of the Earth's Core. *Geophysical Journal of the Royal Astronomical Society*, **4**, 276-281. http://dx.doi.org/10.1111/j.1365-246X.1961.tb06819.x

[8] Morgan, W.J. (1971) Convection Plumes in the Lower Mantle. *Nature*, **230**, 42-43. http://dx.doi.org/10.1038/230042a0

[9] Gubbins, D. (1977) Energetics of the Earth's Core. *J. Geophys.*, **43**, 453-464.

[10] Loper, D.E. (1978) The Gravitationally Powered Dynamo. *Geophysical Journal of the Royal Astronomical Society*, **54**, 389-404. http://dx.doi.org/10.1111/j.1365-246X.1978.tb04265.x

[11] Loper, D.E. and Roberts, P.H. (1978) On the Motion of an Iron-Alloy Core Containing a Slurry. I. General Theory. *Geophysical & Astrophysical Fluid Dynamics*, **9**, 289-321. http://dx.doi.org/10.1080/03091927708242333

[12] Loper, D.E. and Roberts, P.H. (1981) A Study of Conditions at the Inner Core Boundary of the Earth. *Physics of the Earth and Planetary Interiors*, **24**, 302-307. http://dx.doi.org/10.1016/0031-9201(81)90117-5

[13] Loper, D.E. and Stacey, F.D. (1983) The Dynamical and Thermal Structure of Deep Mantle Plumes. *Physics of the Earth and Planetary Interiors*, **34**, 304-317. http://dx.doi.org/10.1016/0031-9201(83)90047-X

[14] Moffatt, H.K. (1989) Liquid Metal MHD and the Geodynamo. In: Lielpeteris, J. and Moreau, R., Eds., *Liquid Metal Magneto-Hydrodynamics*, Kluwer Academic Publishers, Dordrecht, 403-412.

[15] Eltayeb, I.A. and Loper, D.E. (1991) On the Stability of Vertical Double-Diffusive Interfaces. Part 1. A Single Plane Interface. *Journal of Fluid Mechanics*, **228**, 149-181.

[16] Eltayeb, I.A. and Loper, D.E. (1994) On the Stability of Vertical Double-Diffusive Interfaces. Part 2. Two Parallel Interfaces. *Journal of Fluid Mechanics*, **267**, 251-273. http://dx.doi.org/10.1017/S0022112094001175

[17] Eltayeb, I.A. and Loper, D.E. (1997) On the Stability of Vertical Double-Diffusive Interfaces. Part 3. Cylindrical Interfaces. *Journal of Fluid Mechanics*, **353**, 45-66. http://dx.doi.org/10.1017/S0022112097007374

[18] Eltayeb, I.A. and Hamza, E.A. (1998) Compositional Convection in the Presence of Rotation. *Journal of Fluid Mechanics*, **354**, 277-299. http://dx.doi.org/10.1017/S0022112097007696

[19] Eltayeb, I.A. (1999) The Stability of Compositional Plumes in a Rotating Magnetic Fluid. *Physics of the Earth and Planetary Interiors*, **110**, 1-19.

[20] Eltayeb, I.A. (2006) The Stability of a Compositional Plume Rotating in the Presence of a Magnetic Field. *Geophysical & Astrophysical Fluid Dynamics*, **100**, 429-455. http://dx.doi.org/10.1080/03091920600799541

[21] Kumagai, I. (2002) On the Anatomy of Mantle Plumes: Effect of the Viscosity Ratio on Entrainment and Stirring. *Earth and Planetary Science Letters*, **198**, 211-224. http://dx.doi.org/10.1016/S0012-821X(02)00513-7

[22] Shimizu, H., Poirier, J.P. and Le Mouel, J.L. (2005) On Crystallization at the Inner Core Boundary. *Physics of the Earth and Planetary Interiors*, **151**, 37-51. http://dx.doi.org/10.1016/j.pepi.2005.01.001

[23] Kerr, R.C., Mériaux, C. and Lister, J.R. (2008) Effect of Thermal Diffusion on the Stability of Strongly Tilted Mantle Plume Tails. *Journal of Geophysical Research*, **113**, Article ID: B09401. http://dx.doi.org/10.1029/2007JB005510

[24] Al-Lawatia, M.A., Elbashir, T.B.A., Eltayeb, I.A., Rahman, M.M. and Balakrishnan, E. (2011) The Dynamics of Two Interacting Compositional Plumes in the Presence of a Magnetic Field. *Geophysical & Astrophysical Fluid Dynamics*, **105**, 586-615. http://dx.doi.org/10.1080/03091929.2010.518316

[25] Eltayeb, I.A. and Elbashir, T.B.A. (2013) The Stability of Two Interacting Compositional Plumes under the Influence of Inclined Rotation.

[26] Stern, M.E. (1960) The "Salt Fountain" and Thermohaline Convection. *Tellus*, **12**, 172-175.

http://dx.doi.org/10.1111/j.2153-3490.1960.tb01295.x

[27] Bennington, K.O. (1963) Some Crystal Growth Features of Sea Ice. *Journal of Glaciology*, **4**, 669-689.

[28] Lake, R. and Lewis, E. (1970) Salt Rejection by Sea Ice during Growth. *Journal of Geophysical Research*, **75**, 583-597. http://dx.doi.org/10.1029/JC075i003p00583

[29] Holyer, J.Y. (1984) The Stability of Long, Steady, Two-Dimensional Salt Fingers. *Journal of Fluid Mechanics*, **147**, 169-185. http://dx.doi.org/10.1017/S0022112084002044

[30] Howard, L.N. and Veronis, G. (1992) Stability of Salt Fingers with Negligible Salt Diffusivity. *Journal of Fluid Mechanics*, **239**, 511-522. http://dx.doi.org/10.1017/S0022112092004518

[31] Wettlaufer, J.S., Worster, M.G. and Huppert, H.E. (1997) Natural Convection during Solidification of an Alloy from above with Application to the Evolution of Sea Ice. *Journal of Fluid Mechanics*, **344**, 291-316. http://dx.doi.org/10.1017/S0022112097006022

[32] Schmitt, R.W. (2003) Observational and Laboratory Insights into Salt-Finger Convection. *Progress in Oceanography*, **56**, 419-433. http://dx.doi.org/10.1016/S0079-6611(03)00033-8

[33] Feltham, D.L., Untersteiner, N., Wettlaufer, J.S. and Worster, M.G. (2006) Sea Ice Is a Mushy Layer. *Geophysical Research Letters*, **33**, Article ID: L14501. http://dx.doi.org/10.1029/2006GL026290

[34] Notz, D. and Worster, M.G. (2008) *In Situ* Measurements of the Evolution of Young Sea Ice. *Journal of Geophysical Research*, **113**, Article ID: C03001. http://dx.doi.org/10.1029/2007JC004333

[35] Sample, A.K. and Hellawell, A. (1984) The Mechanisms of Formation and Prevention of Channel Segregation during Alloy Solidification. *Metallurgical Transactions A*, **15**, 2163-2173. http://dx.doi.org/10.1007/BF02647099

[36] Chen, C.F. and Chen, F. (1991) Experimental Study of Directional Solidification of Aqueous Ammonium Chloride Solution. *Journal of Fluid Mechanics*, **227**, 567-586. http://dx.doi.org/10.1017/S0022112091000253

[37] Hellawell, A., Sarazin, J.R. and Steube, R.S. (1993) Channel Convection in Partly Solidified Systems. *Philosophical Transactions of the Royal Society A. London*, **345**, 507-544. http://dx.doi.org/10.1098/rsta.1993.0143

[38] Classen, S., Heimpel, M. and Christensen, U. (1999) Blob Instability in Rotating Compositional Convection. *Geophysical Research Letters*, **26**, 135-138. http://dx.doi.org/10.1029/1998GL900227

## Appendix A: The Derivation of the Dispersion Relation

In this appendix, we apply the boundary conditions (26)-(28) when the subscript is 0 for the variables (41), (44) and (45) to evaluate the constants of the solution. Application of the boundary conditions at the interface $x = x_0$ gives

$$\sum_{j=1}^{3}\mu_j^3 E_j = 0, \quad \sum_{j=1}^{3}\mu_j^2 E_j = 0, \quad \sum_{j=1}^{3}\left(1-\tau^2\right)\mu_j E_j = 0, \tag{A:1}$$

$$\sum_{j=1}^{3}\lambda_j\mu_j^3 F_j = 1, \quad \sum_{j=1}^{3}\lambda_j\mu_j^2 F_j = 0, \quad \sum_{j=1}^{3}\lambda_j\mu_j\left\{\mu_j^2+\left(1-\tau^2\right)\right\}F_j = 0, \tag{A:2}$$

$$\sum_{j=1}^{3}\frac{\mu_j^2}{\mu_j+n^2}\left\{-m\mu_j E_j + \mathrm{i}\tau\lambda_j F_j\right\} + E = 0, \tag{A:3}$$

$$\sum_{j=1}^{3}\frac{\lambda_j\mu_j^2}{\mu_j+n^2}\left\{-m\mu_j F_j + \mathrm{i}\tau\lambda_j E_j\right\} + mF = 0, \tag{A:4}$$

where $E_j$, $F_j$, $E$ and $F$ are defined by

$$E_j = \left(A_j^{(2)} - A_j^{(3)}\right)\cosh\left(\lambda_j x_0\right) + \left(B_j^{(2)} - B_j^{(3)}\right)\sinh\left(\lambda_j x_0\right), \tag{A:5}$$

$$F_j = \left(A_j^{(2)} - A_j^{(3)}\right)\sinh\left(\lambda_j x_0\right) + \left(B_j^{(2)} - B_j^{(3)}\right)\cosh\left(\lambda_j x_0\right), \tag{A:6}$$

$$E = \left(C^{(2)} - C^{(3)}\right)\cosh\left(m x_0\right) + \left(D^{(2)} - D^{(3)}\right)\sinh\left(m x_0\right), \tag{A:7}$$

$$F = \left(C^{(2)} - C^{(3)}\right)\sinh\left(m x_0\right) + \left(D^{(2)} - D^{(3)}\right)\cosh\left(m x_0\right). \tag{A:8}$$

We use the Equations (A:5)-(A:8) to obtain

$$E_j\cosh\left(\lambda_j x_0\right) - F_j\sinh\left(\lambda_j x_0\right) = A_j^{(2)} - A_j^{(3)}, \tag{A:9}$$

$$E_j\sinh\left(\lambda_j x_0\right) - F_j\cosh\left(\lambda_j x_0\right) = B_j^{(3)} - B_j^{(2)}, \tag{A:10}$$

$$E\cosh\left(m x_0\right) - F\sinh\left(m x_0\right) = C^{(2)} - C^{(3)}, \tag{A:11}$$

$$E\sinh\left(m x_0\right) - F\cosh\left(m x_0\right) = D^{(3)} - D^{(2)}. \tag{A:12}$$

Similarly, at the interface $x = -x_0$, we find

$$\sum_{j=1}^{3}\mu_j^3 K_j = 0, \quad \sum_{j=1}^{3}\mu_j^2 K_j = 0, \quad \sum_{j=1}^{3}\left(1-\tau^2\right)\mu_j K_j = 0, \tag{A:13}$$

$$\sum_{j=1}^{3}\lambda_j\mu_j^3 L_j = -\eta_1, \quad \sum_{j=1}^{3}\lambda_j\mu_j^2 L_j = 0, \quad \sum_{j=1}^{3}\lambda_j\mu_j\left\{\mu_j^2+\left(1-\tau^2\right)\right\}L_j = 0, \tag{A:14}$$

$$\sum_{j=1}^{3}\frac{\mu_j^2}{\mu_j+n^2}\left\{-m\mu_j K_j + \mathrm{i}\tau\lambda_j L_j\right\} + K = 0, \tag{A:15}$$

$$\sum_{j=1}^{3}\frac{\lambda_j\mu_j^2}{\mu_j+n^2}\left\{-m\mu_j L_j + \mathrm{i}\tau\lambda_j K_j\right\} + mL = 0, \tag{A:16}$$

in which $K_j$, $L_j$, $K$ and $L$ are given by

$$K_j = \left(A_j^{(1)} - A_j^{(2)}\right)\cosh\left(\lambda_j x_0\right) - \left(B_j^{(1)} - B_j^{(2)}\right)\sinh\left(\lambda_j x_0\right), \tag{A:17}$$

$$L_j = -\left(A_j^{(1)} - A_j^{(2)}\right)\sinh\left(\lambda_j x_0\right) + \left(B_j^{(1)} - B_j^{(2)}\right)\cosh\left(\lambda_j x_0\right), \tag{A:18}$$

$$K = \left(C^{(1)} - C^{(2)}\right)\cosh\left(mx_0\right) - \left(D^{(1)} - D^{(2)}\right)\sinh\left(mx_0\right), \quad \text{(A:19)}$$

$$L = -\left(C^{(1)} - C^{(2)}\right)\sinh\left(mx_0\right) + \left(D^{(1)} - D^{(2)}\right)\cosh\left(mx_0\right). \quad \text{(A:20)}$$

Equations (A:17)-(A:20) give

$$K_j \cosh\left(\lambda_j x_0\right) + L_j \sinh\left(\lambda_j x_0\right) = A_j^{(1)} - A_j^{(2)}, \quad \text{(A:21)}$$

$$K_j \sinh\left(\lambda_j x_0\right) + L_j \cosh\left(\lambda_j x_0\right) = B_j^{(1)} - B_j^{(2)}, \quad \text{(A:22)}$$

$$K \cosh\left(mx_0\right) + L \sinh\left(mx_0\right) = C^{(1)} - C^{(2)}, \quad \text{(A:23)}$$

$$K \sinh\left(mx_0\right) + L \cosh\left(mx_0\right) = D^{(1)} - D^{(2)}. \quad \text{(A:24)}$$

The boundary conditions at the boundaries $x = a_1$ and $x = -a_2$ give

$$\sum_{j=1}^{3} \mu_j^3 P_j^{(3)} = 0, \quad \sum_{j=1}^{3} \mu_j^2 P_j^{(3)} = 0, \quad \sum_{j=1}^{3} \mu_j^3 P_j^{(1)} = 0, \quad \sum_{j=1}^{3} \mu_j^2 P_j^{(1)} = 0, \quad \text{(A:25)}$$

$$\sum_{j=1}^{3} \frac{\mu_j^2}{\mu_j + n^2}\left[-m\mu_j P_j^{(3)} + \mathrm{i}\tau\lambda_j Q_j^{(3)}\right] + C^{(3)} \cosh\left(ma_1\right) + D^{(3)} \sinh\left(ma_1\right) = 0, \quad \text{(A:26)}$$

$$\sum_{j=1}^{3} \frac{\mu_j^2}{\mu_j + n^2}\left[-m\mu_j P_j^{(1)} + \mathrm{i}\tau\lambda_j Q_j^{(1)}\right] + C^{(1)} \cosh\left(ma_2\right) - D^{(1)} \sinh\left(ma_2\right) = 0, \quad \text{(A:27)}$$

$$\sum_{j=1}^{3} \lambda_j \mu_j \left\{\mu_j^2 + 1 - \tau^2\right\} Q_j^{(3)} = 0, \quad \sum_{j=1}^{3} \lambda_j \mu_j \left\{\mu_j^2 + 1 - \tau^2\right\} Q_j^{(1)} = 0, \quad \text{(A:28)}$$

such that $P_j^{(3)}$, $P_j^{(1)}$, $Q_j^{(3)}$ and $Q_j^{(1)}$ are defined by

$$P_j^{(3)} = A_j^{(3)} \cosh\left(\lambda_j a_1\right) + B_j^{(3)} \sinh\left(\lambda_j a_1\right), \quad \text{(A:29)}$$

$$P_j^{(1)} = A_j^{(1)} \cosh\left(\lambda_j a_2\right) - B_j^{(1)} \sinh\left(\lambda_j a_2\right), \quad \text{(A:30)}$$

$$Q_j^{(3)} = A_j^{(3)} \sinh\left(\lambda_j a_1\right) + B_j^{(3)} \cosh\left(\lambda_j a_1\right), \quad \text{(A:31)}$$

$$Q_j^{(1)} = -A_j^{(1)} \sinh\left(\lambda_j a_2\right) + B_j^{(1)} \cosh\left(\lambda_j a_2\right). \quad \text{(A:32)}$$

The equations (A:1) and (A:2) can be solved to find

$$E_j = 0, \quad F_j = \frac{-\left(\mu_j + n^2\right)}{\lambda_j \mu_j \left\{2\mu_j + 3n^2\left(1 - \tau^2\right)\right\}}, \quad j = 1,2,3, \quad \text{(A:33)}$$

since the roots $\mu_j$ are distinct. Hence Equations (A:3)-(A:5) give

$$\left(A_j^{(2)} - A_j^{(3)}\right)\cosh\left(\lambda_j x_0\right) + \left(B_j^{(2)} - B_j^{(3)}\right)\sinh\left(\lambda_j x_0\right) = 0, \quad \text{(A:34)}$$

$$\mathrm{i}\tau \sum_{j=1}^{3} \frac{\lambda_j \mu_j^2}{\mu_j + n^2} F_j + E = 0, \quad \text{(A:35)}$$

$$-\sum_{j=1}^{3} \frac{\lambda_j \mu_j^3}{\mu_j + n^2} F_j + F = 0. \quad \text{(A:36)}$$

Using the expression for $F_j$ in (A:33), and the properties of the 3 roots of the cubic Equation (42) to simplify the expressions (A:35), (A:36). The simplifications lead to

$$E = F = 0. \quad \text{(A:37)}$$

Hence the expressions (A:9)-(A:12) reduce to

$$A_j^{(2)} - A_j^{(3)} = -F_j \sinh\left(\lambda_j x_0\right), \tag{A:38}$$

$$B_j^{(2)} - B_j^{(3)} = F_j \cosh\left(\lambda_j x_0\right), \tag{A:39}$$

$$C^{(2)} = C^{(3)}, \quad D^{(2)} = D^{(3)}. \tag{A:40}$$

The same method can be applied to the conditions at $x = -x_0$. The Equations (A:13)-(A:16) leads to

$$K_j = 0, \quad L_j = -\eta_1 F_j, \quad K = L = 0, \tag{A:41}$$

thus the expressions (A:21)-(A:24) give

$$A_j^{(1)} - A_j^{(2)} = L_j \sinh\left(\lambda_j x_0\right), \tag{A:42}$$

$$B_j^{(1)} - B_j^{(2)} = L_j \cosh\left(\lambda_j x_0\right), \tag{A:43}$$

$$C^{(1)} = C^{(2)}, \quad D^{(1)} = D^{(2)}. \tag{A:44}$$

The expressions $P_j^{(3)}$, $P_j^{(1)}$, $Q_j^{(3)}$ and $Q_j^{(1)}$, given in (A:29)-(A:32), can be simplified by using the relations (A:38), (A:39) and (A:42), (A:43) to find

$$P_j^{(3)} = V_j + F_j \sinh\left\{\lambda_j\left(x_0 - a_1\right)\right\}, \tag{A:45}$$

$$Q_j^{(3)} = W_j - F_j \cosh\left\{\lambda_j\left(x_0 - a_1\right)\right\}, \tag{A:46}$$

$$P_j^{(1)} = R_j + L_j \sinh\left\{\lambda_j\left(x_0 - a_2\right)\right\}, \tag{A:47}$$

$$Q_j^{(1)} = S_j + L_j \cosh\left\{\lambda_j\left(x_0 - a_2\right)\right\}, \tag{A:48}$$

where we have defined the notation

$$V_j = A_j^{(2)} \cosh\left(\lambda_j a_1\right) + B_j^{(2)} \sinh\left(\lambda_j a_1\right), \tag{A:49}$$

$$W_j = A_j^{(2)} \sinh\left(\lambda_j a_1\right) + B_j^{(2)} \cosh\left(\lambda_j a_1\right), \tag{A:50}$$

$$R_j = A_j^{(2)} \cosh\left(\lambda_j a_2\right) - B_j^{(2)} \sinh\left(\lambda_j a_2\right), \tag{A:51}$$

$$S_j = -A_j^{(2)} \sinh\left(\lambda_j a_2\right) + B_j^{(2)} \cosh\left(\lambda_j a_2\right). \tag{A:52}$$

The Equations (A:49)-(A:52) give

$$A_j^{(2)} = \frac{1}{\sinh\left(\lambda_j d\right)}\left\{R_j \sinh\left(\lambda_j a_1\right) + V_j \sinh\left(\lambda_j a_2\right)\right\}, \tag{A:53}$$

$$B_j^{(2)} = \frac{-1}{\sinh\left(\lambda_j d\right)}\left\{R_j \cosh\left(\lambda_j a_1\right) - V_j \cosh\left(\lambda_j a_2\right)\right\}, \tag{A:54}$$

$$W_j = \frac{1}{\sinh\left(\lambda_j d\right)}\left\{-R_j + V_j \cosh\left(\lambda_j d\right)\right\}, \tag{A:55}$$

$$S_j = \frac{-1}{\sinh\left(\lambda_j d\right)}\left\{R_j \cosh\left(\lambda_j d\right) - V_j\right\}. \tag{A:56}$$

The boundary conditions at $x = a_1$ and $x = -a_2$, given in (A:25) and (A:28), can be written by using

(A:45)-(A:48) and (A:55) and (A:56) to obtain

$$\sum_{j=1}^{3}\left\{\mu_j^3;\mu_j^2;\bar{d}_j\left(\cosh\left(\lambda_j d\right)-1\right)\right\}\left(R_j+V_j\right)=\left\{x_1;x_2;x_3\right\}, \tag{A:57}$$

$$\sum_{j=1}^{3}\left\{\mu_j^3;\mu_j^2;\bar{d}_j\left(\cosh\left(\lambda_j d\right)+1\right)\right\}\left(R_j-V_j\right)=\left\{x_4;x_5;x_6\right\}, \tag{A:58}$$

in which $x_1$, $x_2$, $x_3$, $x_4$, $x_5$ and $x_6$ are defined by

$$x_1 = Y_3-\eta_1 Y_1,\qquad x_2 = Y_4-\eta_1 Y_2,\qquad x_3 = Y_6-\eta_1 Y_5, \tag{A:59}$$

$$x_4 = -Y_3-\eta_1 Y_1,\qquad x_5 = -Y_4-\eta_1 Y_2,\qquad x_6 = -Y_6-\eta_1 Y_5, \tag{A:60}$$

and we have defined $\bar{d}_j$ and $Y_i$, $i=1-6$ as

$$\bar{d}_j = \frac{\lambda_j\mu_j}{\sinh\left(\lambda_j d\right)}\left\{\mu_j^2+1-\tau^2\right\},\qquad j=1,2,3, \tag{A:61}$$

$$Y_1=\sum_{j=1}^{3}\mu_j^3F_{2j}\sinh\left[\lambda_j\left(a_2-x_0\right)\right],\qquad Y_2=\sum_{j=1}^{3}\mu_j^2F_{2j}\sinh\left[\lambda_j\left(a_2-x_0\right)\right], \tag{A:62}$$

$$Y_3=\sum_{j=1}^{3}\mu_j^3F_{2j}\sinh\left[\lambda_j\left(a_1-x_0\right)\right],\qquad Y_4=\sum_{j=1}^{3}\mu_j^2F_{2j}\sinh\left[\lambda_j\left(a_1-x_0\right)\right], \tag{A:63}$$

$$Y_5=\sum_{j=1}^{3}\bar{d}_jF_{2j}\sinh\left(\lambda_j d\right)\cosh\left[\lambda_j\left(a_2-x_0\right)\right],\quad Y_6=\sum_{j=1}^{3}\bar{d}_jF_{2j}\sinh\left(\lambda_j d\right)\cosh\left[\lambda_j\left(a_1-x_0\right)\right]. \tag{A:64}$$

The algebraic system (A:57) and (A:58) can be solved by elimination to find

$$R_j = R_{j0}+\eta_1 R_{je},\qquad V_j = V_{j0}+\eta_1 V_{je},\quad \text{for } j=1,2,3, \tag{A:65}$$

in which $R_{j0}$, $R_{je}$, $V_{j0}$ and $V_{je}$ are given by

$$R_{10}=\frac{-\left(f_{2+}g_{10-}+f_{2-}g_{10+}\right)}{2f_{2-}f_{2+}},\qquad R_{1e}=\frac{-\left(f_{2+}g_{1e-}+f_{2-}g_{1e+}\right)}{2f_{2-}f_{2+}}, \tag{A:66}$$

$$R_{20}=\frac{\left(f_{2+}f_{10-}+f_{2-}f_{10+}\right)}{2f_{2-}f_{2+}},\qquad R_{2e}=\frac{\left(f_{2+}f_{1e-}+f_{2-}f_{1e+}\right)}{2f_{2-}f_{2+}}, \tag{A:67}$$

$$R_{30}=\frac{-\left(f_{2+}A_{10-}+f_{2-}A_{10+}\right)}{2f_{2-}f_{2+}},\qquad R_{3e}=\frac{-\left(f_{2+}A_{1e-}+f_{2-}A_{1e+}\right)}{2f_{2-}f_{2+}}, \tag{A:68}$$

$$V_{10}=\frac{-\left(f_{2+}g_{10-}-f_{2-}g_{10+}\right)}{2f_{2-}f_{2+}},\qquad V_{1e}=\frac{-\left(f_{2+}g_{1e-}-f_{2-}g_{1e+}\right)}{2f_{2-}f_{2+}}, \tag{A:69}$$

$$V_{20}=\frac{\left(f_{2+}f_{10-}-f_{2-}f_{10+}\right)}{2f_{2-}f_{2+}},\qquad V_{2e}=\frac{\left(f_{2+}f_{1e-}-f_{2-}f_{1e+}\right)}{2f_{2-}f_{2+}}, \tag{A:70}$$

$$V_{30}=\frac{-\left(f_{2+}A_{10-}-f_{2-}A_{10+}\right)}{2f_{2-}f_{2+}},\qquad V_{3e}=\frac{-\left(f_{2+}A_{1e-}-f_{2-}A_{1e+}\right)}{2f_{2-}f_{2+}}, \tag{A:71}$$

and we have defined the following notation

$$g_{10\pm}=\mp\mu_2^2T_{3\pm}\left(\mu_2Y_4-Y_3\right)\mp\mu_3^2T_{2\pm}\left(Y_3-\mu_3Y_4\right)\pm\mu_2^2\mu_3^2\left(\mu_2-\mu_3\right)Y_6, \tag{A:72}$$

$$g_{1e\pm}=\mu_2^2T_{3\pm}\left(-\mu_2Y_2+Y_1\right)+\mu_3^2T_{2\pm}\left(-Y_1+\mu_3Y_2\right)+\mu_2^2\mu_3^2\left(\mu_2-\mu_3\right)Y_5, \tag{A:73}$$

$$f_{10\pm}=\mp\mu_1^2T_{3\pm}\left(\mu_1Y_4-Y_3\right)\mp\mu_3^2T_{1\pm}\left(\mp Y_3-\mu_3Y_4\right)\pm\mu_1^2\mu_3^2\left(\mu_1-\mu_3\right)Y_6, \tag{A:74}$$

$$f_{1e\pm} = \mu_1^2 T_{3\pm}\left(-\mu_1 Y_2 + Y_1\right) + \mu_3^2 T_{1\pm}\left(-Y_1 + \mu_3 Y_2\right) + \mu_1^2 \mu_3^2 \left(\mu_1 - \mu_3\right) Y_5, \tag{A:75}$$

$$A_{10\pm} = \mp \mu_1^2 T_{2\pm}\left(\mu_1 Y_4 - Y_3\right) \mp \mu_2^2 T_{1\pm}\left(Y_3 - \mu_2 Y_4\right) \pm \mu_1^2 \mu_2^2 \left(\mu_1 - \mu_2\right) Y_6, \tag{A:76}$$

$$A_{1e\pm} = \mu_1^2 T_{2\pm}\left(-\mu_1 Y_2 + Y_1\right) + \mu_1^2 \mu_2^2 \left(\mu_1 - \mu_2\right) Y_5 + \mu_2^2 T_{1\pm}\left(-Y_1 + \mu_2 Y_2\right), \tag{A:77}$$

$$f_{2\pm} = \mu_2^2 \mu_3^2 \left(\mu_2 - \mu_3\right) T_{1\pm} - \mu_1^2 \mu_3^2 \left(\mu_1 - \mu_3\right) T_{2\pm} + \mu_1^2 \mu_2^2 \left(\mu_1 - \mu_3\right) T_{3\pm}, \tag{A:78}$$

$$T_{j\pm} = \bar{d}_j \left\{\cosh\left(\lambda_j d\right) \pm 1\right\}. \tag{A:79}$$

The constants $A_j^{(i)}$ and $B_j^{(i)}$ for $i, j = 1,2,3$ can easily evaluated using (A:38), (A:39), (A:42), (A:43), (A:53), (A:54). The next step is to find the expressions of the constants $C^{(i)}$ and $D^{(i)}$. The Equations (A:26), (A:27), (A:40) and (A:44) give

$$C^{(1)} = C^{(2)} = C^{(3)} = \frac{-1}{\sinh(md)}\left\{\sinh(ma_2)\sum_{j=1}^{3} H_j^{(3)} + \sinh(ma_1)\sum_{j=1}^{3} H_j^{(1)}\right\}, \tag{A:80}$$

$$D^{(1)} = D^{(2)} = D^{(3)} = \frac{-1}{\sinh(md)}\left\{\cosh(ma_2)\sum_{j=1}^{3} H_j^{(3)} - \cosh(ma_1)\sum_{j=1}^{3} H_j^{(1)}\right\}, \tag{A:81}$$

where $H_j^{(1)}$ and $H_j^{(3)}$ are defined by

$$H_j^{(1)} = \frac{\mu_j^2}{\mu_j + n^2}\left\{-m\mu_j P_j^{(1)} + \mathrm{i}\lambda_j \tau Q_j^{(1)}\right\}, \qquad H_j^{(3)} = \frac{\mu_j^2}{\mu_j + n^2}\left\{-m\mu_j P_j^{(3)} + \mathrm{i}\lambda_j \tau Q_j^{(3)}\right\}. \tag{A:82}$$

The application of the boundary condition (29) gives

$$\Omega_1 - \mathrm{i}n\bar{w}(x_0) = F_{1+} + \eta_1 F_{e+}, \qquad \left\{\Omega_1 - \mathrm{i}n\bar{w}(-x_0)\right\}\eta_1 = F_{1-} + \eta_1 F_{e-}, \tag{A:83}$$

such that $F_{1\pm}$ and $F_{e\pm}$ are given by

$$F_{1\pm} = G_\pm\left(R_{j0}, V_{j0}\right) + R_{0\pm}, \qquad F_{e\pm} = G_\pm\left(R_{je}, V_{je}\right) + V_{0\pm}, \tag{A:84}$$

where $G_\pm\left(R_{j\alpha}, V_{j\alpha}\right)$ for $\alpha = \{0, e\}$, $R_{0\pm}$ and $V_{0\pm}$ are defined by

$$G_\pm\left(R_{j\alpha}, V_{j\alpha}\right) = \frac{\mathrm{i}}{n}\sum_{j=1}^{3} \bar{d}_j \left[-R_{j\alpha}\cosh\left\{\lambda_j\left(a_1 \mp x_0\right)\right\} + V_{j\alpha}\cosh\left\{\lambda_j\left(a_2 \pm x_0\right)\right\}\right] + \frac{\tau}{n} Z_\pm, \tag{A:85}$$

$$R_{0\pm} = \frac{\tau\sinh\left[m\left(a_2 \pm x_0\right)\right]}{n\sinh(md)}\sum_{j=1}^{3}\frac{\mu_j^2 a_{1j}}{\mu_j + n^2} F_j, \qquad V_{0\pm} = \frac{\tau\sinh\left[m\left(a_1 \mp x_0\right)\right]}{n\sinh(md)}\sum_{j=1}^{3}\frac{\mu_j^2 a_{2j}}{\mu_j + n^2} F_j, \tag{A:86}$$

and we have used the notation $Z_\pm$, $a_{1j}$, $a_{2j}$, $a_{3j\pm}$, $a_{4j\pm}$, $E_{1s\pm}$ and $E_{2s\pm}$

$$\begin{aligned} Z_\pm = m\sum_{j=1}^{3}\frac{\mu_j^2}{\sinh\left(\lambda_j d\right)}\left\{R_j \sinh\left[\lambda_j\left(a_1 \mp x_0\right)\right] + V_j \sinh\left[\lambda_j\left(a_2 \pm x_0\right)\right]\right\} \\ + \sum_{j=1}^{3}\frac{\mu_j^2\left(-a_{3j\mp}R_j + a_{4j\pm}V_j\right)}{\left(\mu_j + n^2\right)\sinh\left(\lambda_j d\right)} + \frac{1}{\sinh(md)}\sum_{j=1}^{3}\frac{\mu_j^2\left(E_{1s\mp}R_j + E_{2s\pm}V_j\right)}{\mu_j + n^2}, \end{aligned} \tag{A:87}$$

$$a_{1j} = -m\mu_j \sinh\left\{\lambda_j\left(a_1 - x_0\right)\right\} + \mathrm{i}\tau\lambda_j \cosh\left\{\lambda_j\left(a_1 - x_0\right)\right\}, \tag{A:88}$$

$$a_{2j} = m\mu_j \sinh\left\{\lambda_j\left(a_2 - x_0\right)\right\} + \mathrm{i}\tau\lambda_j \cosh\left\{\lambda_j\left(a_2 - x_0\right)\right\}, \tag{A:89}$$

$$a_{3j\pm} = m\mu_j \sinh\left\{\lambda_j\left(a_1 \pm x_0\right)\right\} + \mathrm{i}\tau\lambda_j \cosh\left\{\lambda_j\left(a_1 \pm x_0\right)\right\}, \tag{A:90}$$

$$a_{4j\pm} = -m\mu_j \sinh\left\{\lambda_j\left(a_2 \pm x_0\right)\right\} + \mathrm{i}\tau\lambda_j \cosh\left\{\lambda_j\left(a_2 \pm x_0\right)\right\}, \tag{A:91}$$

$$E_{1s\pm} = E_{j+} \sinh\left\{m\left(a_1 \pm x_0\right)\right\} + \mathrm{i}\tau\lambda_j \frac{\sinh\left\{m\left(a_2 \mp x_0\right)\right\}}{\sinh\left\{\lambda_j d\right\}}, \tag{A:92}$$

$$E_{2s\pm} = E_{j-} \sinh\left\{m\left(a_2 \pm x_0\right)\right\} - \mathrm{i}\tau\lambda_j \frac{\sinh\left\{m\left(a_1 \mp x_0\right)\right\}}{\sinh\left\{\lambda_j d\right\}}, \tag{A:93}$$

and $E_{j\pm}$ has the form

$$E_{j\pm} = m\mu_j \pm \mathrm{i}\tau\lambda_j \coth\left(\lambda_j d\right). \tag{A:94}$$

Now the growth rate $\Omega_1$ and the displacement $\eta_1$ can be evaluated by solving the Equation (A:83) to find

$$\eta_1 = \frac{F_{1-}}{\Omega_1 - \mathrm{i}n\overline{w}(-x_0) - F_{e-}}, \tag{A:95}$$

$$\left(\Omega_1/\mathrm{i}n\right)^2 + S_1\left(\Omega_1/\mathrm{i}n\right) + S_2 = 0, \tag{A:96}$$

where $S_1$ and $S_2$ are defined by

$$S_1 = \frac{\mathrm{i}}{n}\left(F_{1+} + F_{e-}\right) - \overline{w}\left(x_0\right) - \overline{w}\left(-x_0\right), \tag{A:97}$$

$$S_2 = \left(\frac{\mathrm{i}}{n}F_{1+} - \overline{w}\left(x_0\right)\right)\left(\frac{\mathrm{i}}{n}F_{e-} - \overline{w}\left(-x_0\right)\right) + \frac{1}{n^2}F_{e+}F_{1-}. \tag{A:98}$$

# 6

# A Seven-Dimensional System of the Navier-Stokes Equations for a Two-Dimensional Incompressible Fluid on a Torus

**Heyuan Wang[1], Yan Gao[2]**

[1]College of Sciences, Liaoning University of technology, Jinzhou, China
[2]Liaoning Petro-chemical Vocational Technology College, Jinzhou, China
Email: wangheyuan6400@sina.com, gaoyan67830@126.com

## Abstract

**A seven-mode truncation system of the Navier-Stokes equations for a two-dimensional incompressible fluid on a torus is considered. Its stationary solutions and stability are presented; the existence of the attractor and the global stability of the system are discussed. The whole process, which shows a chaos behavior approached through instability of invariant tori, is simulated numerically by computers with the changing of Reynolds number. Based on numerical simulation results of bifurcation diagram, Lyapunov exponent spectrum, Poincare section, power spectrum and return map of the system, some basic dynamical behaviors of the new chaos system are revealed.**

## Keywords

**The Navier-Stokes Equations, Strange Attractor, Lyapunov Function, Bifurcation, Chaos**

## 1. Introduction

In recent years much attention has been devoted to the study of simple differential or difference equations, which although deterministic, exhibit a transition as some parameters go through certain values to a chaos behavior. The equations which are studied often arise in a natural way as simplified models in fluid dynamics and in ecology. The best known examples are perhaps the models of [1]-[3]. In these models a chaos behavior arises as a consequence of the appearance of an attractor of complicated structure which is called "strange attractor". Trajectories in a neighborhood of the attractor appear to move in a completely erratic way. Phenomena of this kind are predicted under certain hypothesis, by the mathematical theory of turbulence of Ruelle and Takens [4]. In the following we consider a model obtained by a suitable seven-mode truncation of the Navier-Stokes equations for a 2-dimensional incompressible fluid on a torus [5]. The paper [5] has been published in Proceedings of the IWCFTA 2012, but the published paper is too rough and several deficiencies in the study of the published paper can be found. Only local results of dynamic behavior are given, and some simulation results are wrong; the transition to chaos is confused. This paper gives more accurate simulation results, discusses the dynamic behavior of the system at the high values of the Reynolds number, and analyses the evolution of the dynamic behavior of the system. This paper is extension and modification of the published article [5] and the content is more abundant. We argue on the basis of numerical simulation results analysis where in a certain range of the Reynolds number our system behaves in a way similar to that predicted by Ruelle and Takens. Dynamical behaviors of this new chaotic system, including some basic dynamical properties, bifurcations and routes to chaos, etc., have been investigated both theoretically and numerically by changing Reynolds number. Our purpose is to study how the phenomena of the model change when the number of modes in the truncation is slightly increased. The existence of the attractor and the global stability of the equations have been firmly verified and these theories can be used in other similar systems. Furthermore, some basic dynamical behaviors of the new chaos system, such as bifurcation diagram, Lyapunov exponent spectrum, Poincare section, power spectrum and return map, are presented.

## 2. Seven-Mode Lorenz-Like Equations

Consider the incompressible Navier-Stokes equations:

$$\frac{\partial u}{\partial t}+(u\cdot\nabla)u=-\nabla p+f+\nu\Delta u, \tag{1}$$

$$\nabla\cdot u=0, \tag{2}$$

$$\int_{T^2} u\mathrm{d}X=0, \tag{3}$$

on the torus $T^2=[0,2\pi]\times[0,2\pi]$, where $u$ is the velocity field, $p$ is the pressure and $f$ is a (periodic) volume force.

We expanded $u$, $f$, $p$ in Frourier series:

$$u(X,t)=\sum_{K\neq 0}\mathrm{e}^{iK\cdot X}r_K\frac{K^\perp}{|K|} \tag{4}$$

$$f(X,t)=\sum_{K\neq 0}\mathrm{e}^{iK\cdot X}f_K\frac{K^\perp}{|K|} \tag{5}$$

$$p(X,t)=\sum_{K\neq 0}\mathrm{e}^{iK\cdot X}p_K\frac{K^\perp}{|K|} \tag{6}$$

where $K=(k_1,k_2)$ is a "wave vector", with integer components, $K^\perp=(k_2,-k_1)$, $r_K=r_K(t)$ is a function of $t$, and the reality condition $r_K=-\overline{r}_{-K}$ holds. Substituting (4)-(6) into (1), we get formally the following equations for $\{r_K\}_{K\neq 0}$

$$\dot{r}_K=-i\sum_{K_1+K_2+K=0}\frac{K_1^\perp\cdot K_2\left(K_2^2-K_1^2\right)}{2|K||K_1||K_2|}\overline{r}_{K_1}\overline{r}_{K_2}-\nu|K|^2 r_K+f_K \tag{7}$$

where $L$ is a set of wave vectors such that if $K\in L$, also $-K\in L$.

We take as the set of vectors $K_1=(1,-1)$, $K_2=(0,3)$, $K_3=(1,2)$, $K_4=(1,-2)$, $K_5=(0,1)$, $K_6=(1,0)$, $K_7=(2,1)$, and their opposites, namely $L=\{\pm K_1,\pm K_2,\pm K_3,\pm K_4,\pm K_5,\pm K_6,\pm K_7\}$. Let $\nu=1$, and make the following transform

$$\begin{cases} r_{K_1} = 2\sqrt{10}x_1, r_{K_2} = 2\sqrt{10}ix_2, r_{K_3} = 2\sqrt{10}x_3, r_{K_4} = 2\sqrt{10}ix_4, \\ r_{K_5} = 2\sqrt{10}x_5, r_{K_6} = 2\sqrt{10}ix_6, r_{K_7} = -2\sqrt{10}ix_7. \end{cases}$$

Taking the force acting on the mode $K_3$, and let $r = \sum_{k=K_1}^{K_7} |f_K| - |f_{K_3}| - Re$ (Reynolds number), with a lot of Calculation we obtain the following system

$$\begin{cases} \dot{x}_1 = -2x_1 + 4x_2x_3 + 4x_4x_5 & (8.1) \\ \dot{x}_2 = -9x_2 + 3x_1x_3 & (8.2) \\ \dot{x}_3 = -5x_3 - 7x_1x_2 + \frac{9\sqrt{5}}{5}x_1x_7 + r & (8.3) \\ \dot{x}_4 = -5x_4 - x_1x_5 & (8.4) \\ \dot{x}_5 = -x_5 - 3x_1x_4 + \sqrt{5}x_1x_6 & (8.5) \\ \dot{x}_6 = -x_6 - \sqrt{5}x_1x_5 & (8.6) \\ \dot{x}_7 = -5x_7 - \frac{9\sqrt{5}}{5}x_1x_3 & (8.7) \end{cases} \tag{8}$$

## 3. The Stationary Solution and Their Stability Properties

In this section we discuss the stationary solution and their stability properties of the system (2.8). Let

$$F(X,r) = \begin{pmatrix} -2x_1 + 4x_2x_3 + 4x_4x_5 \\ -9x_2 + 3x_1x_3 \\ -5x_3 - 7x_1x_2 + \frac{9\sqrt{5}}{5}x_1x_7 + r \\ -5x_4 - x_1x_5 \\ -x_5 - 3x_1x_4 + \sqrt{5}x_1x_6 \\ -x_6 - \sqrt{5}x_1x_5 \\ -5x_7 - \frac{9\sqrt{5}}{5}x_1x_3 \end{pmatrix},$$

setting $F(X,r) = 0$ we can find out stationary solutions of the system (8). In the following we present stability properties

(a) For $0 \le r < R_1 = \frac{5\sqrt{6}}{2}$ there is only one stationary solution $P_0 = \left(0,0,\frac{r}{5},0,0,0,0\right)$, which turns out to be stable for $r$ small enough (this is a particular case of general results on the theory of the Navier-Stokes equations [6]), and numerical evidence suggests that the above solution is a global attractor.

(b) For $R_1 < r < 67.54$ there are 3 stationary solutions: the old one $P_0$, which has become unstable (as a consequence of the crossing of the imaginary axis by one of the eigenvalues of the Lyapunov matrix) and two additional $P_\pm$ as follow

$$\left(\sigma 5\sqrt{\frac{\sqrt{6}\left(2r-5\sqrt{6}\right)}{836}}, \sigma\frac{5\sqrt{6}}{6}\sqrt{\frac{\sqrt{6}\left(2r-5\sqrt{6}\right)}{836}}, \frac{\sqrt{6}}{2}, 0, 0, 0, -\sigma\frac{9\sqrt{30}}{10}\sqrt{\frac{\sqrt{6}\left(2r-5\sqrt{6}\right)}{836}}\right)$$

where $\sigma = \pm 1$, and they are stable. Numerical evidences indicate that any randomly chosen initial data is attracted by them, so they are global attractors. When $r \ge 114.685$, a pair of complex conjugate eigenvalues crosses the imaginary axis, so we have the following conclusion.

(c) For $r > 114.685$, all the stationary solutions of (8) become unstable.

## 4. The Existence of Attractor and Analysis of Global Stability

In the following we prove the existence of attractor of the system (8).

By calculating $(8.1)\times x_1+(8.2)\times x_2+(8.3)\times x_3+(8.4)\times x_4+(8.5)\times x_5+(8.6)\times x_6+(8.7)\times x_7$ we get

$$\begin{aligned}&\dot{x}_1x_1+2x_1^2+\dot{x}_2x_2+9x_2^2+\dot{x}_3x_3+5x_3^2+\dot{x}_4x_4+5x_4^2\\&+\dot{x}_5x_5+x_5^2+\dot{x}_6x_6+x_6^2+\dot{x}_7x_7+5x_7^2=x_3r\end{aligned}$$

accordingly,

$$\frac{1}{2}\frac{\mathrm{d}}{\mathrm{d}t}\left(x_1^2+x_2^2+x_3^2+x_4^2+x_5^2+x_6^2+x_7^2\right)+\left(2x_1^2+9x_2^2+5x_3^2+5x_4^2+x_5^2+x_6^2+5x_7^2\right)=x_3r.$$

letting $|u(t)|^2=x_1^2+x_2^2+x_3^2+x_4^2+x_5^2+x_6^2+x_7^2$, and using Young Inequality we get

$$\frac{1}{2}\frac{\mathrm{d}}{\mathrm{d}t}\left(x_1^2+x_2^2+x_3^2+x_4^2+x_5^2+x_6^2+x_7^2\right)+\left(2x_1^2+9x_2^2+5x_3^2+5x_4^2+x_5^2+x_6^2+5x_7^2\right)=x_3r\le\frac{r^2}{4}+x_3^2,$$

as a result, $\frac{\mathrm{d}}{\mathrm{d}t}|u|^2+2|u|^2\le\frac{r^2}{2}$. Using the Gronwall Inequality [7] we get

$$|u|^2\le|u(0)|^2\,\mathrm{e}^{-2t}+\frac{r^2}{4}\left(1-\mathrm{e}^{-4t}\right)$$

, then

$$\limsup_{t\to\infty}|u(t)|^2\le\frac{r^2}{4}.$$

From above we have

$$\limsup_{t\to\infty}|u(t)|\le\frac{r}{2}=\rho_0.$$

If $\rho$ big enough, $B(0,\rho)$ is an not only functional invariant set but also absorbing set. As a result the system (8) has the global attractor [7] [8].

When the system is a global stability,its orbits contract into a domain called the trapping region. Therefore, if the existence of the trapping region is proved, the system has the global stability, though the stationary solutions are unstable. We construct a following Liapunov function of the system (8)

$$V\left(x_1,x_2,x_3,x_4,x_5,x_6,x_7\right)=x_1^2+x_2^2+x_3^2+x_4^2+x_5^2+x_6^2+x_7^2>0 \tag{9}$$

Setting $V\left(x_1,x_2,x_3,x_4,x_5,x_6,x_7\right)=k$, obviously, when $k$ is a positive constant, the Equation (9) represents a sphere,which is labeled as $E$. By calculating we obtain the following derivative

$$\begin{aligned}\frac{\mathrm{d}V}{\mathrm{d}t}&=2x_1\dot{x}_1+2x_2\dot{x}_2+2x_3\dot{x}_3+2x_4\dot{x}_4+2x_5\dot{x}_5+2x_6\dot{x}_6+2x_7\dot{x}_7\\&=-2\left(2x_1^2+9x_2^2+5x_3^2+5x_4^2+x_5^2+x_6^2+x_7^2-x_3r\right)\\&=-2\left[2x_1^2+9x_2^2+5\left(x_3-\frac{r}{10}\right)^2+5x_4^2+x_5^2+x_6^2+x_7^2-\frac{r^2}{20}\right]\end{aligned} \tag{10}$$

Obviously $2x_1^2+9x_2^2+5\left(x_3-\frac{r}{10}\right)^2+5x_4^2+x_5^2+x_6^2+x_7^2-\frac{r^2}{20}=0$ represents an ellipsoid in $R^7$, which is labeled as $C$. From (10) We know that $\frac{\mathrm{d}V}{\mathrm{d}t}<0$ on outside of $C$, $\frac{\mathrm{d}V}{\mathrm{d}t}=0$ on $C$, and $\frac{\mathrm{d}V}{\mathrm{d}t}>0$ inside of $C$. If $k$

is big enough, $E$ will include $C$. Therefore, from (10) we know that $\frac{dV}{dt}<0$, $V\frac{dV}{dt}<0$ on outside of $C$. From the Liapunov theory we know that the orbits out of system (8) will enter E. Namely E is the trapping region of the Equations (8). Though the stationary solutions $P_0, P_\pm$ are all unstable, the system (8) still has the global stability. orbits of system contract into the trapping region and oscillates in the trapping region. Finally the orbits form an invariant set in the trapping region, which is called the attractor.

## 5. Numerical Simulation

With the increasing of Reynolds number $r$, the stability of the Equation (8) will change, and some nonlinear phenomena appear, such as the Hopf bifurcation and the chaos. In this section, we present the numerical simulation results of dynamical behavior of the system (8).

1) At $r<114.685$ the stationary solutions $P_\pm$ of (8) is stable, Numerical evidences indicate that any randomly chosen initial data is attracted by one of them, so they are global attractors(**Figures 1-4**).

2) When $r=114.685$, the stationary solutions $P_\pm$ of (8) become unstable because a pair of complex conjugate eigenvalues of Jacobian matrix at stationary point $P_\pm$ cross the imaginary axis, and the stable periodic orbits $\zeta_+$ and $\zeta_-$ around the fixed points $P_\pm$ arise via a Hopf bifurcation, and they are stable up to $r=143.463$ and numerical results shows that they attract any point chosen at random (**Figure 5** and **Figure 6**).

3) At $r=143.463$, the periodic orbits lose stability. As predicted by the general theory of bifurcation [3], the Numerical Simulation shows that two attracting tori $T(\zeta_+)$ and $T(\zeta_-)$ arise from the two periodic orbits $\zeta_+$ and $\zeta_-$ (**Figure 7** and **Figure 8**).

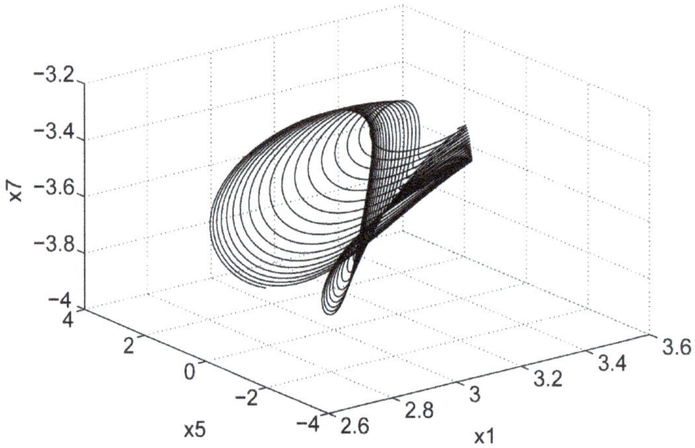


**Figure 1.** ($P_\pm$ is stable for $r < 114.685$).

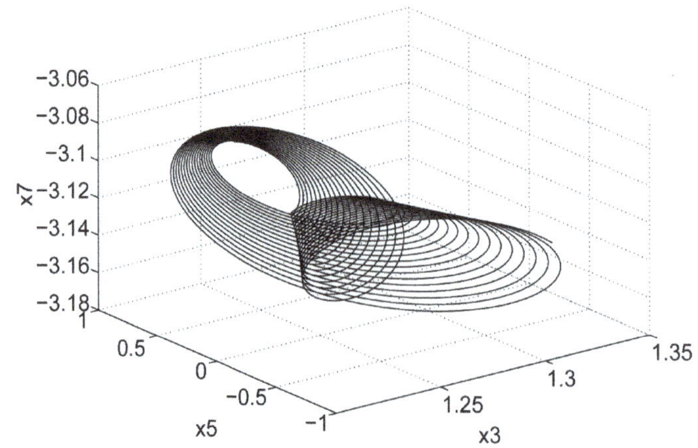


**Figure 2.** ($P_\pm$ is stable for $r < 114.685$).

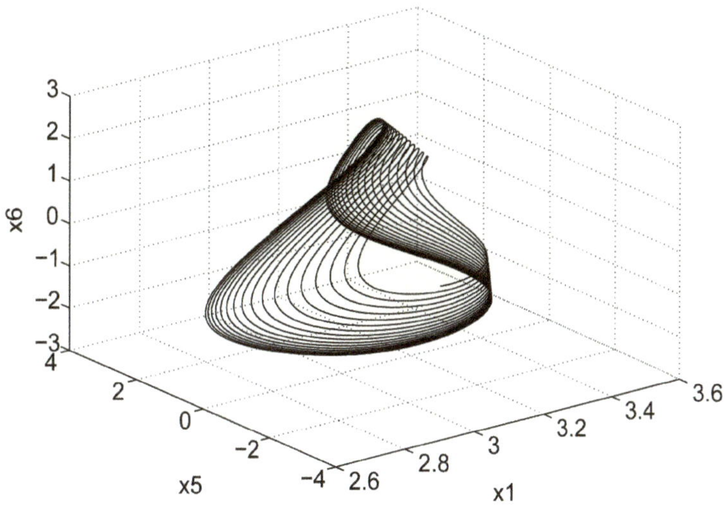


**Figure 3.** ($P_{\pm}$ is stable for $r < 114.685$).

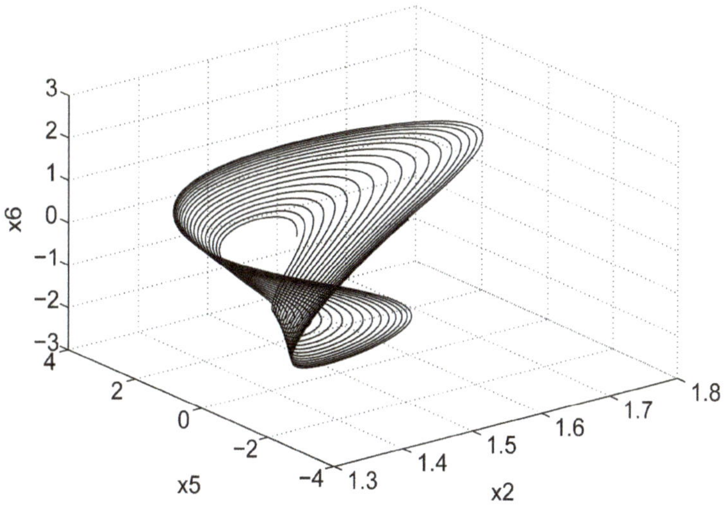


**Figure 4.** ($P_{\pm}$ is stable for $r < 114.685$).

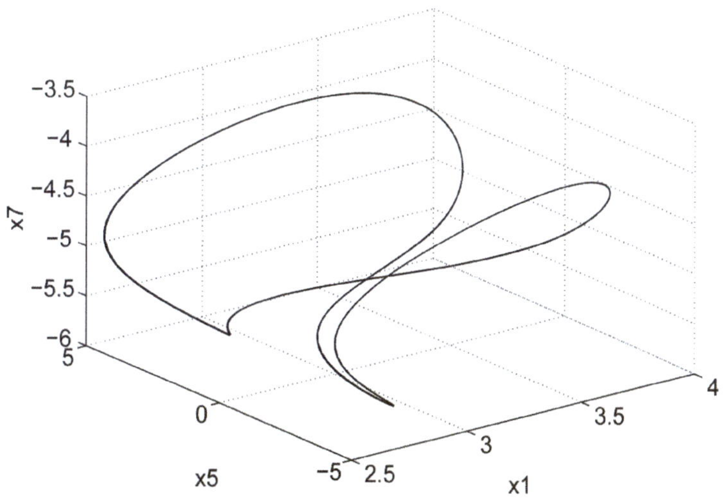


**Figure 5.** (Periodic orbit for $r = 114.685$).

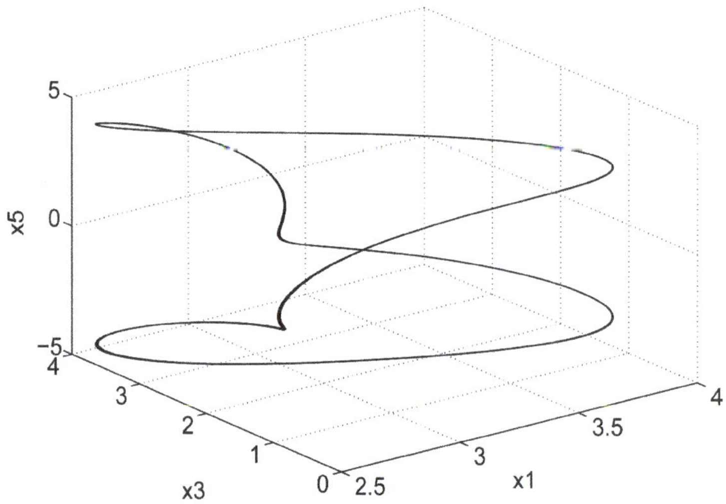


**Figure 6.** (Periodic orbit for $r = 114.685$).

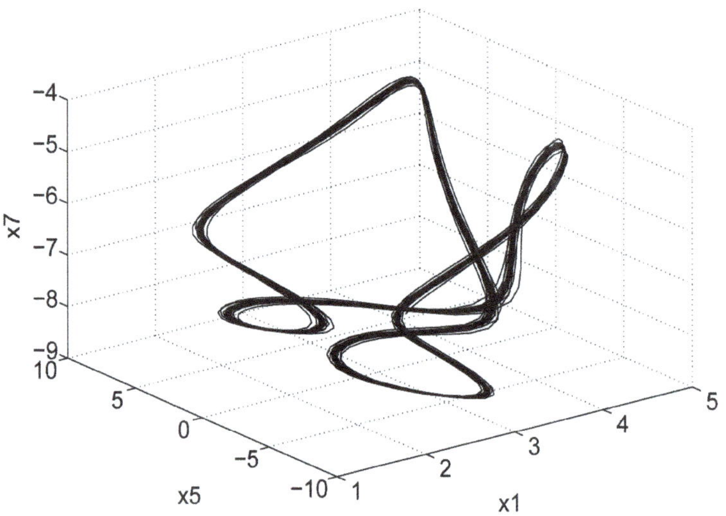


**Figure 7.** (The attracting tori for $r = 143.463$).

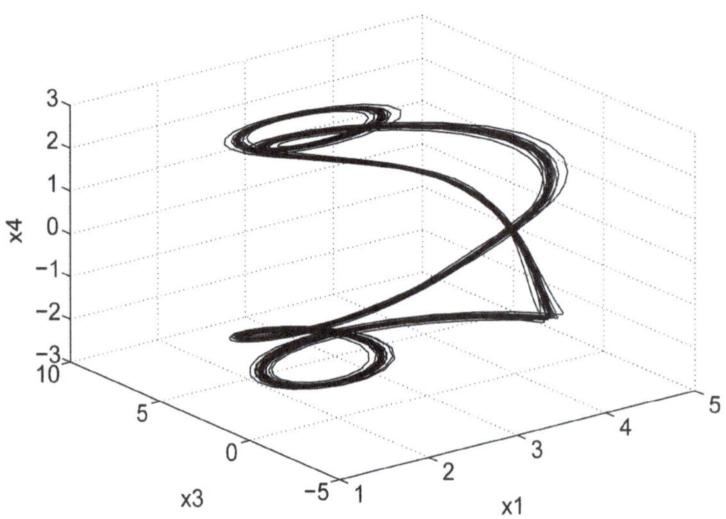


**Figure 8.** (The attracting tori for $r = 143.463$).

4) With the increasing of the Reynolds number $r$, a strong hysteresis phenomenon(*i.e.*, coexistence of stable attractors) appears, in some intervals hysteresis takes place between closed orbits and tori (**Figures 9-19**).

5) At $r = 158.631$, these attracting toris lose stability, the strange attractor appears (**Figures 20-25**).

6) **Figure 26** and **Figure 27** show Bifurcation diagrams and the largest Lyapunov exponents of the system (2.8).

7) **Figures 28-30** show Poincare section, return map and power spectrum of the system (2.8) when $r = 184$, they indicate chaos behavior feature of the new chaos system.

8) For $r > 158.631$ all stable orbits disappears, by studying the flow of a randomly chosen initial point, trajectories are observed which all appear completely chaotic and sensitively dependent on initial conditions. The entire process repeats itself indefinitely. This situation appears analogous to that found by Curry in [4], where the flow in the turbulent parameter range is driven by a similar mechanism, When $r$ increases, the behavior of our system becomes more complicated (**Figures 20-25**).

## 6. Conclusions

In this work we have reported the results of our theoretical and numerical investigation on a model of seven nonlinear ordinary differential equations. Such a model, obtained by a suitable seven-mode truncation of the Navier-Stokes equations for an incompressible fluid on a torus, exhibits a very varied phenomenology, with an

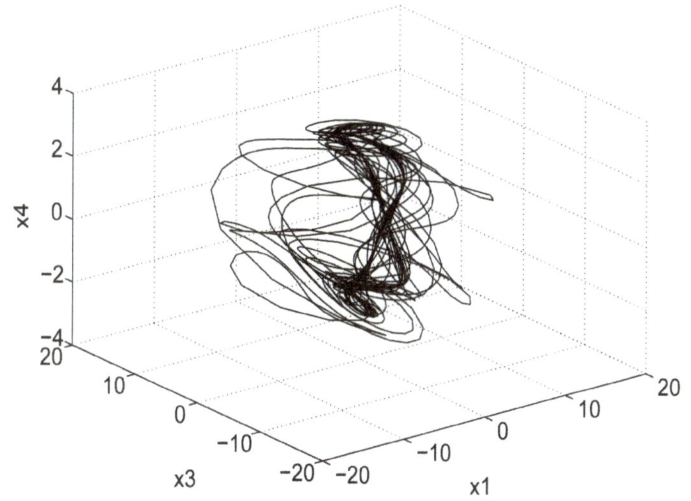


**Figure 9.** (Quasi-periodic orbit for $r = 148.685$).

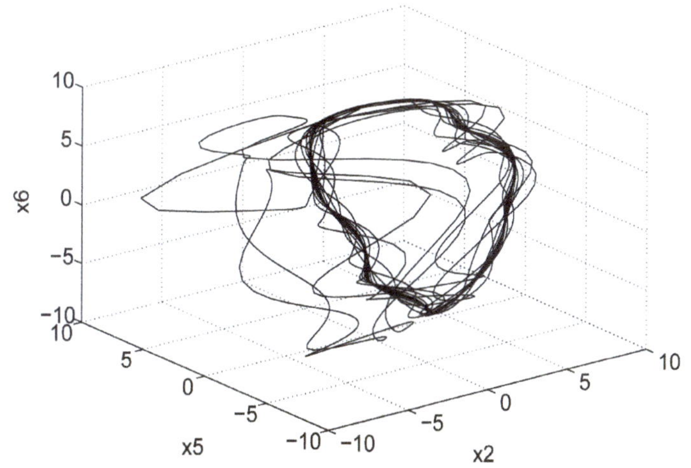


**Figure 10.** (Quasi-periodic orbit for $r = 148.685$).

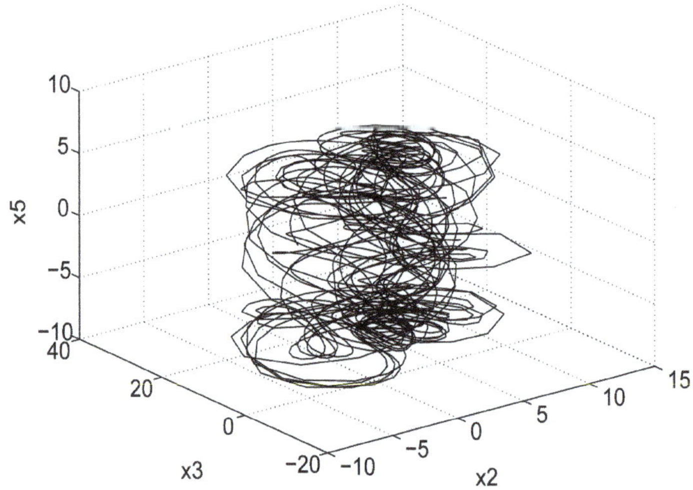


**Figure 11.** (Quasi-periodic orbit for $r = 149.35$).

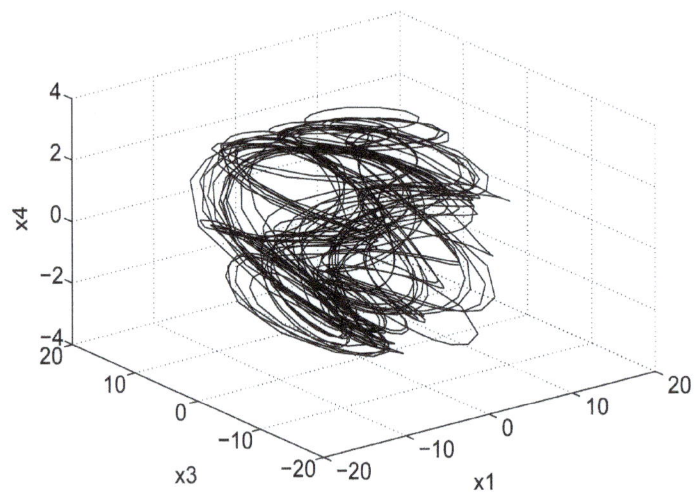


**Figure 12.** (Quasi-periodic orbit for $r = 151.35$).

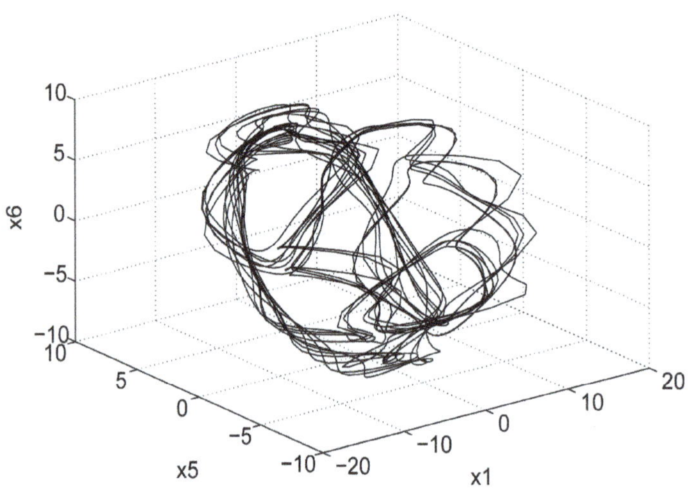


**Figure 13.** (Quasi-periodic orbit for $r = 152.65$).

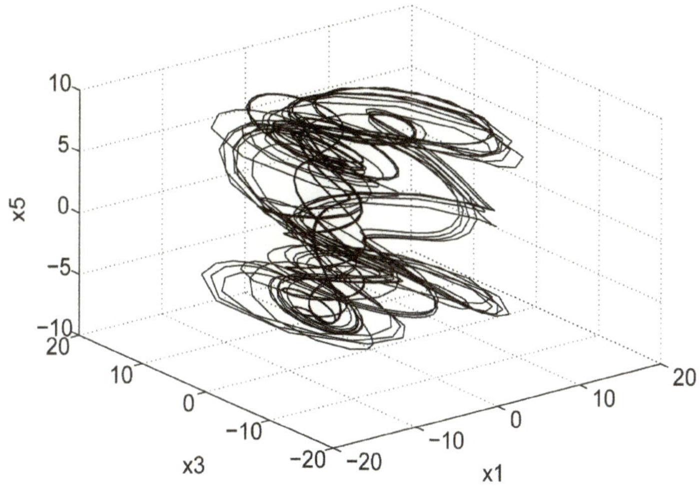


**Figure 14.** (Quasi-periodic orbit for $r = 154.35$).

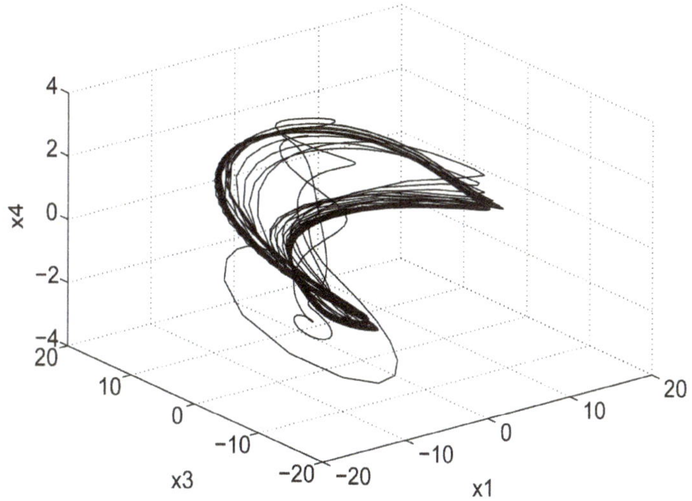


**Figure 15.** (Quasi-periodic orbit for $r = 155.12$).

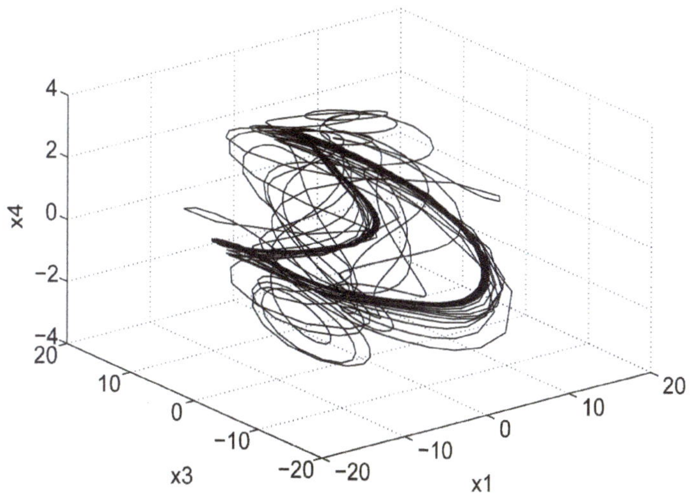


**Figure 16.** (Quasi-periodic orbit for $r = 156.43$).

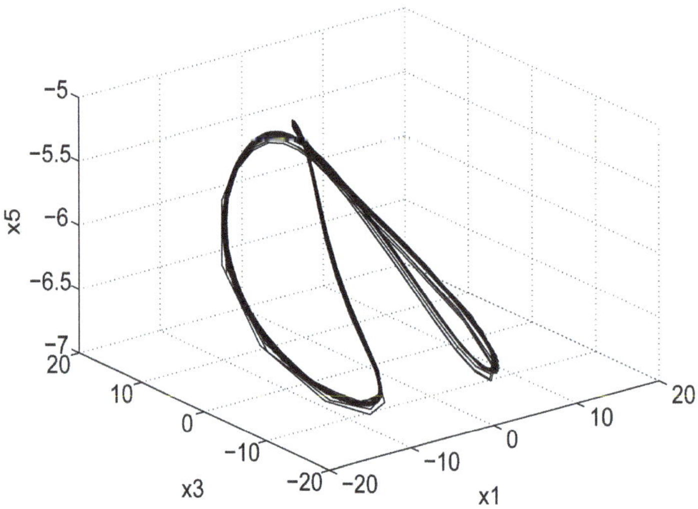


**Figure 17.** (The attracting tori for $r = 156.43$).

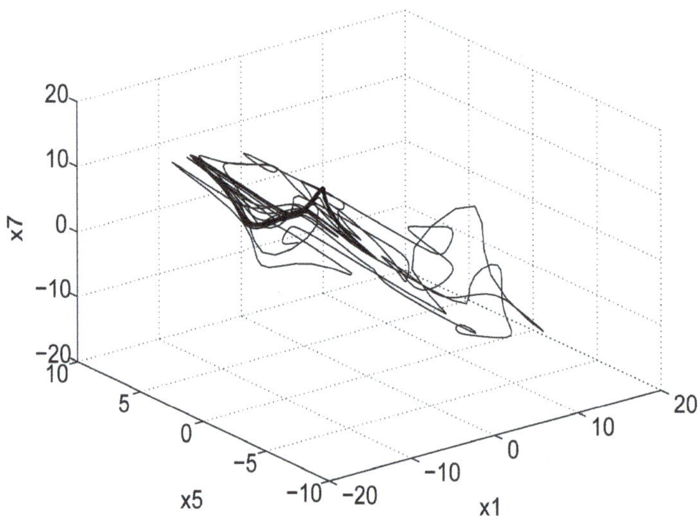


**Figure 18.** (Quasi-periodic orbit for $r = 156.83$).

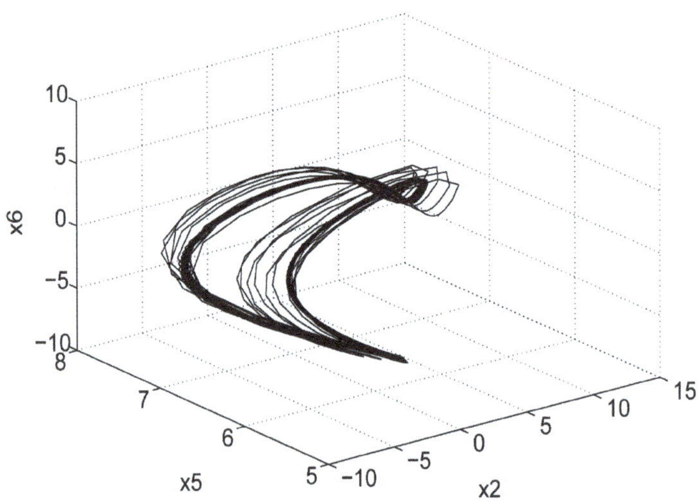


**Figure 19.** (The attracting tori for $r = 157.23$).

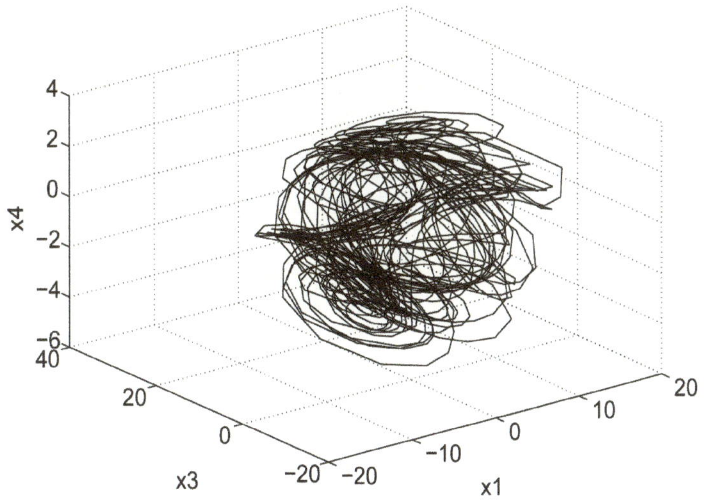


**Figure 20.** (The strange attractor for $r = 158.5$).

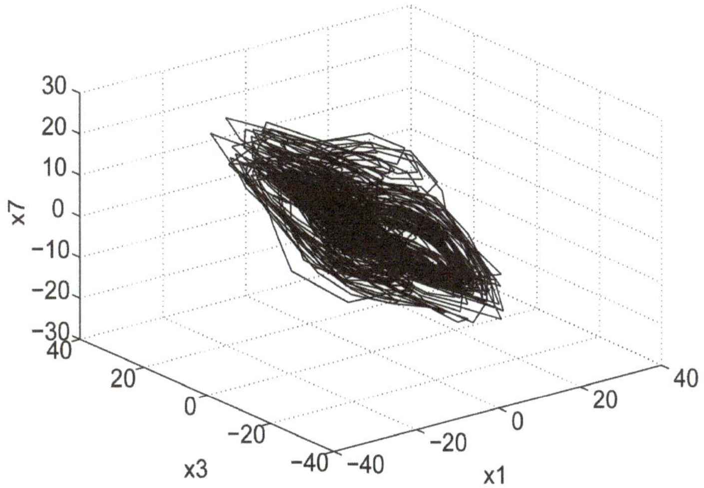


**Figure 21.** (The strange attractor for $r = 159.5$).

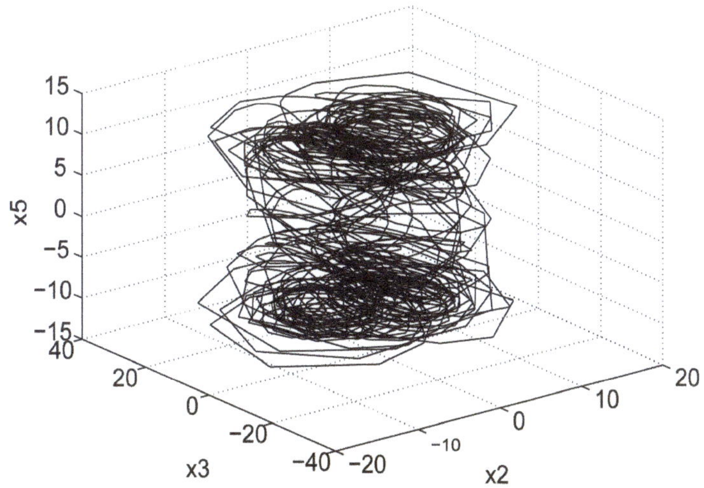


**Figure 22.** (The strange attractor for $r = 159.5$).

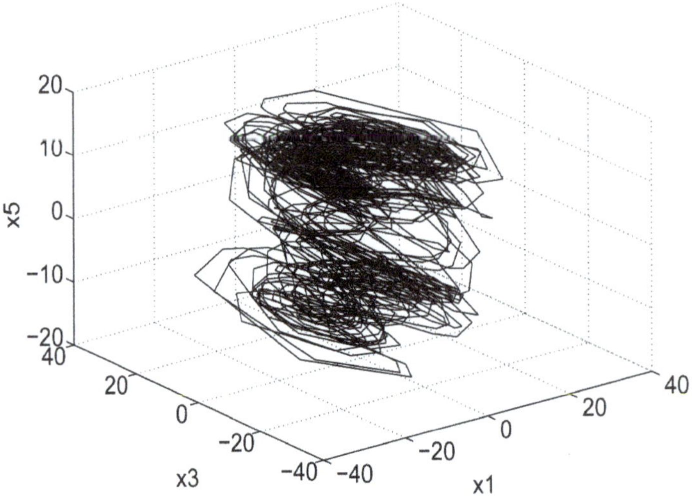


**Figure 23.** (The strange attractor for $r = 160.5$).

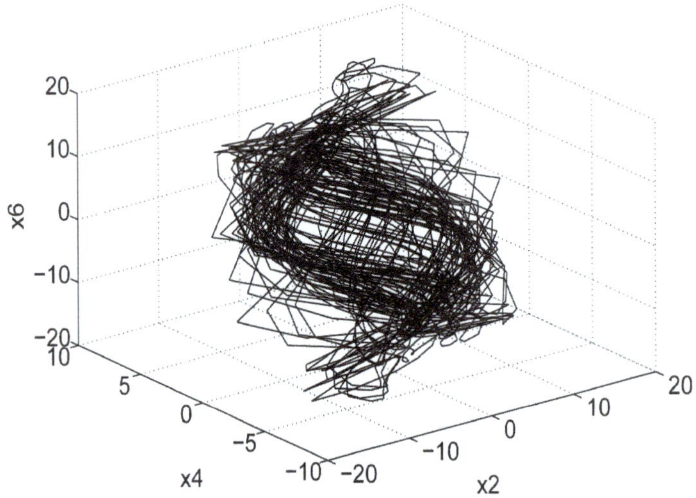


**Figure 24.** (The strange attractor for $r = 250.5$).

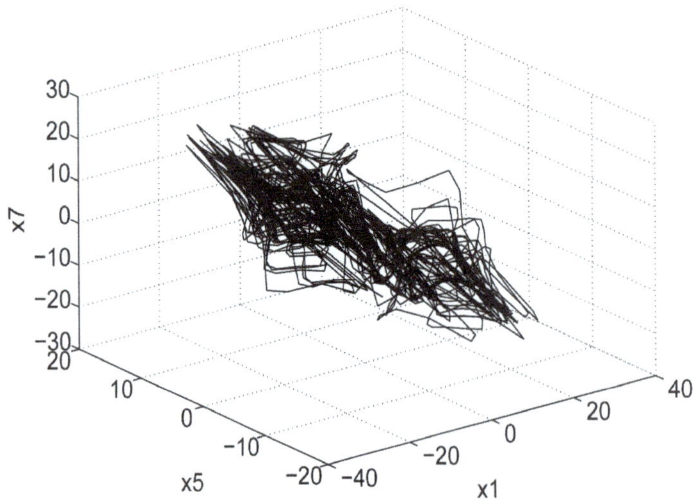


**Figure 25.** (The strange attractor for $r = 360.5$).

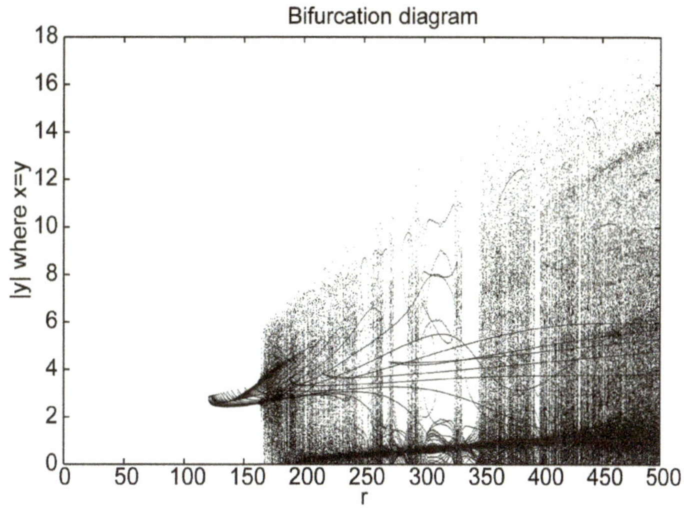


**Figure 26.** (Bifurcation diagrams).

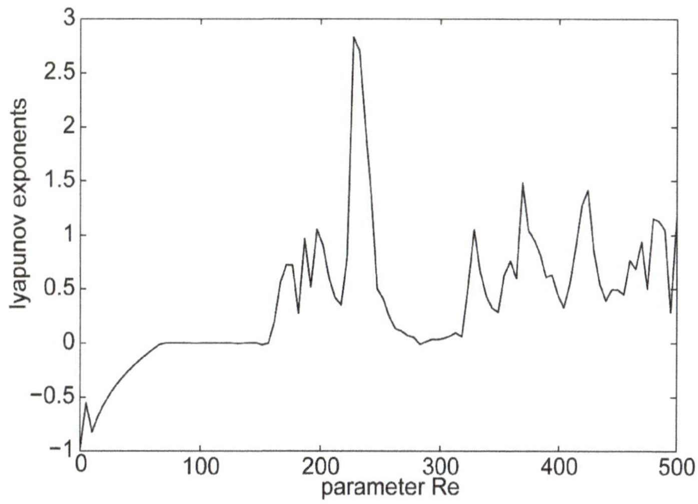


**Figure 27.** (The largest Lyapunov exponents).

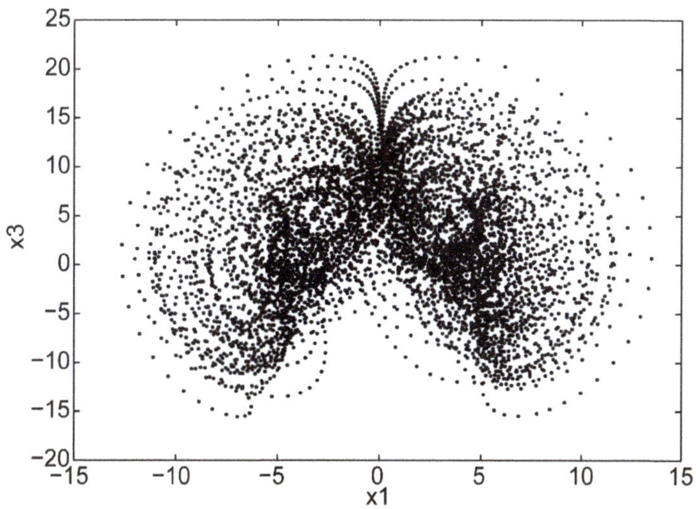


**Figure 28.** (The Poincare section for $r = 184$).

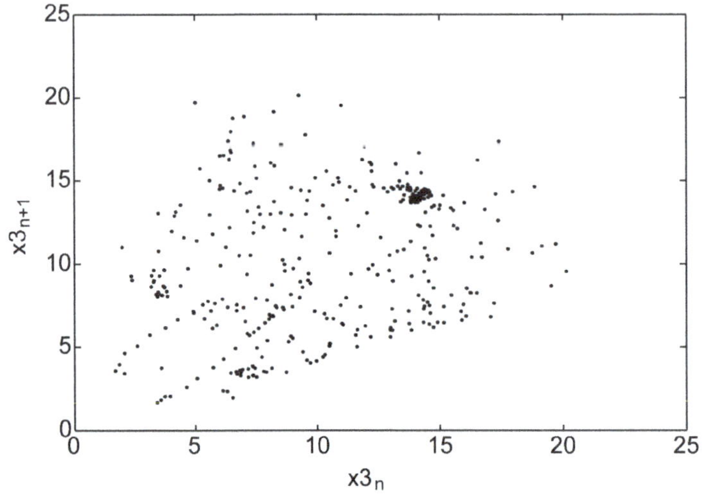


**Figure 29.** (The return map for $r = 184$).

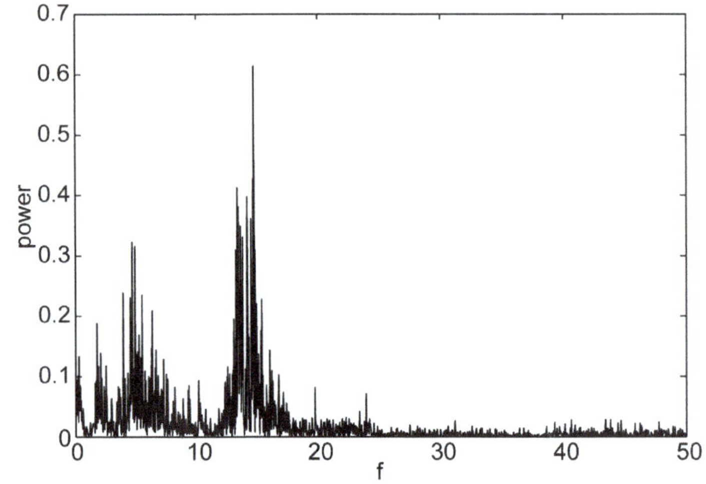


**Figure 30.** (The power spectrum for $r = 184$).

interesting sequence of bifurcations. From numerical results we present four different and independent stories describing the complete phenomenology of the model. The first story consists of a sequence of a bifurcations very similar to the one found by Curry in [4]: The fixed point $P_0$ bifurcates to the two fixed $P_{\pm}$; via a direct Holf bifurcation $P_+$ and $P_-$ bifurcate to the periodic orbits $\zeta_+$ and $\zeta_-$, which on their turn bifurcate to the tori $T(\zeta_+)$ and $T(\zeta_-)$. The further three stories show three interesting examples of "life" of orbit (**Figure 9**, **Figure 14** and **Figure 18**), each one with its own characteristics. A remark feature concerns a strong phenomenon of hysteresis (*i.e*, coexistence of stable attractors) characterizing the model (**Figures 9-19**); three different stable orbits are present in some intervals, and hysteresis takes place between closed orbits and tori in the intervals.

For all the values of the Reynolds number $r$ larger than 158.631, when no stable periodic orbits or tori are present any more, the model exhibits a turbulent behavior. In fact any randomly chosen point describes trajectories which appear to be completely random and sensitively dependent on initial conditions. Since all the numerical investigations carried up to $r = 5000$ keep showing a stochastic behavior, we think that turbulence might also persist for $r$ tending to infinity and that no stable attracting periodic orbit exists at the high values of the Reynolds number. Our seven-mode system does not reproduce the qualitative features of the five-mode model from which it has been obtained as an extension. This result makes more striking what appears already in [4], where a 14-mode generalization of the three-mode Lorenz system is presented: new modes can change quite completely the phenomenology of a model.

## Acknowledgements

We thank the editor and the referee for their comments. Research work is funded by the funds for education department of Liaoning Province (L2013248) and science and technology funds of Jinzhou city (13A1D32).

## References

[1] Wang, H.Y. (2012) Dynamical Behaviors and Numerical Simulation of Lorenz Systems for the Incompressible Flow between Two Concentric Rotating Cylinders. *International Journal of Bifurscation and Chaos*, **22**, 1-11.

[2] Wang, H.Y. and Jiang, H.B. (2007) Bifurcation Analysis of the Model System Similar to the Lorenz Equations of the Flow between Two Concentric Rotating Spheres. *Numerical Mathematics*: *A Journal of Chinese Universities English Series*, **29**, 278-288.

[3] Franceschini, V. and Zanasi, R. (1992) Three-Dimensional Navier-Stokes Equations Truncated on a Torus. *Nonlinearity*, **4**, 189-209. http://dx.doi.org/10.1088/0951-7715/5/1/008

[4] Curry, J.H. (1978) A Generalized Lorenz System. *Communications in Mathematical Physics*, **60**, 193-204. http://dx.doi.org/10.1007/BF01612888

[5] Wang, H.Y. and Gao, Y. (2012) The Dynamical Behaviors and Numerical Simulation of a Seven-Mode Truncation System of the Navier-Stokes Equations for a Two-Dimensional Incompressible Fluid on a Torus. *Proceedings of the IWCFTA* 2012, 54-57.

[6] Ladyzhenskaya, O.A. (1969) The Mathematical Theory of Viscous Incompressible Flows. Gordon and Breach, New York.

[7] Li, K.T. and Ma, Y.C. (1992) The Hilbert Space Method of Math and Physics Equations. Xi'an Jiaotong University Press, Xi'an. (In Chinese)

[8] Liu, B.Z. and Peng, J.H. (2004) Nolinear Dynamic. Higher Education Publishing House, Beijing. (In Chinese)

7

# Numerical Analysis of Turbulent Fluid Flow and Drag Coefficient for Optimizing the AUV Hull Design

**João Victor Nunes de Sousa[1*], Antônio Roberto Lins de Macêdo[2], Wanderley Ferreira de Amorim Junior[1], Antonio Gilson Barbosa de Lima[1]**

[1]Department of Mechanical Engineering, Center of Sciences and Technology, Federal University of Campina Grande, Campina Grande, Brazil
[2]ARMTEC Technology in Robotics, Fortaleza, Brazil
Email: [*]joao.vns@hotmail.com

## Abstract

**Autonomous Underwater Vehicles (AUVs) are robots able to perform tasks without human intervention (remote operators). Research and development of this class of vehicles has growing, due to the excellent characteristics of the AUVs to operate in different situations. Therefore, this study aims to analyze turbulent single fluid flow over different geometric configurations of an AUV hull, in order to obtain test geometry that generates lower drag force, which reduces the energy consumption of the vehicle, thereby increasing their autonomy during operation. In the numerical analysis was used *ANSYS-CFX*® 11.0 software, which is a powerful tool for solving problems involving fluid mechanics. Results of the velocity (vectors and streamlines), pressure distribution and drag coefficient are showed and analyzed. Optimum hull geometry was found. Lastly, a relationship between the geometric parameters analyzed and the drag coefficient was obtained.**

## Keywords

**AUV, Hull, Drag Coefficient, Numerical Simulation**

## 1. Introduction

An Autonomous Underwater Vehicle, or AUV, is defined as a robot which travels underwater without physical communication with the land and without the necessity of the human operator. The AUVs are included in the

*Corresponding author.

group of Unmanned Underwater Vehicles, better known as UUVs.

During the last years, several AUVs have been developed and researches in the area are becoming more frequent, due to the extremely favorable characteristics that these robots have, like the ability to operate autonomously in hostile environments, such as unexplored areas, enemy water territories (in wartime), contaminated or deepwater areas, etc. All these features make the use of AUVs very interesting for military, scientific and industrial sectors.

Most existing AUVs use batteries as an energy source for the propulsion system. High value of drag force generated during the displacement of the robot increases the energy consumption of the system and therefore the AUV autonomy will be lowered, which is undesirable for any engineer.

Viscous fluid passing around axisymmetric body such a cylinder, sphere and spheroid has been reported in many studies [1]-[4]. The behavior of viscous fluid flow is described by *Navier-Stokes Equations*, which are particular forms of *Newton's Laws of Motion*, supplemented by an equation describing the conservation of mass, subjected to the prevailing boundary conditions. Except for very simple conditions, these equations need to be solved numerically with the aid of computers. For solution, the previously defined continuous flow domain is converted in a discrete domain by a numerical mesh. The flow governing differential equations are then transformed, using numerical discretisation schemes, in sets of algebraic equations. The coupled algebraic equations are then solved, by using linear methods, on a computer to yield discrete values of velocity and pressure at mesh nodes. The collection of theoretical, numerical and computational techniques that facilitate this process is called *Computational Fluid Dynamics* (*CFD*). Within the area of *CFD*, some of the most challenging flows are turbulent. Particular interest in turbulence derives from the fact that it is largely responsible for dispersion, mixing, vibrations, frictional losses and noise. Being a highly effective mechanism for momentum redistribution, turbulence has a major impact on hydrodynamic characteristics [5]-[8].

Many authors uses *CFD* in submersible vehicle projects and optimizations due to the large amount of information obtained with reduced cost and time compared with experimental tests. Barros *et al.* [9] compared the analytical and semi-empirical results with numerical results for normal force and momentum coefficient of an AUV. It was shown that the *CFD* approach allows for a good prediction of the coefficients and shows qualitative information from flow visualizations; Juong *et al.* [10] used *CFD* to optimize the design of an AUV hull. They used a commercial software and they obtained an optimum value of nozzle angle, drag force and pressure and velocity fields; Malik *et al.* [11] used *CFD* to calculate the hydrodynamic derivates for a submersible vehicle in transient flow regime. It was concluded that the *CFD* method is well capable of economically evaluating the hydrodynamic derivatives of submersible platforms such as submarines, torpedoes and autonomous underwater vehicles; Dantas *et al.* [12] used *CFD* to analyze the influence of control surfaces in manoeuvrability of an AUV. It was verified that the occurrence of the control surface stall depends on a linear relationship between the angle of attack and the control surface deflection.

Therefore this paper aims to predict fluid flow around the hull of an AUV. Here is created a numerical model analyzing some geometric configurations for the hull of an AUV, to obtain geometry that generates low drag and is suitable for application to this class of vehicles.

## 2. Methodology

### 2.1. The Geometry of Hull

The hull of AUV is torpedo type (cylindrical body with a large ratio between the length and the diameter), due to its good features, which explains the widespread use of this type of hull by all major manufacturers of AUVs. To model the profiles of the bow and stern of the vehicle it was used the *Myring Equations*. These theoretical equations describe curves for bow and stern of the torpedo bodies which generate the smallest possible drag coefficient [13]. The equations are as follows:

- *Bow*,

$$r_1(x) = \frac{1}{2}D\left[1-\left(\frac{x-a}{a}\right)^2\right]^{1/n}, \tag{1}$$

- *Stern*,

$$r_2(x) = \frac{1}{2}D - \left[\frac{3D}{2c^2} - \frac{\text{tg}\theta}{c}\right](x-a-b)^2 + \left[\frac{D}{c^3} - \frac{\text{tg}\theta}{c^2}\right](x-a-b)^3, \tag{2}$$

where all parameters of these equations, with the exception of the parameter $n$, are geometric, and are shown in **Figure 1**.

In this paper the geometric parameters considered are shown in **Table 1**. These parameters were based on dimensions of AUVs existing in the market.

The influence of the parameters $n$ and $\theta$ in the drag of the hull was the focus of this paper. Profiles with $n$ equal to 1, 2, 3 and 4, and $\theta$ equal to 15°, 20°, 25° and 30° were examined. The choice of these ranges of the bow and stern was based in previous works found in the literature [9] [10] [12] [14] and images of others torpedo bodies.

**Figure 2** and **Figure 3** show the profiles (with the altered scale for clarity) to bow and stern, respectively, which were used in the numerical simulation.

## 2.2. The Physical Domain and the Numerical Mesh

The dimensions of the fluid domain were obtained after various computational tests until to reach an optimal

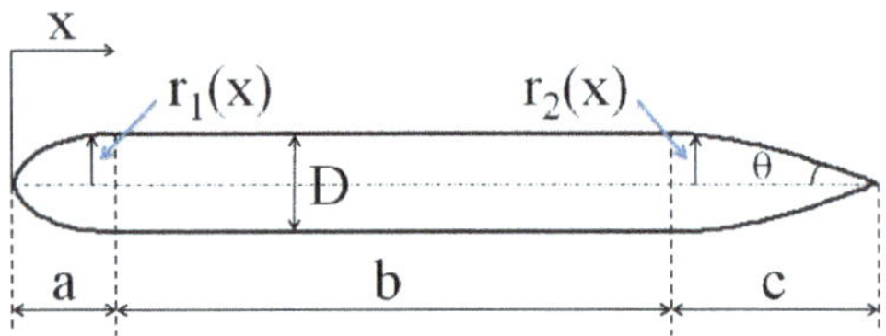


**Figure 1.** Squematic figure of the AUV hull and geometric parameters of *Myring Equations*.

**Table 1.** AUV geometric parameters used in the simulations.

| *Parameter* | *Value (mm)* |
|---|---|
| a | 215 |
| b | 1155 |
| c | 430 |
| D | 200 |

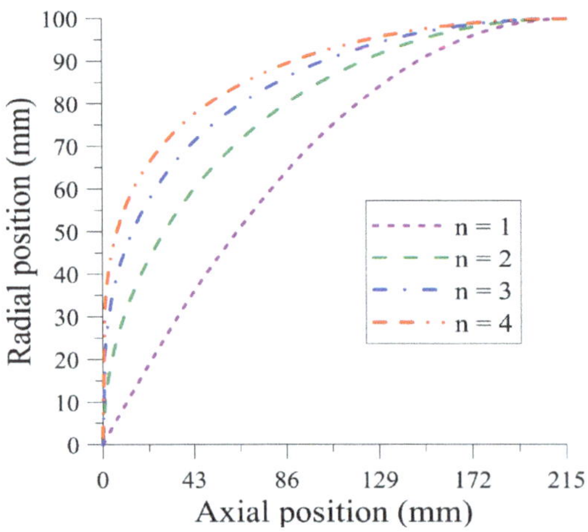


**Figure 2.** Bow profiles analyzed for different parameters $n$.

size of the field, allied to good results with good accuracy and low computational time. **Figure 4** shows the dimensions of the domain as a function of the maximum diameter, $D = 200$ mm, and the total length of AUV, $L = 1800$ mm. It was opted for a semi-cylindrical domain, seeking a smaller number of control volumes, when compared to a quadrilateral domain, thus reducing the computational cost.

**Figure 5** illustrates the mesh representing the study domain, which was built with the support of *ICEM-CFD*® 11.0 software. A mesh was made for each variation in curves of the bow and stern. These non-structured meshes were obtained after various refinements, and have approximately 550,000 tetrahedral elements.

**Figure 6** shows details of the mesh, especially the region in front of the hull, the region of the boundary layer and wake region.

Much care must be taken when constructing numerical meshes aimed at solving problems of flow around immersed bodies. Here, many factors were taken into account [15]. It was used the Equation (3) that provides the average thickness of the kinetic boundary layer (region where viscous flow effects are important) around immersed bodies, $\delta$, as follows:

$$\delta = 0.035 L Re^{-1/7}, \tag{3}$$

where $Re$ is *Reynolds number* of the flow, calculated as follows:

$$Re = \frac{\rho |\boldsymbol{U}| D}{\mu}, \tag{4}$$

where $\rho$ is the fluid density, $|\boldsymbol{U}|$ is the norm of flow velocity vector and $\mu$ is the fluid viscosity. The meshes were built with great refinement near the surface of the AUV (boundary layer region), in order to encompass precisely the viscous effects of the flow around the AUV.

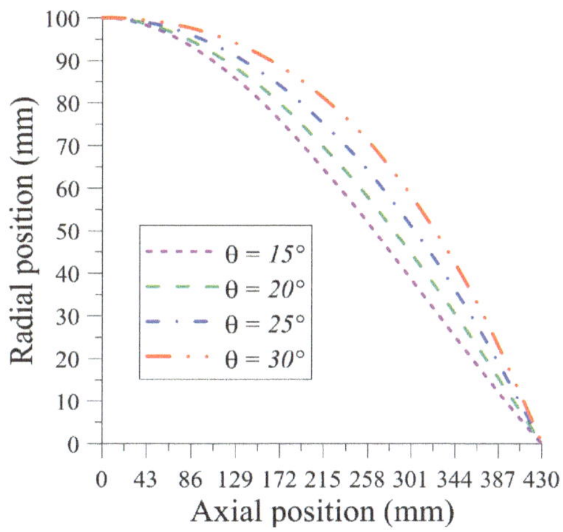


**Figure 3.** Stern profiles for different parameters $\theta$.

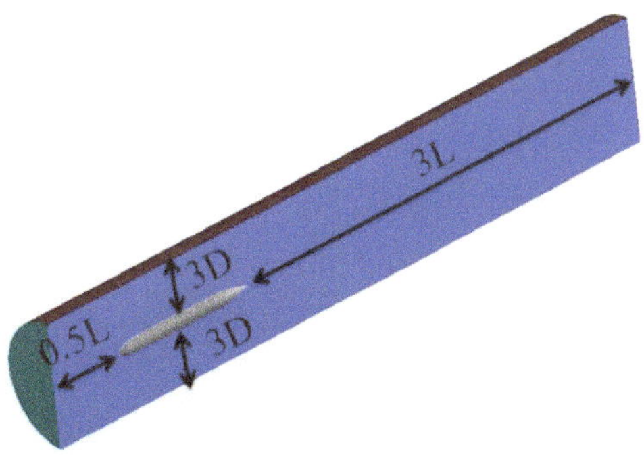


**Figure 4.** Fluid domain.

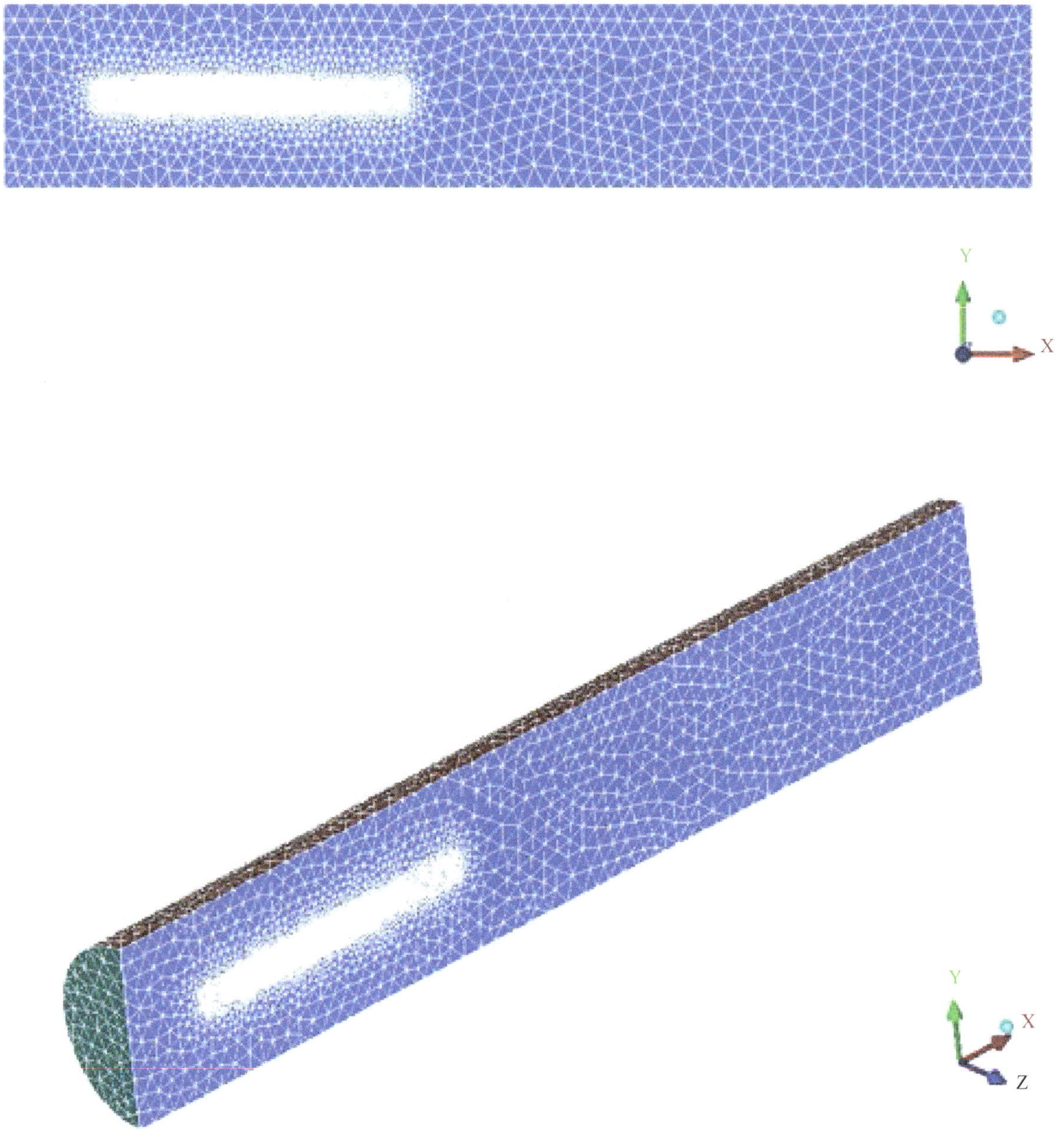


**Figure 5.** Mesh created in the studied domain.

## 2.3. Mathematical Modeling

To investigate the single-phase flow of sea water around the hull of AUV, it was considered three-dimensional, permanent, incompressible and isothermal turbulent flow.

The general equations used in this work are:

- *Mass conservation equation,*

$$\frac{\partial \rho}{\partial t} + \nabla \cdot (\rho \boldsymbol{U}) = 0, \tag{5}$$

where $\boldsymbol{U} = (u, v, w)$ is the velocity vector,

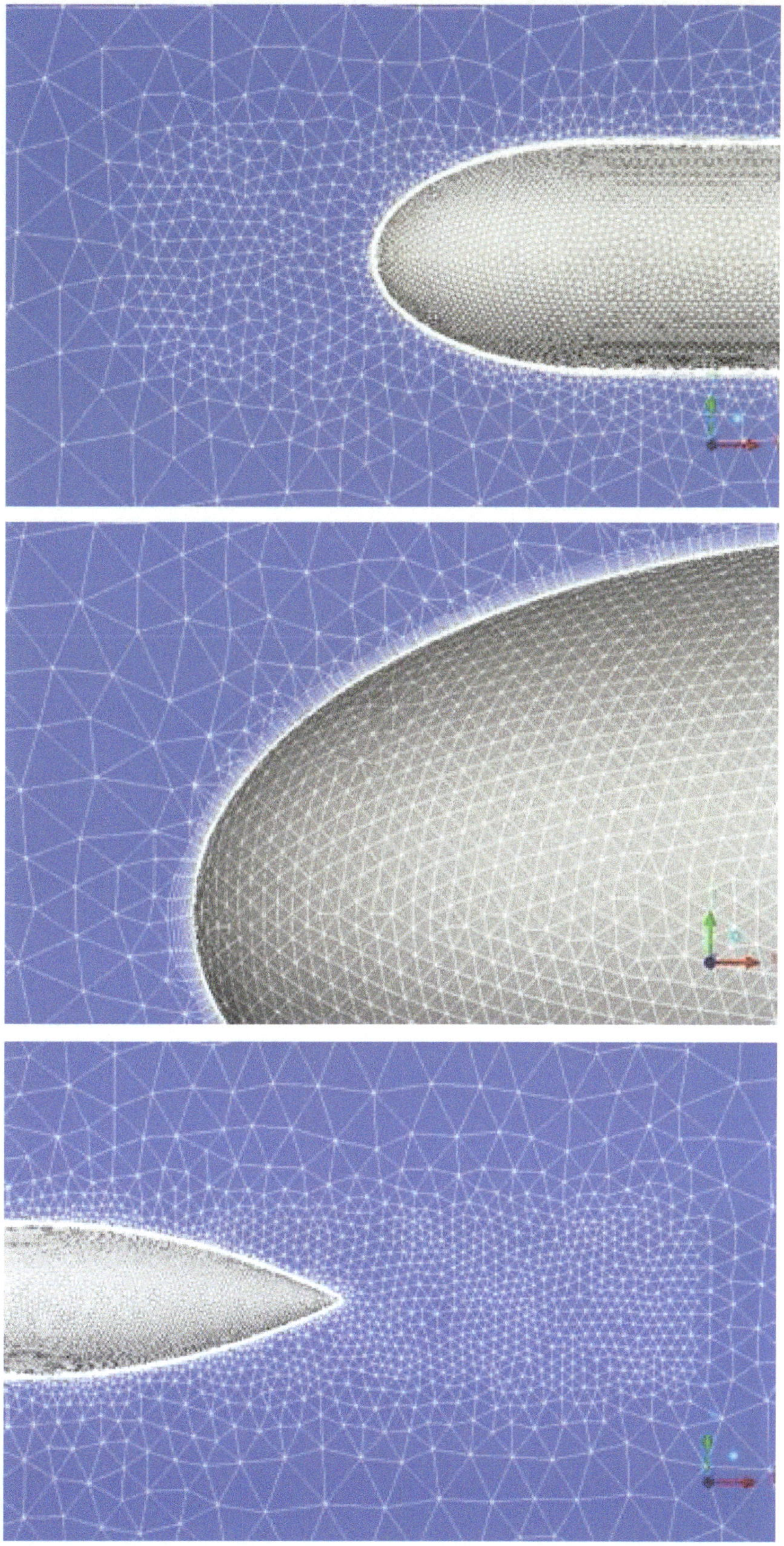

**Figure 6.** Details of the mesh: region in front of the hull (up), region of the boundary layer (center) and wake region (down).

- *Momentum conservation equation,*

$$\frac{\partial(\rho \boldsymbol{U})}{\partial t}+\nabla\cdot(\rho \boldsymbol{U}\otimes \boldsymbol{U})-\nabla\cdot(\mu_{eff}\nabla \boldsymbol{U})=-\nabla p'+\nabla\cdot(\mu_{eff}\nabla \boldsymbol{U})^{\mathrm{T}}+\rho \boldsymbol{g}, \quad (6)$$

where $p'$ is the corrected pressure, which depends of turbulent model to be used, $\boldsymbol{g}$ is the local gravity acceleration vector (adopted value 9.81 m/s$^2$) and $\mu_{eff}$ is the effective viscosity, calculated by:

$$\mu_{eff}=\mu+\mu_t, \quad (7)$$

where $\mu_t$ is the turbulent viscosity.

It is necessary to add in the model new equations governing the phenomenon of turbulence that is present in the flow. The turbulence consists of fluctuations in the flow field in time and space (time-dependent velocity and pressure fields). It is a complex process and can have a significant effect on the flow characteristics. Turbulence occurs when the inertial forces acting on the fluid becomes significantly higher than the viscous forces and is characterized by a high *Reynolds Number* of the flow. Turbulence can also be caused by surface roughness, which induces secondary flow (*i.e.* vortices) [16].

The turbulence model adopted in this work is the *Shear Stress Transport model* (*SST model*), which is based on the turbulence *k-ω model*. This model was used because of its good treatment of external flow with high *Reynolds numbers*. It was considered flow fully turbulent around the AUV.

The *SST model* introduces two new variables in the problem, which are *k*, which is the turbulent kinetic energy, and *ω*, which is the turbulent frequency. These variables are calculated by:

- *Turbulent kinetic energy equation,*

$$\frac{\partial(\rho k)}{\partial t}+\nabla\cdot(\rho \boldsymbol{U}k)=\nabla\cdot\left[\left(\mu+\frac{\mu_t}{C_{k1}}\right)\nabla k\right]+P_k-C_{k2}\rho k\omega, \quad (8)$$

where $P_k$ is the turbulence production, $C_{k1}=2.000$ and $C_{k2}=0.009$.

- *Turbulence frequency equation,*

$$\frac{\partial(\rho\omega)}{\partial t}+\nabla\cdot(\rho \boldsymbol{U}\omega)=\nabla\cdot\left[\left(\mu+\frac{\mu_t}{C_{\omega1}}\right)\nabla \omega\right]+\frac{\omega}{k}(C_{\omega2}P_k-C_{\omega3}\rho\omega k), \quad (9)$$

where $C_{\omega1}=2.000$, $C_{\omega2}=0.556$ and $C_{\omega3}=0.075$. In the *SST model* the parameters $\mu_t$, $p'$ and $P_k$ are given as follows:

$$\mu_t=\rho\frac{k}{\omega}, \quad (10)$$

$$p'=p_d+\frac{2}{3}\rho k+\frac{2}{3}\mu_t\nabla\cdot \boldsymbol{U}, \quad (11)$$

$$P_k=\mu_t\nabla \boldsymbol{U}\cdot(\nabla \boldsymbol{U}+\nabla \boldsymbol{U}^{\mathrm{T}})-\frac{2}{3}\nabla\cdot \boldsymbol{U}(3\mu_t\nabla\cdot \boldsymbol{U}+\rho k), \quad (12)$$

where $p_d$ is the dynamic pressure, calculated by the equation

$$p_d=\frac{1}{2}\rho|\boldsymbol{U}|^2. \quad (13)$$

Total pressure, $p_t$, is calculated by:

$$p_t=p_s+p_d, \quad (14)$$

where $p_s$ is the static pressure.

According to Warsi [6] the action of viscosity is diminish the velocity of fluid past a surface and, thus, decreasing the fluid momentum within the boundary layer. It strongly affects the overall flow behavior past the surface. Since, fluid flow is governed by the pressure distribution impressed on the boundary layer. We must analyze both the retarding action of viscosity and the imposed pressure distribution. Thus is done calculated the drag coefficient in the flow. The total drag force is generated by the friction and pressure forces acting on a body immersed in a flowing fluid.

The friction drag is due to the boundary layer surface shear stress, while pressure drag is due to pressure difference in the flow direction resulting from formation of the wake in the downstream region.

The drag on a body is usually expressed in terms of a dimensionless drag coefficient. The total drag coefficient, $C_d$, is calculated by:

$$C_d = C_{df} + C_{dp} = \frac{F_{df}}{\frac{1}{2}\rho U_*^2 A_f} + \frac{F_{dp}}{\frac{1}{2}\rho U_*^2 A_f}, \tag{15}$$

where $C_{df}$ is the friction drag coefficient, $C_{dp}$ is the pressure drag coefficient (form drag), $F_{df}$ is the friction drag force, $F_{dp}$ is the pressure drag force, $U_*$ is the free stream fluid velocity and $A_f$ is the frontal area of AUV. $F_{df}$ and $F_{dp}$ are calculated by:

$$F_{df} = \left(\int_A \tau \mathrm{d}A\right)\hat{\imath}, \tag{16}$$

$$F_{dp} = \left(\int_A p_d \mathrm{d}A\right)\hat{\imath}, \tag{17}$$

where $\tau$ is the shear stress, $A$ is the AUV superficial area and $\hat{\imath}$ is the unitary vector in parallel flow direction.

## 2.4. Boundary Conditions and Fluid Properties

**Figure 7** shows domain with indicative information related to boundary conditions and **Table 2** shows specified values of these boundary conditions.

The fluid adopted in all simulations was sea water with 35 g/L of salinity, on the depth of 1000 m (temperature of 15˚C and static pressure 10.2 MPa). The properties of the fluid are showed in **Table 3**. The flow analyzed has $Re = 4.2 \times 10^6$.

**Table 4** shows the considerations adopted for the numerical solver.

## 2.5. Validation of the Results

To validate the methodology for constructing the numerical mesh, as well as the mathematical model used in this

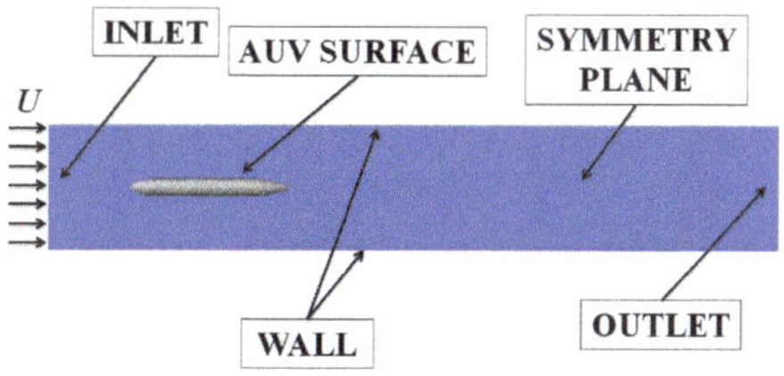


**Figure 7.** Physical specification of the domain with the boundary conditions.

**Table 2.** Boundary conditions used in the simulations.

| | *Condition* | *Value* |
|---|---|---|
| Inlet | Prescribed velocity | 2.83 m/s (5.5 knot) |
| Wall | Prescribed pressure[c] | 10.2 MPa |
| AUV surface[a] | Prescribed velocity | 0 m/s |
| Outlet[b] | Prescribed pressure[c] | 10.2 MPa |
| Symmetry Plane | Symmetry | - |

a. *No slip and smooth wall*; b. *Opening condition*; c. *Pressure on the depth of* 1000 *m*.

**Table 3.** Fluid properties.

| *Fluid* | $\rho$ [kg/m$^3$] | $\mu$ [mPa.s] |
|---|---|---|
| Sea water | 1027 | 1.25 |

**Table 4.** Considerations adopted in numerical solver.

| *Characteristic* | *Consideration* |
|---|---|
| Flow regime | Permanent |
| Convergence criterion | Variation in $C_d$ less than $10^{-5}$ |
| Advection scheme | High resolution |
| Interpolation scheme for pressure | Trilinear |
| Interpolation scheme for velocity | Trilinear |

study, we analyzed the axial flow on two well-known geometries: a *cylinder* and *ellipsoid of revolution* (**Figure 8**). Was chose these cases because of the similarity of these geometries with the AUV hull.

The simulations for validation are subject to the same considerations, boundary conditions and mathematical model described in the previous sections.

The drag coefficient, $C_d$, as well as corresponding friction drag coefficient, $C_{df}$, and pressure drag coefficient, $C_{dp}$, obtained numerically will be compared with those obtained experimentally.

The following is detailed for geometry was conducted to evaluate how the experimental coefficient of drag:

a) *Cylinder:* Length $L$ = 1800 mm and diameter $D$ = 200 mm. Reference [17] shows $C_d$, $C_{df}$ and $C_{dp}$, obtained experimentally for $Re \geq 10^4$, as a function of ratio $L/D$. For $L/D = 9$, $C_d \approx 0.85$, where approximately 90% is referent of $C_{dp}$;

b) *Ellipsoid:* Length of major axis $L'$ = 1800 mm and length of minor axis $L''$ = 200 mm. Reference [17] shows $C_d$ (Equation (18)), $C_{df}$ and $C_{dp}$, obtained experimentally for $Re \geq 10^5$, as a function of ratio $L'/L''$. For the analyzed case, $C_d \approx 0.117$, where approximately 100% is referent of $C_{df}$.

$$C_d = 0.02 + 0.01\left(L'/L''\right) + 0.6\left(L''/L'\right)^2 . \tag{18}$$

## 3. Results and Discussion

**Table 5** and **Table 6** show the comparison between the numerical and experimental results of different drag coefficients cited in the text for the cases of flow around cylinder and ellipsoid. The difference between the results is small, being of the order of 7.1% for the case of the cylinder and 12.0% for the ellipsoid. Moreover, there was also a good approximation between the corresponding pressure and friction drag fractions. The low deviations between the results validate the methodology used in this work, showing that it describe well the studied phenomenon.

**Figures 9-12** illustrate, respectively, the velocity field around the cylinder (streamlines), the velocity vector field around the cylinder, the velocity field around the ellipsoid (streamlines) and the velocity vector field around the ellipsoid by vectors, in the longitudinal symmetry plane for the analyzed cases. Because the shape of cylinder is very different of ellipsoid, in the flow pattern more intense turbulence zones are found, specially in the wake zone. In cylinder case is visible more intense recirculation zones in the flow behind the cylinder (wake region), in comparison with the ellipsoid, thus, increasing $C_{dp}$ factor. For this reason $C_d$ for cylinder is dramatically higher than the ellipsoid.

For analyze the drag of AUV hull by varying the shape of the bow (**Figure 13**) was used stern profile with $\theta = 25^\circ$.

**Table 7** shows the complete comparison between the results of the drag coefficients obtained by varying the parameter $n$. Further, the average shear stress at the AUV surface, $\tau_{ave}$, and the difference of pressure between the point $x = 0$ and the point $x = 1.8$ m in longitudinal axis of AUV (**Figure 1**), $\Delta p$, are showed too.

It was found that with increasing the parameter $n$ we have an increase in the total drag, $C_d$, and particularly in

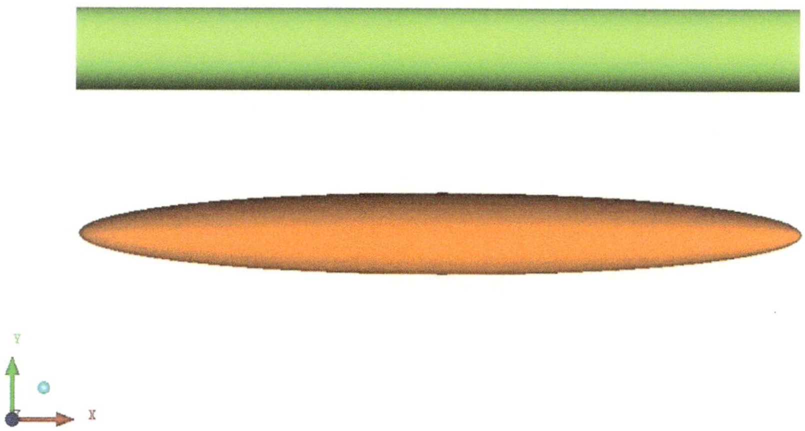


**Figure 8.** Cylinder and ellipsoid of revolution used in the validation.

**Table 5.** Comparison between experimental and numerical results of the drag coefficients for flow around the cylinder.

| *Results* | $C_{df}/C_d$ | $C_{dp}/C_d$ | $C_d$ |
|---|---|---|---|
| Experimental[a] | 10% | 90% | 0.85 |
| Numerical | 5% | 95% | 0.91 |

a. *Reference* [17].

**Table 6.** Comparison between experimental and numerical results of the drag coefficients for flow around the ellipsoid.

| *Results* | $C_{df}/C_d$ | $C_{dp}/C_d$ | $C_d$ |
|---|---|---|---|
| Experimental[a] | 100% | 0% | 0.117 |
| Numerical | 91% | 9% | 0.103 |

a. *Reference* [17].

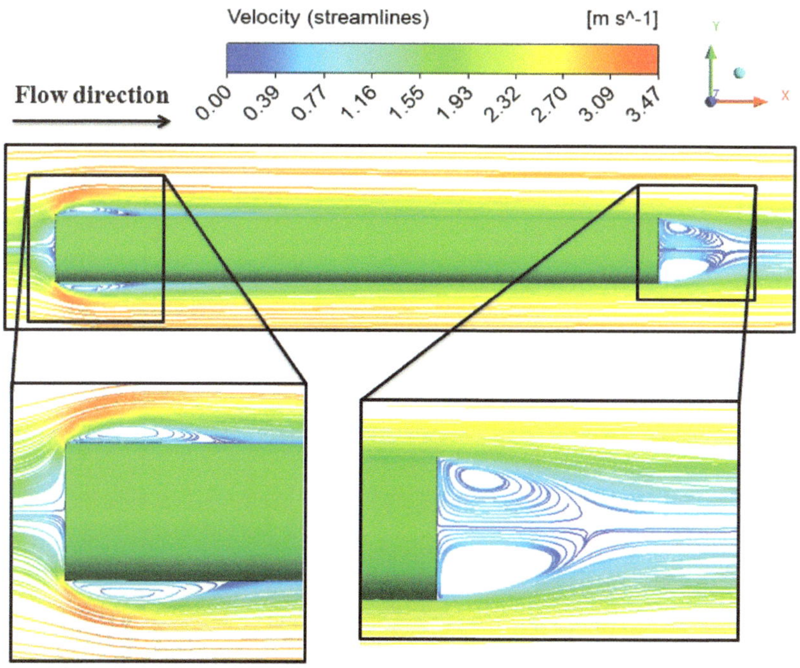


**Figure 9.** Streamline showing the velocity field around the cylinder, with emphasis for recirculation zones.

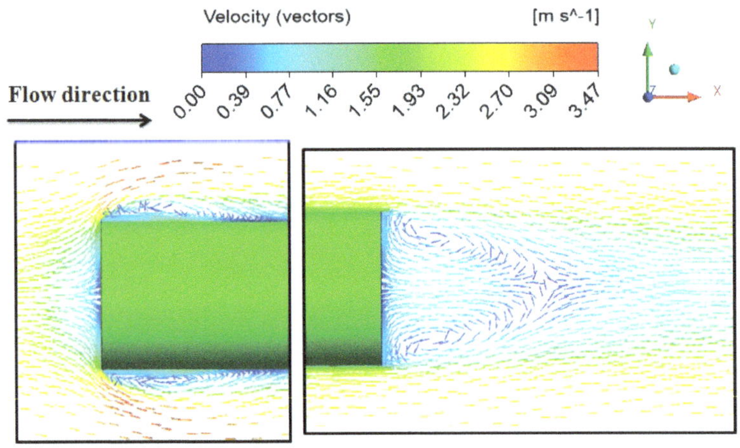


**Figure 10.** Vectors showing the velocity field in the recirculation zones around the cylinder.

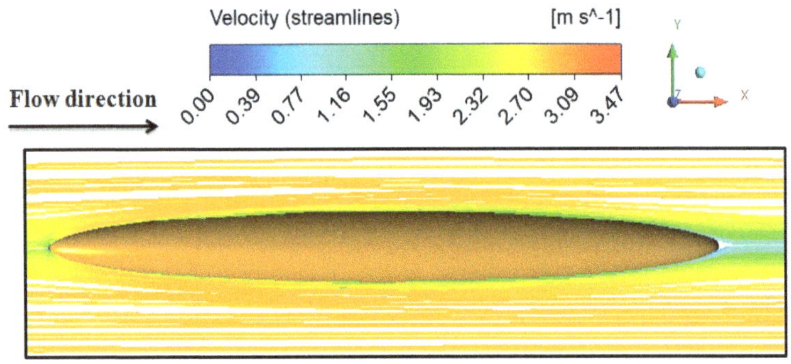


**Figure 11.** Streamline showing the velocity field around the ellipsoid.

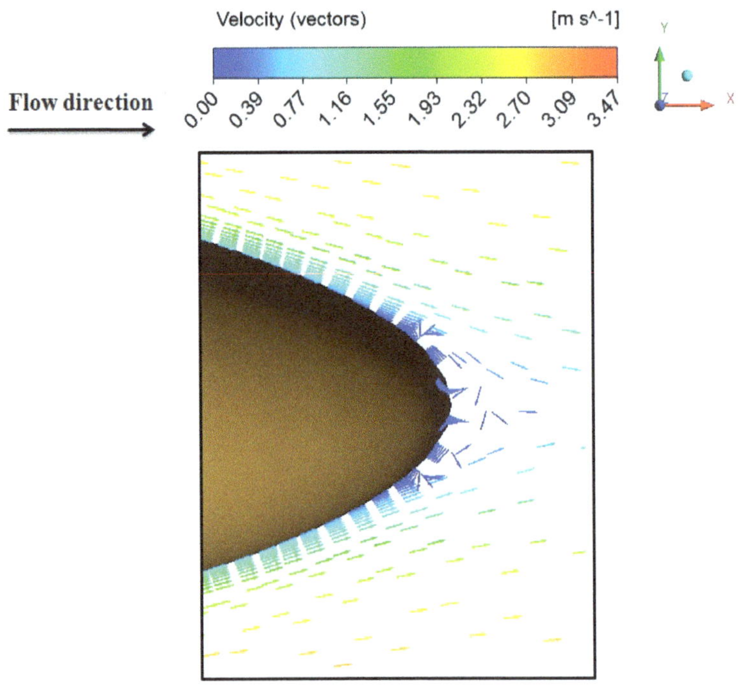


**Figure 12.** Vectors showing the velocity field in the small recirculation zone around the ellipsoid.

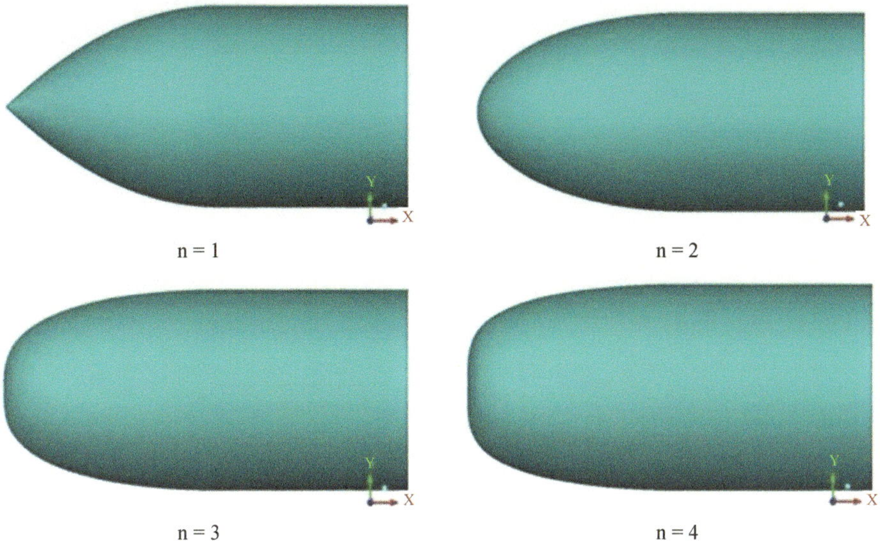


**Figure 13.** Bow profiles used in the simulations ($\theta = 25°$).

**Table 7.** Results for bow profiles ($\theta = 25°$).

| $n$ | $C_{df}/C_d$ [-] | $C_{dp}/C_d$ [-] | $\tau_{ave}$ [Pa] | $\Delta p$ [Pa] | $C_d$ [-] |
|---|---|---|---|---|---|
| 1 | 86.2% | 13.8% | 13.7356 | 3356 | 0.1227 |
| 2 | 86.0% | 14.0% | 13.5922 | 3486 | 0.1238 |
| 3 | 85.1% | 14.9% | 13.5411 | 3486 | 0.1254 |
| 4 | 83.6% | 16.4% | 13.4463 | 3449 | 0.1271 |

the portion corresponding to the drag pressure, $C_{dp}$. However, it is seen that these parameters does not vary much over the range computed. In all cases, the predominant fraction of the drag is due to friction, $C_{df}$, accounting for over 80% of total drag. It is verified that the increasing the parameter $n$ the parameters $\tau_{ave}$ increase and $\Delta p$ oscillate. Based on data obtained was defined the profile with $n = 2$ as ideal shape for the bow. The profile with $n = 2$ is the most efficient because it gives to the hull a lower drag (only 0.9% above the drag of the hull profile when $n = 1$) and a good internal volume of the bow (25.0% above of the hull profile when $n = 1$), which would facilitate the accommodation of all internal systems of the vehicle.

With the bow profile with parameter $n = 2$ fixed, were simulated cases changing the parameter $\theta$ which alters the stern shape (**Figure 14**).

**Table 8** shows the complete comparison between the results obtained for the drag AUV hull by varying the parameter $\theta$. **Figure 15** shows the results graphically.

It is found that the best performance profile is that when parameter $\theta = 20°$, giving to the hull the lowest drag coefficient, which is 2.5% less than the hull with the worst profile analyzed in this section, which is the profile with $\theta = 30°$. Is verified that the increasing the parameter $\theta$ the parameters $\tau_{ave}$ increase and $\Delta p$ decrease.

The ideal geometry among the analyzed cases is shown in **Figure 15**. The geometry has parameters $n = 2$ and $\theta = 20°$.

Based on the simulations results it was obtained a correlation between the parameters $n$ and $\theta$ with $C_d$ (Equation (19)). This equation were obtained by the method of the least squares, with determination coefficient of 0.99694, and valid for the ranges $1 \le n$ [-] $\le 4$ and $15 \le \theta$ [degree] $\le 30$.

$$C_d = 0.12 + 0.000096\left((3.2 + 1.22n)^2 + (10.5 - 0.54\theta)^2\right). \tag{19}$$

The optimal design gives the hull a drag coefficient of 0.1230, which is very close to the drag coefficient of the ellipsoid (just 5.1% upper), with the advantages of easier construction and approximately 10.3% more in volume.

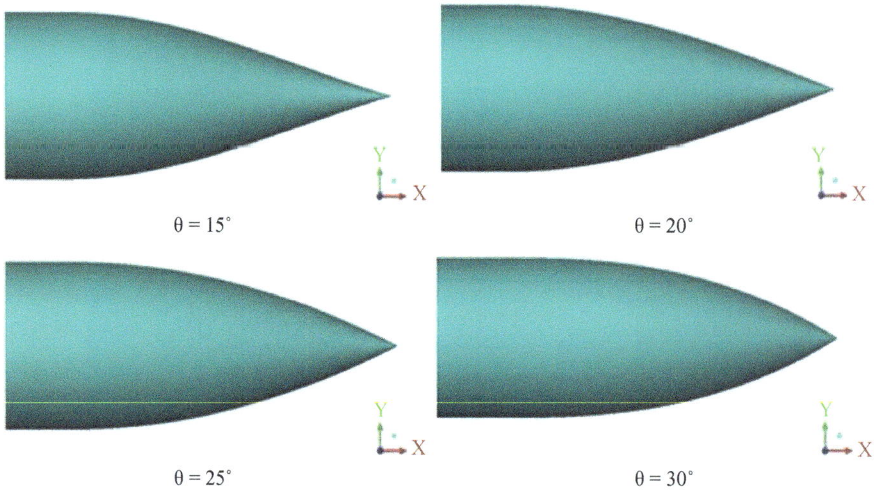


**Figure 14.** Stern profiles used in the simulations ($n$ = 2).

**Table 8.** Results for the stern profiles ($n$ = 2).

| $\Theta$ | $C_{df}/C_d$ [-] | $C_{dp}/C_d$ [-] | $\tau_{ave}$ [Pa] | $\Delta p$ [Pa] | $C_d$ [-] |
|---|---|---|---|---|---|
| 15° | 83.6% | 16.4% | 12.8694 | 3621 | 0.1235 |
| 20° | 84.2% | 15.8% | 13.3103 | 3544 | 0.1230 |
| 25° | 86.0% | 14.0% | 13.5922 | 3486 | 0.1238 |
| 30° | 83.6% | 16.4% | 13.2253 | 3454 | 0.1262 |

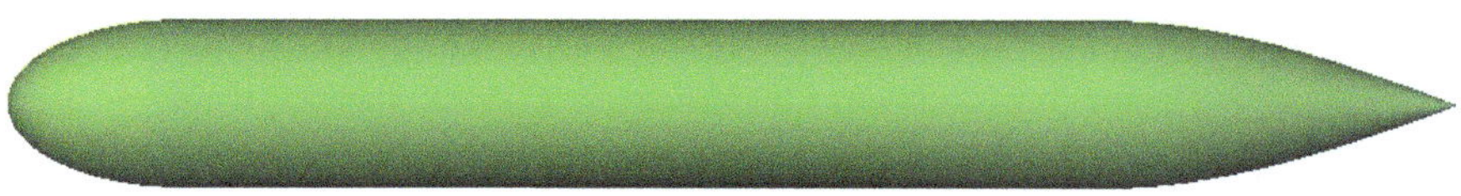

**Figure 15.** Ideal AUV hull design, among the studied cases.

**Figure 16** shows the total pressure field around the ideal AUV hull, in the longitudinal symmetry plane. It is verified the higher pressure in the stagnation point of the flow. Further, it is verified the high pressure values and low range. This is due the high contribution of static pressure, $p_s$, in the total pressure, due the big depth (1000 m) in that the study was performed.

**Figure 17** and **Figure 18** show the velocity field around the ideal AUV hull. By analyzing these figures we can see the low velocity in the wake region and the null velocity in stagnation point. Is visualized yet the velocity profile in kinetic boundary layer in stern region.

## 4. Conclusion

In this paper the hydrodynamic single-phase flow around the AUV hull is discussed. The study is related to sea water flow in the turbulent regime by using the *ANSYS-CFX*® 11.0 commercial software. The simulations revealed the good mathematical treatment used, with good precision between numerical and experimental results. It reached an optimized design for AUV hull that has a drag coefficient of 0.1230, which is very close to the drag coefficient of the ellipsoid, with 10.3% more volume, and is about 7 times smaller than the cylinder, proving the efficiency of the use of bow and stern smoothed profiles.

## Acknowledgements

The authors would like to express their thanks to Brazilian Research Agencies *CNPq*, *CAPES* and *FINEP* and *ARMTEC Technology in Robotics* for supporting this work, and are also grateful to the authors of the references

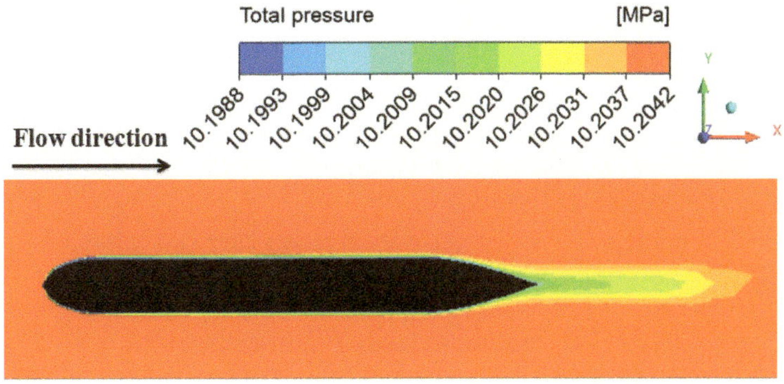


**Figure 16.** Total pressure field for ideal AUV hull ($n$ = 2 and $\theta$ = 20°).

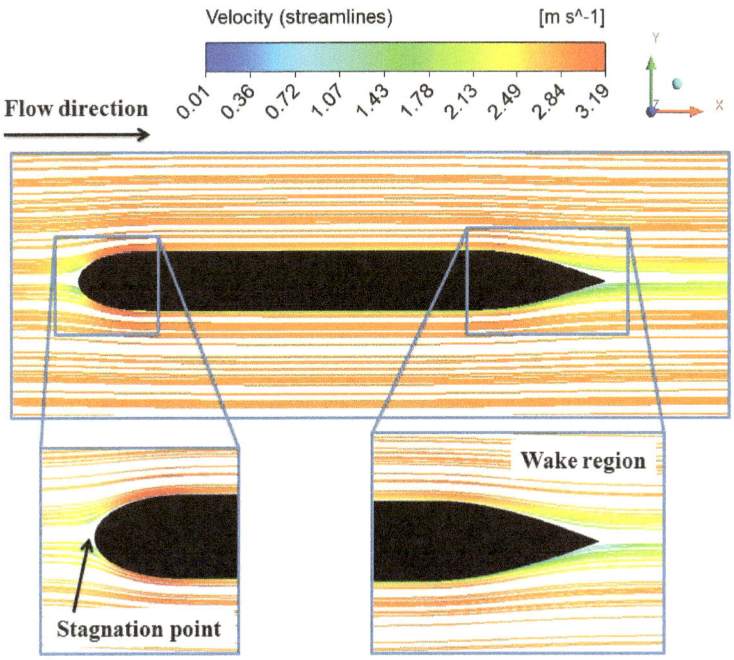


**Figure 17.** Velocity field around the AUV, in terms of streamlines.

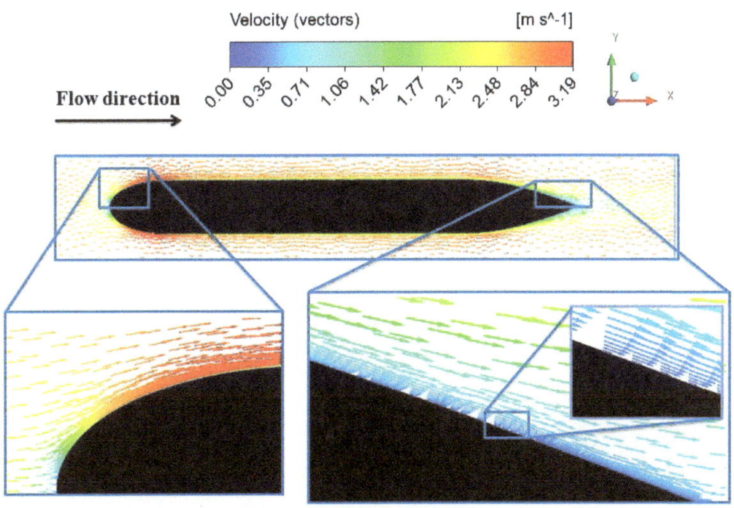


**Figure 18.** Velocity vector field around the AUV.

in this paper that helped in the improvement of quality

## References


[1] Proudman, I. and Pearson, J.R.A. (1957) Expansions at Small Reynolds Numbers for the Flow Past a Sphere and a Circular Cylinder. *Journal of Fluid Mechanics*, **2**, 237-262. http://dx.doi.org/10.1017/S0022112057000105

[2] Masliyah, J.H. and Epstein, N. (1970) Numerical Study of Steady Flow Past Spheroids. *Journal of Fluid Mechanics*, **44**, 493-512. http://dx.doi.org/10.1017/S0022112070001957

[3] Masliyah, J.H. and Epstein, N. (1971) Steady Symmetric Flow Past Elliptical Cylinder. *Industrial & Engineering Chemistry Fundaments*, **10**, 293-299. http://dx.doi.org/10.1021/i160038a017

[4] Rosenfeld, M., Wolfshtein, M. and Israeli, M. (1992) A Numerical Study of the Laminar Incompressible Flow Over a 6:1 Prolate Spheroid at 10° Incidence. *International Journal for Numerical Methods in Fluids*, **15**, 147-173. http://dx.doi.org/10.1002/fld.1650150203

[5] White, F.M. (1991) Viscous Fluid Flow. 2nd Edition, McGraw-Hill, New York.

[6] Warsi, Z.U.A. (1999) Fluid Dynamics: Theoretical and Computational Approaches. 2nd Edition, CRC Press, Boca Raton.

[7] Ferziger, J.H. and Peric, M. (1999) Computational Methods for Fluid Dynamics. 2nd Edition, Springer-Verlag, Berlin. http://dx.doi.org/10.1007/978-3-642-98037-4

[8] Spurk, J.H. and Aksel, N. (2008) Fluid Mechanics. 2nd Edition, Springer-Verlag, Berlin.

[9] Barros, E.A., Dantas, J.L.D., Pascoal, A.M. and Sá, E. (2008) Investigation of Normal Force and Moment Coefficients for an AUV at Nonlinear Angle of Attack and Sideslip Range. *IEEE Journal of Oceanic Engineering*, **33**, 538-549. http://dx.doi.org/10.1109/JOE.2008.2004761

[10] Juong, T., Sammut, K., He, F. and Lee, S.K. (2009) A Study on the Design Optimization of an AUV by Using Computational Fluid Dynamic Analysis. *Proceedings of the* 19*th International Offshore and Polar Engineering Conference*, Osaka, 21-26 June 2009, 696-702.

[11] Malik, S.A. and Guang, P. (2013) Transient Numerical Simulation for Hydrodynamic Derivates Predictions of an Axisymmetric Submersible Vehicle. *Research Journal of Applied Sciences, Engineering and Technology*, **5**, 5003-5011.

[12] Dantas, J.L.D. and Barros, E.A. (2013) Numerical Analysis of Control Surface Effects on AUV Manoeuvrability. *Applied Ocean Research*, **42**, 168-181. http://dx.doi.org/10.1016/j.apor.2013.06.002

[13] Myring, D.F. (1976) A Theoretical Study of Body Drag in Subcritical Axisymmetric Flow. *The Aeronautical Quartely*, **27**, 186-194.

[14] Jun, B.H., Park, J.Y., Lee, F.Y., Lee, P.M., Lee, C.M., Kim, K., Lim, Y.K. and Oh, J.H. (2009) Development of the AUV "ISiMI" and a Free Running Test in an Ocean Engineering Basin. *Ocean Engineering*, **36**, 2-14. http://dx.doi.org/10.1016/j.oceaneng.2008.07.009

[15] ANSYS (2006) ANSYS CFX-Solver Theory Guide (Release 11.0). ANSYS, Inc., Canonsburg.

[16] Kaviany, M. (2001) Principles of Convective Heat Transfer. 2nd Edition, Springer, New York. http://dx.doi.org/10.1007/978-1-4757-3488-1

[17] Hoerner, S.F. (1965) Fluid-Dynamic Drag. Published by the Author, New York.

8

# Dynamics of Drainage of Power-Law Liquid into a Deformable Porous Material

**Javed I. Siddique[1], Forrest A. Landis[2], Muhammad R. Mohyuddin[3,4]**

[1]Department of Mathematics, Pennsylvania State University, York Campus, York, USA
[2]Department of Chemistry, Pennsylvania State University, York Campus, York, USA
[3]Department of Mathematics, FAST University, Islamabad, Pakistan
[4]NCBA & E, Gujrat, Pakistan
Email: jis15@psu.edu, fal100@psu.edu, m_raheel@yahoo.com

## Abstract

**In this study we explore the one-dimensional drainage of a power-law fluid into a deformable porous material. Initially, the fluid is imbibed into the dry undeformed material due to capillary suction which in turn deforms the porous material and forms liquid and solid interfaces. Mixture theory is employed to study the movement of the liquid and solid phases. The zero-gravity model contains the similarity solution that is solved numerically. The stress gradient within the deformable porous material is induced from a pressure gradient that produces an evolving solid fraction and hence deformation. In the absence of gravity effects, the deformation of the solid seems in the same direction of imbibition. This is because of attraction of gravity. Note that these liquid and solid dynamics depend on both the power-law indexes $n$ and $\mu$. We performed the experiments to measure the drainage and deformations of deformable porous materials for two samples of silicon oil (polydimethylsiloxane) in a polyurethane foam. Our experiments show that the silicon with high viscosity drains slower than silicon oil with low viscosity. The theoretical and experimental results show the same qualitative trend.**

## Keywords

**Power Law Fluid, Mixture Theory, Deformable Porous Material**

## 1. Introduction

In this study we develop a model for the drainage of a power law liquid into deformable porous materials. The motivation of this work is connected to many scientific fields, such as oil recovery, inkjet printing, textile engineering, soil consolidation, reservoir engineering and biomechanics. The movement of fluid through the porous materials swells the porous material which in turn affects the flow of fluid. This alteration in fluid flow and deformation of the porous material identifies the complexity and importance of these flows. Interestingly enough, most of the fluids involved in these processes are not Newtonian in nature and hence should be incorporated in the complex dynamics of deformation. To accommodate this need, we present a coupled model for deformation of porous material and drainage of power law fluids.

An overview of literature shows that the deformation of porous materials couple with fluid flow goes back to Terzaghi [1]. This idea was later extended by Biot [2] [3] to study soil consolidation. Later on, mixture theory [4] [5] was introduced to enhance the understanding of material deformation causes by fluid imbibition. Some of the biological applications that use the mixture theories include articular cartilage [6]-[12], arterial tissue [13]-[16] and skin [17]. Note that most of the fluids involved in the above referenced biological studies possess non-Newtonian properties and therefore, non-Newtonian properties must be incorporated with the mixture theory in future studies.

Most industrial applications (e.g. composite materials, paper and inkjet printing, and dyeing of colored fabrics) involve fluids that are non-Newtonian. In an effort to address this behavior, Sommer and Mortensen [18] studied a forced unidirectional infiltration of deformable porous materials. They considered a constant pressure driven flow in an initially dry sponge like material and an agreement between theory and experiment was reported. A similar model of an infiltration of an incompressible liquid into an initially dry porous material was developed by Preziosi *et al.* [19], where they allowed the porous material to deform and relax. Another study that uses mixture theory to model the imbibition of a liquid droplet on a deformable substrate in a one dimensional setting was presented by Anderson [20]. In the absence of gravity effects, the imbibition causes swelling, swelling relaxation and shrinking of a porous material.

The history of capillary rise into porous media goes back to the classical model first presented by Washburn [21]. This model shows that the amount of liquid imbibed into a porous material is proportional to $\sqrt{t}$ . To validate the Washburn model, Zhmud *et al.* [22] and Lago and Araujo [23] presented an experimental and theoretical model of capillary rise into porous materials. Their results are consistent with the Washburn model for initial times and deviate from this trend for long times. Following these studies, Siddique *et al.* [24] presented an analog of the Washburn model of capillary rise. Mixture theory has been used to take into account the deformation of porous materials.

The non-Newtonian modeling along with the mixture theory was studied by Siddique and Anderson [25]. In particular, they studied the capillary rise of a power law fluid into a deformable porous material, where they assumed imbibition occurring from an infinite bath of power law fluid. When gravity effects are present, both liquid and solid interface positions reach equilibrium heights depending upon the power law index $n$. This study mimics only one particular aspect of capillary rise; however the need is to extend the current power law model along with mixture theory to explore many other physical settings. In this study we explore the drainage of a finite amount of liquid into deformable porous materials in the presence and absence of gravity effects.

There are many types of non-Newtonian fluid models, e.g. Herschel-Bulkley fluid, power law and differential fluid type, etc. These fluid models have been widely used in a variety of settings to analyze various physical aspects. Most relevant ones are Christopher and Middleman [26], Sadowski [27] and Hayes *et al.* [28]. These studies explained some of the important aspects of porous medium flow based on power law model but there are still important phenomena to explain and clarify. The current study is an effort to model the power law fluid with mixture theory (for details see [25]).

In this paper, we study the drainage of a power law fluid into deformable porous material. The initial height $H(t)$ of fluid is modeled via an Equation. Our experiments of drainage of silicon oil with low and high viscosity into deformable porous material encouraged us to present a mathematical model that can be used to predict the similar dynamics of drainage. We do this by following Siddique *et al.* [24] [25]. We hope that our preliminary efforts in this regard may lead to further investigate the additional features of flows in complex porous media.

## 2. Experiment

A sample of polyurethane foam (pore size ca. 50 - 150 μm (**Figure 1**)) was cut into a $1\times1\times2$ cm piece. The foam was then inserted into a 1 cm$^2$ plastic cuvette with a hole drilled in the bottom to allow for drainage. The foam fit just snuggly in the cuvette so that liquid could not flow around the foam, but little compression of the

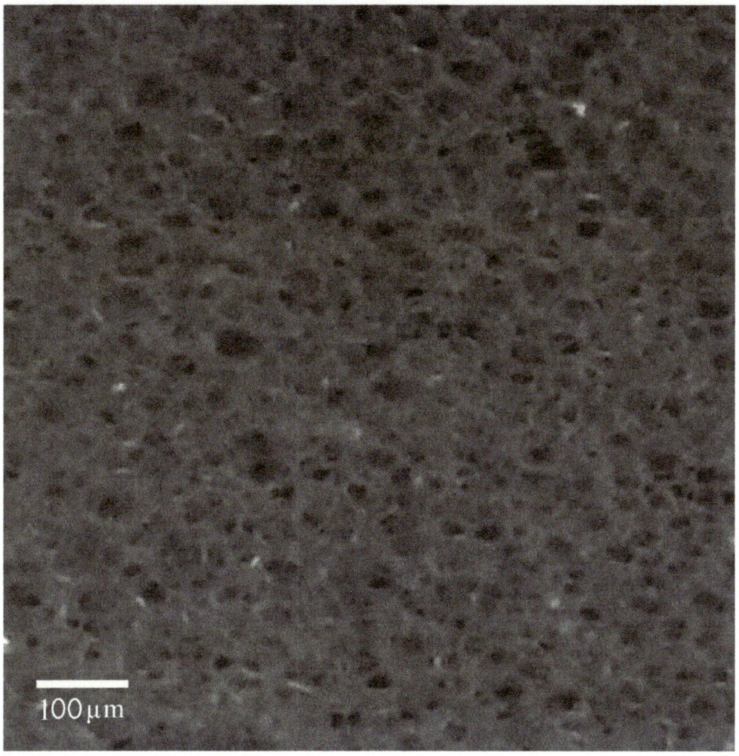


**Figure 1.** Optical micrograph of a cross-section of the polyurethane foam.

foam occurred which would alter the drainage of the liquid (**Figure 2**). Two samples of silicon oil (polydimethylsiloxane) were used as the drainage fluid: a low viscosity sample (10 cPs) and a high viscosity sample (1000 cPs). Since these non-Newtonian silicon oils have identical chemical structure, their molecular interaction with the foam should also be identical; however, they have different molar masses, which is manifested in their differences in viscosity. Only these differences in viscosity should affect their drainage through the foam. A single grain of solid iodine was dissolved in both silicon oils to provide contrast with the foam and aid in the determination of the movement of the oils through the foam. It is not expected that the presence of the iodine will alter the drainage of the oils. 1 mL of the dyed silicon oil was injected into the top of the cuvette and it began to drain into the foam. A digital camera operating at 30 frames per second and a resolution of $1920\times1080$ pixels was used to record the flow of the silicon oils through the foam. Using the images produced by the camera, the position (in pixels) of the oil above the foam, $H(t)$, the foam interface position, $h_s(t)$, and the liquid interface, $h_\ell(t)$, were determined as a function of time (**Figure 2**). The distance in pixels was converted to distance in mm using the length of the cuvette (1 cm) as a scale. The experiment was concluded when the level of the silicon above the foam reached the level of the foam.

## Experimental Results

**Figure 2** shows a representative drainage experiment using the high viscosity silicon oil as the fluid. The beginning point (0 s in **Figure 2**) was chosen as the image where all of the 1 mL volume of silicon oil had been added to the cuvette. As the experiment continued, the height of the liquid above the foam, $H(t)$, and the liquid interface position, $h_\ell(t)$, both dropped as the oil absorbed into the foam (10 s and 30 s in **Figure 2**). This experiment was terminated when all of the silicon oil had penetrated into the foam (when $H(t)=h_s(t)$ at ca. 100 s). The positions of each of the three interfaces are plotted as a function of time in **Figure 3** for both the low and high viscosity silicon oils. It took approximately 4 s for the low viscosity oil to completely drain into the foam while the higher viscosity oil took significantly longer at 100 s.

It should be noted that no measurable expansion of the foam was observed (*i.e.*, $h_s(t)$ was constant) for these foams when imbibed with either silicon oil. This is in contrast to drainage experiments where low molar

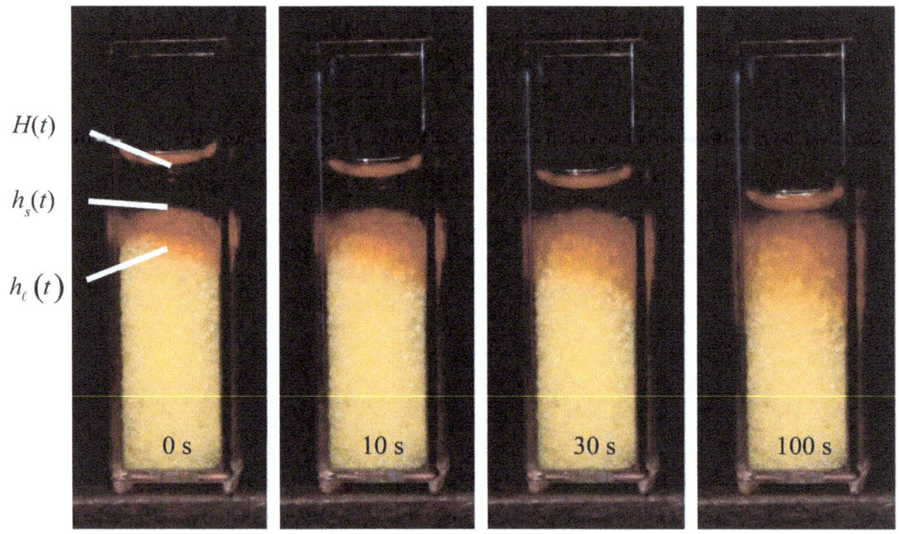


**Figure 2.** Representative photographs showing the drainage of the high viscosity silicon oil into the polyurethane foam at different time intervals. The liquid height above the foam, the foam height, and the liquid interface are labeled ( $H(t), h_s(t)$ , and $h_\ell(t)$ , respectively).

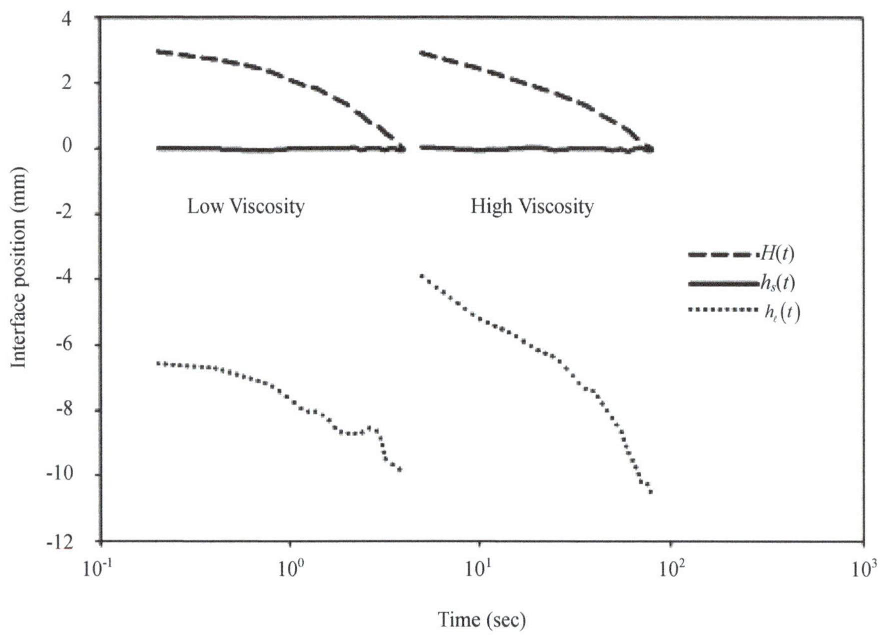


**Figure 3.** This figure shows the dynamics behavior where the curves are shown in dimensional form. The dashed curves shows the initial height as a function of time $H(t)$ , the solid line shows the deformation of foam as function of time $h_s(t)$ and drainage of liquid as a function of time $h_\ell(t)$ .

mass alcohols such as ethanol were used. While ethanol has similar rapid drainage characteristics to the low viscosity silicon oil, it differed in that it also caused a significant expansion of the foam. Ethanol is a polar liquid that can interact strongly with the polyurethane foam causing expansion. In contrast, the silicon oil is essentially nonpolar and will not interact strongly with the foam resulting in little expansion. Clearly the expansion of the foam is very dependent on the nature of the intermolecular interactions between the foam and the liquid and will be examined in future research efforts.

**Figure 3** shows the dynamics behavior of the low and high viscosity silicon oils into the polyurethane foam. Here the curves are shown in dimensional form. The dashed curves shows the initial height as a function of time $H(t)$, the solid line shows the deformation of foam as function of time $h_s(t)$ and drainage of liquid as a function of time $h_\ell(t)$.

## 3. Mathematical Modeling

The basic geometrical description for our problem is shown in **Figure 4**. We consider a finite amount of a non-Newtonian liquid with an initial height defined as $H_0$ in contact with a deformable porous material at time $t = 0$. In the figure this initial contact position between the liquid and the deformable porous material is specified by $z = 0$. We assume the pressure at the positions $z = H(t)$ is atmospheric pressure. Later, the height of the liquid will be represented by the Equation for $H(t)$. After time $t > 0$, the non-Newtonian liquid starts to imbibe into the deformable porous material which in turn deforms the porous material and forms two interfaces: the upper interface $z = h_s(t)$, and lower interface $h_\ell(t)$, as can be seen from **Figure 4**. In order to define the dimensionless system of equations, we use the following set of scaling parameters

$$z' = \frac{z}{L}, \quad t' = \frac{t}{T}, \quad u_s' = \frac{T}{L}u_s, \quad u_\ell' = \frac{T}{L}u_\ell, \quad p' = \frac{p}{\Sigma_0},$$
$$h_s' = \frac{h_s}{L}, \quad h_\ell' = \frac{h_\ell}{L}, \quad \sigma'(\phi) = \frac{\sigma(\phi)}{\Sigma_0}, \quad K'(\phi) = \frac{K(\phi)}{K_0}.$$

The above dimensionless quantities suggest the following time and length scales after balancing the terms in both momentum equations $T^n = \frac{L^{n+1}\mu^*}{md^{n+1}}$ and $L = \frac{m\rho_\ell}{g}$. Note that this choice of time scale depends on power-law index $n$. We will present the details on how to make this choice of time scale independent of power law index $n$ for numerical simulation in the result section. In the above defined set of scaling parameters we use $\Sigma_0$ and $K_0$ are representative scales for solid stress and the permeability respectively. We follow Siddique and Anderson [25] where they defined the permeability $K(\phi)$ in terms of the permeability of bed particles of diameter $d$ and solid volume fraction $\phi$ (see Hayes *et al.* [28] and references therein for details)

$$K(\phi) = W(\phi)d^2, \qquad W(\phi) = \frac{(1-\phi)^3}{150\phi^2} \tag{1}$$

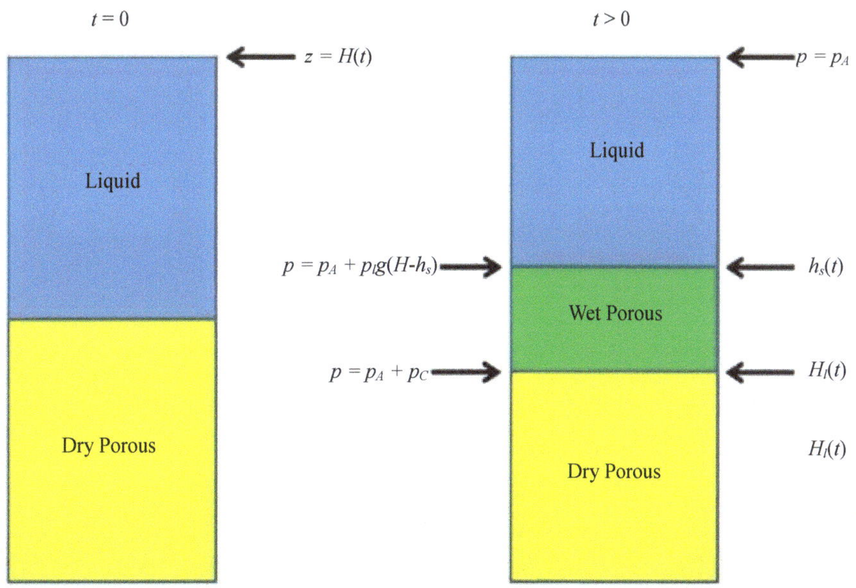


**Figure 4.** Problem configuration at $t = 0$ and $t > 0$.

and $\mu_{eff}$ for the power law fluid is given by

$$\mu_{eff} = \mu^* Y(\phi,n) d^{1-n} \tag{2}$$

where

$$Y(\phi,n) = \left[ \left( \frac{3n+1}{n(1-\phi)} \right)^n \left( \frac{1-\phi}{4} \right) \left( \frac{1-\phi}{3\phi} \right)^{1-n} \right].$$

Note, setting $n=1$ in $Y(\phi,n)$ relation and Equation (2) yields $\mu_{eff} = \mu^*$, that represents the Newtonian viscosity. The resulting set of Equations describing the one-dimensional drainage of non-Newtonian liquid (e.g. power law) into deformable porous material after dropping the primes is

$$\phi_t + (\phi u_s)_z = 0 \tag{3}$$

$$\phi_t - \left[ (1-\phi) u_\ell \right]_z = 0 \tag{4}$$

$$(u_\ell - u_s)^{n-1} (u_\ell - u_s) = -\frac{K(\phi)}{(1-\phi)} (p_z + 1) \tag{5}$$

$$p_z = \sigma_z - (\rho\phi + 1), \tag{6}$$

where $\rho = \rho_s^{\mathrm{T}} / \rho_\ell^{\mathrm{T}} - 1$ and stress is a function of solid volume fraction $\sigma = \sigma(\phi)$. Following Anderson 2005, we assume $\sigma(\phi) = \phi_r - \phi$. This choice is suitable for one dimensional deformation and needs to be generalized to take into account effects such as shear deformation in a higher dimension. It is worthwhile mentioning that subscript $t$ and $z$ represent the derivative with respect to corresponding variables.

Equations (3) and (4) represent mass balances and Equations (5) and (6) are momentum balances for the solid and liquid phases (see [25]). The set of parameters in these Equations can be divided into two categories, first set of parameters inside the wet material regions such as the solid volume fraction $\phi$, the liquid $u_\ell$ and solid $u_s$ component of velocities, the liquid pressure $p$, the solid stress $\sigma(\phi)$, and the permeability of porous material and second set represents liquid $h_\ell$ and solid $h_s$ boundary positions.

We follow the same procedure as was followed in Anderson (2005), Siddique *et al.* [24], and Siddique and Anderson [25] to obtain the partial differential equation for the solid volume fraction $\phi$ in the wet sponge region $h_\ell \le z \le h_s$

$$\phi_t + c(t)\phi_z = \partial_z \left[ -\frac{W(\phi)\phi^n (1-\phi)^{n-1}}{Y(\phi,n)} \left( \sigma'(\phi)\phi_z - \rho\phi \right) \right]^{1/n} \tag{7}$$

The appropriate boundary conditions for the solid volume fraction are

$$\phi = \phi_r, \quad \text{at } z = h_s \tag{8}$$

$$\phi = \phi_\ell^* + H - h_s + (h_s - h_\ell) \int_0^1 (\rho\phi + 1)\,\mathrm{d}z \quad \text{at } z = h_\ell. \tag{9}$$

In the derivation of boundary condition (8), we integrate Equation (6) after substituting $\sigma(\phi)$ and applying the pressure boundary conditions given below in (12) and (15). The liquid height Equation is derived using a conservation of liquid argument (see Siddique *et al.* [24])

$$H(t) = H_0 + h_s - (h_s - h_\ell) \int_{h_\ell}^{h_s} (1-\phi)\,\mathrm{d}z \tag{10}$$

where $H_0$ represents the initial height of the non-Newtonian liquid before it starts to drain into deformable porous material.

The boundary conditions at the liquid-wet material interface $z = h_s(t)$ are

$$u_s\left(h_s^-,t\right)=\frac{\partial h_s}{\partial t}, \tag{11}$$

$$p\left(h_s^-,t\right)=\frac{p_A}{\Sigma_0}-\left[h_s(t)-H(t)\right] \tag{12}$$

$$\sigma\left(h_s^-,t\right)=0 \tag{13}$$

where $p_A/\Sigma_0$ is the dimensionless constant atmospheric pressure. Here, Equations (11)-(13) represent the kinematic, hydrostatic, and zero stress conditions respectively. While modeling hydrostatic boundary condition (12) we have neglected the inertial effects.

The kinematic and pressure boundary conditions at wet-material-dry material interface $z=h_\ell(t)$ are

$$u_\ell\left(h_\ell^+,t\right)=\frac{\partial h_\ell}{\partial t}, \tag{14}$$

$$p\left(h_\ell^+,t\right)=p_A/\Sigma_0+p_c/\Sigma_0 \tag{15}$$

where $p_c/\Sigma_0$ represents the dimensionless constant capillary pressure.

Boundary conditions (11) and (14) transform to ordinary differential Equations for the solid and liquid interfaces

$$\frac{\mathrm{d}h_s}{\mathrm{d}t}=c(t)-\left[\frac{(1-\phi)^{n-1}W(\phi)}{Y(\phi,n)}\left(\frac{\partial\phi}{\partial z}+\rho\phi\right)\bigg|_{z=0}\right]^{\frac{1}{n}} \tag{16}$$

$$\frac{\mathrm{d}h_\ell}{\mathrm{d}t}=c(t)+\left[\frac{W(\phi)\phi^n}{Y(\phi,n)(1-\phi)}\left(\frac{\partial\phi}{\partial z}+\rho\phi\right)\bigg|_{z=1}\right]^{\frac{1}{n}} \tag{17}$$

where $c(t)$

$$c(t)=\frac{1-\phi_0}{\phi_0}\left[\frac{W(\phi)\phi^n}{Y(\phi,n)(1-\phi)}\left(\frac{\partial\phi}{\partial z}+\rho\phi\right)\bigg|_{z=1}\right]^{\frac{1}{n}} \tag{18}$$

If we take $n=1$ in Equation (7) we recover Equation (9) of Siddique *et al.* [24] in dimensionless form and if we take $n=1$ and $g=0$ in (7) we recover Equation (20) of Anderson [20] in dimensionless form and Equation (44) of Prezoisi *et al.* [5]. The steady state solution of the above system is the same as for the Newtonian case (see Siddique *et al.* [24]). Below we will summarize the solution procedure in the absence of gravity and in the presence of gravity effects.

## 4. Gravity Independent Solution

In the absence of gravity effects, Equation (7) admits the solution in terms of a similarity variable $\eta=\dfrac{z}{(n+1/n)t^{\frac{n}{n+1}}}$ that yields the following ordinary differential Equation

$$\left(-\frac{n}{n+1}\right)^{1/n}\eta\frac{\mathrm{d}\phi}{\mathrm{d}\eta}+\frac{1-\phi_0}{\phi_0}\left[\frac{W(\phi)\phi^n}{Y(\phi,n)(1-\phi)}\frac{\mathrm{d}\phi}{\mathrm{d}\eta}\bigg|_{\lambda^+}\right]^{\frac{1}{n}}\frac{\mathrm{d}\phi}{\mathrm{d}\eta}=\frac{\mathrm{d}}{\mathrm{d}\eta}\left[\frac{W(\phi)\phi^n(1-\phi)^{n-1}}{Y(\phi,n)}\frac{\mathrm{d}\phi}{\mathrm{d}\eta}\right] \tag{19}$$

subject to the boundary conditions $\phi(\lambda_s)=\phi_r$, and $\phi(\lambda_\ell)=\phi_\ell^*$, where $\phi_\ell^*=\phi_r-p_c/\Sigma_0$. The boundary condition, $\phi(\lambda_\ell)=\phi_\ell^*$, is obtained by substituting $\sigma=\sigma(\phi)$ in Equation (6), integrating, and using boundary conditions (12) and (15) in the absence of gravity. Note that $\lambda_s$ and $\lambda_\ell$ represents the interface positions in terms of the similarity variables given below. Similarly, introducing $h_s(t)=(n+1/n)\lambda_s t^{n/n+1}$ and $h_\ell=(n+1/n)\lambda_\ell t^{n/n+1}$ yields this following relations for $\lambda_s$ and $\lambda_\ell$

$$\lambda_s=\left(\frac{n}{n+1}\right)^{1/n}\left[\frac{1-\phi_0}{\phi_0}\left(\frac{\phi^n W(\phi)}{(1-\phi)Y(\phi,n)}\frac{\mathrm{d}\phi}{\mathrm{d}\eta}\bigg|_{\lambda_\ell^+}\right)^{1/n}-\left(\frac{(1-\phi)^n W(\phi)}{Y(\phi,n)}\frac{\mathrm{d}\phi}{\mathrm{d}\eta}\bigg|_{\lambda_s^-}\right)^{1/n}\right] \tag{20}$$

$$\lambda_\ell=\left(\frac{n}{n+1}\right)^{1/n}\left[\frac{1}{\phi_0}\left\{\frac{\phi^n W(\phi)}{(1-\phi)Y(\phi,n)}\frac{\mathrm{d}\phi}{\mathrm{d}\eta}\bigg|_{\lambda_\ell^+}\right\}^{1/n}\right] \tag{21}$$

The non-linear ODE (19) along with the non-linear Equations (20) and (21) is solved numerically. The ODE and non-linear Equations are discretized using finite difference and midpoint rule yielding a system of non-linear Equations. This coupled system of non-linear Equations is solved numerically. We will present the solution of zero gravity case in the results section. It is worthwhile highlighting that this numerically computed zero gravity solution will be used as an initial condition for non-zero gravity case which we will discuss in the section below.

## 5. Non-Zero Gravity Solution

In the presence of gravity effects, we will first transform the moving domain problem given by (7) along with (16) and (17) to fixed domain using the following transformation

$$z=\frac{z-h_\ell(t)}{h_s(t)-h_\ell(t)}. \tag{22}$$

This helps transforming the moving domain problem $h_\ell\le z\le h_s$ to a fixed domain problem $0\le z\le 1$. We use $h_s(t=0)=0$ and $h_\ell(t=0)=0$ initial conditions for solid and liquid interface positions respectively. We use method of lines along with boundary conditions given in (8) and (9). Again, we use the zero gravity solution as initial conditions for the non-zero gravity case. We approximate the spatial derivatives using second order accurate finite difference and mid-point discretizations. This reduces the system of partial differential Equations (PDEs) to system of ordinary differential Equations (ODEs) in time which is solved using Matlab's solver ode23s. We start our numerical integration for $0<t<t_{\text{start}}$, where $t_{\text{start}}$ is a numerical small value.

## 6. Results

Our results are based on a specific set of test fluids for which power-law index $n$ and consistency index $\mu^*$ values are available (see Missirlis *et al.* [29]). The calculation of the $t^*$ values is based on the assumption that the capillary pressure $|p_c|=(\gamma\cos\theta)/d$, where $\gamma$ is the surface tension assumed to be same and non-Newtonian case and $\theta$ is the wetting angle, that we also assumed to be $\theta=0$. We introduce the Newtonian time scale $T=\mu_N/|p_c|$ in such a way that computed simulation results are independent of the power law index $n$. Note that the time scales are related through capillary pressure $p_c$ that depends on particle diameter $d$. When capillary pressure increases the drainage of fluid increases. Although we do not explore the dependence on $\mu^*$ here, increasing $\mu^*$ means increasing friction force. In other words, $\mu^*$ is directly proportional to the drainage process.

In the beginning of the process, a finite amount of liquid is supplied, whose height is shown by $H(t)$ in **Figure 5**. It is important to note that when gravity effects are not present, the solution is independent of $\rho$, and both interface positions $h_s(t)$ and $h_\ell(t)$ admits similarity solution until the entire fluid is drained into the deformable porous material (*i.e.*, $h_s(t)=H(t)$). **Figure 5** summarizes the drainage comparison between the Newtonian and different power law fluids. Each of these curves corresponds to set of parameters $(\phi_0,\phi_r,\phi_\ell^*)$ used in sponge experiment in Siddique *et al.* [24] along with power law consistency index $\mu^*$ and the power law index $n$ values from Missirlis *et al.* [29]. However it is important to note that equilibrium heights $h_s(t)$ and $h_\ell(t)$ attained depend on the values of $\phi_0,\phi_r,\phi_\ell^*$ and $\rho$. The bottom right of **Figure 5** shows power law

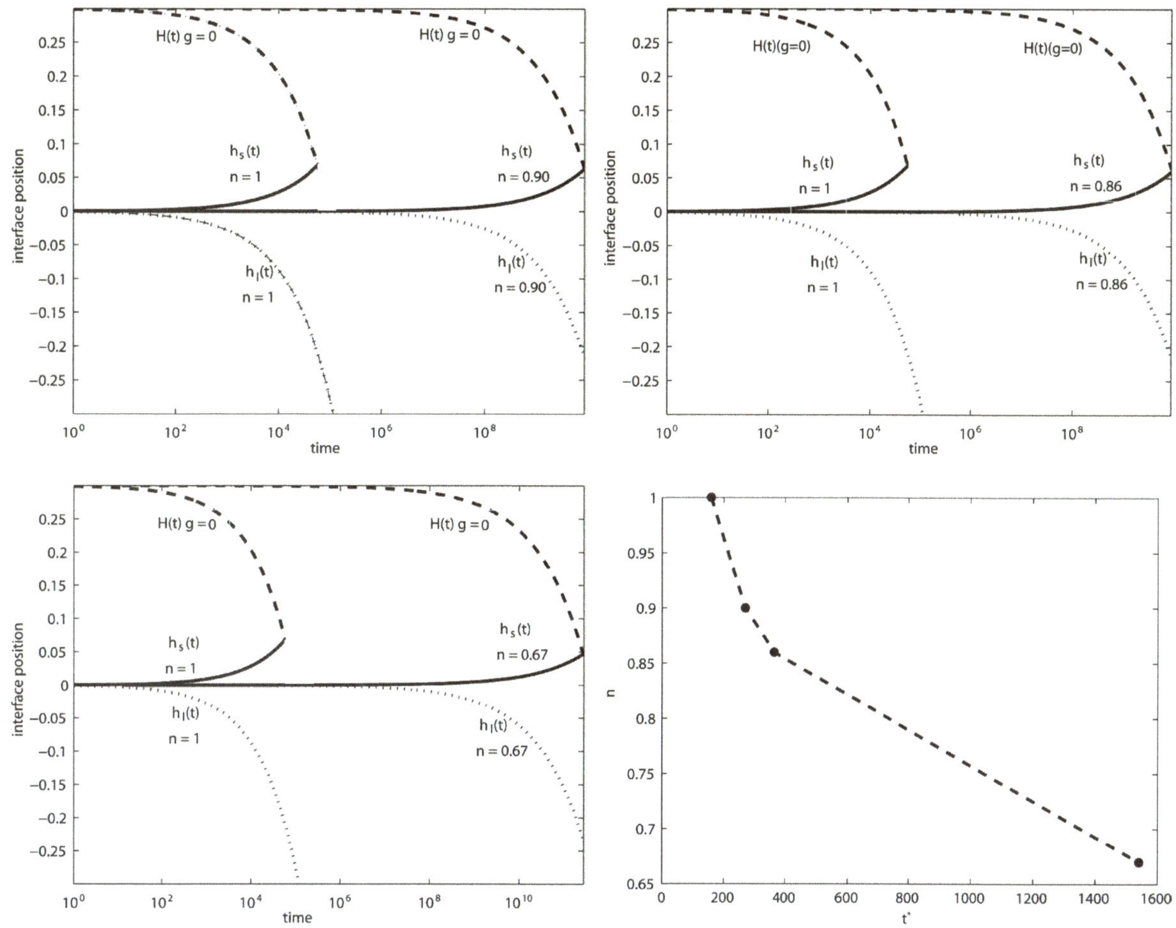


**Figure 5.** The plots show the drainage of Newtonian and non-Newtonian liquids. Top left shows the comparison between $n=1$ and $n=0.90$, top right $n=1$ and $n=0.86$ and bottom left $n=1$ and $n=0.67$. Bottom right plot shows that drainage time $t^*$ for different for different fluids. The other parameter used in these figures are $\rho_\ell = 1000\ \mathrm{kg/m^3}$, $g = 9.8\ \mathrm{m/sec^2}$, $d = 0.001$, $\phi_0 = 0.33$, and surface tension $63.57\times10^{-3}\ \mathrm{N/m}$.

index $n$ as a function of $t$. As $n$ decrease the drainage time increases. We denote this drainage time as $t^*$. It is interesting to note that both the Newtonian and non-Newtonian fluids follow the similarity behavior but as $n$ decreases, $t^*$ increases which means decreasing the $n$ values slows down the dynamics. **Figure 6** shows the drainage of liquid into a deformable porous material in the presence of gravity effects. The time scales used in **Figure 6** allow us to present a direct comparison for different values of $n$. Similar to the Newtonian case, a finite amount of liquid is supplied, shown by $H(t)$ in **Figure 6**. For this case, both curves $h_s(t)$ and $h_\ell(t)$ follow similarity solution for very short time and then depart from this behavior until the fluid is entirely drained into porous material. The drainage time for the nonzero gravity case is faster than the zero gravity case.

In order to connect our theoretical predictions with experiments, we performed drainage of silicon oil of two different viscosities with identical chemical structure into deformable sponge like materials. Our experimental predictions show a similar trend for drainage and deformation of deformable porous materials. The drainage of water into deformable porous material follow approximately $t^{1/2}$ which is expected for the case of Newtonian fluids. Our experimental curves shows a modified power law of $t^{0.48}$ and $t^{0.70}$ for two different set of silicon oils which suggests $n=0.92$ and $n=2.3$ respectively. Although we are unable to present a direct qualitative comparison between theory and experiments due to limited access to experimental facilities but our finding shows that any fluid having different properties than water require different power law behavior. We hope our findings with these preliminary comparison will encourage further experimental investigation leading to capture the realistic drainage phenomena.

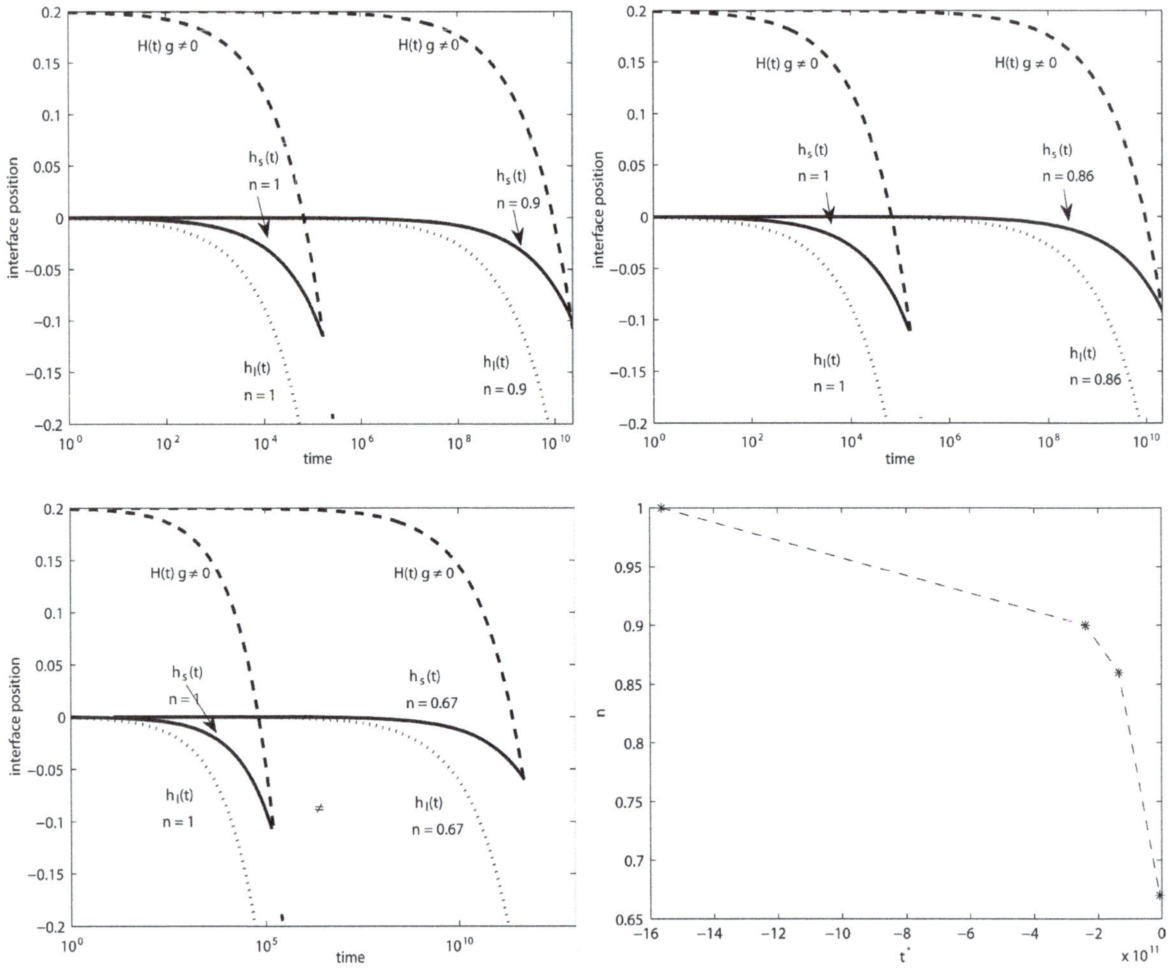


**Figure 6.** This plot shows the comparison of drainage of Newtonian fluid with non-Newtonian fluid for different power-law index. The curves as labeled $H(t)$, $h_s(t)$ and $h_\ell(t)$ for hight of liquid, height of solid, and height of liquid. Drainage time is noted as the time it takes for all of the above liquid to drain into the deformable porous material.

## 7. Discussion

In this present work we have presented a basic model for predicting the drainage of a power law liquid into a deformable porous material. More specifically, we transform the capillary rise theory developed for power law liquids into deformable porous materials [24] to the drainage setting by specifying liquid height [25] via an equation based on conservation of liquid argument. In addition to theoretical predictions, this work includes a set of experiments showing the similar behavior as shown in our numerically simulations. In this study we use the mixture theory [15] [16] [18]-[20] [24] [25] to model the problem of drainage of power law liquid into deformable porous material.

In the absence of gravity effects, the deformation of the porous material and the penetration of liquid follows a different power law behavior $t^{\frac{n}{n+1}}$ compared to the Newtonian case. Our findings show that the dynamics for both $h_s$ and $h_\ell$ follow different dynamics compared to the Newtonian case. We were able to compute the drainage time for different fluids. These findings are consistent with our experimental findings. When gravity effects are present in our model, the drainage of power law liquid depends on $\rho$. Note that both curves, $h_s(t)$ and $h_\ell(t)$, follow similarity solution initially and then depart from this trend depending upon their power law consistency index $\mu^*$ and power law index $n$. In short, the dynamics for fluid with different properties follow different dynamics.

The present study is preliminary, but suggests some interesting possibilities in the theory of drainage of power law fluid. An interesting area to explore in future investigations would be the use of a more generalized function for solid stress $\sigma$ and a complicated permeability function $K(\phi)$ and for exploring the deformation in higher dimension. Alternatively, keeping a simplified functions for $\sigma$ and $K(\phi)$ and extending the ability to directly compare the theoretical and experimental predictions could lead to a better understanding of the basic mechanisms of drainage

## Acknowledgements

The author J. I. Siddique greatly appreciate the support of Simons Foundation Grant No. 281839.

## References

[1] Terzaghi, K. (1925) Erdbaumechanik auf Bodenphysikalischen Grundlagen. Deuticke, Wien.

[2] Biot, M.A. (1955) Theory of Elasticity and Consolidation for a Porous Anisotropic Solid. *Journal of Applied Physics*, **26**, 182-185. http://dx.doi.org/10.1063/1.1721956

[3] Biot, M.A. (1962) Mechanics of Deformation and Acoustic Propogation in Porous Media. *Journal of Applied Physics*, **33**, 1482-1498.

[4] Atkin, R.J. and Crain, R.E. (1976) Continuum Theories of Mixture: Basic Theory and Historical Development. *Quarterly Journal of Mechanics and Applied Mathematics*, **29**, 209-244. http://dx.doi.org/10.1093/qjmam/29.2.209

[5] Bowen, R.M. (1980) Incompressible Porous Media Models by Use of the Theory of Mixtures. *International Journal of Engineering Science*, **18**, 1129-1148. http://dx.doi.org/10.1016/0020-7225(80)90114-7

[6] Lai, W.M. and Mow, V.C. (1980) Drag Induced Compression of Articular Cartilage during a Permeation Experiment. *Biorheology*, **17**, 111-123.

[7] Holmes, M.H. (1983) A Nonlinear Diffusion Equation Arising in the Study of Soft Tissue. *Quarterly of Applied Mathematics*, **41**, 209.

[8] Holmes, M.H. (1984) Comparison Theorems and Similarity Solution Approximations for a Nonlinear Diffusion Equation Arising in the Study of Soft Tissue. *SIAM Journal on Applied Mathematics*, **44**, 545-556. http://dx.doi.org/10.1137/0144037

[9] Holmes, M.H. (1985) A Theoretical Analysis for Determining the Nonlinear Hydraulic Permeability of a Soft Tissue from a Permeation Experiment. *Bulletin of Mathematical Biology*, **47**, 669-683. http://dx.doi.org/10.1007/BF02460132

[10] Holmes, M.H. (1986) Finite Deformation of Soft Tissue: Analysis of a Mixture Model in Uni-Axial Compression. *Journal of Biomechanical Engineering*, **108**, 372-381. http://dx.doi.org/10.1115/1.3138633

[11] Holmes, M.H. and Mow, V.C. (1990) The Nonlinear Characteristic of Soft Gels and Hydrated Connective Tissue in Ultrafiltration. *Journal of Biomechanics*, **23**, 1145-1156. http://dx.doi.org/10.1016/0021-9290(90)90007-P

[12] Hou, J.S., Holmes, M.H., Lai, W.M. and Mow, V.C. (1989) Boundary Conditions at the Cartilage-Synovial Fluid Interface for Joint Lubrication and Theoretical Verifications. *Journal of Biomechanical Engineering*, **111**, 78-87. http://dx.doi.org/10.1115/1.3168343

[13] Kenyon, D.E. (1976) The Theory of an Incompressible Solid-Fluid Mixture. *Archive for Rational Mechanics and Analysis*, **62**, 131-147.

[14] Klanchar, M. and Tarbell, J.M. (1987) Modelling Water Flow through Arterial Tissue. *Bulletin of Mathematical Biology*, **49**, 651-669. http://dx.doi.org/10.1007/BF02481766

[15] Barry, S.I. and Aldis, G.K. (1992) Flow Induced Deformation from Pressurized Cavities in Absorbing Porous Tissues. *Bulletin of Mathematical Biology*, **54**, 977-997. http://dx.doi.org/10.1007/BF02460662

[16] Barry, S.I., Parker, K.H. and Aldis, G.K. (1991) Fluid Flow over a Thin Deformable Porous Layer. *Journal of Applied Mathematics and Pysics* (*ZAMP*), **42**, 633-648.

[17] Oomens, C.W.J., Van Campen, D.H. and Grootenboer, H.J. (1987) A Mixture Approach to the Mechanics of Skin. *Journal of Biomechanics*, **20**, 877-885. http://dx.doi.org/10.1016/0021-9290(87)90147-3

[18] Sommer, J.L. and Mortensen, A. (1996) Forced Unidirectional Infiltration of Deformable Porous Media. *Journal of Fluid Mechanics*, **311**, 193-217. http://dx.doi.org/10.1017/S002211209600256X

[19] Preziosi, L., Joseph, D.D. and Beavers, G.S. (1996) Infiltration of Initially Dry, Deforamable Porous Media. *International Journal of Multiphase Flow*, **22**, 1205-1222. http://dx.doi.org/10.1016/0301-9322(96)00035-3

[20] Anderson, D.M. (2005) Imbibition of a Liquid Droplet on a Deformable Porous Substrate. *Physics of Fluids*, **17**, Ar-

ticle ID: 087140. http://dx.doi.org/10.1063/1.2000247

[21] Washburn, E.W. (1921) The Dynamics of Capillary Flow. *Physical Review*, **17**, 273-283. http://dx.doi.org/10.1103/PhysRev.17.273

[22] Zhmud, B.V., Tiberg, F. and Hallstensson, K. (2000) Dynamic of Capillary Rise. *Journal of Colloid and Interface Science*, **228**, 263-269. http://dx.doi.org/10.1006/jcis.2000.6951

[23] Lago, M. and Araujo, M. (2001) Capillary Rise in Porous Media. *Journal of Colloid and Interface Science*, **234**, 35-43. http://dx.doi.org/10.1006/jcis.2000.7241

[24] Siddique, J.I., Anderson, D.M. and Bondarev, A. (2009) Capillary Rise of Liquid into Deformable Porous Material. *Physics of Fluids*, **21**, Article ID: 013106. http://dx.doi.org/10.1063/1.3068194

[25] Siddique, J.I. and Anderson, D.M. (2011) Capillary Rise of Non-Newtonian Liquid into Deformable Porous Material. *Journal of Porous Media*, **14**, 1087-1102. http://dx.doi.org/10.1615/JPorMedia.v14.i12.40

[26] Christopher, R.H. and Middlemen, S. (1965) Power-Law Flow through a Packed Tube. *Industrial & Engineering Chemistry Fundamentals*, **4**, 422-426. http://dx.doi.org/10.1021/i160016a011

[27] Sadowski, T.J. (1963) Non-Newtonian Flow through Porous Media. Ph.D. Thesis, University of Winconsin, Madison.

[28] Hayes, R.E., Afacan, A., Boulanger, B. and Shenoy, A.V. (1996) Modeling the Flow of Power Law Fluids in a Packed Bed Using a Volume-Averaged Equations of Motion. *Transport in Porous Media*, **41**, 175-196.

[29] Missirlis, K.A., Assimacopoulos, D., Mitsoulis, E. and Chhabra, R.P. (2001) Wall Effects for Motion of Spheres in Power-Law Fluids. *Journal of Non-Newtonian Fluid Mechanics*, **96**, 459-471.

# 9

# Numerical Calculation of the Horizontal Stress Difference of Viscous-Elastic Fluid Acting on the Residual Oil in Micro Pore

**Lili Liu[1], Shuren Yang[1], Lihui Wang[2], Di Xu[1], Liang Zhao[3], Yue Cui[1]**

[1]Petroleum Engineering Department, Northeast Petroleum University, Daqing, China
[2]Daqing Petroleum Equipment Group, Daqing, China
[3]Daqing Oilfield Tianyu Engineering Design Co., Ltd., Daqing, China
Email: dqliull2009@163.com

## Abstract

**In order to analyze the microscopic theory of viscous-elastic fluid flooding residual oil, the flow equation of polymer solution in the micro pore can be derived by selecting upper-convected Maxwell constitutive equation, continuity equation and motion equation. Then, the flow velocity field and stress field can be calculated under the boundary condition, and with the theory of stress tensor, the horizontal stress difference of polymer solution acting on the residual oil can be calculated. The results show that the greater the elasticity of viscous-elastic fluid is, the wider the flow channel is, the greater the horizontal stress difference is. The force acting on residual oil by viscous-elastic fluid can be increased by increasing the concentration of the polymer solution.**

## Keywords

**Viscous-Elastic Fluid, Upper-Convected Maxwell Constitutive Equation, Horizontal Stress Difference, *We***

## 1. Introduction

The practice of enhanced oil recovery shows that the recovery can be improved by the viscous-elastic polymer solution flooding. The conclusion that the greater the elasticity of viscous-elastic fluid is, the higher the sweep efficiency is, which has been derived through using the modified upper-convected Maxwell constitutive model by Yin Hongjun [1] [2] *et al.* The flow field of viscous-elastic fluid on the dead ends has been calculated and analyzed by Zhang Lijuan [3] *et al.* And conclusion that viscous-elastic fluid can improve oil displacement efficiency has been received by Wang Demin [4] [5], Xia Huifen [6] [7], *et al.* and they qualitative analyzed the microscopic force acting on residual oil in detail. And they pointed out that in the microscopic pore space, it is microscopic force that made the residual oil deform and separate.

In this paper, the microscopic force acting on the residual oil of viscous-elastic fluid has been calculated quantitative. And the microscopic force which made the residual oil deform is the horizontal stress difference acting on the residual oil at the same level, which lays the basis for the further analysis of deformation and separation of residual oil.

## 2. The Mathematical Model of Upper-Convected Maxwell Viscous-Elastic Fluid in the Micro Pore

### 2.1. Flow Channel Model

In order to analyze the force of the residual oil in micro pore of different flow channel width, the calculation model has been established in **Figure 1**. $H$ is width of the flow channel, taking 20 μm, 40 μm, 60 μm respectively. The length of the flow channel is 80 μm. The oil film height is 10 μm.

### 2.2. The Flow Equation of Upper-Convected Maxwell Viscous-Elastic Fluid

The viscous-elastic fluid has complicated rheological characteristics. After the practical research, it is suitable for upper-convected Maxwell constitutive equation to describe the rheological characteristics of the viscous-elastic fluid. And the continuity equation, motion equation and constitutive equation are as follows:

Continuity equation:

$$\frac{\partial u}{\partial x}+\frac{\partial v}{\partial y}=0 \tag{1}$$

where: $u$, $v$ is the velocity in the $x$, $y$ direction respectively, m/s;

Motion equation:

$$Re\left(u\frac{\partial u}{\partial x}+v\frac{\partial u}{\partial y}\right)=-\frac{1}{\rho}\frac{\partial p}{\partial x}+\frac{\partial T^{xx}}{\partial x}+\frac{\partial T^{xy}}{\partial y} \tag{2}$$

$$Re\left(u\frac{\partial v}{\partial x}+v\frac{\partial v}{\partial y}\right)=-\frac{1}{\rho}\frac{\partial p}{\partial x}+\frac{\partial T^{xy}}{\partial x}+\frac{\partial T^{yy}}{\partial y} \tag{3}$$

Upper-convected Maxwell constitutive equation [8]:

$$T^{ik}+\lambda \overset{\nabla}{T^{ik}}=\eta_0 A^{ik} \tag{4}$$

where: $\overset{\nabla}{T^{ik}}$ is the upper-convected derivative, the expression is

$$\overset{\nabla}{T^{ik}}=\frac{\partial T^{ik}}{\partial t}+v^m\frac{\partial T^{ik}}{\partial x^m}-T^{mk}\frac{\partial v^i}{\partial x^m}-T^{im}\frac{\partial v^k}{\partial x^m} \tag{5}$$

where: $A$ is the First Rivlin-Ericksen deformation tensor, the expression is

$$A^{ik}=v^{i,j}+v^{j,i} \tag{6}$$

Constitutive equation:

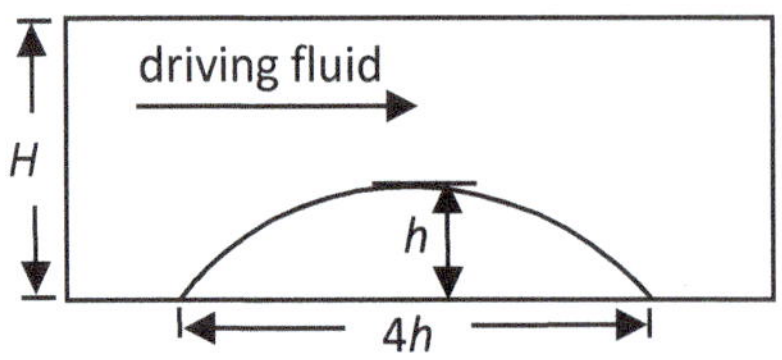


**Figure 1.** Calculation model.

$$T^{xx} + We\left(u\frac{\partial T^{xx}}{\partial x} + v\frac{\partial T^{xx}}{\partial y} - 2T^{xx}\frac{\partial u}{\partial x} - 2T^{xy}\frac{\partial u}{\partial y}\right) = 2\frac{\partial u}{\partial x} \tag{7}$$

$$T^{xy} + We\left(u\frac{\partial T^{xy}}{\partial x} + v\frac{\partial T^{xy}}{\partial y} - T^{xy}(\frac{\partial u}{\partial x} + \frac{\partial v}{\partial y}) - T^{yy}\frac{\partial u}{\partial y} - T^{xx}\frac{\partial v}{\partial x}\right) = \frac{\partial u}{\partial y} + \frac{\partial v}{\partial x} \tag{8}$$

$$T^{yy} + We\left(u\frac{\partial T^{yy}}{\partial x} + v\frac{\partial T^{yy}}{\partial y} - 2T^{xy}\frac{\partial v}{\partial x} - 2T^{yy}\frac{\partial v}{\partial y}\right) = 2\frac{\partial v}{\partial y} \tag{9}$$

Make the motion equation and constitutive equation dimensionless and introduce the dimensionless parameters as follows: take $U$ (the average velocity of inlet) as the characteristic velocity; take $h$ (the width of inlet) as the characteristic length.

$$x_* = \frac{x}{h},\ y_* = \frac{y}{h},\ u_* = \frac{u}{U},\ v_* = \frac{v}{U},\ T_*^{xx} = \frac{hT^{xx}}{\eta U},\ T_*^{xy} = \frac{hT^{xy}}{\eta U}$$

$$T_*^{yy} = \frac{hT^{yy}}{\eta U},\ p_* = \frac{hp}{\eta U},\ Re = \frac{\rho U h}{\eta},\ We = \lambda\frac{U}{h}$$

For the sake of simplicity, "$_*$" in the lower right corner is omitted.

### 2.3. Boundary Condition

1) The solid boundaries are all in the no-slip condition. And the boundary condition of wall is

$$u = v = \frac{\partial u}{\partial y} = \frac{\partial u}{\partial x} = \frac{\partial v}{\partial x} = \frac{\partial v}{\partial y} = 0 \tag{10}$$

2) The inlet and outlet are in the same pressure difference. The pressure gradient is 0.02 MPa/m. And the boundary condition of the inlet is

$$v = \frac{\partial u}{\partial x} = \frac{\partial v}{\partial x} = \frac{\partial v}{\partial y} = 0, \frac{\partial T^{xx}}{\partial x} = \frac{\partial T^{xy}}{\partial x} = \frac{\partial T^{yy}}{\partial x} = 0 \tag{11}$$

The boundary condition of outlet is: normal stress is 0, tangential velocity is 0.

## 3. Calculation Result

The oil micro pore of different permeability is calculated with the change of *We*, the variation curves of the horizontal stress difference are all shown in **Figure 2**.

From **Figure 2**, with the increase of *We*, the horizontal stress difference increases; And with the increase of the flow channel width, the horizontal stress difference increases, too. Because of the limitation of the calculation condition, the maximum value of *We* is 0.6. In the practical reservoir, the *We* of polymer solution is greater than 0.6. In order to reflect the affection to horizontal stress difference of *We* in the real situation, the curves of the maximum of horizontal stress difference are shown in **Figure 3**.

Three conclusions can be got from **Figure 3**. First, the wider the flow channel (the higher the permeability) is, the greater the maximum of displacing force of residual oil is, vice versa. Second, the greater the *We* is, the greater the maximum curve slope of horizontal stress difference is. Namely, with the increasing of *We*, the increasing of horizontal stress difference maximum is faster. Third, in order to obtain the same force of horizontal displacement force, *We* is large when the flow channel is 20 μm. *We* is in the middle value when the flow channel is 40 μm. And it is lower when the flow channel is 60 μm. The third conclusion shows that it is easy to derive that the same displacing effect needs great *We* in low permeability reservoir. Although there are many ways to increasing *We*, but in the limitation of field condition, the most effective way is increasing the polymer concentration.

## 4. Conclusions

1) The wider the flow channel is, the greater the permeability is. The horizontal stress difference acting on

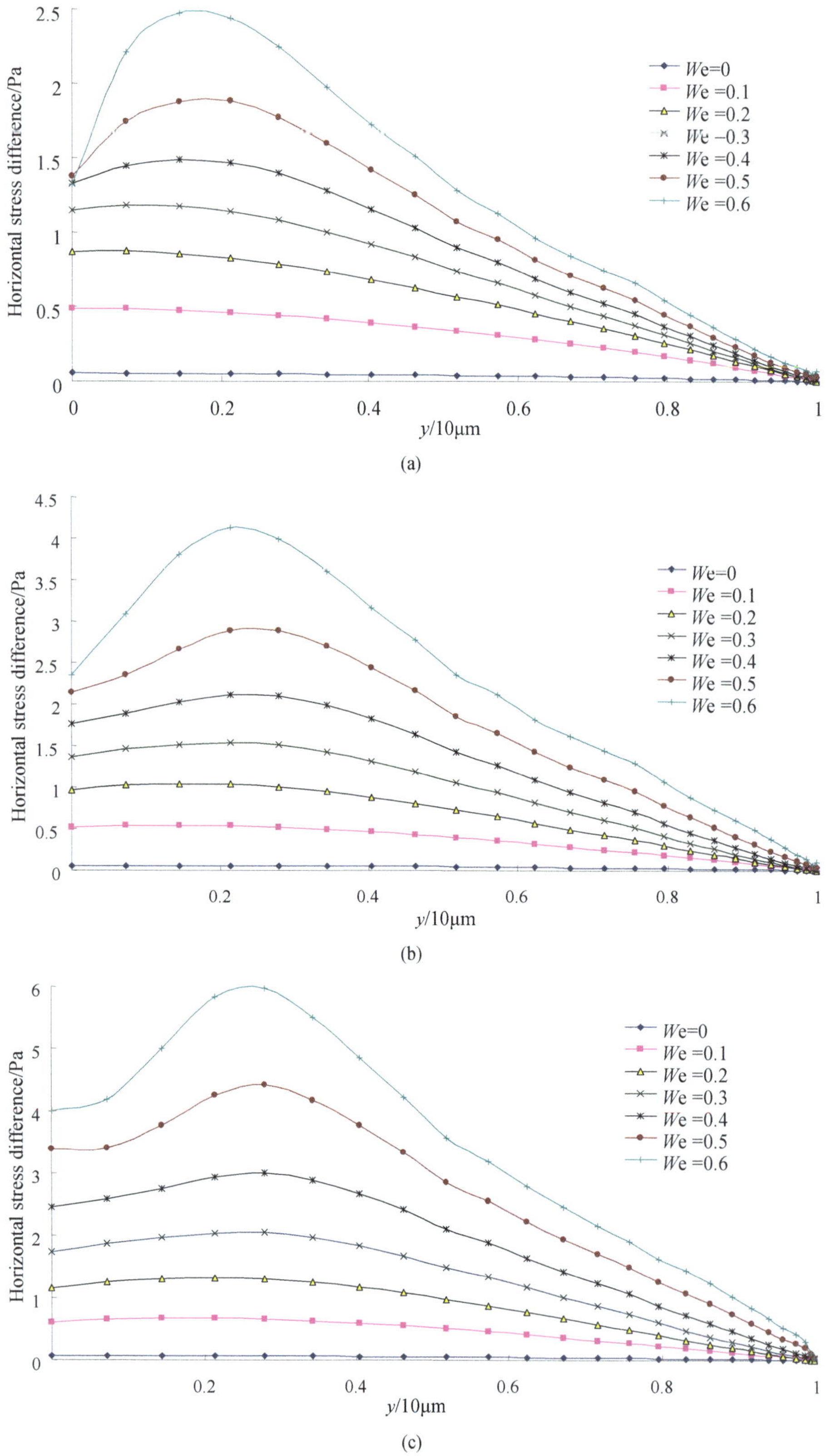


**Figure 2.** The horizontal stress difference of residual oil on the surface of oil wet rock. (a) The flow channel width $H$ = 20 μm; (b) The flow channel width $H$ = 40 μm; (c) The flow channel width $H$ = 60 μm.

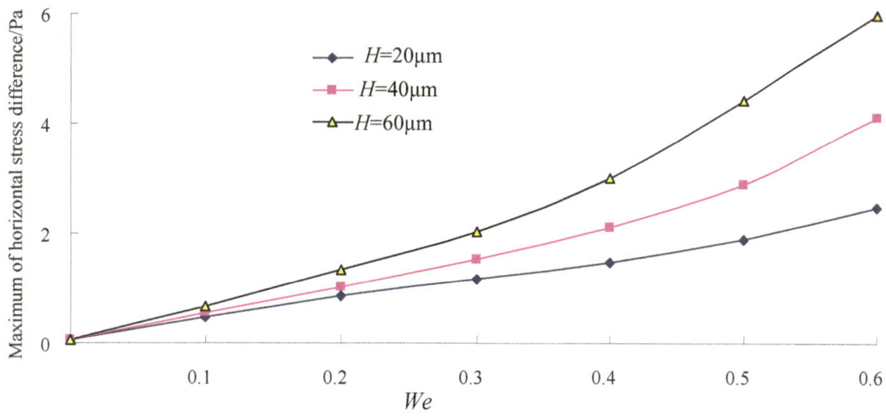


**Figure 3.** The relationship curve of the maximum of horizontal stress difference and *We*.

the residual oil will be greater.

2) The greater *We* (namely, the greater the elasticity of polymer solution) is, the greater the horizontal stress difference acting on the residual oil is.

For class II reservoirs, the concentration of polymer solution can be increased, and the *We* of the polymer fluid will be increased. Then the force of residual oil can be increased. That is in favor of the deformation of residual oil in low permeability reservoir, which lays the basis for the analysis of residual oil deformation and the separation of the small oil drop from the residual oil.

## References


[1] Yin, H.J., Wang, D.M., Zhong, H.Y., *et al.* (2012) Flow Characteristics of Viscous-Elastic Polymer. *Solution in Micro-Pores*, 1-9.

[2] Yin, H.J., Wang, D.M. and Zhong, H.Y., *et al.* (2006) Study on Flow Behaviors of Viscous-Elastic Polymer Solution in Micro Pore with Dead End. SPE 101950.

[3] Zhang, L.J., Yue, X.G., Liu, Z.C. and Hou, J.R. (2002) Flow Field of Viscous-Elastic Fluid in Pores with Dead Ends. *Journal of Hydrodynamics*, *Ser A*, **17**, 748-755.

[4] Wang, D.M., Cheng, J.C., Yang, Q.Y., Gong, W.C., Li, Q. and Chen, F.M. (2000) Viscous-Elastic Polymer Can Increase Displacement Efficiency in Cores. *SPE Annual Technical Conference and Exhibition*, Dallas, 1-4 October 2000.

[5] Wang, D.M., Wang, G., Wu, W.X., *et al.* (2008) Influence of the Micro-Force Produced by Viscous-Elastic Displacement Liquid on Displacement Efficiency. *Journal of Xi'an Shiyou University* (*Natural Science Edition*), **23**, 43-55.

[6] Xia, H.F., Ye, J. and Kong, F.S. (2004) Effect of Elastic Behavior of HPAM Solutions on Displacement Efficiency on Mixed Wettability Conditions. *SPE Asia Pacific Oil and Gas Conference and Exhibition*, Perth, 18-20 October 2004.

[7] Xia, H.F., Wang, D.M., Wang, G., *et al.* (2009) Effect of Micro Forces Caused by Driving Fluid with Viscous-Elasticity on Residual Oil in Chemical Flooding. *Journal of China University of Petroleum* (*Natural Science Edition*), **33**, 151-155.

[8] Han, S.F. (2000) Non-Newtonian Fluid Constitutive Equation and Calculative Analytic Theory. Science Press, Beijing, 102.

# 10

# 3D Thermo-Fluid Dynamic Simulations of High-Speed-Extruded Starch Based Products

**Alessandro Cubeddu[1,2], Cornelia Rauh[1,2], Antonio Delgado[1]**

[1]Institute of Fluid Mechanics, Technical Faculty, University of Erlangen-Nuremberg, Erlangen, Germany
[2]Department of Food Biotechnology and Food Process Engineering, Faculty III—Process Science and Engineering, Berlin Institute of Technology, Berlin, Germany
Email: alessandro.cubeddu@tu-berlin.de

## Abstract

**This paper aims to investigate a method to perform non-isothermal flow simulations in a complex geometry for generalised Newtonian fluids. For this purpose, 3D numerical simulations of starch based products are performed. The geometry of a co-rotating twin-screw extruder is considered. Process conditions concern high rotational speed (up to 1800 rpm), different flow rates (30, 40 and 60 kg/h) and water contents (22% and 36%), for a total of 54 simulations. To cope with the geometry complexity a Mesh Superposition Technique (MST) was adopted. The pseudoplastic behaviour of the fluid is taken into account by considering viscosity as function of shear rate (Ostwald-de Waele relationship) and temperature (Arrhenius law). Simulated temperature variations are compared with measurements at same process conditions for validation. Qualitative behaviour of temperature T and shear stress τ along the screw are analysed and comparisons of different process conditions are presented. By these simulations a database is formed to develop a process control strategy for novel extruder operating points in food technology.**

## Keywords

**High Speed Extrusion; Non-Isothermal Numerical Simulation; Ostwald-De Waele Model**

## 1. Introduction

Since their massive development in the first half of 20th century, extruders have been experiencing a non-stop growth in all industrial fields, not only in plastic industry but also in food processing. The need to develop new products and increase the existent production encourages manufacturers to develop new geometrical screw configurations and to promote research to better understand the complex nature of extrusion processes. Especially in the area of breakfast cereals and convenience food many works have been carried out in order to explain the rheological properties of starch based extrudates. Senouci and Smith [1] as well as Xie *et al.* [2] characterised shear stress and melt viscosity as function of product temperature, moisture content and melt amylose/amylopectin content and investigated the regression power law parameters. The influence of the specific mechanical energy (SME) on cornmeal viscosity was illustrated by Chang *et al.* [3] while product degradation and expansion mechanisms were widely studied in many other works [4] [5] [6] [7]. An exhaustive review of the state of the art about the rheology of starch polymers and related rheometric techniques can be found in [8]. Engineering aspects of extruder performances from an energy point of view have been also investigated [9] [10] and assessment of twin-screw extrusion in producing snack food [11] as well as effects on product characteristics [12] have been experimentally studied. Some of the most important independent variables in the extrusion process are: preconditioning, screw speed $N$, water content $x_w$, flow rate $\dot{Q}$, barrel temperature $T_W$, die geometry and extruder configuration [13]. All manipulable variables influence in a non-linear way the mechanical and thermal stresses and are responsible for the quality of the final product. Due to the complexity of the extrusion process and high number of design variables, most research has been historically carried out on an experimental basis and only recently, with the growth in computer science, the empirical approach has been supported by numerical contributions. Since the begin of the 1990s theoretical models have been used to simulate the transformation of starchy twin-screw extrudates [14], but only with the advent of the new millennium first attempts have been done to solve the governing equations of fluid dynamics in 3D twin-screw geometries by means of Finite Element Method (FEM). Ishikawa *et al.* [15] modelled six discrete angular positions of the screw geometry and performed a non-isothermal quasi-steady-state simulation of a polypropylene material. Avalosse *et al.* [16] reproduced the results of Ishikawa using a Mesh Superposition Technique (MST). The MST as well as the Mesh Partitioning Technique [17] allows simulating the extrusion process without regenerating the finite element mesh as the screws rotate. Use of MST to perform isothermal fluid dynamic simulations of extruded starch based matrix can be found in [18].

The present work is intended to provide a method to set non-isothermal flow simulations in a complex geometry for generalised Newtonian fluids. The new contribution is the analysis and discussion of the flow behaviour of the studied extrudate comparing a wide range of process conditions, which cover new and extremely high operating points, up to 1800 rpm. For this aim the MST implemented in the POLYFLOW® software is applied to a twin-screw co-rotating geometry and non-isothermal 3D numerical simulations of starch based products are preformed. The analysis of the extrudate flow behaviour is carried out from a thermal and kinematic point of view for screw speeds ranging from 400 rpm to 1800 rpm. Three different feed rates (30, 40 and 60 kg/h) and two water contents (22% and 36%) are also considered for a total of 54 simulations. The modelled zone is the melt-conveying zone of the extruder, before the die, where the mixture flows as molten monophasic dough. Counter pressure is taken into account. Material parameters were provided by the Department of Food Process Engineering and Department of Applied Mechanics of Karlsruhe Institute of Technology, that performed the experiments and measurements. Validation is made with experimental data for the temperature across the simulated area. Profiles of temperature and shear stress along the simulated area are qualitatively analysed. Quantitative comparisons of all process conditions are also discussed. The performed simulations investigate novel extrusion process conditions in food technology and offer a base of numerical data which can be used to develop a control strategy for optimisation and control of the extrusion process.

## 2. Materials and Methods

### 2.1. Geometry and Simulated Area

In **Figure 1** the extruder cross section is schematically represented. The modelled area is found at the end of the extruder (melt-conveying zone). The advantages of simulating the flow just before the die are a filling degree of 100%, the possibility to take into account the counter pressure and to compare the calculated temperature variations with the measurements. In fact the temperature transducer plugs are situated across the considered elements. The simulated area consists of three screw elements in a row. Frontal and lateral view with dimensions (in mm) of a single screw and the lateral view of the whole simulated domain are illustrated in **Figure 2**.

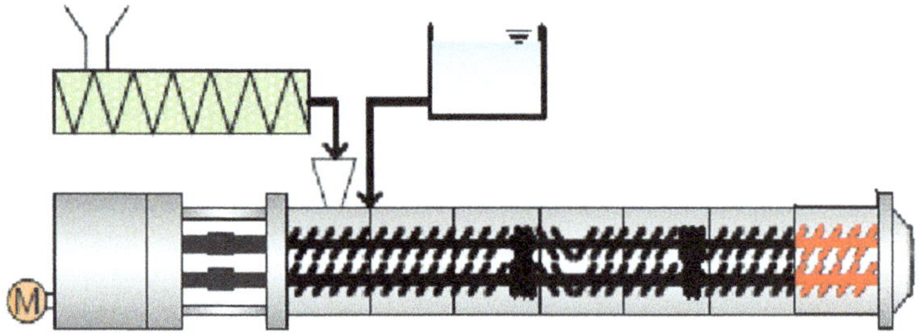

Figure 1. Schematic draw of the extruder and location of the modeled area.

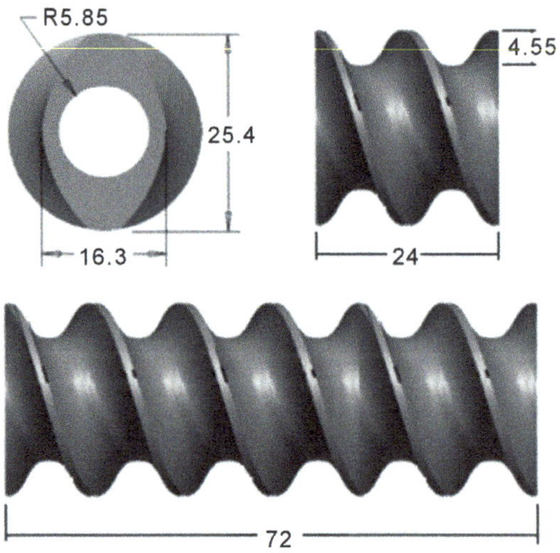


**Figure 2.** Frontal and lateral view of a single screw element and lateral view of the modeled area (3 elements). Dimensions are in mm.

## 2.2. Mesh Superposition Technique

Even in a steady state operating condition (*i.e.* constant screw speed and far away from start-up or shut-down transitories), due to the rotation of the helicoidal screw geometry, the fluid inside the extruder flows at a non-steady periodic condition. The state variables are time dependent with a periodicity depending on the screw speed. Thus, the discretisation of the fluid domain is in principle also time dependent. In a classical approach [19] the mesh is created only on the volume occupied by the fluid, following the geometrical shape of the screw surface at different angles. As the screws rotate, the mesh must be regenerated to take into account the new relative position between the screw volume and the fluid domain. In this study the use of MST allows avoiding a re-meshing algorithm. The screw meshes and the fluid mesh are independently generated with the mesh generator program ICEM CFD®. The fluid mesh is made up of hexahedra bricks for a total of 25,230 elements. The screws are discretised with tetrahedral, for a total of 21,846 elements. The created meshes are then merged using the tool POLYFUSE, which allows to combine multiple meshes. In the resulting merged mesh some portions of the fluid overlap the screw cells (**Figure 3**). As the screw rotate new portions of the fluid domain will overlap partially or totally the screws. Making use of the MST requires a modification of the fundamental equations of fluid dynamics.

For an incompressible fluid, mass (Equation (1)), momentum (Equation (2)) and energy (Equation (3)) conservation equations can be written as follows[1] [20]:

$$\nabla \cdot v + \frac{\beta}{\eta} \nabla^2 P = 0 \tag{1}$$

where $v$ is the velocity, $\beta$ is the relative compression factor, artificially added to guarantee the mass

[1]Variables without subscription are referred to the fluid. The subscription *el* refers to the solid moving elements (screws).

**Figure 3.** Domain discretisation and mesh superposition.

conservation and avoid peaks of pressure near to the moving elements, $\eta$ is the shear viscosity and $P$ the pressure.

$$H(\boldsymbol{v}-\boldsymbol{v}_{el})+(1-H)\left(\rho\frac{\partial \boldsymbol{v}}{\partial t}+\rho\boldsymbol{v}\cdot(\nabla\boldsymbol{v})+\nabla P-\nabla\cdot\vec{\vec{\tau}}+\rho\boldsymbol{g}\right)=0 \tag{2}$$

where $H \in \{0,1\}$ is a step function introduced to distinguish if a fluid mesh node is inside or outside the screw subdomain. If H = 0 then the node is considered part of the fluid, else it is considered to be inside of the moving parts and takes all physical properties of the screws. The parameter $\rho$ is the fluid density, $\vec{\vec{\tau}}$ is shear stress tensor and $\boldsymbol{g}$ is the acceleration of volume forces,

$$H\left[\rho_{el}c_{p_{el}}\frac{DT}{Dt}-\nabla(k_{el}\nabla T)\right]+(1-H)\left[\rho c_p\frac{DT}{Dt}-\vec{\vec{\tau}}:\nabla\boldsymbol{v}-\nabla\cdot(k\nabla T)\right]=0 \tag{3}$$

where $c_p$ is the specific heat capacity, $T$ is the temperature and $k$ is the thermal conductivity. $\frac{D}{Dt}$ is the substantial derivative.

## 2.3. Kinematic and Thermal Boundary Conditions

The kinematic and thermal boundary conditions are shown in **Figure 4**. Rotation of the screws and main flow direction are indicated by the arrows. Flow rate $\dot{Q}$ is imposed at the inlet domain. Simulations at 30, 40 and 60 kg/h have been performed. The chosen values of flow rate guarantee a filling degree of 100% in the considered area. Temperature at the inlet is in principle different for each process condition. Indeed, the inlet temperature decreases with the increase in water content and screw speed. However, for simplicity, we used two constant values of the inlet temperature: 418 K for all simulations at $x_w = 22\%$ and 373 K for all simulations at $x_w = 36\%$. These values are experimental averages for each of the considered water contents, respectively.

On the outlet of the domain a counter pressure is imposed. The values of the counter pressure have been measured at the Karlsruhe Institute of Technology for each considered operating point. The screws are insulated solids which rotate at constant angular velocity. Screw speed varies from 400 rpm to 1800 rpm. A non-slip condition is imposed on the screw surface. On the barrel, there is a non-slip condition and a constant temperature of 373 K. The Newton´s law of cooling has been used on the boundary layer of the internal barrel surface in order to take into account convective effects:

$$-k\frac{\partial T}{\partial r}=h(T-T_W) \tag{4}$$

where $r$ is the radial direction and $h$ is the heat transfer coefficient. Common value of $h$ for starch based products is 500 W/(m$^2$·K) [21].

## 2.4. Material Parameters

The fluid is a non-Newtonian shear thinning fluid. It means that the higher the shear rate is, the smaller is the viscosity. A pure viscous model can be written as:

$$\eta = f(\dot{\gamma},T) \tag{5}$$

where $\dot{\gamma}$ is the shear rate. The model used in this study is the Ostwald-de Waele relationship:

$$\eta(\dot{\gamma},T)=K(T)\dot{\gamma}^{n-1} \tag{6}$$

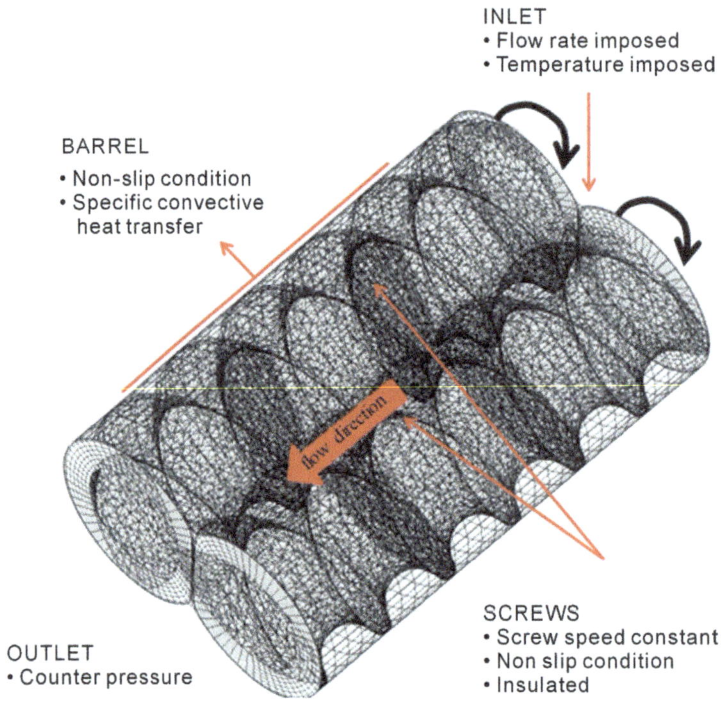


**Figure 4.** Kinematic and thermal boundary conditions.

where $K(T)$ is the Arrhenius law and $n$ is the flow index, representing the slope of the viscosity curve in a ln-ln diagram. It takes a value <1 in case of a shear thinning fluid like the one considered in this study. The Arrhenius law can be written as follows:

$$K(T) = Ae^{\alpha\left(\frac{1}{T}-\frac{1}{T_W}\right)} \tag{7}$$

where $A$ is the consistency related to the magnitude of the average viscosity. It represents the apparent viscosity at unit shear rate. The parameter $\alpha$ is the ratio of the activation energy to the universal gas constant. In this case $\alpha$ takes the value of 2407 K, as the measured average activation energy is 20 kJ/mol. The complete equation for the viscosity can be written as follows:

$$\eta(\dot{\gamma},T) = Ae^{\alpha\left(\frac{1}{T}-\frac{1}{T_W}\right)}\dot{\gamma}^{n-1} \tag{8}$$

The flow function $\eta(\dot{\gamma},T)$ is univocal for each process condition. The values of the consistency $A$ and flow index $n$ depend strongly on the process conditions. In general $A$ decreases when screw speed $N$ or water content $x_w$ are increased, whereas $n$ has the opposite behaviour.

The fluid density $\rho$ is constant during the extrusion process and takes the value of 1351 kg/m$^3$ for $x_w = 22\%$ and 1288 kg/m$^3$ for $x_w = 36\%$.

Inertial terms and acceleration of gravity have been neglected: the inertial terms because the Reynolds number is less than one, due to the high fluid viscosity and small dimensions of the extruder section, and the acceleration of gravity because it does not play a relevant role in the extrusion process which takes place in a horizontal plane.

The thermal conductivity of the fluid $k$ is considered constant with a value of 0.6 W/(m·K) [22]. The fluid specific heat capacity $c_p$ has a constant value of 2300 J/(kg·K) [23].

## 2.5. Numerical Solution

Due to the high number of finite volumes and the complexity of the numerical problem, the matrices of the coefficients are very large sparse matrices whose factorisation requires high computational resources. For that reason a domain parallelisation was adopted, in order to reduce the computational load on the CPU. Therefore a mesh decomposition was made by using the Chaco decomposition method. A number of 512 sub-parts was specified to partition the whole domain in smaller and simpler to handle sub-domains. The resulting sub-matrices are then solved and reassembled on a parallel basis making use of a High Performance Computing (HPC) technology. To take advantage of the HPC parallelisation capabilities, a multifrontal method [24] has been adopted as solver. To avoid convergence problems due to the dependence of the viscosity on both shear rate and temperature, two simulations for each process condition were performed. In the first simulation only the viscosity dependence on shear rate is considered, while $K(T)=1$. Momentum and energy equations are hence decoupled. In the second simulation the results of the first simulation are used as initial conditions and only the viscosity dependence on temperature is taken into account. In this case momentum and energy are coupled and therefore an iterative one-way coupled solution is adopted. The one-way method ensures convergence of the solution even if at a convergence rate generally lower than a full coupled method. Convergence difficulties can be caused also by the high value of Péclet number. It is defined as the ratio of the advection heat transfer to the conduction heat transfer:

$$Pe = \frac{\rho c_p d v}{k} \tag{9}$$

where $d$ is a characteristic dimension of the geometry (in this case the depth of the channel, $d = 4.55$ mm) and $v$ is the modulus of the flow velocity. In polymer extrusion the advection heat transfer term is much higher than the conduction heat transfer. The Péclet number is of the order of $10^3$ to $10^5$ [25]. In this study it varies from $3 \times 10^3$ to $3 \times 10^4$, depending on the process conditions. A high Péclet number means high non-linearity of the energy equation and causes instability of the numerical solution. To solve this problem the Péclet number was varied iteratively during the second step simulation from a very small value $(\ll 1)$ to its actual value. Another dimensionless number common in extrusion science is the Brinkman number. It is defined as the ratio of viscous heat dissipation to heat conduction transfer through the barrel:

$$Br = \frac{\eta v^2}{k\left|T_W - T_{ref}\right|} \tag{10}$$

where $T_{ref}$ is a reference temperature, e.g. the bulk temperature of the fluid. High values of Brinkman number are very common in the extrusion process, due to the high viscosity dissipation. This is a cause of numerical instability. In this study, the Brinkman number ranges from 10 to 30 and therefore also for this number an iterative evolution (from a very small value to its actual value) was adopted.

## 3. Results and Discussion

In **Table 1** are summarised the Ostwald-de Waele parameters $A$ and $n$ of 8 numerical simulations used to validate the temperature field. A constant flow rate of 40 kg/h has been used. Of the 8 validation points, the first 4 are referred to a fluid with 22% of water content and the remaining 4 points to a fluid with 36% of water content. The comparison between measured $\Delta T$ and simulated $\Delta T$ across the modelled domain are diagrammed in **Figure 5**. Simulation results show a good agreement with the measured values for both considered water contents. The source of error is due to the simplifications adopted in the simulations. For instance, the inlet temperature is considered function of the only water content because this variable influences considerably the temperature at the inlet. The influence of the screw speed on the inlet temperature is neglected. Furthermore, thermal parameters like specific heat capacity, thermal conductivity and convective heat transfer coefficient are kept constant, for simplicity. However, the simulations can reproduce accurately the temperature variation trend, demonstrating the suitability of the Ostwald-de Waele model to reproduce the extrudate temperature.

### 3.1. State Variables along the Screw

It is interesting to focus the attention on one simulation ($N = 800$ rpm, $x_w = 36\%$, $\dot{Q} = 40$ kg/h) and have a look at the qualitative variation of the temperature and shear stress in the simulated area. In **Figure 6** it is represented the temperature contour over two revolution surfaces located at 11 mm of radius from the rotational axes and over the plane of symmetry of the extruder. The screws rotate anticlockwise and the fluid is pumped in z-direction. The temperature contour shows an increasing in temperature in z-direction from the

**Table 1.** Ostwald-de Waele parameters $A$ and $n$ used to validate the model.

| $\dot{Q}$ (kg/h) | $x_w$ (%) | $N$ (rpm) | $A$ (Pa·s$^n$) | $n$ |
|---|---|---|---|---|
| 40 | 22 | 600 | 5757 | 0.404 |
| 40 | 22 | 1000 | 3058 | 0.465 |
| 40 | 22 | 1400 | 1635 | 0.556 |
| 40 | 22 | 1800 | 809 | 0.656 |
| 40 | 36 | 600 | 1570 | 0.398 |
| 40 | 36 | 1000 | 850 | 0.532 |
| 40 | 36 | 1400 | 436 | 0.648 |
| 40 | 36 | 1800 | 197 | 0.763 |

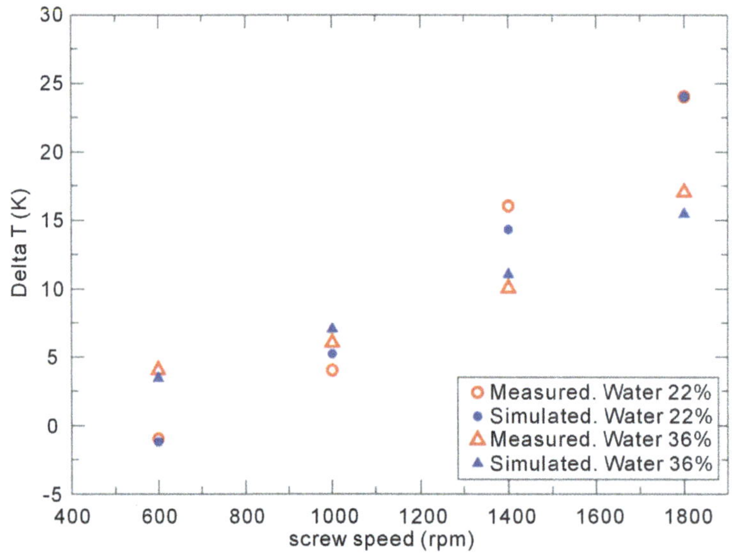


**Figure 5.** Comparison between simulated and measured values of temperature.

**Figure 6.** Temperature contour in the simulated area.

initial value of 373 K to its final value at the outlet of the domain. The viscous dissipation is the cause of the temperature increase. As the temperature increases, the viscosity decreases (shear thinning behaviour) resulting in a reduction of viscous dissipation and then in a lower increase of the temperature. The equilibrium is reached when the viscous heating is balanced by the heat flux through the barrel. The dissipative effects are homogeneously distributed and the temperature profile shows an almost linear increasing trend along the screw (**Figure 7**). The shear stress contour is represented in **Figure 8**. Highest values of shear stress are found in the intermeshing zone (between the two screws) and especially close to the barrel. Here the stresses

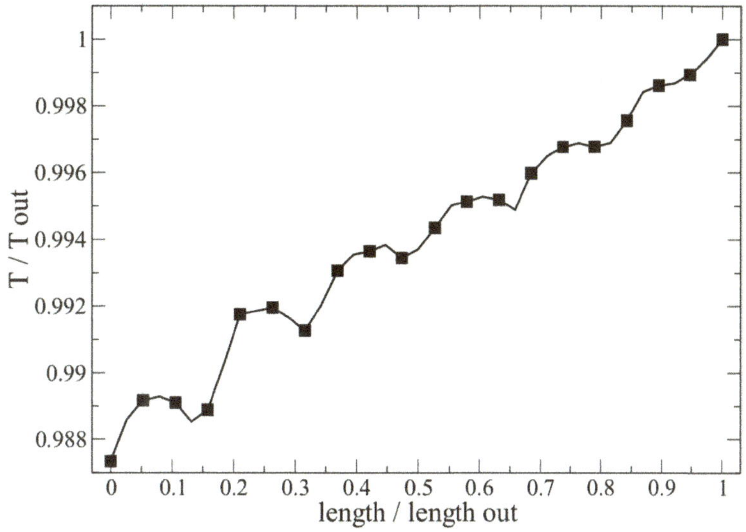


**Figure 7.** Normalised temperature variation along the screw.

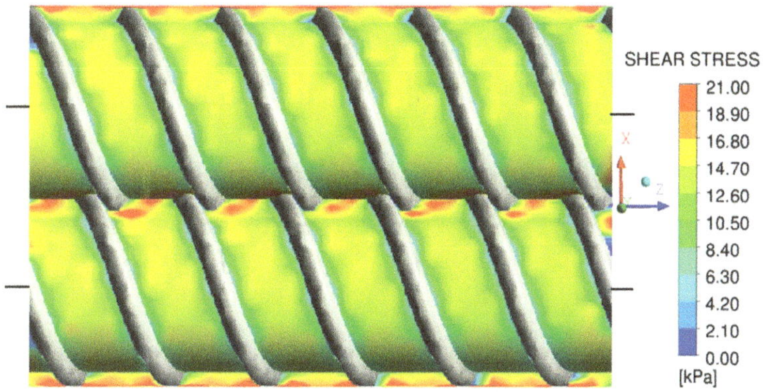


**Figure 8.** Shear stress contour in the simulated area.

concentrate because of the high relative speed between screws (intermeshing zone) and between screws and barrel (close to the barrel wall). In z-direction the shear stress presents a periodic behaviour. This can be seen in **Figure 9** that shows the normalised shear stress variation in a line parallel to the rotation axes. In the middle of each channel where the velocity gradient is higher, the shear stress has a higher value in comparison with its value near to the flights.

## 3.2. Comparison of Different Process Conditions

The fluid dynamic variables $\Delta T$ and average shear stress in the simulated area in function of screw speed are diagrammed in figures 10 to 13. **Figure 10** and **Figure 11** show a comparison of the temperature variations for the three flow rates considered in this study, *i.e.* 30, 40 and 60 kg/h. The water content of the dough is 22% and 36% for **Figure 10** and **Figure 11**, respectively. For all curves there is an increase of $\Delta T$ with the screw speed. This is due to the higher mechanical energy transferred to the fluid at higher screw speed. The curves show in general a reduction of the $\Delta T$ while the flow rate is increasing. In fact, the higher the flow rate is, the lower is the permanence time of the dough inside the pumping area and then the temperature increase. Comparing **Figure 10** with **Figure 11**, it is to be noted that, fixing the other process conditions, the temperature increase is smaller for dough with higher percentage of water content in comparison with a fluid with a lower water content. This is the consequence of the smaller average viscosity of a fluid with higher water content and therefore a minor viscous dissipation. The curve at 60 kg/h shows a smaller slope in respect to the curves at lower flow rates. It may be caused from the fact that a high reduction of the permanence time of the dough inside the pumping area makes the fluid less sensitive to the other process conditions.

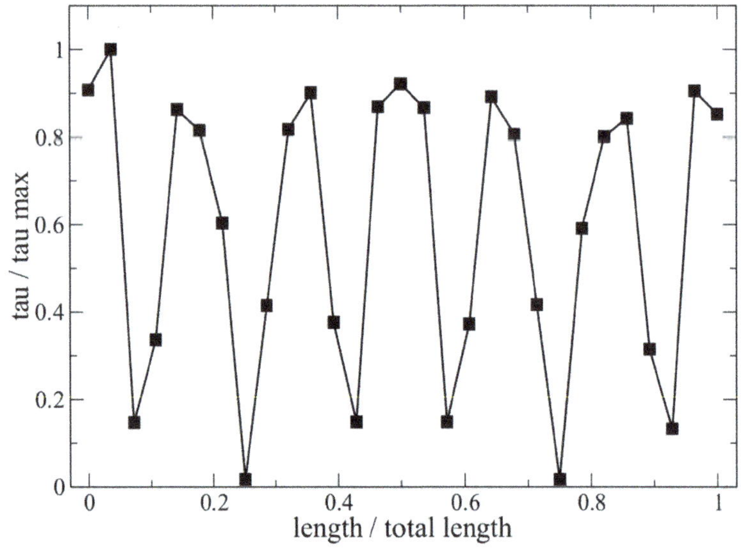


**Figure 9.** Normalised shear stress along the screw.

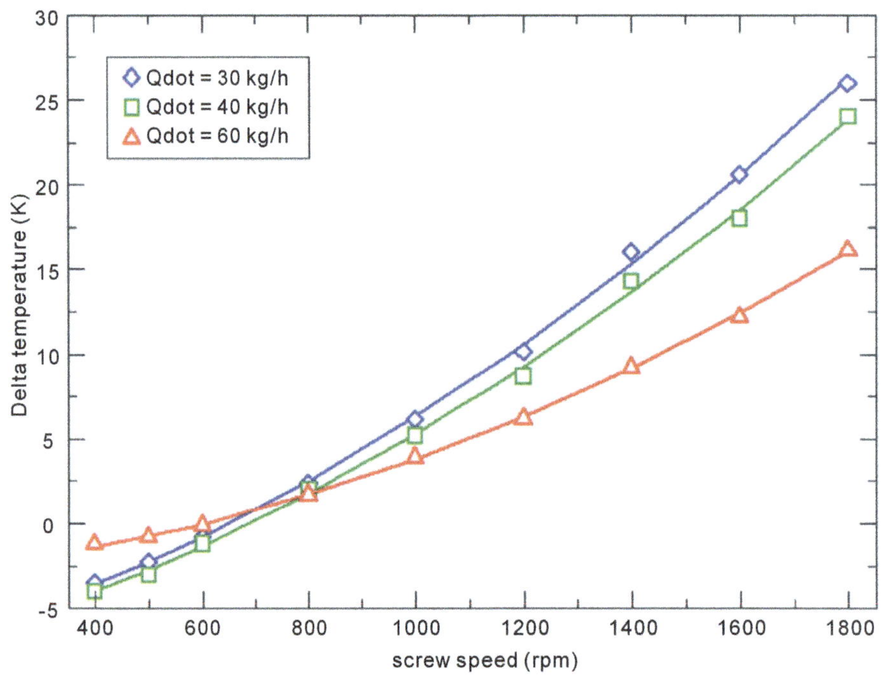


**Figure 10.** $\Delta T$ in function of screw speed for three different flow rates. Water content is 22%.

Therefore an increase in screw speed results in a smaller percentage of temperature increase, which causes a flattening of the curve.

In **Figure 12** and **Figure 13** the average shear stresses at different process conditions are compared for an extrudate with 22% and 36% of water content respectively. In case of a dough with medium water content (**Figure 12**), the shear stress decreases while the screw speed is increasing, whereas in case of a dough with high water content (**Figure 13**) the shear stress increases while the screw speed is increasing. This can be explained with a different behaviour of the viscous fluid at different water contents. Increasing the water content results in fact in a more Newtonian behaviour of the fluid. Therefore, the effects of mechanical stress on the fluid, caused by the increasing in screw speed, overtakes the effects of the viscosity reduction due to the shear thinning nature of the fluid. Thus the shear stress increases with the screw speed. At medium and

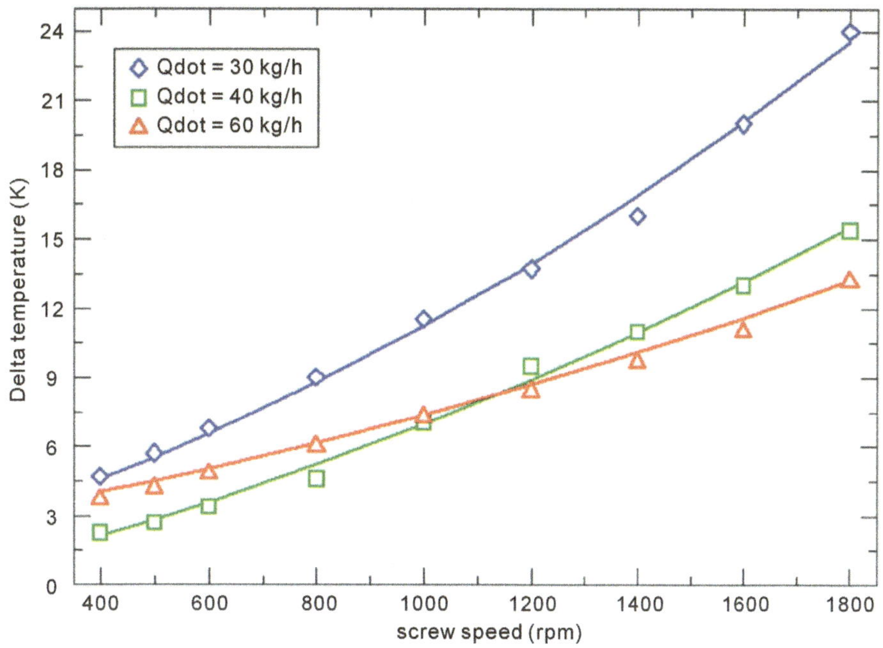


**Figure 11.** $\Delta T$ in function of screw speed for three different flow rates. Water content is 36%.

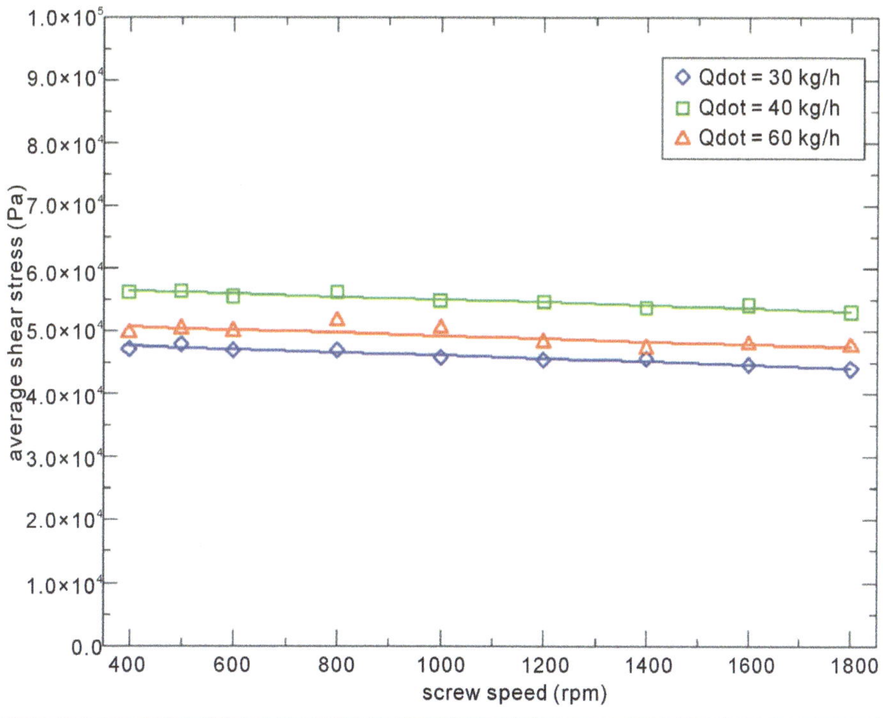


**Figure 12.** Average shear stress in function of screw speed for three different flow rates. Water content is 22%.

low water contents the shear thinning behaviour of the polymer is vice versa more relevant and the shear stress reduces while the screw speed is increasing. The influence of the flow rate on the shear stress seems not to be very relevant. The absolute values of shear stresses remain of the order of 50 kPa and 20 kPa for a fluid with 22% and 36% of water content respectively. The higher absolute values for the extrudate with at 22% of water content is due to its higher average viscosity in comparison with a fluid at 36% of water content.

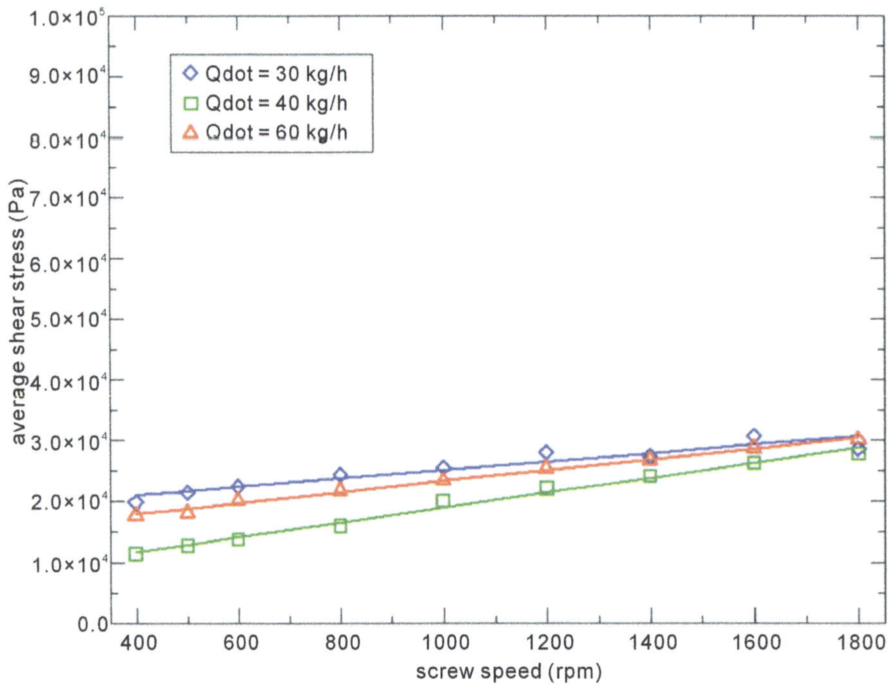


**Figure 13.** Average shear stress in function of screw speed for three different flow rates. Water content is 36%.

## 4. Conclusion

3D thermo fluid dynamic simulations of starch based extruded products have been performed. 54 numerical simulations have been carried out for a wide range of process conditions considering novel food properties, different flow rates and very high screw speeds. The fluid has been modelled with the Ostwald-de Waele relationship and the Arrhenius law. The study shows that the chosen viscous model coupled with the temperature dependence of the viscosity is able to describe the behaviour of the studied dough inside the pumping area of an industrial extruder. Temperature variation as well as shear stress is discussed for a fixed operating point in the simulated area and along the pumping direction. Comparison of the simulated variables is also presented for all studied process conditions and interpretations are proposed. Results of the simulations can be used in the industry and for design of a process control strategy which allows to optimise the process of extrusion and to improve the product quality.

## Acknowledgements

This research was supported by the German Ministry of Economics and Technology (via AiF) and the FEI (Forschungskreis der Ernährungsindustrie e.V., Bonn) Project AiF 332 ZN. Thanks to the company Coperion GmbH for providing the screw geometry.

## References

[1] Senouci, A. and Smith, A.C. (1988) An Experimental Study of Food Melt Rheology. I. Shear Viscosity Using a Slit Die Viscometer and Capillary Rheometer. *Acta Rheologica*, **27**, 546-554. http://dx.doi.org/10.1007/BF01329355

[2] Xie, F., Yu, L., Su, B., Liu, P., Wang, J. and Chen, L. (2009) Rheological Properties of Starches with Different Amylose/Amylopectin Ratios. *Journal of Cereal Science*, **49**, 371-377. http://dx.doi.org/10.1016/j.jcs.2009.01.002

[3] Chang, Y.K., Martinez-Bustos, F., Park, T.S. and Kokini, J.L. (1999) The Influence of Specific Mechanical Energy on Cornmeal Viscosity Measured by an On-Line System during Twin-Screw Extrusion. *Brazilian Journal of Chemical Engineering*, **3**, 285-295.

[4] Willet, J.L., Millard, M.M. and Jasberg, B.K. (1997) Extrusion of Waxy Maize Starch: Melt Rheology and Molecular Weight Degradation of Amylopectin. *Polymer*, **38**, 5983-5989. http://dx.doi.org/10.1016/S0032-3861(97)00155-9

[5] Della Valle, G., Boché, Y., Colonna, P. and Vergnes, B. (1995) The Extrusion Behaviour of Potato Starch. *Carbohydrate Polymers*, **28**, 255-264. http://dx.doi.org/10.1016/0144-8617(95)00111-5

[6] Della Valle, G., Vergnes, B., Colonna, P. and Patria, A. (1997) Relations between Rheological Properties of Molten Starches and their Expansion Behaviour in Extrusion. *Journal of Food Engineering*, **31**, 277-296. http://dx.doi.org/10.1016/S0260-8774(96)00080-5

[7] Launay, B. and Lisch, J.M. (1983) Twin-Screw Extrusion Cooking of Starches: Flow Behaviour of Starch Pastes, Expansion and Mechanical Properties of Extrudates. *Journal of Food Engineering*, **2**, 259-280. http://dx.doi.org/10.1016/0260-8774(83)90015-8

[8] Xie, F., Halley, P.J. and Avérous, L. (2012) Rheology to Understand and Optimize Processibility, Structures and Properties of Starch Polymeric Materials. *Progress in Polymer Science*, **37**, 595-623. http://dx.doi.org/10.1016/j.progpolymsci.2011.07.002

[9] Janssen, L.P.B.M., Mościcki, L. and Mitrus, M. (2002) Energy Aspects in Food Extrusion-Coocking. *International Agrophysics*, **16**, 191-195.

[10] Liang, M., Huff, H.E. and Hsieh, F.H. (2001) Evaluating Energy Consumption and Efficiency of a Twin-Screw Extruder. *Journal of Food Science*, **67**, 1803-1807. http://dx.doi.org/10.1111/j.1365-2621.2002.tb08726.x

[11] Shi, C., Wang, L., Wu, M., Adhikari, B. and Li, L. (2011) Optimization of Twin-Screw Extrusion Process to Produce Okara-Maize Snack Foods Using Response Surface Methodology. *Journal of Food Engineering*, **7**, 1-24.

[12] Guha, M., Ali, S.Z. and Bhattacharya, S. (1997) Twin-Screw Extrusion of Rice Flour without a Die: Effect of Barrel Temperature and Screw Speed on Extrusion and Extrudate Characteristics. *Journal of Food Engineering*, **32**, 251-267. http://dx.doi.org/10.1016/S0260-8774(97)00028-9

[13] Hubner, G.R. (2000) Twin-Screw Extruders. In Riaz, M.N., Ed., *Extruders in Food Applications*, CRC Press, Taylor & Francis Group, Boca Raton, 8-114.

[14] Della Valle, G., Barrès, C., Plewa, J., Tayeb, J. and Vergnes, B. (1993) Computer Simulation of Starchy Products' Transformation by Twin-Screw Extrusion. *Journal of Food Engineering*, **19**, 1-31. http://dx.doi.org/10.1016/0260-8774(93)90059-S

[15] Ishikawa, T., Kihara, S.I. and Funatsu, K. (2000) 3-D Numerical Simulations of Nonisothermal Flow in Co-Rotating Twin Screw Extruders. *Polymer Engineering and Science*, **40**, 357-364. http://dx.doi.org/10.1002/pen.11169

[16] Avalosse, T., Rubin, Y. and Fondin, L. (2002) Non-Isothermal Modeling of Co-Rotating and Contra-Rotating Twin Screw Extruders. *Journal of Reinforced Plastics and Composites*, **21**, 419-429. http://dx.doi.org/10.1177/0731684402021005442

[17] Gupta, M. (2008) Non-Isothermal Simulation of the Flow in Co-Rotating and Counter-Roteting Twin-Screw Extruders using Mesh Partitioning Technique. In: *ANTEC* 2008 *Plastics*: *Annual Technical Conference Proceedings*, ANTEC, Milwaukee, 316-320.

[18] Emin, M.A. and Schuchmann, H.P. (2013) Analysis of the Dispersive Mixing Efficiency in a Twin-Screw Extrusion Processing of Starch Based Matrix. *Journal of Food Engineering*, **115**, 132-143. http://dx.doi.org/10.1016/j.jfoodeng.2012.10.008

[19] Bravo, V.L., Hrymak, A.N. and Wright, J.D. (2000) Numerical Simulation of Pressure and Velocity Profiles in Kneading Elements of a Co-Rotating Twin Screw Extruder. *Polymer Engineering and Science*, **40**, 525-541. http://dx.doi.org/10.1002/pen.11184

[20] (2009) Ansys Polyflow 12.1 User's Guide. http://docsfiles.com/pdf_ansys_polyflow_12_1_user_s_guide.html

[21] Yacu, W.A. (1985) Modeling of a Twin Screw Co-Rotating Extruder. *Journal of Food Engineering*, **8**, 1-21. http://dx.doi.org/10.1111/j.1745-4530.1985.tb00095.x

[22] Sakiyama, T., Han, S., Kincal, N.S. and Yano, T. (1993) Intrinsic Thermal Conductivity of Starch: A Model-Independent Determination. *Journal of Food Science*, **58**, 413-415. http://dx.doi.org/10.1111/j.1365-2621.1993.tb04287.x

[23] Noel, R.T. and Ring, S.G. (1992) A Study of the Heat Capacity of Starch/Water Mixtures. *Carbohydrate Research*, **227**, 203-213. http://dx.doi.org/10.1016/0008-6215(92)85072-8

[24] Liu, J.W.H. (1992) The Multifrontal Method for Sparse Matrix Solution: Theory and Practice. *Society for Industrial and Applied Mathematics*, **34**, 82-109.

[25] Rauwendaal, C. (2001) Polymer Extrusion. Carl Hanser Verlag, Munich.

# 11

# Nonlinear Vortex Structures in Obliquely Rotating Fluid

**Michael Kopp[1,2], Anatoly Tur[3], Vladimir Yanovsky[1,2]**

[1]Institute for Single Crystals, National Academy of Sciences of Ukraine, Kharkov, Ukraine
[2]V.N. Karazin Kharkiv National University 4 Svobody Sq., Kharkov, Ukraine
[3]Université de Toulouse [UPS], CNRS, Institut de Recherche en Astrophysique et Planétologie, Toulouse Cedex, France
Email: anatoly.tour@irap.omp.eu

## Abstract

**In this paper, we find a new large scale instability which appears in obliquely rotating flow with the small scale turbulence, generated by external force with small Reynolds number. The external force has no helicity. The theory is based on the rigorous method of multi-scale asymptotic expansion. Nonlinear equations for instability are obtained in the third order of the perturbation theory. In this article, we explain in detail the nonlinear stage of the instability and we find the nonlinear periodic vortices and the vortex kinks of Beltrami type.**

## Keywords

**Large Scale Vortex Instability, Coriolis Force, Multi-Scale Asymptotic Development, Small Scale Turbulence, Vortex Kinks**

## 1. Introduction

It is well known that the rotating effects play an important role in many theoretical and practical applications for fluid mechanics [1] and are especially important for geophysics and astrophysics [2]-[4] when one has to deal with rotating objects such as the Earth, Jupiter, the Sun, etc. Rotating fluids could generate different wave and vortex motions, for example, gyroscopic waves, Rossbywaves, internal waves, located vortices and coherent vortex structures [4]-[7]. Among the vortex structures, the most interesting are the large scale ones since they carry out the efficient transport of energy and impulse. The structures which have characteristic scale much more than the scale of turbulence or the scale of external force which generates this turbulence are understood as large scale ones. In this paper we find a new large scale instability in obliquely rotating flow which is influenced by the small scale external force with zero helicity. Its axis of rotation does not coincide with the Z axis. This force supports small scale turbulent fluctuations in fluid. The nonlinear large scale helical vortex structures such as Beltrami vortices or localized kinks appear as a result of the development of this instability in rotating fluid. This supposes that the external mall-scale force substitutes the action of small-scale turbulence. Further we consider that the external force acts in the plane (X, Y). Instability occurs only when the vector of angular velocity of rotation $\Omega$ is inclined relatively to the plane (X, Y), as shown in **Figure 1**. If the fluid is rotating around the axis Z strictly, then instability does not occur. The helical 2D velocity field $W_x, W_y$ turns around the axis Z when Z changes in the periodic wave (**Figure 2**) and makes one turn in the kink (**Figure 3**). The found instability belongs to the class of instabilities called hydrodynamic $\alpha$-effects. For these instabilities the positive feedback between velocity components is typical:

$$\partial_T W_x - \Delta W_x - \alpha_y \frac{\partial}{\partial z} W_y = 0,$$

$$\partial_T W_y - \Delta W_y + \alpha_x \frac{\partial}{\partial z} W_x = 0,$$

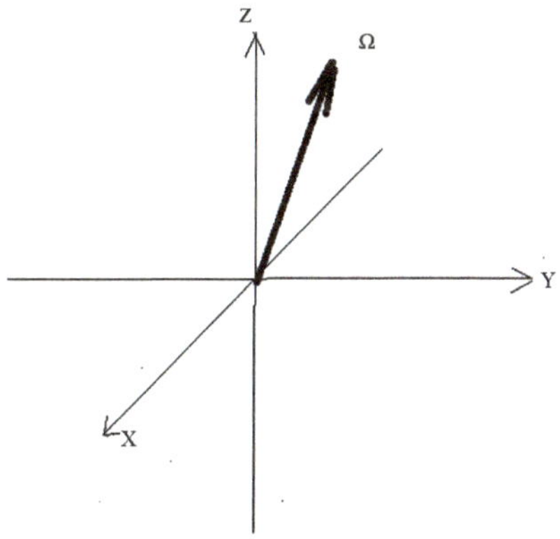


**Figure 1.** In general, the angular velocity $\Omega$ is inclined relatively to the plane (X, Y) in which there is an external force $F_{0\perp}$.

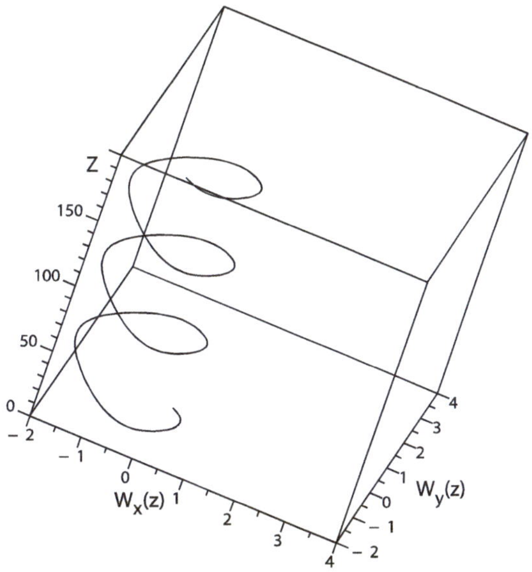


**Figure 2.** Nonlinear helical Beltramiwave, which corresponds to the closed trajectory in the phase plane ($C_1 = 0.1$, $C_2 = 0.1$). The spiral is oriented along Z axis and inclined relatively to the axis of rotation.

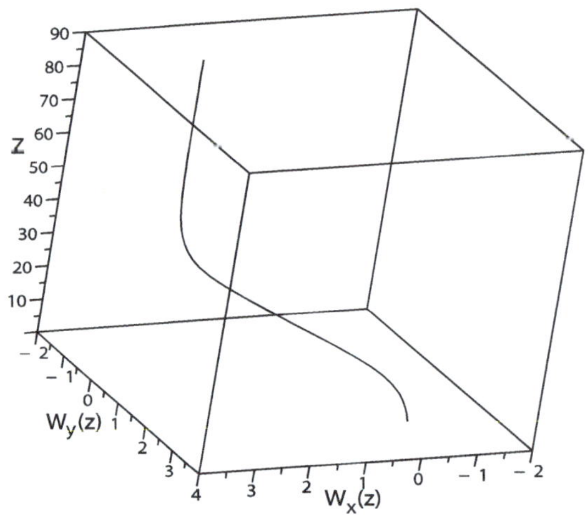


**Figure 3.** Localized solution (kink), which corresponds to the separatrice in the phase plane ($C_1 = 0.1$, $C_2 = 0.1$).

and leads to the instability. $\alpha$-effect origins from magnetic hydrodynamics where it engenders the increase of large scale magnetic fields (see for example, [8]). Later it was extended to ordinary hydrodynamics. Several examples of hydrodynamics $\alpha$-effect [9]-[16] are known for today. From this point of view, in this study we found a new example of the $\alpha$-effect. The theory of this instability is based on a rigorous method of multi-scale development, which was proposed by Frisch, She and Sulem for the theory of the AKA effect [14]. This method allows finding the equations for large scale perturbations in the form of secular equations of the asymptotic theory, to calculate the Reynolds stress tensor and to find the instabilities. The small parameter of asymptotical development is the number of Reynolds $R, R \le 1$. Our paper is organized as follows: in Section 2 we formulate the problem and the main equations in rotating system coordinates; in Section 3 we discuss the concept of multi-scale development and we give the secular equations. In Section 4 we calculate the velocity field of zero approximation. In Section 5 we describe the calculation of the Reynolds stress and find the large scale instability. In Section 6 we discuss the saturation of the instability and find the nonlinear stationary vortex structures. The results obtained are discussed in the conclusions given in Section 7.

## 2. The Main Equations and Formulation of the Problem

Let us examine the equations of motion for non-compressible rotating fluid with the external force $\boldsymbol{F}_0$ in rotating coordinates system:

$$\frac{\partial \boldsymbol{V}}{\partial t} + (\boldsymbol{V}\nabla)\boldsymbol{V} + 2\boldsymbol{\Omega} \times \boldsymbol{V} = -\frac{1}{\rho_0}\nabla P + \nu \Delta \boldsymbol{V} + \boldsymbol{F}_0 \tag{1}$$

$$div\boldsymbol{V} = 0 \tag{2}$$

The external force $\boldsymbol{F}_0$ is divergence-free. Here $\boldsymbol{\Omega}$ is angular velocity of fluid rotation, $\nu$ is viscosity and $\rho_0$ is constant fluid density. Let us designate the characteristic amplitude of force as $f_0$, and its characteristic space and time scale as $\lambda_0$ and $t_0$ respectively.

Then $\boldsymbol{F}_0 = f_0 \boldsymbol{F}_0\left(\frac{\boldsymbol{x}}{\lambda_0}, \frac{t}{t_0}\right)$. We will designate the characteristic amplitude of velocity, generated by external force as $v_0$. Further we choose the dimensionless variables $(t, \boldsymbol{x}, \boldsymbol{V})$:

$$\boldsymbol{x} \to \frac{\boldsymbol{x}}{\lambda_0}, \quad t \to \frac{t}{t_0}, \quad \boldsymbol{V} \to \frac{\boldsymbol{V}}{v_0}, \quad \boldsymbol{F}_0 \to \frac{\boldsymbol{F}_0}{f_0}, \quad P \to \frac{P}{\rho_0 P_0},$$

$$t_0 = \frac{\lambda_0^2}{\nu}, \quad P_0 = \frac{v_0 \nu}{\lambda_0}, \quad f_0 = \frac{v_0 \nu}{\lambda_0^2}, \quad v_0 = \frac{f_0 \lambda_0^2}{\nu}.$$

Then, in dimensionless variables the equation (1) takes the form:

$$\frac{\partial \boldsymbol{V}}{\partial t}+R(\boldsymbol{V}\cdot\nabla)\boldsymbol{V}+\boldsymbol{D}\times\boldsymbol{V}=-\nabla P+\Delta\boldsymbol{V}+\boldsymbol{F}_0, \tag{3}$$

$R=\frac{\lambda_0 v_0}{\nu}, |D|=\sqrt{Ta}$. Where $R$ and $Ta=\frac{4\Omega^2\lambda_0^4}{\nu^2}$ are respectively the Reynolds number and the Taylor number on scale $\lambda_0$. Further we will consider the Reynolds number as small $R\le 1$ and will construct on this small parameter the asymptotical development. Concerning the parameter *D*, we do not choose any range of values for the moment. Let us examine the following formulation of the problem. We consider the external force as being of small scale and of high frequency. This force leads to small scale fluctuations in velocity. After averaging, these rapidly oscillating fluctuations vanish. Nevertheless, due to small nonlinear interactions in some orders of perturbation theory, nonzero terms can occur after averaging. This means that they are not oscillatory, that is to say, they are large scale. From a formal point of view, these terms are secular, *i.e.*, they create the conditions for the solvability of large-scale asymptotic development. So, the purpose of this paper is to find and study the solvability equations, *i.e.*, the equations for the large scale perturbations. Let us denote the small scale variables by $x_0=(\boldsymbol{x}_0,t_0)$, and the large scale ones by $X=(\boldsymbol{X},T)$. The small scale partial derivative operation $\frac{\partial}{\partial x_0^i},\frac{\partial}{\partial t_0}$, and the large scale ones $\frac{\partial}{\partial \boldsymbol{X}},\frac{\partial}{\partial T}$ are written, respectively, as $\partial_i,\partial_t,\nabla_i$ and $\partial_T$. To construct a multi-scale asymptotic development we follow the method which is proposed in [8].

## 3. The Multi-Scale Asymptotic Development

Let us search the solution to equations (2) and (3) in following form:

$$V(\boldsymbol{x},t)=\frac{1}{R}\boldsymbol{W}_{-1}(X)+\boldsymbol{v}_0(x_0)+R\boldsymbol{v}_1+R^2\boldsymbol{v}_2+R^3\boldsymbol{v}_3+\cdots, \tag{4}$$

$$P(\boldsymbol{x},t)=\frac{1}{R^3}P_{-3}(X)+\frac{1}{R^2}P_{-2}(X)+\frac{1}{R}P_{-1}(X)+P_0(x_0)+R\left(P_1+\overline{P}_1(X)\right)+R^2P_2+R^3P_3+\cdots. \tag{5}$$

We introduce the slow variables $\boldsymbol{X}=R^2\boldsymbol{x}_0$ and $T=R^4t_0$ which lead to the following expressions for the spatial and temporal derivatives:

$$\frac{\partial}{\partial x^i}=\partial_i+R^2\nabla_i, \tag{6}$$

$$\frac{\partial}{\partial t}=\partial_t+R^4\partial_T, \tag{7}$$

$$\frac{\partial^2}{\partial x^j\partial x^j}=\partial_{jj}+2R^2\partial_j\nabla_j+R^4\partial_{jj}. \tag{8}$$

Using initial notation, the system of equations can be written as:

$$\begin{aligned}&\left(\partial_t+R^4\partial_T\right)V^i+R\left(\partial_j+R^2\nabla_j\right)\left(V^iV^j\right)+D^j\varepsilon_{ijk}V^k\\&=\left(\partial_j+R^2\nabla_j\right)P+\left(\partial_{jj}+2R^2\partial_j\nabla_j+R^4\nabla_{jj}\right)V^i+F_0^i,\end{aligned} \tag{9}$$

$$\left(\partial_i+R^2\nabla_i\right)V^i=0. \tag{10}$$

Substituting these expressions into the initial equations (2) and (3) and then gathering together the terms of the same order, we obtain the equations of the multi-scale asymptotic development and write down the obtained equations up to order $R^3$ including. In the order $R^{-3}$ there is only one equation:

$$\partial_i P_{-3}=0\Rightarrow P_{-3}=P_{-3}(X). \tag{11}$$

In order $R^{-2}$ we have the equation:

$$\partial_i P_{-2} = 0 \Rightarrow P_{-2} = P_{-2}(X). \tag{12}$$

In order $R^{-1}$ we get a system of equations:

$$\partial_t W_{-1}^i - \partial_{jj} W_{-1}^i + D^j \varepsilon_{ijk} W_{-1}^k = -(\partial_i P_{-1} + \nabla_i P_{-3}) - \partial_j W_{-1}^i W_{-1}^j, \tag{13}$$

$$\partial_i W_{-1}^i = 0. \tag{14}$$

The system of equations (13) and (14) gives secular terms

$$-\nabla_i P_{-3} = D^j \varepsilon_{ijk} W_{-1}^k, \tag{15}$$

which corresponds to a geostrophic equilibrium equation. In zero order $R^0$, we have the following system of equations:

$$\partial_t v_0^i - \partial_{jj} v_0^i + \partial_j \left(W_{-1}^i v_0^j + v_0^i W_{-1}^j\right) + D^j \varepsilon_{ijk} v_0^k = -(\partial_i P_0 + \nabla_i P_{-2}) + F_0^i, \tag{16}$$

$$\partial_i v_0^i = 0. \tag{17}$$

These equations give the following secular equation:

$$\nabla P_{-2} = 0 \Rightarrow P_{-2} = \text{Const.} \tag{18}$$

Let us consider the equations of the first approximation $R$:

$$\begin{aligned} &\partial_t v_1^i - \partial_{jj} v_1^i + D^j \varepsilon_{ijk} v_1^k + \partial_j \left(W_{-1}^i v_1^j + v_1^i W_{-1}^j + v_0^i v_0^j\right) \\ &= -\nabla_j \left(W_{-1}^i W_{-1}^j\right) - (\partial_i P_1 + \nabla_i P_{-1}), \end{aligned} \tag{19}$$

$$\partial_i V_1^i + \nabla_i W_{-1}^i = 0. \tag{20}$$

Secular equations follow from this system of equations:

$$\nabla_i W_{-1}^i = 0, \tag{21}$$

$$\nabla_j \left(W_{-1}^i W_{-1}^j\right) = -\nabla_i P_{-1}, \tag{22}$$

Secular equations (21) and (22) are satisfied by choosing the following geometry for the velocity field (Beltrami field):

$$\boldsymbol{W} = \left(W_{-1}^x(Z), W_{-1}^y(Z), 0\right);\ T_{-1} = T_{-1}(Z); \tag{23}$$

$$\nabla P_{-1} = 0 \Rightarrow P_{-1} = \text{Const.}$$

In the second order $R^2$, we obtain the equations:

$$\begin{aligned} &\partial_t v_2^i - \partial_{jj} v_2^i - 2\partial_j \nabla_j v_0^i + \partial_j \left(W_{-1}^i v_2^j + v_2^i W_{-1}^j + v_0^i v_1^j + v_1^i v_0^j\right) + D^j \varepsilon_{ijk} v_2^k \\ &= -\nabla_j \left(W_{-1}^i v_0^j + v_0^i W_{-1}^j\right) - (\partial_i P_2 + \nabla_i P_0), \end{aligned} \tag{24}$$

$$\partial_i v_2 + \nabla_i v_0 = 0. \tag{25}$$

It is easy to see that there are no secular terms in this order.

Let us come now to the most important order $R^3$. In this order we obtain the equations:

$$\begin{aligned} &\partial_t v_3^i + \partial_T W_{-1}^i - \left(\partial_{jj} v_3^i + 2\partial_j \nabla_j v_1^i + \nabla_{jj} W_{-1}^i\right) + \nabla_j \left(W_{-1}^i v_1^j + v_1^i W_{-1}^j + v_0^i v_0^j\right) \\ &+ \partial_j \left(W_{-1}^i v_3^j + v_3^i W_{-1}^j + v_0^i v_2^j + v_2^i v_0^j + v_1^i v_1^j\right) + D^j \varepsilon_{ijk} v_3^k = -\left(\partial_i P_3 + \nabla_i \overline{P}_1\right), \end{aligned} \tag{26}$$

$$\partial_i v_3 + \nabla_i v_1 = 0.$$

From this we get the main secular equation:

$$\partial_T W_{-1}^i - \Delta W_{-1}^i + \nabla_k \left( \overline{v_0^k v_0^i} \right) = -\nabla_i \overline{P}_1, \tag{27}$$

There is also an equation to find the pressure $P_{-3}$:

$$-\nabla_i P_{-3} = D^j \varepsilon_{ijk} W_{-1}^k. \tag{28}$$

## 4. The Velocity Field in Zero Approximation

It is clear that the most important is the equation (27). In order to obtain these equations in closed form, we need to calculate the Reynolds stress $\nabla_k \left( \overline{v_0^k v_0^i} \right)$. First of all, we have to calculate the fields of the zero approximation $v_0^k$. From the asymptotic development in zero order we have:

$$\partial_t v_0^i - \partial_{jj} v_0^i + W_{-1}^k \partial_k v_0^i + D^j \varepsilon_{ijk} v_0^k = -\partial_i P_0 + F_0^i, \tag{29}$$

Let us introduce the operator $\widehat{D_0}$:

$$\widehat{D_0} \equiv \partial_t - \partial_{jj} + W^k \partial_k. \tag{30}$$

Using $\widehat{D_0}$, we rewrite equation (29) in the form:

$$\widehat{D_0} v_0^i + D^j \varepsilon_{ijk} v_0^k = -\partial_i P_0 + F_0^i, \tag{31}$$

Pressure $P_0$ can be found from condition $divV = 0$.

$$P_0 = \frac{[\boldsymbol{D} \times \partial]_i v_0^i}{\partial^2} \tag{32}$$

Let us introduce the designations for the operators:

$$P_{ij} = \partial_j \frac{[\boldsymbol{D} \times \partial]_i}{\partial^2} \tag{33}$$

and for velocities: $v_0^x = u_0, v_0^y = v_0, v_0^z = w_0$. Then excluding pressure from (31), we obtain the system of equations to find the velocity field of zero approximation:

$$\begin{aligned}
&\left(\widehat{D_0} + \widehat{P_{xx}}\right) u_0 + \left(\widehat{P_{yx}} - D_z\right) v_0 + \left(\widehat{P_{zx}} + D_y\right) w_0 = F_0^x, \\
&\left(\widehat{P_{xy}} + D_z\right) u_0 + \left(\widehat{D_0} + \widehat{P_{yy}}\right) v_0 + \left(\widehat{P_{zy}} - D_x\right) w_0 = F_0^y, \\
&\left(\widehat{P_{xz}} - D_y\right) u_0 + \left(\widehat{P_{yz}} + D_x\right) v_0 + \left(\widehat{D_0} + \widehat{P_{zz}}\right) w_0 = F_0^z.
\end{aligned} \tag{34}$$

In order to solve this system of equations we have to set the force in the explicit form. Let us choose now the external force in the rotating system of coordinates in the following form:

$$\begin{aligned}
&F_0^z = 0,\ \boldsymbol{F}_{0\perp} = f_0 \left( \boldsymbol{i} \mathrm{Cos} \varphi_2 + \boldsymbol{j} \mathrm{Cos} \varphi_1 \right);\ \varphi_1 = \boldsymbol{k}_1 \boldsymbol{x} - \omega_0 t,\ \varphi_2 = \boldsymbol{k}_2 \boldsymbol{x} - \omega_0 t, \\
&\boldsymbol{k}_1 = k_0 (1,0,0),\ \boldsymbol{k}_2 = k_0 (0,1,0).
\end{aligned}$$

It is obvious that divergence and helicity of this force us equal to zero: $\boldsymbol{F}_0 rot \boldsymbol{F} = 0$. Thus, the external force is given in the plane (x, y), which is orthogonal to the projection of angular velocity $\boldsymbol{\Omega}$.

The solution for equations system (34) can be found easily in accordance with Cramer's Rule:

$$u_0 = \frac{\Delta_1}{\Delta}, v_0 = \frac{\Delta_2}{\Delta}, w_0 = \frac{\Delta_3}{\Delta}. \tag{35}$$

Here $\Delta$ is the determinant of the system (34):

$$\Delta = \begin{vmatrix} \widehat{D_0} + \widehat{P_{xx}} & \widehat{P_{yx}} - D_z & \widehat{P_{zx}} + D_y \\ \widehat{P_{xy}} + D_z & \widehat{D_0} + \widehat{P_{yy}} & \widehat{P_{zy}} - D_x \\ \widehat{P_{xz}} \quad D_y & \widehat{P_{yz}} + D_x & \widehat{D_0} + \widehat{P_{zz}} \end{vmatrix}, \tag{36}$$

$$\Delta_1 = \begin{vmatrix} F_0^x & \widehat{P_{yx}} - D_z & \widehat{P_{zx}} + D_y \\ F_0^y & \widehat{D_0} + \widehat{P_{yy}} & \widehat{P_{zy}} - D_x \\ 0 & \widehat{P_{yz}} + D_x & \widehat{D_0} + \widehat{P_{zz}} \end{vmatrix}, \tag{37}$$

$$\Delta_2 = \begin{vmatrix} \widehat{D_0} + \widehat{P_{xx}} & F_0^x & \widehat{P_{zx}} + D_y \\ \widehat{P_{xy}} + D_z & F_0^y & \widehat{P_{zy}} - D_x \\ \widehat{P_{xz}} - D_y & 0 & \widehat{D_0} + \widehat{P_{zz}} \end{vmatrix}, \tag{38}$$

$$\Delta_3 = \begin{vmatrix} \widehat{D_0} + \widehat{P_{xx}} & \widehat{P_{yx}} - D_z & F_0^x \\ \widehat{P_{xy}} + D_z & \widehat{D_0} + \widehat{P_{yy}} & F_0^y \\ \widehat{P_{xz}} - D_y & \widehat{P_{yz}} + D_x & 0 \end{vmatrix}. \tag{39}$$

Expanding the determinant, we obtain:

$$\begin{aligned} u_0 = &\frac{1}{\Delta}\Big[\big(\widehat{D_0}+\widehat{P_{yy}}\big)\big(\widehat{D_0}+\widehat{P_{zz}}\big)-\big(\widehat{P_{yz}}+D_x\big)\big(\widehat{P_{zy}}-D_x\big)\Big]F_0^x \\ &+\frac{1}{\Delta}\Big[\big(\widehat{P_{zx}}+D_y\big)\big(\widehat{P_{yz}}+D_x\big)-\big(\widehat{P_{yx}}-D_z\big)\big(\widehat{D_0}+\widehat{P_{zz}}\big)\Big]F_0^y, \end{aligned} \tag{40}$$

$$\begin{aligned} v_0 = &\frac{1}{\Delta}\Big[\big(\widehat{P_{xz}}-D_y\big)\big(\widehat{P_{zy}}-D_x\big)-\big(\widehat{P_{xy}}+D_z\big)\big(\widehat{D_0}+\widehat{P_{zz}}\big)\Big]F_0^x \\ &+\frac{1}{\Delta}\Big[\big(\widehat{D_0}+\widehat{P_{xx}}\big)\big(\widehat{D_0}+\widehat{P_{zz}}\big)-\big(\widehat{P_{xz}}-D_y\big)\big(\widehat{P_{zx}}+D_y\big)\Big]F_0^y, \end{aligned} \tag{41}$$

$$\begin{aligned} w_0 = &\frac{1}{\Delta}\Big[\big(\widehat{P_{xy}}+D_z\big)\big(\widehat{P_{yz}}+D_x\big)-\big(\widehat{P_{xz}}-D_y\big)\big(\widehat{D_0}+\widehat{P_{yy}}\big)\Big]F_0^x \\ &+\frac{1}{\Delta}\Big[\big(\widehat{P_{xz}}-D_y\big)\big(\widehat{P_{yx}}-D_z\big)-\big(\widehat{D_0}+\widehat{P_{xx}}\big)\big(\widehat{P_{yz}}+D_x\big)\Big]F_0^y. \end{aligned} \tag{42}$$

$$\begin{aligned} \Delta = &\big(\widehat{D_0}+\widehat{P_{xx}}\big)\Big[\big(\widehat{D_0}+\widehat{P_{yy}}\big)\big(\widehat{D_0}+\widehat{P_{zz}}\big)-\big(\widehat{P_{yz}}+D_x\big)\big(\widehat{P_{zy}}-D_x\big)\Big] \\ &-\big(\widehat{P_{yx}}-D_z\big)\Big[\big(\widehat{P_{xy}}+D_z\big)\big(\widehat{D_0}+\widehat{P_{zz}}\big)-\big(\widehat{P_{xz}}-D_y\big)\big(\widehat{P_{zy}}-D_x\big)\Big] \\ &+\big(\widehat{P_{zx}}+D_y\big)\Big[\big(\widehat{P_{xy}}+D_z\big)\big(\widehat{P_{yz}}+D_x\big)-\big(\widehat{D_0}+\widehat{P_{yy}}\big)\big(\widehat{P_{xz}}-D_y\big)\Big]. \end{aligned} \tag{43}$$

In order to calculate the expressions (40)-(43) we present the external force in complex form:

$$F_0^x = \frac{f_0}{2}\left(e^{i\varphi_2}+e^{-i\varphi_2}\right),\ F_0^y = \frac{f_0}{2}\left(e^{i\varphi_1}+e^{-i\varphi_1}\right). \tag{44}$$

Then all operators in formulae (40) - (42) act from the left on their eigenfunction. In particular:

$$\begin{aligned} &\widehat{D_0}e^{i\varphi_2} = e^{i\varphi_2}\,\widehat{D_0}\left(k_2,-\omega_0\right),\ \widehat{D_0}e^{i\varphi_1} = e^{i\varphi_1}\,\widehat{D_0}\left(k_1,-\omega_0\right), \\ &\Delta e^{i\varphi_2} = e^{i\varphi_2}\Delta\left(k_2,-\omega_0\right),\ \Delta e^{i\varphi_1} = e^{i\varphi_1}\Delta\left(k_1,-\omega_0\right) \end{aligned} \tag{45}$$

To simplify the formulae, let us choose $k_0 = 1, \omega_0 = 1, f_0 = 1$.
Now let us designate:

$$\widehat{D_0}(k_2,-\omega_0)=1-i(1-w_y)=A_y, \widehat{D_0}(k_1,-\omega_0)=1-i(1-w_x)=A_x. \quad (46)$$

Before doing further calculations, we have to note that some components of tensors $\widehat{P_{ij}}(k_1)$ and $\widehat{P_{ij}}(k_2)$ vanish. Let us write the non-zero components only:

$$\widehat{P_{yx}}(k_1)=D_z,\ \widehat{P_{zx}}(k_1)=-D_y,\ \widehat{P_{xy}}(k_2)=-D_z,\ \widehat{P_{zy}}(k_2)=D_x. \quad (47)$$

Taking into account the formulae (45)-(47), we can find the determinant:

$$\Delta(k_1)=A_x^3+D_x^2A_x,\ \Delta(k_2)=A_y^3+D_y^2A_y. \quad (48)$$

In a similar way we find velocity field of zero approximation:

$$u_0=\frac{1}{2}\frac{e^{i\varphi_2}A_y}{A_y^2+D_y^2}+C.C., \quad (49)$$

$$v_0=\frac{1}{2}\frac{e^{i\varphi_1}A_x}{A_x^2+D_x^2}+C.C., \quad (50)$$

$$w_0=\frac{1}{2}\frac{e^{i\varphi_2}D_y}{A_y^2+D_y^2}-\frac{1}{2}\frac{e^{i\varphi_1}D_x}{A_x^2+D_x^2}+C.C.. \quad (51)$$

We note that the angular velocity $D_z$ component disappears from the expression for the velocity field of zero approximation, which is a consequence of the properties of an external force.

## 5. Reynolds Stress and Large Scale Instability

To close the equations (27) we have to calculate the Reynolds stresses $\overline{w_0u_0}$ and $\overline{w_0v_0}$. These terms are easily calculated with the help of formulae (49)-(51). As a result we obtain:

$$\overline{w_0u_0}=\frac{1}{2}\frac{D_y}{\left|A_y^2+D_y^2\right|^2},\ \overline{w_0v_0}=-\frac{1}{2}\frac{D_x}{\left|A_x^2+D_x^2\right|^2}. \quad (52)$$

Now equations (27) are closed and take form:

$$\begin{aligned}&\partial_T W_x-\Delta W_x+\frac{\partial}{\partial z}\frac{1}{2}\frac{D_y}{\left|A_y^2+D_y^2\right|^2}=0,\\&\partial_T W_y-\Delta W_y-\frac{\partial}{\partial z}\frac{1}{2}\frac{D_x}{\left|A_x^2+D_x^2\right|^2}=0.\end{aligned} \quad (53)$$

We calculate the modules and write the equations (53) in the explicit form:

$$\begin{aligned}&\partial_T W_x-\Delta W_x+\frac{1}{2}\frac{\partial}{\partial z}\frac{D_y}{4(1-w_y)^2+\left[D_y^2+w_y(2-w_y)\right]^2}=0,\\&\partial_T W_y-\Delta W_y-\frac{1}{2}\frac{\partial}{\partial z}\frac{D_x}{4(1-w_x)^2+\left[D_x^2+w_x(2-w_x)\right]^2}=0.\end{aligned} \quad (54)$$

With small $W_x, W_y$ we obtain the linearized equations (54):

$$\begin{aligned}&\partial_T W_x-\Delta W_x-\alpha_y\frac{\partial}{\partial z}W_y=0,\\&\partial_T W_y-\Delta W_y+\alpha_x\frac{\partial}{\partial z}W_x=0.\end{aligned} \quad (55)$$

$$\alpha_y = 2\frac{D_y\left(D_y^2-2\right)}{\left(4+D_y^2\right)^2},\quad \alpha_x = 2\frac{D_x\left(D_x^2-2\right)}{\left(4+D_x^2\right)^2}.$$

The system (55) describes the positive feedback between the components of velocity. We will look for the solution of linear system (55) in the following form:

$$W_x, W_y \sim \exp\left(\gamma T + ikZ\right). \tag{56}$$

Substituting (56) in equation (55), we obtain the dispersion equation:

$$\gamma = \pm\sqrt{\alpha_x\alpha_y}k - k^2. \tag{57}$$

The dispersion equation (57) shows the existence at $\alpha_x\alpha_y \succ 0$ of the large scale instability with maximum growth rate $\gamma_{\max} = \frac{\alpha_x\alpha_y}{4}$, at the wave vector $k_{\max} = \frac{1}{2}\sqrt{\alpha_x\alpha_y}$. As a result of the development of instability the large scale helical Beltrami vortices are generated in the system. When $\alpha_x\alpha_y \prec 0$, damped oscillations with a frequency $\omega_0 = \sqrt{\alpha_x\alpha_y}k$ arise instead of instability. In fact the behavior of $\gamma$ depends on how is located the external force $F_0^x, F_0^y$ with respect to the perpendicular projections of the angular velocity of rotation and the values of components $D_x, D_y$. If one of the component $D_x, D_y$ is zero or equals to $\sqrt{2}$, then the instability is absent. Instability exists in the following cases:

$$1. D_x \succ \sqrt{2}, D_y \succ \sqrt{2};$$
$$2. D_x, D_y \succ 0, D_x \prec \sqrt{2}, D_y \prec \sqrt{2};$$
$$3. D_x \prec 0, D_y \prec 0, D_x^2 \succ 2, D_y^2 \succ 2;$$
$$4. D_x \prec 0, D_y \prec 0, D_x^2 \prec 2, D_y^2 \prec 2;$$
$$5. D_x \prec 0, D_y \succ 0, D_y^2 \succ 2, D_x^2 \prec 2; \text{ or } D_y^2 \prec 2, D_x^2 \succ 2;$$
$$6. D_x \succ 0, D_y \prec 0, D_y^2 \succ 2, D_x^2 \prec 2; \text{ or } D_y^2 \prec 2, D_x^2 \succ 2;$$

In all other cases damped oscillations occur.

## 6. Saturation of Instability and Nonlinear Vortex Structures

It is clear that with increasing of amplitude nonlinear terms decrease and instability becomes saturated. Consequently stationary nonlinear vortex structures are formed. To find these structures let us choose equations (54) $\frac{\partial}{\partial T} = 0$ and integrate equations one time over Z. We obtain the system of equations:

$$\begin{aligned}\frac{d}{dZ}W_x &= \frac{1}{2}\frac{D_y}{4\left(1-w_y\right)^2+\left[D_y^2+w_y\left(2-w_y\right)\right]^2}+C_1,\\ \frac{d}{dZ}W_y &= -\frac{1}{2}\frac{D_x}{4\left(1-w_x\right)^2+\left[D_x^2+w_x\left(2-w_x\right)\right]^2}+C_2.\end{aligned} \tag{58}$$

Let's take for this system new variables: $1-w_x = u_x, 1-w_y = u_y$. Then we obtain:

$$\begin{aligned}\frac{du_x}{dZ} &= -\frac{1}{2}\frac{D_y}{\left(D_y^2+1\right)^2+2\left(1-D_y^2\right)u_y^2+u_y^4}+C_1,\\ \frac{du_y}{dZ} &= \frac{1}{2}\frac{D_x}{\left(D_x^2+1\right)^2+2\left(1-D_x^2\right)u_x^2+u_x^4}+C_2.\end{aligned} \tag{59}$$

The system of equations (59) can be written in Hamiltonian form:

$$\frac{du_x}{dZ} = -\frac{\partial H}{\partial u_y},$$
$$\frac{du_y}{dZ} = \frac{\partial H}{\partial u_x}.$$

Where Hamiltonian $H$ has the form:

$$H = h(D_x, u_x) + h(D_y, u_y), \tag{60}$$

with function $h(D,u)$:

$$h(D,u) = \frac{D}{2}\int \frac{du}{\left(D^2+1\right)^2 + 2\left(1-D^2\right)u^2 + u^4} + Cu. \tag{61}$$

Integral in expression (61) is calculated in elementary functions [17]. Let us choose for simplicity $D_x = D_y = D = 1$. In this case, the function (61) is equal to [17]:

$$h(u) = \frac{1}{16}\left\{\ln\frac{u^2+2u+2}{u^2-2u+2} + \text{arctg}\frac{2u}{2-u^2}\right\} + Cu. \tag{62}$$

The sum $h(u_x) + h(u_y)$ can be written down as one formula. Then Hamiltonian is equal to:

$$H = \frac{1}{16}\ln\frac{\left(u_x^2+2u_x+2\right)\left(u_y^2+2u_y+2\right)}{\left(u_x^2-2u_x+2\right)\left(u_y^2-2u_y+2\right)} + \frac{1}{16}\text{arctg}\frac{2u_y\left(u_x^2-2\right)+2u_x\left(u_y^2-2\right)}{2\left(u_x+u_y\right)^2 - u_x^2u_y^2 - 4} + C_1u_x + C_2u_y \tag{63}$$

It is easy to construct the phase portrait of **Figure 4** for Hamiltonian (63) and specific values $C_1 = 0.1$, $C_2 = 0.1$. The phase portrait shows the presence of closed trajectories in the phase plane around elliptic points and separatrices that connect hyperbolic points. It is obvious that the closed trajectories correspond to nonlinear periodic solutions. The separatrices correspond to localized solutions of kink type.

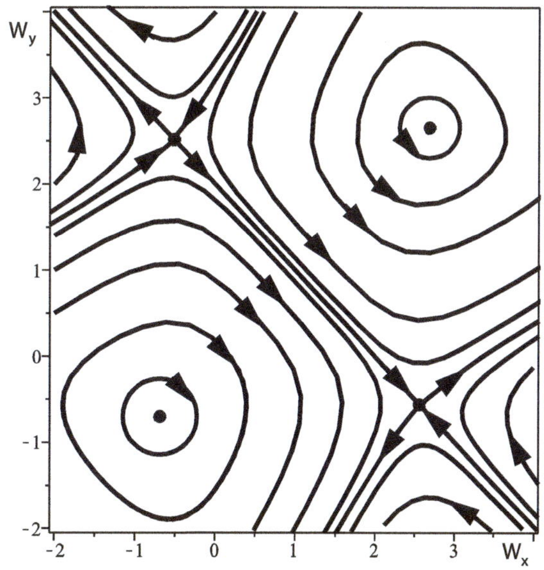


**Figure 4.** Phase plane for Hamiltonian (63) ($C_1 = 0.1$, $C_2 = 0.1$). We see the presence of closed trajectories around the elliptic points and separatrices which connect the hyperbolic points. Phase portrait is typical for Hamiltonian systems.

## 7. Conclusions and Discussion of the Results

In this work we found new large scale instability in rotating fluid. It is supposed that the small scale vortex external force in rotating coordinates system acts on fluid which maintains the small velocity field fluctuations (small-scale turbulence with low Reynolds number $R, R \ll 1$). For the real applications this Reynolds number should be calculated with the help of the turbulent viscosity. The asymptotic development of motion equations by small Reynolds number allows obtaining motion equations for the large scale. These equations are of the hydrodynamic $\alpha$-effect type, in which velocity components $W_x, W_y$ are connected by the positive feedback. This may result in the appearance of the large scale vortex instability. This instability is responsible for the formation of large scale Beltrami vortices in rotating fluid with small scale external force. With further increase of amplitude the instability stabilizes and passes to a stationary mode. In this mode the nonlinear stationary vortex structures are formed. The most interesting structures belong to a variety of vortex kinks. These kinks connect stationary hyperbolic points of the dynamical system (58).

Note that in contrast to previous work on the hydrodynamic $\alpha$-effect in rotating fluid, the method enables us to construct an asymptotic development in a natural way and to explore non-linear theory of nonlinear stationary vortex kinks.

## References

[1] Grinspen, H.P. (1990) The Theory of Rotating Fluids. Brookline Press, Brooklyn, MA.

[2] Roberts P.H. and Soward, A.M., Eds. (1978) Rotating Fluids in Geophysics. Acad. Press Inc., London.

[3] Clarke, C. and Carswell, B. (2007) Principles of Astrophysical Fluid Dynamics. Cambridge University Press, Cambridge, UK. http://dx.doi.org/10.1017/CBO9780511813450

[4] Vallis, G.K. (2010) Atmospheric and Oceanic Fluid Dynamics. Cambridge University Press, Cambridge, MA.

[5] Abramowicz, M.A., Lanza, A., Spigel, E.A. and Szuszkiewicz, E. (1992) Vortices on Accretion Disks. *Nature*, **356**, 41-43. http://dx.doi.org/10.1038/356041a0

[6] Brandt, P.N., Scharmer, G.B., Ferguson, S., Shine, R.A., Tarbell, T.D. and Title, A.M. (1988) Vortex Flow in the Solar Photosphere. *Nature*, **335**, 238-240. http://dx.doi.org/10.1038/335238a0

[7] Dritschel, G. and Legras, B. (1993) Modeling Oceanic and Atmospheric Vortices. *Physics Today*, **46**, 44. http://dx.doi.org/10.1063/1.881375

[8] Moffat, H.K. (1978) Magnetic Field Generation in Electrically Conducting Fluids. Cambridge University Press, Cambridge, UK.

[9] Moiseev, S.S., Sagdeev, R.Z., Tur, A.B., Khomenko, G.A. and Yanovsky, V.V. (1983) Theory of Large-Scale Structures in Hydrodynamic Turbulence. *ZHETF*, **58**, 1149.

[10] Moiseev, S.S., Rutkiewicz, P.B., Tur, A.B. and Yanovsky, V.V. (1988) Spiral Vortex Dynamo in Turbulent Convection. *ZHETF*, **67**, 294.

[11] Loupian, E.A., Mazurov, A.A., Rutkiewicz, P.B. and Tur, A.B. (1992) Generation of Large-Scale Eddies as a Result of the Spiral Turbulence Convective Nature. *ZHETF*, **75**, 833.

[12] Khomenko, G.A., Moiseev, S.S. and Tur, A.V. (1991) The Hydrodynamic Alpha-Effect in a Compressible Fluid. *Journal of Fluid Mechanics*, **225**, 355-369. http://dx.doi.org/10.1017/S0022112091002082

[13] Levina, G.V., Moiseev, S.S. and Rutkevich, P.B. (2000) Hydrodynamic Alpha-Effect in a Convective System. *Advances in Fluid Mechanics*, **25**, 111-162.

[14] Frisch, U., She, Z.S. and Sulem, P.L. (1987) Large-Scale Flow Driven by the Anisotropic Kinetic Alpha Effect. *Physica D: Nonlinear Phenomena*, **28**, 382-392. http://dx.doi.org/10.1016/0167-2789(87)90026-1

[15] Tur, A.V. and Yanovsky, V.V. (2013) Non Linear Vortex Structures in Stratified Fluid Driven by Small-Scale Helical Force. *Open Journal of Fluid Dynamics*, **3**, 64-74. http://dx.doi.org/10.4236/ojfd.2013.32009

[16] Kitchatinov, L.L., Rudiger, G. and Khomenko, G. (1994) Large-Scale Vortices in Rotating Stratified Disks. *Astronomy and Astrophysics*, **287**, 320-324.

[17] Gradshteyn, I.S. and Ryzhik, I.M. (1974) Tables of Integrals, Sums, Series and Products. Academic Press, San Diego, San Francisco, New York, Boston, London, Sydney, Tokyo.

# 12

# A Twin Unidirectional Impulse Turbine for Wave Energy Conversion—Effect of Fluidic Diode on the Performance

**Shinya Okuhara[1], Manabu Takao[2], Hideki Sato[3], Akiyasu Takami[2], Toshiaki Setoguchi[4]**

[1]Support Center for Practical Education, National Institute of Technology, Matsue College, Matsue, Japan
[2]Department of Mechanical Engineering, National Institute of Technology, Matsue College, Matsue, Japan
[3]Advanced Engineering Faculty, National Institute of Technology, Matsue College, Matsue, Japan
[4]Institute of Ocean Energy, Saga University, Saga, Japan
Email: takao@matsue-ct.jp

## Abstract

**As a system using a conventional unidirectional air turbine in oscillating water column (OWC) based on a wave energy plant, a twin unidirectional impulse turbine topology has been suggested in previous studies. However, the average efficiency of the suggested twin turbine is considerably lower than that of a conventional unidirectional turbine in this topology because reciprocating air flow can't be rectified adequately by a unidirectional turbine. In order to improve the efficiency, using fluidic diode is discussed. In this study, two different fluidic diodes were discussed by computational fluid dynamics (CFD) and a wind tunnel test. Further, its usefulness is discussed from a view point of the turbine efficiency. The fluidic diode was shown to improve rectification of the topology. However, it needs more improvement in regards to its energy loss in order to enhance the turbine efficiency.**

## Keywords

**Fluidic Diode, Twin Unidirectional Turbine, Wave Energy Conversion, Oscillating Water Column**

## 1. Introduction

As an air turbine equipped with oscillating water column (OWC) based wave energy plant, a rectification-valve system using a conventional unidirectional turbine has been invented. However, this system has problems with the durability of the valves and the complexity of the mechanism. Moreover, it has a major fault that the valves must be large so as to produce high output. Therefore, a twin unidirectional impulse turbine topology has been suggested in previous studies in order to use conventional unidirectional turbines without a rectification valve [1] [2]. The topology is composed of two unidirectional impulse turbines coupled with an electrical generator, as shown in **Figure 1**. However, the past study indicated that the average efficiency of the topology was shown to be low, when the performance prediction of the topology in oscillating airflow was carried out by means of quasi-steady analysis [1]. Further, the cause of the low efficiency is that part of the air flow gets through the unidirectional impulse turbine in the direction of low efficiency [1].

In this study, a fluidic diode [2] [3] is adopted in order to suppress the air flow rate into the inefficient turbine in a twin unidirectional impulse turbine topology for wave energy plant, and the effect of the fluidic diodes on the performance of twin unidirectional impulse turbine topology is investigated by a wind tunnel test and computational fluid dynamics (CFD). Further, its usefulness is discussed from a view point of the turbine efficiency.

## 2. Twin Unidirectional Impulse Turbine Topology and Fluidic Diode

In a twin unidirectional impulse turbine topology, two unidirectional impulse turbines coupled with an electric generator are installed as shown in **Figure 1**. In this study, when air flow gets through a turbine or fluidic diode, the flow direction that indicates smaller flow resistance is described as "forward flow", and the flow direction that indicates larger flow resistance is described as "reverse flow". Index "f" stands for "forward flow" and index "r" stands for reverse flow. A turbine in forward flow and it in reverse flow are respectively described as "forward turbine" and "reverse turbine".

The principle of a twin unidirectional impulse turbine topology is as followed. Oscillating air flow is produced in an air chamber. Then, it is rectified into a forward turbine that has the high efficiency by a reverse turbine. Consequently, electricity is produced with high efficiency. However, the average efficiency of the topology is considerably lower than that of a unidirectional impulse turbine because part of the air flow gets through the reverse turbine resulting in the efficiency of energy conversion being very low [1]. As a way of improving this fault, rectifying an oscillating air flow by installing fluidic diode downstream from a forward turbine is suggested [2].

Fluidic diodes indicate different pressure differences between the front and back against different flow directions. **Figure 2** shows the fluidic diodes adopted in this study. **Figure 2(a)** shows the fluidic diode which has a shape of a conical nozzle (Type I). The projection length is 125 mm and the larger diameter is 240 mm. A wind tunnel test and CFD were carried out by varying the taper angle $\theta$. Then, both results were compared. **Figure 2(b)** shows the special type fluidic diode (Type II) which is designed by referring a fluidic diode suggested in references [2]. The flow passage is composed of a bluff body (B) that has a hollow (H) rear, a toroidal region (T) and a conical nozzle region (N). In this study, CFD for Type II was carried out by changing the taper angle $\theta$ of the conical nozzle region.

## 3. Experiments and Computational Method

For CFD software, SCRYU/Tetra of Software Cradle Co., Ltd. was used. The governing equation was Reynolds averaged Navier-Stokes equation and turbulence model was a standard $k$-$\varepsilon$ model. The working fluid was incompressible air at 20°C. The analyzed region was a flow passage composed of tube and fluidic diodes. The analysis region of Type I consisted of approximately 1,500,000 lattices and the analysis region of Type II consisted of approximately 6,470,000 lattices. As a boundary condition, the inner wall of the tube and fluidic diode were in a non-slip boundary condition, and the inlet had an inflow in a constant flow rate. The outlet was opened to the atmosphere. The forward-flow and reverse-flow pressure differences ($\Delta p_f$ and $\Delta p_r$) were examined in a position far enough from the fluidic diode. In order to calculate the effects of the taper angles $\theta$ on $R_D$, CFD was executed by varying $\theta$ from 10 to 60 in 10 degree intervals in Type I and $\theta$ from 0 to 60 in 10 degree intervals in Type II.

In this study, the rectification property of the Type I in steady flow was examined by a wind tunnel test so as to indicate the validity of CFD. In the wind tunnel test, the suction test apparatus was composed of a centrifugal blower, the fluidic diode Type I and a tube of which the internal diameter was 240 mm as shown in **Figure 3**. As an experimental method, the blower generates a steady flow in the fluidic diode. The flow rate $Q$ was measured by a Pitot tube (OKANO WORKS, LK-0). The pressure in the settling chamber was measured by a pressure detector (OKANO WORKS, POP202) in order to examine the pressure differences $\Delta p_f$ and $\Delta p_r$ between the front

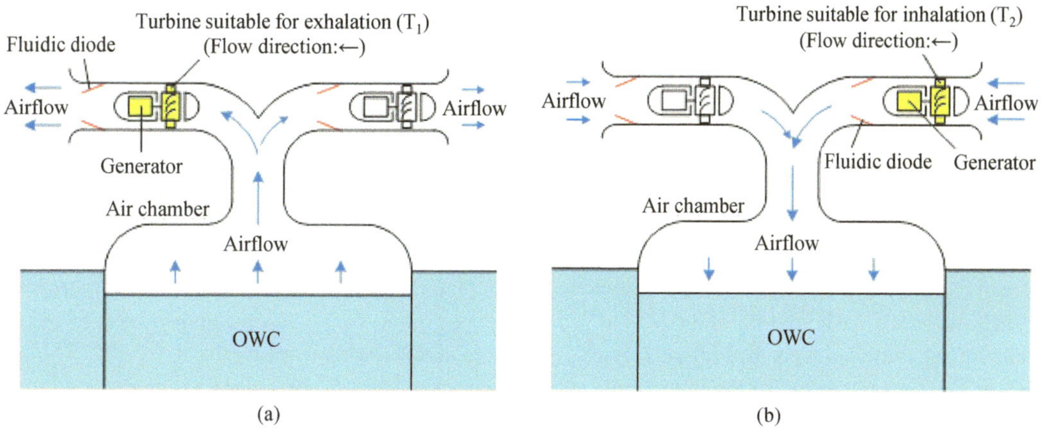


**Figure 1.** Principle of twin impulse turbine for wave energy conversion: (a) Exhalation; (b) Inhalation.

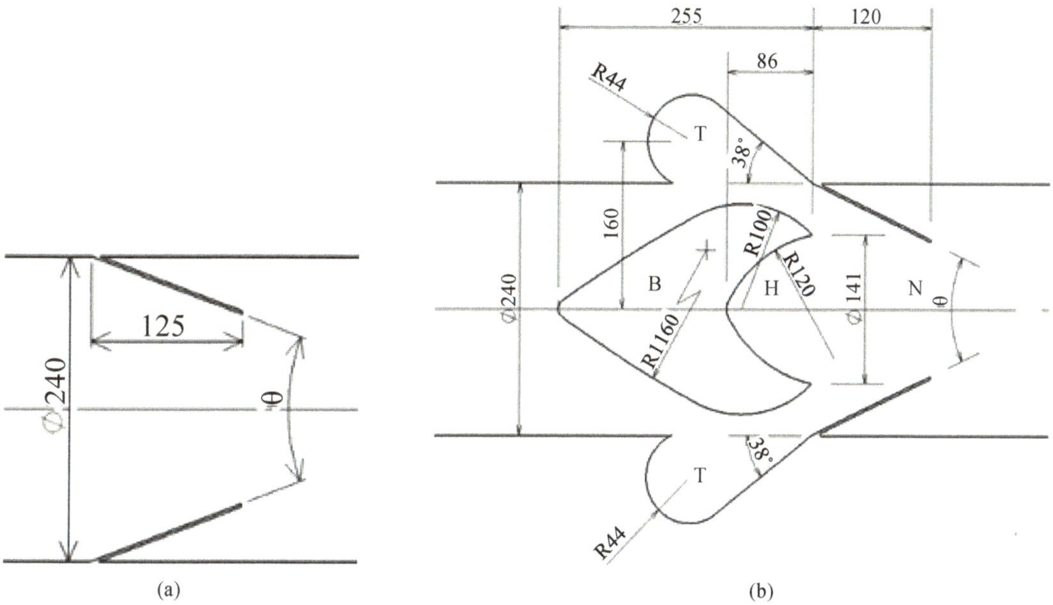


**Figure 2.** Tested fluidic diode: (a) Type I; (b) Type II.

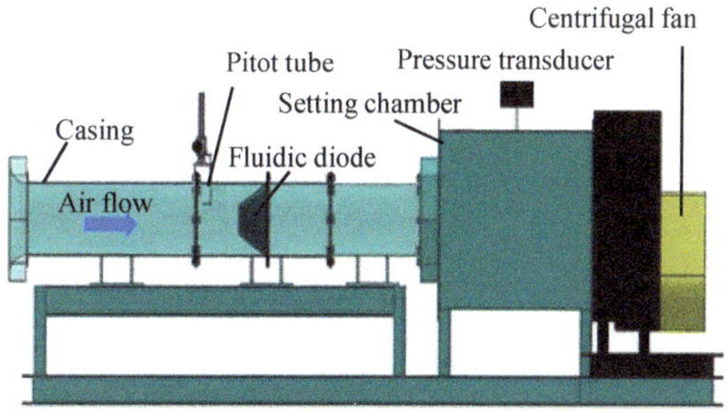


**Figure 3.** Test apparatus.

and the back of the fluidic diode in both the forward and reverse flow.

The performance of the fluidic diode is evaluated by the pressure ratio $R_D$ and the pressure coefficients $C_{pf}$, $C_{pr}$. The definitions of these parameters are as follows:

$$R_D = \Delta p_r / \Delta p_f \tag{1}$$

$$C_{pf} = \Delta p_f / \left( \rho v^2 / 2 \right) \quad (2)$$

$$C_{pr} = \Delta p_r / \left( \rho v^2 / 2 \right) \quad (3)$$

where $\rho$ and $v$ denote the density of air and axial flow velocity respectively.

## 4. Results and Discussion

**Figure 4** shows the effect of the taper angle $\theta$ on the pressure coefficients $C_{pf}$ and $C_{pr}$, respectively in both forward and reverse flow in Type I, as the results to the CFD and the wind tunnel test. In the figure, flow rates $Q$ are 0.320 $m^3/s$ in the calculation and 0.308 $m^3/s$ in the experiment. In the figure, both results of calculation and experiment are matched well. In **Figure 4**, both $C_{pf}$ and $C_{pr}$ increase as $\theta$ increases.

**Figure 5** shows the effect of the taper angle $\theta$ on the pressure ratio $R_D$ of Type I as the result to the CFD. The $R_D$ becomes the maximum $R_D = 2.00$ after it increases with $\theta$. The rectification of cascades geometry profitable for twin unidirectional impulse turbine topology [4] [5] in the maximum efficiency was calculated by the $C_A$-$\phi$ curve ($C_A$: input coefficient, $\phi$: flow coefficient). The value is $R_D = 2.63$. The maximum $R_D$ of Type I is shown to be lower than that of the unidirectional impulse turbine in **Figure 7**. Therefore, this indicates that the rectification of Type I was lower than that of the unidirectional impulse turbine.

On the other hand, **Figure 6** shows the effect of the taper angle $\theta$ on the pressure ratio $R_D$, the pressure coefficients $C_{pf}$ and $C_{pr}$, respectively in both forward and reverse flow in Type II. The tendency of the graph is similar to that of Type I. The maximum $R_D$ of Type II is 5.03 in the taper angle $\theta = 40^\circ$. The maximum $R_D$ of Type II is much higher than that of the unidirectional impulse turbine, as shown in **Figure 7**. Therefore, rectification of the twin unidirectional impulse turbine topology can be improved by installing Type II. However, the efficiency of a forward turbine coupled with Type II is lower than that of the only forward turbine because the additional pressure loss is caused by Type II in the forward flow. Therefore, Type II must be improved to decrease the flow resistance in the forward flow.

**Figure 8** and **Figure 9** show the velocity vector and pressure distribution respectively around Type II in both the forward and reverse flow. Forward flow goes from left to right. Reverse flow goes from right to left. In the forward flow, the cause of main pressure loss is the narrower flow passage of the conical nozzle. In the reverse flow, the causes of main pressure loss are the narrower flow passage of the conical nozzle, the hollow of the bluff body and the toroidal part. **Figure 6** shows that Type II doesn't have sufficient rectification function without the part of conical nozzle. However, it has main flow resistance in the forward flow. On the other hand, the toroidal part doesn't have a large flow resistance in the forward flow while it has large flow resistance in the reverse flow. The route of the forward flow is different from that of the reverse flow because of the each flow separation. Therefore, changing the flow route by making use of flow separation, so that forward flow avoids stagnation zones which reverse flow goes to, is effective.

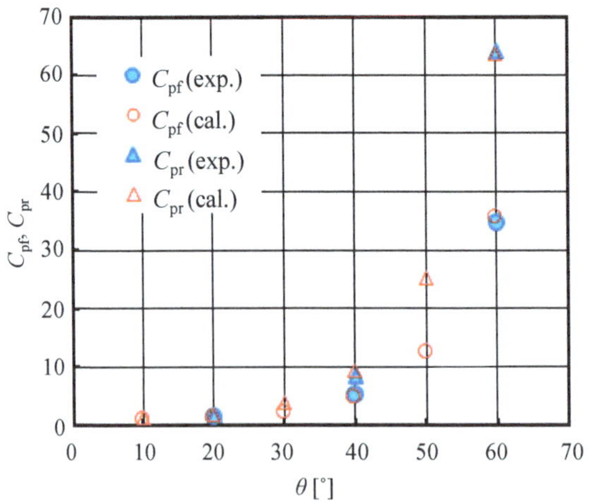


**Figure 4.** Effect of taper angle on pressure coefficient (Type I).

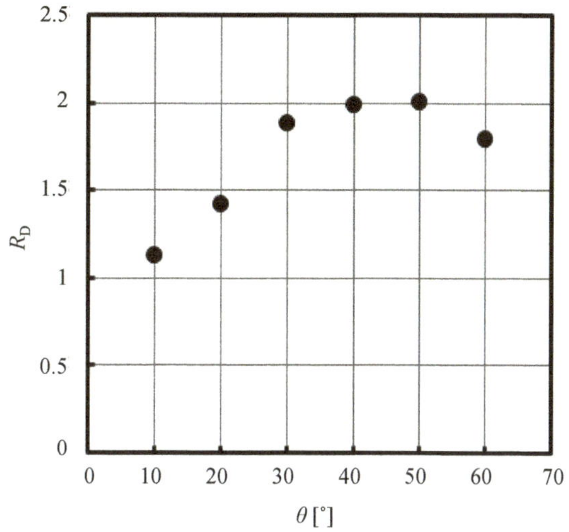


**Figure 5.** Effect of taper angle on pressure ratio (Type I, calculation).

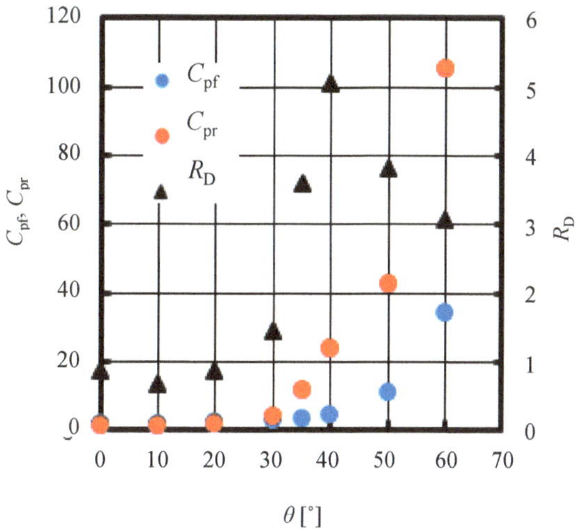


**Figure 6.** Effect of taper angle on pressure coefficient and pressure ratio (Type II, calculation).

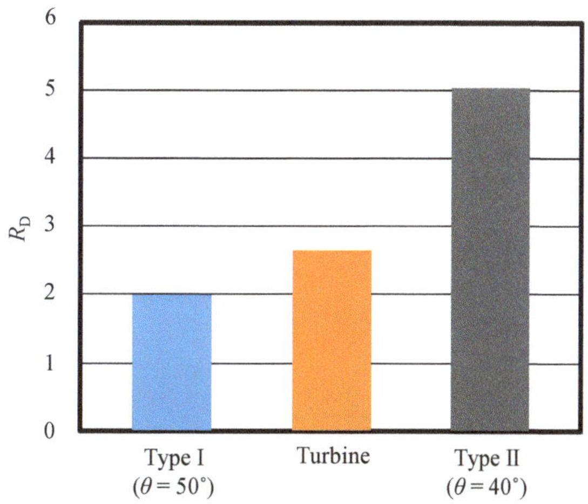


**Figure 7.** Comparison of $R_D$.

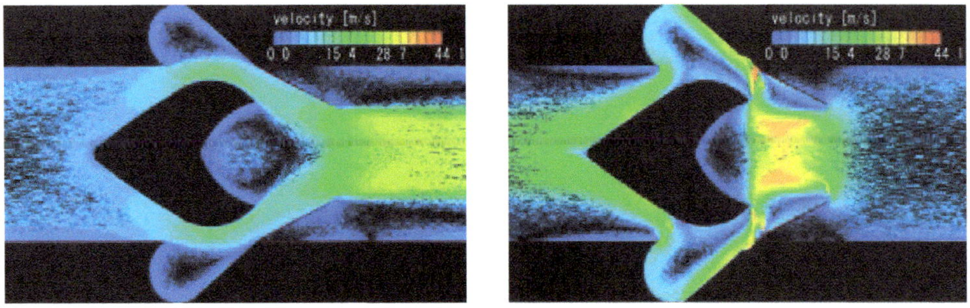


(a) (b)

**Figure 8.** Flow velocity in fluidic diode of Type II ($\theta = 40^{\circ}$): (a) Forward flow; (b) Reverse flow.

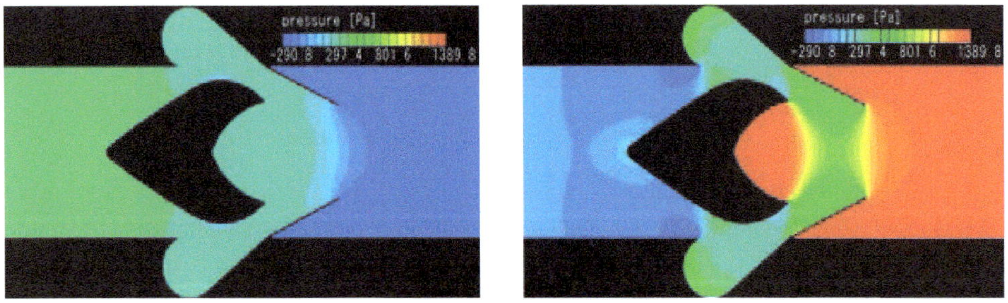


(a) (b)

**Figure 9.** Pressure distribution in fluidic diode of Type II ($\theta = 40^{\circ}$): (a) Forward flow; (b) Reverse flow.

## 5. Conclusion

In this study, the effect of the fluidic diodes on the performance of twin unidirectional impulse turbine topology was discussed by using CFD and wind tunnel testing. As a result, fluidic diodes, which have a shape of conical nozzle (Type I) are shown to be inadequate for this turbine topology. On the other hand, a special type of fluidic diode (Type II), which is designed from the original shape by the authors, was shown to have the possibility of improving performance of this turbine topology. However, from a view point of the turbine efficiency, it needs improvement in order to diminish the pressure difference in the forward flow.

## Acknowledgements

This study was performed as Grant-in-Aid for Scientific Research (B) (No. 24360362) supported by Japan Society for the Promotion of Science (JSPS). The second author wishes to thank JSPS for their financial help in conducting this study.

## References

[1] Takao, M., Takami, A., Okuhara, S. and Setoguchi, T. (2011) A Twin Unidirectional Impulse Turbine for Wave Energy Conversion. *Journal of Thermal Science*, **20**, 394-397. http://dx.doi.org/10.1007/s11630-011-0486-1

[2] Dudhgaonkar, P.V., *et al.* (2011) Fluidic Components for Oscillating Water Column Based Wave Energy Plants. *Proceedings of ASME-JSME-KSME Joint Fluids Engineering Conference* 2011, Paper No. AJK2011-07035.

[3] Takebayashi, M., Akazume, A., Yokoyama, K. and Iwata, H. (1984) Discharge Characteristics of a Oil Feeder Pump with Nozzle Type Fluidic Diodes. *Transactions of JSME* (*Series B*), **51**, 255-261. (In Japanese)

[4] Okuhara, S., Takao, M., Takami, A. and Setoguchi, T. (2012) A Twin Unidirectional Impulse Turbine for Wave Energy Conversion—Effect of Guide Vane Solidity on the Performance. *Open Journal of Fluid Dynamics*, **2012**, 343-347. http://dx.doi.org/10.4236/ojfd.2012.24A043

[5] Takao, M., Setoguchi, T., Kaneko, K., Kim, T.H., Maeda, H. and Inoue, M. (2002) Impulse Turbine for Wave Power Conversion with Air Flow Rectification System. *International Journal of Offshore and Polar Engineering*, **12**, 142-146.

## Nomenclature

$C_A$: Input coefficient
$C_p$: Pressure coefficient $\{C_p = \Delta p/(\rho v^2/2)\}$
$Q$: Flow rate ($m^3/s$)
$R_D$: Ratio of pressure difference ($= \Delta p_r / \Delta p_f$)
$v$: Axial flow velocity (m/s)
$\Delta p$: Pressure difference (Pa)
$\phi$: Flow coefficient
$\theta$: Taper angle of conical nozzle (°)
$\rho$: Air density ($kg/m^3$)

## Subscripts

f: Forward flow
r: Reverse flow

13

# Effects of Thermal Radiation and Radiation Absorption on Flow Past an Impulsively Started Infinite Vertical Plate with Newtonian Heating and Chemical Reaction

**Swetha Ravi[1], Jagdish Prakash[2*], Viswanatha Reddy Gottam[3], Vijaya Kumar Varma Sibyala[3]**

[1]Department of Mathematics, Gudlavalleru Engineering College, Gudlavalleru, India
[2]Department of Mathematics, University of Botswana, Gaborone, Botswana
[3]Department of Mathematics, S. V. University, Tirupati, India
Email: swetha.salini@gmail.com, *prakashj@mopipi.ub.bw, gvr.svu@gmail.com, svijayakumarvarma@yahoo.co.in

## Abstract

**A perfect solution to the present natural convective flow problem of a vertical transfinite plate owing to the impulsive motion in the ubiety of first ordered chemical reaction, radiation absorption, radiation, Newtonian heating and species concentration in its plane is evolved by applying the method of Laplace transforms in closed form at the plate. Exact results for velocity, temperature, concentration fields are prevailed and expressions for heat and mass transfer rates are also found. The effects are analyzed for the respective invariables for both ammonia and water vapor.**

## Keywords

**Newtonian Heating, Natural Convection, Chemical Reaction, Incompressible Fluid, Radiation Absorption and Radiation**

## 1. Introduction

On chemical reaction, the field of mass and heat transfer is of good pragmatic importance to applied scientists

*Corresponding author.

owing to its general occurrence in various fields of engineering and science. Especially, the subject of mass and heat transfer with heat radiation, chemical reaction has significant role in hydrometallurgical and chemical industries. For a moving plate, a chemical reaction takes place in legion chemical processes between a fluid and foreign mass. This sue is involved in many industrial usages such as glassware or ceramics manufacturing, food processing and production of polymers. The convection study with mass and heat transfer plays a major role in the dispersion and formation of fog, design of chemical processing equipment's, temperature distribution, and moisture over agricultural fields and in the paper drying process.

Ahmed *et al.* [1] have identified the analysis for MHD rotating heat or mass transport phenomenon bounded by a vertical oscillating surface in the Mein of Darcian porous regime by using Numerical/Laplace transform. Characteristics of the heat and mass transfer in the Mien of chemical reaction and thermal radiation for a Newtonian incompressible fluid across an extending vertical surface having temperature dependent viscosity was studied by Kandasamy *et al.* [2]. Makinde [3] examined the free transient convection interaction of an absorbing, emitting plate with thermal radiation. Mukhopadhyay [4] performed an investigation on the results of heat transfer and thermal radiation on a mixed unsteady convective flow across an extending porous surface in porous medium.

Heat transfer analysis of a forced convective flow of the fluid past an embedded plate in a porous medium for an incompressible fluid was examined by Mukhopadhyay and Layek [5]. Muthucumaraswamy and Ganesan [6] looked at the impulsively started transient radiation-convection flow with vertical temperature consequences. An analysis of the chemical reaction, theoretically a result, with variable temperature on a vertical oscillating plate was given by Muthucumaraswamy [7]. Reddy *et al.* [8] investigated the consequences of unsteady natural MHD convective flow in a porous medium with constant mass diffusion and Newtonian heating. The effects of MHD radiating and chemically reacting fluid past a non-isothermal impulsively started vertical surface adjacent to a porous regime by using numerical analysis was discussed by Sahin Ahmed [9].

The importance of the present flow problem is to analyze the effects of thermal radiation and radiation absorption on the flow past an impulsively started infinite vertical plate with Newtonian heating and chemical reaction.

## 2. Mathematical Analysis

Free convective unsteady flow of the fluid for a vertical transfinite plate with Newtonian heating, past an impulsively started incompressible viscous fluid in the Mien of radiation and radiation absorption is studied. Along the plate and in the vertical upward direction, axis $\bar{x}^*$ is chosen and normal to the plate, axis $\bar{y}^*$ is considered. Initially the fluid and the plate are having same temperature $\bar{T}_\infty^*$ and the concentration $\bar{C}_\infty^*$ at all points in a stationary state for time $\bar{t}^* \le 0$. The coordinate system and the flow model are shown in **Figure 1**. The plate is fixed with a velocity $u_0$ in the vertical direction into impulsive motion versus the gravitational field at time $\bar{t}^* > 0$. We assumed that i) heat transfer rate and the local surface temperature $\bar{T}^*$ are proportional to one another from the surface, and near the plate concentration rises to $\bar{C}_w^*$ and ii) the consequences of viscous dissipation are negligible in the energy equation. Among the fluid and diffusing species, there is a first order chemical reaction. Since all the physical quantities are expressed in terms of $\bar{y}^*$, $\bar{t}^*$ only and are free from $\bar{x}^*$ and in the direction of $\bar{x}^*$, the plate is considered transfinite.

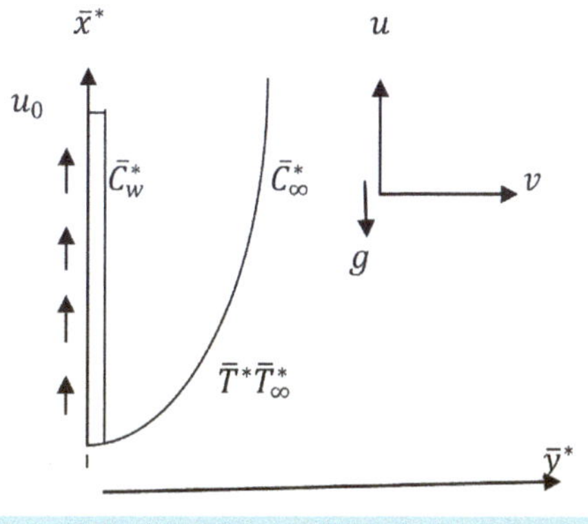


**Figure 1.** A sketch of flowmodel and coordinate system.

The equations for this present flow, by the Boussinesq estimation are as follows

$$\frac{\partial \overline{u}^*}{\partial \overline{t}^*} = g\beta\left(\overline{T}^* - \overline{T}_\infty^*\right) + g\overline{\beta}\left(\overline{C}^* - \overline{C}_\infty^*\right) + \nu\frac{\partial^2 \overline{u}^*}{\partial \overline{y}^{*2}} \tag{1}$$

$$\rho C_p \frac{\partial \overline{T}^*}{\partial \overline{t}^*} = \kappa\frac{\partial^2 \overline{T}^*}{\partial \overline{y}^{*2}} + \overline{Q}_1^*\left(\overline{C}^* - \overline{C}_\infty^*\right) - \frac{\partial \overline{q}_r^*}{\partial \overline{y}^*} \tag{2}$$

$$\frac{\partial \overline{C}^*}{\partial \overline{t}^*} = \mathcal{D}\frac{\partial^2 \overline{C}^*}{\partial y^{*2}} - K_r\left(\overline{C}^* - \overline{C}_\infty^*\right) \tag{3}$$

and the connected conditions for this flow are

$$\left.\begin{aligned} &\text{at } \overline{t}^* \le 0: \overline{u}^* = 0, \overline{T}^* = \overline{T}_\infty^*, \overline{C}^* = \overline{C}_\infty^* \quad \forall \overline{y}^* \\ &\text{at } \overline{t}^* > 0: \overline{u}^* = u_0, \frac{\partial \overline{T}^*}{\partial \overline{y}^*} = -\frac{h}{\kappa}T^*, \overline{C}^* = \overline{C}_w^* \quad \text{for } \overline{y}^* = 0 \\ &\overline{u}^* = 0, \overline{T}^* \to \overline{T}_\infty^*, \overline{C}^* \to \overline{C}_\infty^* \quad \text{as } \overline{y}^* \to \infty \end{aligned}\right\} \tag{4}$$

Here $A = \frac{u_0^2}{\upsilon}$

The term of radiative heat flux by the Rosseland estimation is given by

$$\overline{q}_r^* = -\frac{4\sigma_s}{3k_c}\frac{\partial \overline{T}^{*4}}{\partial \overline{y}^*} \tag{5}$$

But here with in the flow, presuming that the deviation in temperatures can be showed as a linear combination of the temperatures and around $\overline{T}_\infty^*$, which is found by expanding $\overline{T}^{*4}$ in a Taylor's series as follows:

$$\overline{T}^{*4} = \overline{T}_\infty^{*4} + 4\overline{T}_\infty^{*3}\left(\overline{T}^* - \overline{T}_\infty^*\right) + 6\overline{T}_\infty^{*2}\left(\overline{T}^* - \overline{T}_\infty^*\right) + \cdots \tag{6}$$

and ignoring the higher ordered terms, beyond the first degree, we get

$$\overline{T}^{*4} \cong 4\overline{T}_\infty^{*3}\overline{T}^* - 3\overline{T}_\infty^{*4} \tag{7}$$

Differentiating Equation (5) with respect to $\overline{y}^*$ and applying Equation (6), we get

$$\frac{\partial \overline{q}_r^*}{\partial \overline{y}^*} = -\frac{16\sigma_s \overline{T}_\infty^{*3}}{3k_c}\frac{\partial^2 \overline{T}^*}{\partial \overline{y}^{*2}} \tag{8}$$

We have inserted the non-dimensional succeeding measures

$$\left.\begin{aligned} &U = \frac{\overline{u}^*}{u_0}, \mathcal{t} = \frac{\overline{t}^* u_0^2}{\nu}, \mathcal{y} = \frac{\overline{y}^* u_0}{\nu}, Pr = \frac{\mu c_p}{\kappa}, Gr = \frac{g\beta\nu\overline{T}_\infty^*}{u_0^3}, \\ &\theta = \frac{\overline{T}^* - \overline{T}_\infty^*}{\overline{T}_\infty^*}, Sc = \frac{\nu}{\mathcal{D}}, Q_1 = \frac{\nu\overline{Q}_1^*\left(\overline{C}_w^* - \overline{C}_\infty^*\right)}{\rho C_p u_0^2 \overline{T}_\infty^*}, Gm = \frac{g\overline{\beta}\nu\left(\overline{C}_w^* - \overline{C}_\infty^*\right)}{u_0^3}, \\ &C = \frac{\overline{C}^* - \overline{C}_\infty^*}{\overline{C}_w^* - \overline{C}_\infty^*}, K = \frac{K_r\nu}{u_0^2}, N_r = \frac{16\sigma_s\overline{T}_\infty^{*3}}{3\kappa k_c} \end{aligned}\right\} \tag{9}$$

The Equations (1)-(3) are reduced into the following forms by using the Equations (8) and (9) as follows

$$\frac{\partial U}{\partial \mathcal{t}} = G_r\theta + G_m C + \frac{\partial^2 u}{\partial \mathcal{y}^2} \tag{10}$$

$$\frac{\partial\theta}{\partial t}=\left(\frac{1+N_r}{P_r}\right)\frac{\partial^2\theta}{\partial y^2}+Q_1C \tag{11}$$

$$\frac{\partial C}{\partial t}=\frac{1}{S_c}\frac{\partial^2 C}{\partial y^2}-KC \tag{12}$$

In non-dimensional form, the conditions reduce to as follows

$$\left.\begin{aligned}&\text{at } t\le 0: U=0, \theta=0, C=0 \quad \forall y\\ &\text{at } t>0: U=1, \frac{\partial\theta}{\partial y}=-(1+\theta), C=1 \text{ for } y=0\\ &U=0, \theta\to 0, C\to 0 \quad \text{as } y\to\infty\end{aligned}\right\} \tag{13}$$

## 3. Solution of the Problem

The non-dimensional Equations (10)-(12) associated with the conditions given by Equations (13) are evolved by the method of Laplace transforms, and therefore the results of concentration, temperature and velocity are given by

$$C(y,t)=B_6 \tag{14}$$

$$\theta(y,t)=-\frac{1}{\sqrt{a}}B_1-B_2+\frac{1}{\sqrt{a}}B_3+N_1(M_1B_1+M_2B_2+M_3B_4+M_4B_5+M_5B_3)+N_2(B_6-B_7) \tag{15}$$

$$\begin{aligned}U(y,t)=&A_1+C_1(A_4+A_3-A_5-A_{27}+A_2-A_{28})\\ &+C_2(A_{12}+A_{13}+A_6+A_{11}+A_7+A_{10}+A_{14}+A_8+A_{15}+A_9)\\ &+C_3\big[M_6(m_1-m_2)+M_7(A_1-A_{12})+M_8(A_{16}-A_{22})\\ &+M_9(A_{17}-A_{23})+M_{10}(A_{18}-A_{24})+M_{11}(A_{19}-A_{25})\\ &+M_{12}(A_{20}-A_{26})\big]+C_4[A_1-A_{29}-A_{30}+A_{31}]\end{aligned} \tag{16}$$

## 4. The Rate of Heat Transfer

In dimensionless form, heat transfer rate from the temperature gradient is

$$Nu=-\left(\frac{\partial\theta}{\partial y}\right)_{y=0} \tag{17}$$

From Equations (15) and (17), we get

$$\begin{aligned}Nu=&\exp(at)\operatorname{erfc}(-\sqrt{at})+N_1\Bigg[\frac{M_2}{\sqrt{\pi a t}}-M_3\frac{b}{\sqrt{a}}\exp(-bt)\operatorname{erfc}(i\sqrt{bt})\\ &+\frac{i\sqrt{b}}{\sqrt{\pi a t}}e^{-2bt}(M_4-M_3)-M_4\frac{b}{\sqrt{a}}\exp(-bt)\operatorname{erfc}(-i\sqrt{bt})\\ &+M_5\sqrt{a}\exp(at)\operatorname{erfc}(-\sqrt{at})+M_5\frac{1}{\sqrt{\pi t}}\Bigg]+N_2\sqrt{KS_c}\operatorname{erf}(\sqrt{Kt})\\ &-N_2\exp(-bt)\sqrt{S_c(K-b)}\operatorname{erf}\left(\sqrt{(K-b)t}\right)\end{aligned} \tag{18}$$

## 5. The Rate of Mass Transfer

In dimensionless form, mass transfer rate from the concentration gradient is

$$Sh = -\left(\frac{\partial c}{\partial y}\right)_{y=0} \tag{19}$$

From Equations (14) and (19), we get

$$Sh = \sqrt{KS_c}\,\mathrm{erf}\left(\sqrt{Kt}\right) + \sqrt{\frac{S_c}{\pi t}}\exp(-Kt) \tag{20}$$

## 6. Deduction

The effects of this analysis are in good agreement with the results given by Rajesh [8] in the absence of parameters radiation absorption ( $Q_1$ ) and radiation ( $N_r$ ),
where

$$B_1 = \frac{1}{\sqrt{\pi t}}\exp\left(-\frac{y^2}{4at}\right),\quad B_2 = \mathrm{erfc}\left(\frac{y}{2\sqrt{at}}\right) A_1 = \mathrm{erfc}\left(\frac{y}{2\sqrt{t}}\right)$$

$$B_3 = \frac{1}{\sqrt{\pi t}}\exp\left(-\frac{y^2}{4at}\right) + \sqrt{a}\exp\left(-y+at\right)\mathrm{erfc}\left(-\sqrt{at}+\frac{y}{2\sqrt{at}}\right)$$

$$B_4 = \frac{1}{\sqrt{\pi t}}\exp\left(-\frac{y^2}{4at}\right) - i\sqrt{b}\exp\left(\frac{iy\sqrt{b}}{\sqrt{a}} - bt\right)\mathrm{erfc}\left(i\sqrt{bt}+\frac{y}{2\sqrt{at}}\right)$$

$$B_5 = \frac{1}{\sqrt{\pi t}}\exp\left(-\frac{y^2}{4at}\right) + i\sqrt{b}\exp\left(-\frac{iy\sqrt{b}}{\sqrt{a}} - bt\right)\mathrm{erfc}\left(-i\sqrt{bt}+\frac{y}{2\sqrt{at}}\right)$$

$$B_6 = \frac{1}{2}\left[\mathrm{erfc}\left(\frac{y\sqrt{S_c}}{2\sqrt{t}} - \sqrt{Kt}\right)\exp\left(-y\sqrt{KS_c}\right) + \mathrm{erfc}\left(\frac{y\sqrt{S_c}}{2\sqrt{t}} + \sqrt{Kt}\right)\exp\left(y\sqrt{KS_c}\right)\right]$$

$$B_7 = \frac{\exp(-bt)}{2}\left\{\mathrm{erfc}\left(\frac{y\sqrt{S_c}}{2\sqrt{t}} - \sqrt{(K-b)t}\right)\exp\left(-y\sqrt{S_c(K-b)}\right)\right.$$
$$\left. + \mathrm{erfc}\left(\frac{y\sqrt{S_c}}{2\sqrt{t}} + \sqrt{(K-b)t}\right)\exp\left(y\sqrt{S_c(K-b)}\right)\right\}$$

$$A_2 = \frac{1}{b(b-d)}\left[\frac{\exp(-bt)}{2}\left\{\mathrm{erfc}\left(\frac{y\sqrt{S_c}}{2\sqrt{t}} - \sqrt{(K-b)t}\right)\exp\left(-y\sqrt{S_c(K-b)}\right)\right.\right.$$
$$\left.\left. + \mathrm{erfc}\left(\frac{y\sqrt{S_c}}{2\sqrt{t}} + \sqrt{(K-b)t}\right)\exp\left(y\sqrt{S_c(K-b)}\right)\right\}\right]$$

$$A_3 = \frac{1}{d(d-b)}\left[\frac{\exp(-dt)}{2}\left\{\mathrm{erfc}\left(\frac{y\sqrt{S_c}}{2\sqrt{t}} - \sqrt{(K-d)t}\right)\exp\left(-y\sqrt{S_c(K-d)}\right)\right.\right.$$
$$\left.\left. + \mathrm{erfc}\left(\frac{y\sqrt{S_c}}{2\sqrt{t}} + \sqrt{(K-d)t}\right)\exp\left(y\sqrt{S_c(K-d)}\right)\right\}\right]$$

$$A_4 = \frac{1}{bd}\left[\frac{1}{2}\left\{\operatorname{erfc}\left(\frac{y\sqrt{S_c}}{2\sqrt{t}} - \sqrt{Kt}\right)\exp\left(-y\sqrt{S_cK}\right) + \operatorname{erfc}\left(\frac{y\sqrt{S_c}}{2\sqrt{t}} + \sqrt{Kt}\right)\exp\left(y\sqrt{S_cK}\right)\right\}\right]$$

$$A_5 = \frac{1}{b(b-d)}\left[\frac{\exp(-bt)}{2}\operatorname{erfc}\left(\frac{y}{2\sqrt{t}} - \sqrt{-bt}\right)\exp\left(-y\sqrt{-b}\right) + \operatorname{erfc}\left(\frac{y}{2\sqrt{t}} + \sqrt{-bt}\right)\exp\left(y\sqrt{-b}\right)\right]$$

$$A_6 = -\frac{1}{\sqrt{\pi a t}}\exp\left(-\frac{y^2}{4t}\right),\quad A_7 = -A_1,\quad A_8 = -\sqrt{a}A_{16},\quad A_9 = -aA_{17},$$

$$A_{10} = \frac{1}{\sqrt{a}}\left[\frac{1}{\sqrt{\pi t}}\exp\left(-\frac{y^2}{4t}\right) + \sqrt{a}\exp\left(-y\sqrt{a} + at\right)\operatorname{erfc}\left(-\sqrt{at} + \frac{y}{2\sqrt{t}}\right)\right]$$

$$A_{11} = \frac{1}{\sqrt{a}}\left[\frac{1}{\sqrt{\pi t}}\exp\left(-\frac{y^2}{4at}\right)\right],\quad A_{12} = \operatorname{erfc}\left(\frac{y}{2\sqrt{at}}\right),\quad A_{13} = \sqrt{a}A_{22},\quad A_{14} = aA_{23},$$

$$A_{15} = \frac{-1}{\sqrt{a}}\left[\frac{1}{\sqrt{\pi t}}\exp\left(-\frac{y^2}{4at}\right) + \sqrt{a}\exp\left(-y + at\right)\operatorname{erfc}\left(-\sqrt{at} + \frac{y}{2\sqrt{at}}\right)\right]$$

$$A_{16} = 2\sqrt{\frac{t}{\pi}}\exp\left(-\frac{y^2}{4t}\right) - y\operatorname{erfc}\left(\frac{y}{2\sqrt{t}}\right)$$

$$A_{17} = \left(t + \frac{y^2}{2}\right)\operatorname{erfc}\left(\frac{y}{2\sqrt{t}}\right) - y\sqrt{\frac{t}{\pi}}\exp\left(-\frac{y^2}{4t}\right)$$

$$A_{18} = \frac{1}{\sqrt{\pi t}}\exp\left(-\frac{y^2}{4t}\right) - i\sqrt{b}\exp\left(i\sqrt{b}y - bt\right)\operatorname{erfc}\left(i\sqrt{bt} + \frac{y}{2\sqrt{t}}\right)$$

$$A_{19} = \frac{1}{\sqrt{\pi t}}\exp\left(-\frac{y^2}{4t}\right) + i\sqrt{b}\exp\left(-i\sqrt{b}y - bt\right)\operatorname{erfc}\left(-i\sqrt{bt} + \frac{y}{2\sqrt{t}}\right)$$

$$A_{20} = \frac{1}{\sqrt{\pi t}}\exp\left(-\frac{y^2}{4t}\right) + \sqrt{a}\exp\left(-\sqrt{a}y + at\right)\operatorname{erfc}\left(-\sqrt{at} + \frac{y}{2\sqrt{t}}\right),$$

$$A_{22} = 2\sqrt{\frac{t}{\pi}}\exp\left(-\frac{y^2}{4at}\right) - \frac{y}{\sqrt{a}}\operatorname{erfc}\left(\frac{y}{2\sqrt{at}}\right) A_{28} = \frac{1}{bd}A_1$$

$$A_{23} = \left(t + \frac{y^2}{2a}\right)\operatorname{erfc}\left(\frac{y}{2\sqrt{at}}\right) - y\sqrt{\frac{t}{a\pi}}\exp\left(-\frac{y^2}{4at}\right)$$

$$A_{24} = \frac{1}{\sqrt{\pi t}}\exp\left(-\frac{y^2}{4at}\right) - i\sqrt{b}\exp\left(i\sqrt{b}\frac{y}{\sqrt{a}} - bt\right) erfc\left(i\sqrt{bt} + \frac{y}{2\sqrt{at}}\right)$$

$$A_{25} = \frac{1}{\sqrt{\pi t}}\exp\left(-\frac{y^2}{4at}\right) + i\sqrt{b}\exp\left(-i\sqrt{b}\frac{y}{\sqrt{a}} - bt\right)\operatorname{erfc}\left(-i\sqrt{bt} + \frac{y}{2\sqrt{at}}\right)$$

$$A_{26} = \frac{1}{\sqrt{\pi t}} \exp\left(-\frac{y^2}{4at}\right) + \sqrt{a} \exp(-y + at) \operatorname{erfc}\left(-\sqrt{at} + \frac{y}{2\sqrt{at}}\right)$$

$$A_{27} = \frac{1}{d(d-b)} \left[ \frac{\exp(-dt)}{2} \operatorname{erfc}\left(\frac{y}{2\sqrt{t}} - \sqrt{-dt}\right) \exp\left(-y\sqrt{-d}\right) + \operatorname{erfc}\left(\frac{y}{2\sqrt{t}} + \sqrt{-dt}\right) \exp\left(y\sqrt{-d}\right) \right]$$

$$A_{29} = \frac{\exp(-dt)}{2} \left\{ \operatorname{erfc}\left(\frac{y}{2\sqrt{t}} - \sqrt{-dt}\right) \exp\left(-y\sqrt{-d}\right) + \operatorname{erfc}\left(\frac{y}{2\sqrt{t}} + \sqrt{-dt}\right) \exp\left(y\sqrt{-d}\right) \right\}$$

$$A_{30} = \frac{1}{2} \left[ \operatorname{erfc}\left(\frac{y\sqrt{S_c}}{2\sqrt{t}} - \sqrt{Kt}\right) \exp\left(-y\sqrt{KS_c}\right) + \operatorname{erfc}\left(\frac{y\sqrt{S_c}}{2\sqrt{t}} + \sqrt{Kt}\right) \exp\left(y\sqrt{KS_c}\right) \right]$$

$$A_{31} = \frac{\exp(-dt)}{2} \left\{ \operatorname{erfc}\left(\frac{y\sqrt{S_c}}{2\sqrt{t}} - \sqrt{(K-d)t}\right) \exp\left(-y\sqrt{S_c(K-d)}\right) \right.$$
$$\left. + \operatorname{erfc}\left(\frac{y\sqrt{S_c}}{2\sqrt{t}} + \sqrt{(K-d)t}\right) \exp\left(y\sqrt{S_c(K-d)}\right) \right\}$$

## 7. Results and Discussion

In this field of study, in order to examine the consequences of velocity field, temperature profile and concentration profiles by allotting numerical values for several arguments for both water vapor and ammonia, Prandtl number (*Pr*) and Schmidt number (*Sc*) values are considered.

For several values of different arguments the velocities are analyzed and are presented in **Figures 2-7** at time $t = 0.4$ respectively, for both the types of heating ($Gr < 0$, $Gm < 0$) plate and cooling ($Gr > 0$, $Gm > 0$) plate. **Figure 2** depicts the result of Schmidt number (*Sc*) at time $t = 0.4$ on the flow. With an increase in Sc, it is noticed that the velocity increases for heating of the plate and decreases for cooling of the plate and as *Sc* increases at $t = 0.4$ from 0.22 - 0.30 and to 0.60, the maximum velocity of the fluid decreases by 5.7% - 10% for ammonia and by 5.6% - 10.9% for water vapor in the case of cooling plate and the minimum velocity of the fluid increases by 10% and 16.58% for ammonia and by 10.26% - 17% for water vapor in the case of heating plate. Prandtl number (*Pr*) effects on the flow are expressed in **Figure 4** at time $t = 0.4$. It is identified that there is an increase in velocity near the plate and then decreases with a point of separation moving far away from the plate in the type of

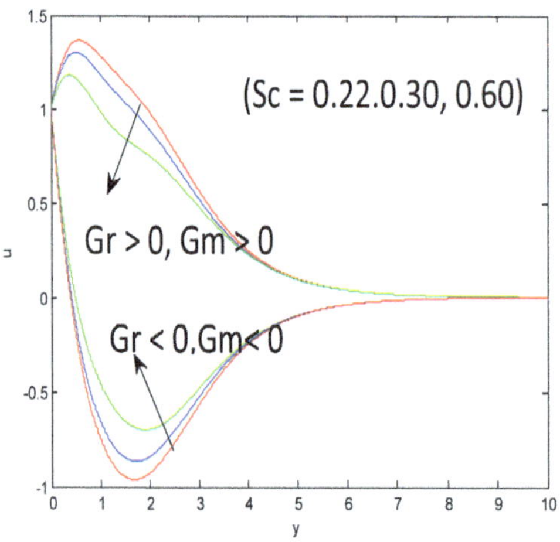


**Figure 2.** Velocity profile shows the effect of *Sc* and $Pr = 0.71$, $K = 0.2$, $Nr = 0.1$, $Q_1 = 0.1$, $t = 0.4$.

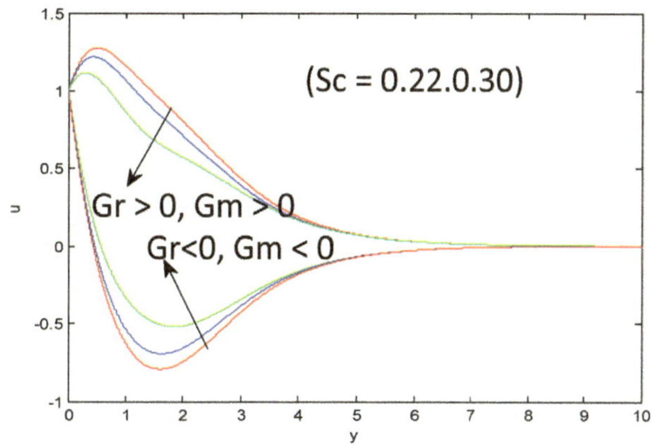


**Figure 3.** Velocity profile displays the result of *Sc* and $Pr = 0.71$, $K = 0.2$, $Nr = 0$, $Q_1 = 0$, $t = 0.4$.

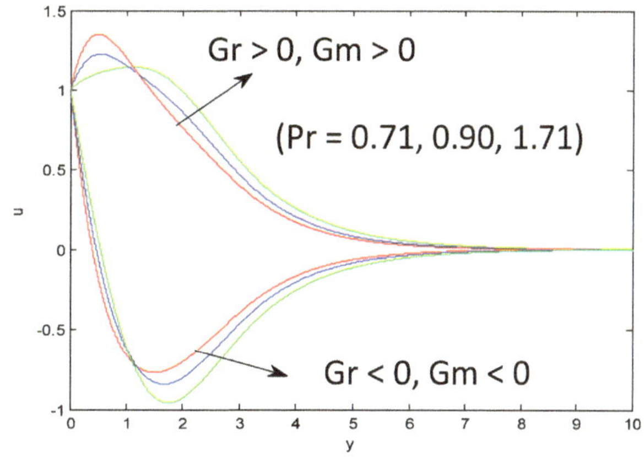


**Figure 4.** Velocity profile shows the effect of *Pr* and $Sc = 0.22$, $K = 0.2$, $Nr = 0.1$, $Q_1 = 0.1$, $t = 0.4$.

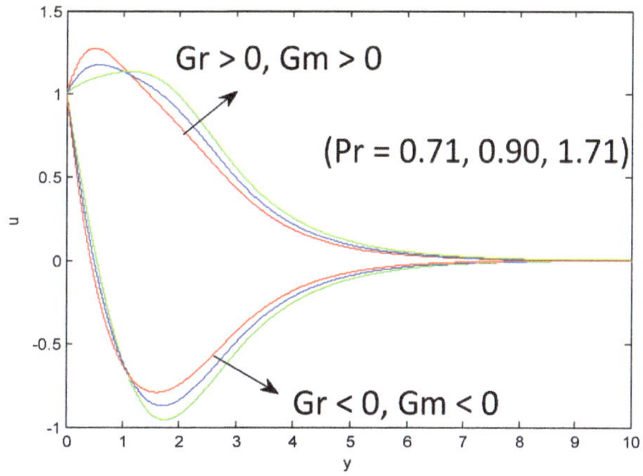


**Figure 5.** Velocity profile displays the result of *Pr* and $Sc = 0.22$, $K = 0.2$, $Nr = 0$, $Q_1 = 0$, $t = 0.4$.

heating and in the instance of cooling plate, the reverse effect is found with the increase of Pr. Owing to the variation in parameter of chemical reaction (*K*), **Figure 6** reveals the consequence of velocity profiles at time $t = 0.4$. As *K* increases, it is found that, there is an increase velocity in the plate of heating type and the velocity decreases in the plate of cooling type and as *K* increases at $t = 0.4$ from 0.2 - 2 and to 5, the maximum velocity of

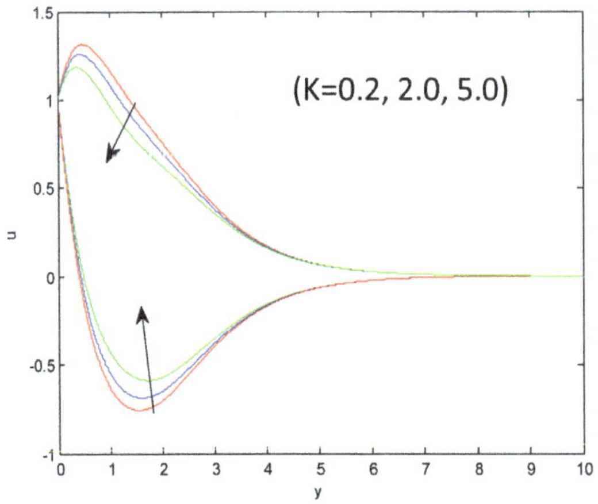


**Figure 6.** Velocity profile shows the effect of $K$ and $Sc = 0.22$, $Pr = 0.71$, $Nr = 0.1$, $Q_1 = 0.1$, $t = 0.4$.

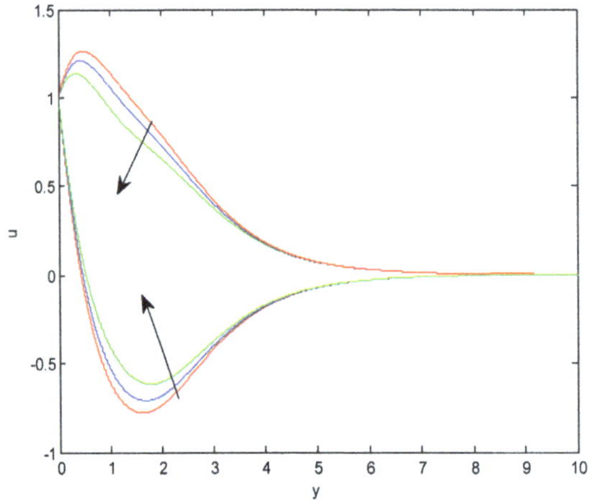


**Figure 7.** Velocity profile displays the result of $K$ and $Sc = 0.22$, $Pr = 0.71$, $Nr = 0$, $Q_1 = 0$, $t = 0.4$.

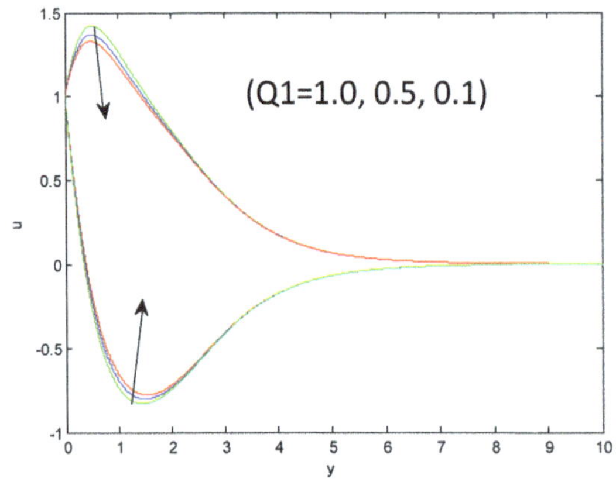


**Figure 8.** Velocity profile shows the effect of $K$ and $Sc = 0.22$, $Pr = 0.71$, $Nr = 0.1$, $Q_1 = 0.1$, $t = 0.4$.

the fluid decreases by 7.4% - 6.8% for ammonia and by 7.1% - 7.4% for water vapor in the case of cooling plate and the minimum velocity of the fluid increases by 8.4% - 9.04% for ammonia and by 5.56% - 9.61% for water vapor in the case of heating plate.

For various values of *Nr* (radiation parameter), the velocity profile is shown in **Figure 8** at $t = 0.4$. It is seen that there is a decrease in velocity near the plate and then increases with a point of separation moving away from the plate in the cooling case and the phenomenon is reversed in the case of heating type with the decrease of *Nr*. **Figure 9** describes the effects of $Q_1$ (radiation absorption parameter) at $t = 0.4$. As $Q_1$ decreases, it is observed that the velocity increases for heating plate and decreases for cooling plate. **Figure 10** reveals the result of velocity profile at several times ($t = 0.4, 0.6, 0.8$). It is identified that there is a considerable decrease in velocity in the heating type and increase when the plate is cooled as time ($t$) increases. The same results are noticed in the absence of thermal radiation and radiation absorption for different values of *Sc*, *Pr*, *K*, *t* which are shown in **Figure 3**, **Figure 5**, **Figure 7**, and **Figure 11**. Hence, these results are in good agreement with the results of Rajesh [10].

The effect of temperature profile for several values of various parameters are studied and shown in **Figures 12-17** at time $t = 0.2$. It is observed from **Figures 12-14** that the temperature rises with the fall in *Sc* and *K*. In **Figures 15-17**, it is observed that the temperature rises with the increase of *Nr*, $Q_1$, and *t*. Moreover **Figure 13** shows the results of velocity for different values of *Pr*. It is noticed from the values that, velocity increases near the

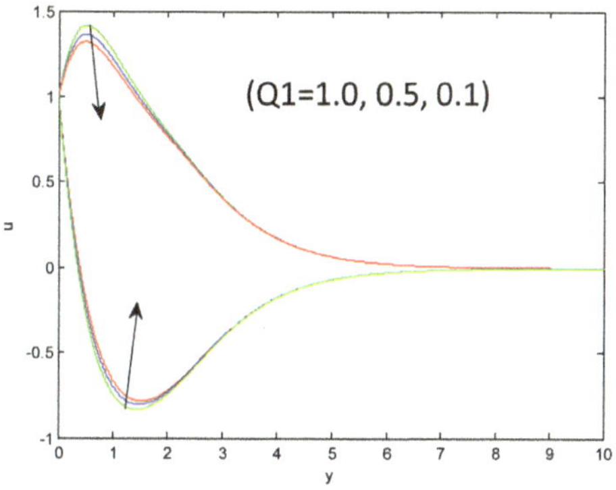


**Figure 9.** Velocity profile displays the result of *K* and $Sc = 0.22$, $Pr = 0.71$, $Nr = 0$, $Q1 = 0$, $t = 0.4$.

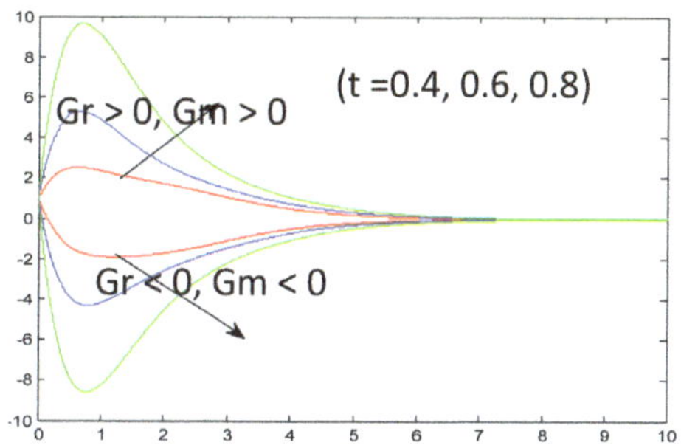


**Figure 10.** The effect of velocity profile fortand $Pr = 0.71$, $K = 0.2$, $Nr = 0.1$, $Sc = 0.22$, $Q_1 = 0.1$.

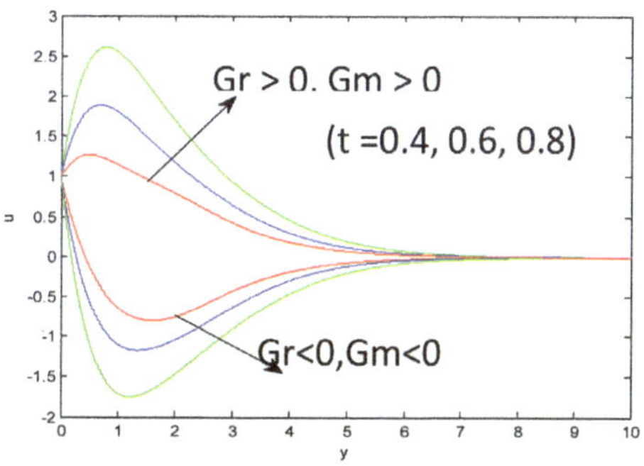


**Figure 11.** The result of velocity profile for tand $Sc = 0.22$, $K = 0.2$, $Nr = 0.1$, $Q_1 = 0.1$, $Pr = 0.71$.

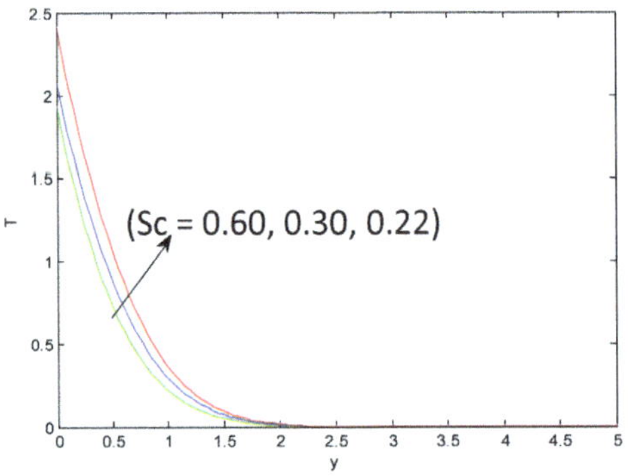


**Figure 12.** Temperature profile presents the effect of *Sc* and $Pr = 0.71$, $K = 0.2$, $Nr = 0.1$, $Q_1 = 0.1$, $t = 0.2$.

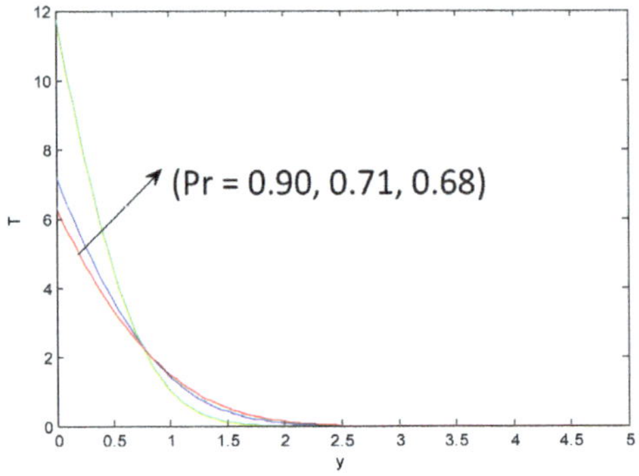


**Figure 13.** Temperature profile shows the resultof *Pr* and $Sc = 0.22$, $K = 0.2$, $Nr = 0.1$, $Q_1 = 0.1$, $t = 0.2$.

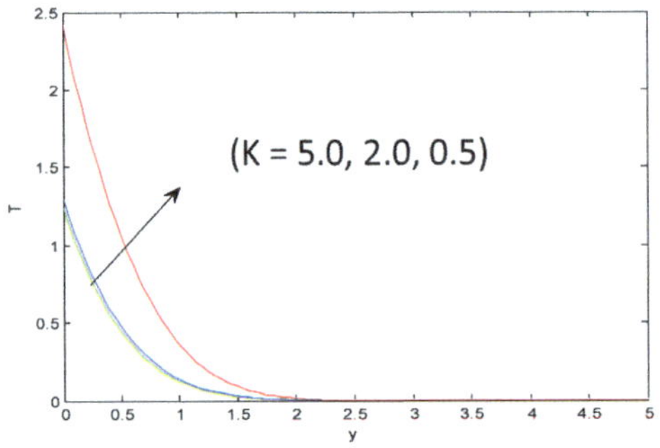


**Figure 14.** Temperature profile displays ensue of *K* and $Sc = 0.22$, $Pr = 0.71$, $Nr = 0.1$, $Q_1 = 0.1$, $t = 0.2$.

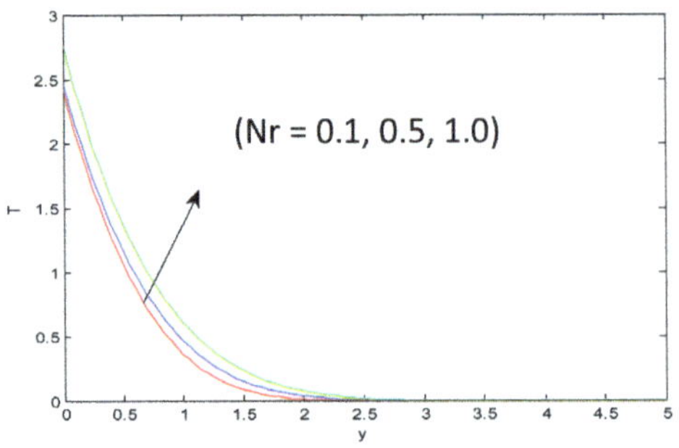


**Figure 15.** Temperature profile indicates the effect of *Nr* and 0.22, $K = 0.2$, $Q_1 = 0.1$, $Pr = 0.71$, $t = 0.2$.

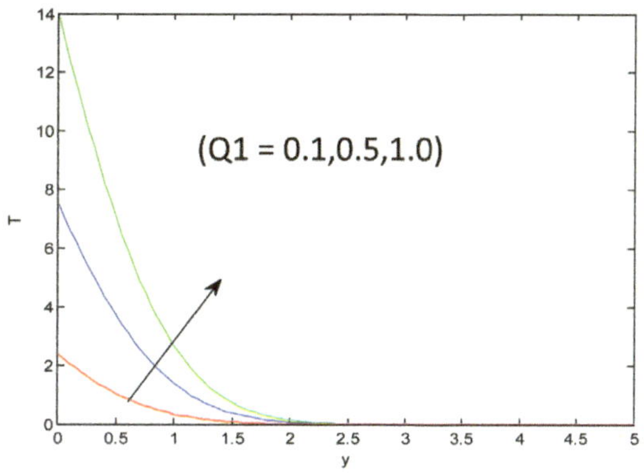


**Figure 16.** The effect of temperature profile for $Q_1$ and $Pr = 0.71$, $Sc = 0.22$, $Nr = 0.1$, $K = 0.2$, $t = 0.2$.

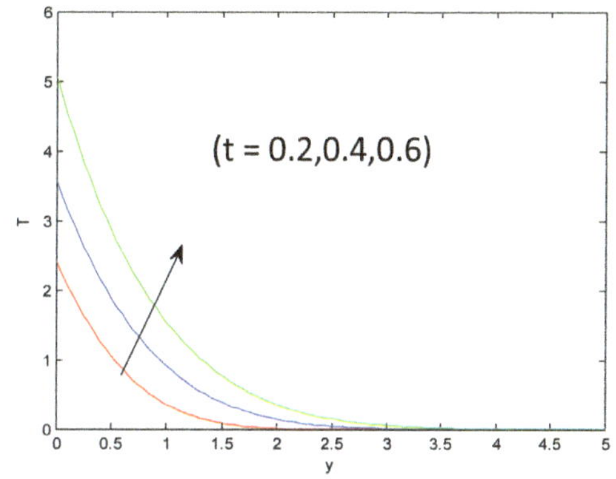


**Figure 17.** Theresult of temperature profile for tand $Sc = 0.22$, $K = 0.2$, $Q_1 = 0.1$, $Pr = 0.1$, $Nr = 0.1$.

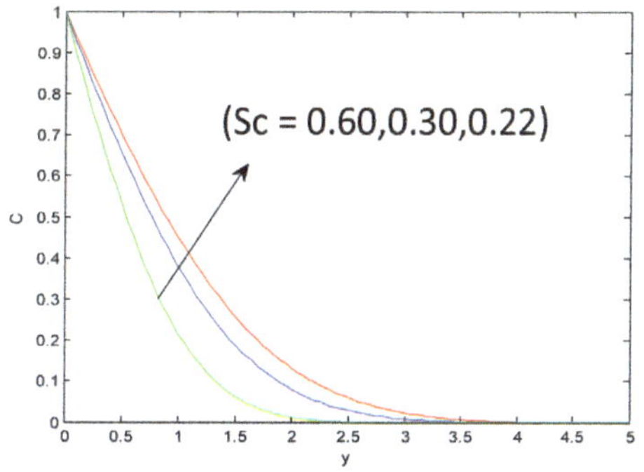


**Figure 18.** Concentration profile displays the effect of $Sc$ and $K = 0.2$, $t = 0.2$.

plate and falls far away from the plate with a point of separation.

The effect of concentration profile for several values of various parameters is studied and is presented in **Figures 18-20** at times 0.4. From **Figure 18** and **Figure 19**, it is identified that with the decrease in *Sc* and *K*, the concentration increases. And from **Figure 20**, it is found that there is a rise in concentration with the rise in time *t*. For several values of various arguments, the Sherwood number versus time is presented in **Figure 21** and **Figure 22**. It is identified that, there is an increase in Sherwood number for both hydrogen and water vapor with the increase of *Sc* and *K*.

For different values of various arguments for both hydrogen and water vapor, the Nusselt number versus time is shown in **Figure 23**. From this figure, it is observed that, there is a rise in Nusselt number with the fall in *Pr*.

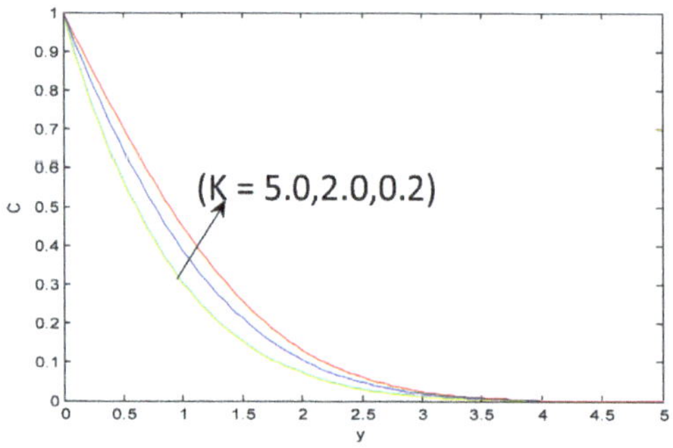


**Figure 19.** Concentration profile shows theresult of $K$ and $Sc = 0.22$, $t = 0.2$.

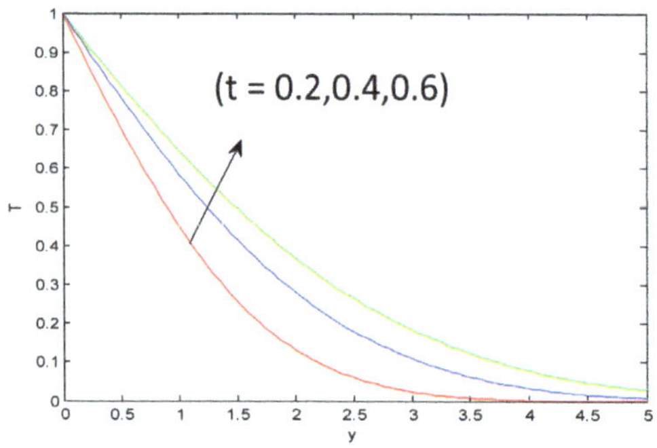


**Figure 20.** Concentration profile presents theensue of $t$ and $Sc = 0.22$, $K = 0.2$.

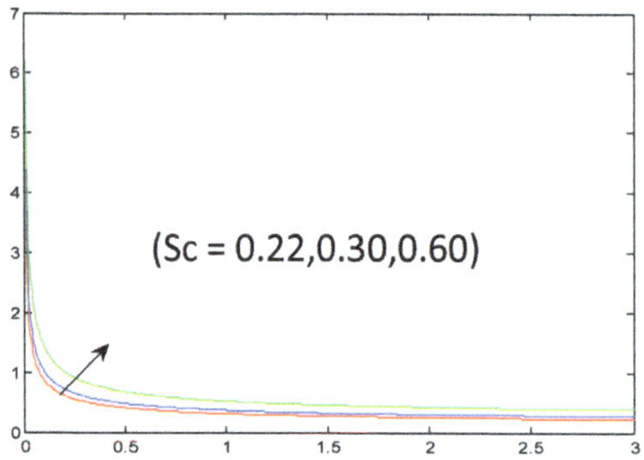


**Figure 21.** Effect of $Sc$ on the real part of Sherwood number and $K = 0.2$.

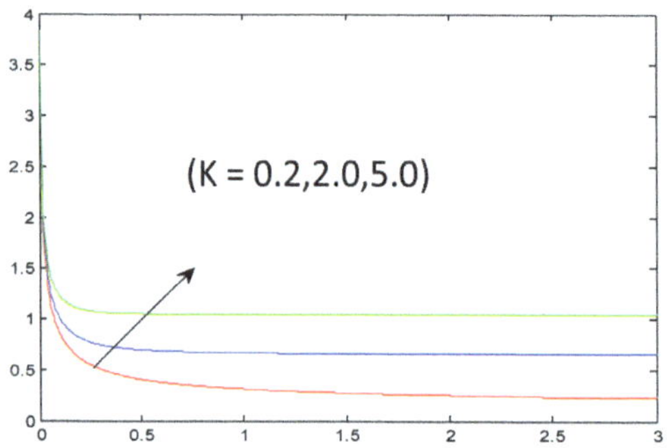


**Figure 22.** Result of $K$ on the real part of Sherwood number and $Sc = 0.22$.

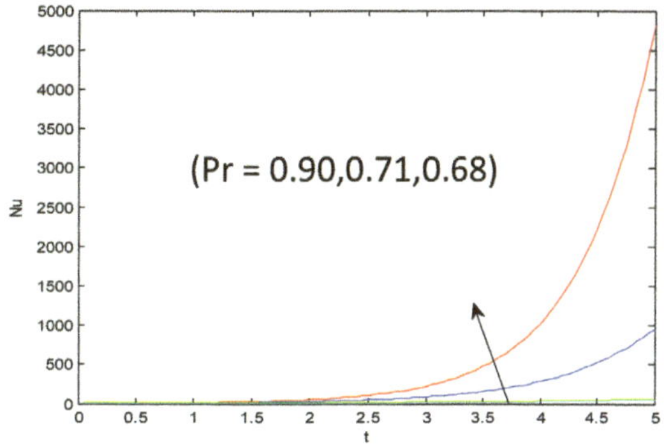


**Figure 23.** Effect of *Pr* on the real part of nusselt number and $K = 0.2$, $Sc = 0.22$, $Nr = 0.1$, $Q_1 = 0.1$.

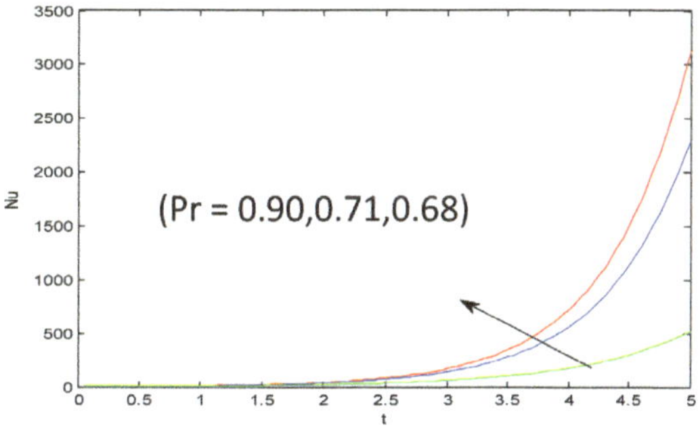


**Figure 24.** Result of *Pr* on the real part of nusselt number and $K = 0.2$, $Sc = 0.22$, $Nr = 0$, $Q_1 = 0$.

And the same results are noticed in the absence of thermal radiation and radiation absorption for different values of *Pr* which is shown in **Figure 24**. Hence, these results are in good agreement with the results of Rajesh [10].

## References

[1] Ahmed, S., Batin, A. and Chamka, A.J. (2015) Numerical/Laplace Transform Analysis for MHD Rotating Heat/ Mass Transport in a Darcian Porous Regime Bounded by an Oscillating Vertical Surface. *Alexandria Engineering Journal*, **54**, 45-54. http://dx.doi.org/10.1016/j.aej.2014.11.006

[2] Kandasamy, R., Muhaimin, I. and Saim, H.B. (2010) Group Analysis for the Effects of the Temperature Dependent Fluid Viscosity and Chemical Reaction on Free Convective Heat and Mass Transfer. *Journal of Applied Mechanics and Technical Physics*, **51**, 887-897. http://dx.doi.org/10.1007/s10808-010-0110-2

[3] Makinde, O.D. (2005) Free Convection Flow with Thermal Radiation and Mass Transfer Past a Moving Vertical Porous Plate. *International Communications in Heat and Mass Transfer*, **32**, 1411-1419. http://dx.doi.org/10.1016/j.icheatmasstransfer.2005.07.005

[4] Mukhopadhyay, S. (2009) Effect of Thermal Radiation on Unsteady Mixed Convection Flow and Heat Transfer over a Porous Stretching Surface in Porous Medium. *International Journal of Heat and Mass Transfer*, **52**, 3261-3265. http://dx.doi.org/10.1016/j.ijheatmasstransfer.2008.12.029

[5] Mukhopadhyay, S. and Layek, G.C. (2009) Radiation Effect on Forced Convective Flow and Heat Transfer over a Porous Plate in a Porous Medium. *Meccanica*, **44**, 587-597. http://dx.doi.org/10.1007/s11012-009-9211-5

[6] Muthucumaraswamy, R. and Ganesan, P. (2003) Radiation Effects on Flow Past an Impulsively Started Infinite Vertical Plate with Variable Temperature. *International Journal of Applied Mechanics and Engineering*, **8**, 125-129.

[7] Muthucumaraswamy, R. (2010) Chemical Reaction Effects on Vertical Oscillating Plate with Variable Temperature. *Chemical Industry & Chemical Engineering Quarterly*, **16**, 167-173.

[8] Hussanan, A., Ismail, Z., Khan, I., Hussein, A.G. and Shafie, S. (2014) Unsteady Boundary Layer MHD Free Convection Flow in a Porous Medium with Constant Mass Diffusion and Newtonian Heating. *The European Physical Journal Plus*, **129**, 46.

[9] Ahmed, S. (2014) Numerical Analysis for MHD Chemically Reacting and Radiating Fluid Past a Non Isothermal Im-

pulsively Started Vertical Surface Adjacent to a Porous Regime. *Ain Shams Engineering Journal*, **5**, 923-933. http://dx.doi.org/10.1016/j.asej.2014.02.005

[10] Rajesh, V. (2012) Effects of Mass Transfer on Flow Past an Impulsively Started Infinite Vertical Plate with Newtonian Heating and Chemical Reaction. *Journal of Engineering Physics and Thermophysics*, **85**, 221-228. http://dx.doi.org/10.1007/s10891-012-0642-9

## Nomenclature

$\bar{C}_{\infty}^{*}$ : concentration far away from the plate;
$\bar{C}^{*}$ : concentration;
$C$ : non-dimensional species concentration;
$\bar{C}_{w}^{*}$ : concentration at the plate;
$D$ : molecular diffusivity;
$C_p$ : specific heat at constant pressure;
$G_r$ : thermal grashoff number;
$G_m$ : mass grashoff number;
$h$ : coefficient of heat transfer;
$g$ : acceleration due to gravity;
$K_r$ : parameter of chemical reaction;
$K$ : non-dimensional parameter of chemical reaction;
$q_r$ : radiative heat flux;
$Q_1$ : parameter of radiation absorption;
$k_c$ : mean absorption coefficient;
$\sigma_s$ : Stefan-Boltzmann constant;
$N_r$ : parameter of radiation;
$Pr$ : Prandtl number;
$Sc$ : Schmidt number;
$Sh$ : Sherwood number;
$Nu$ : Nusselt number;
$\bar{T}^{*}$ : temperature;
$\bar{T}_{\infty}^{*}$, ambient temperature;
$t$ : non-dimensional time;
$\bar{t}^{*}$ : time;
$U$ : non-dimensional velocity of the in the $\bar{x}^{*}$, direction;
$\bar{u}^{*}$ : velocity of the fluid;
$u_0$ : plate velocity;
$\bar{x}^{*}$ and $\bar{y}^{*}$ : Cartesian coordinates along the plate and normal to the plate;
$y$ : non-dimensional coordinate;
$\beta$ : volumetric coefficient of thermal expansion;
$\bar{\beta}$ : volumetric coefficient of expansion with concentration;
$\mu$ : viscosity;
$\rho$ : density;
$\theta$ : non-dimensional temperature;
$\kappa$ : thermal conductivity;
$\nu$ : kinematic viscosity.

## Appendix

$$N_1 = \frac{Q_1 b}{K\sqrt{as_c}},\ N_2 = -\frac{Q_1}{aKS_c},\ d = \frac{KS_c}{S_c - 1},\ a = \frac{1+N_r}{P_r},\ b = \frac{KS_c}{S_c - \frac{1}{a}}$$

$$M_1 = E_{10} - \frac{E_5}{\sqrt{S_c}},\ M_2 = E_6 - \frac{E_4}{\sqrt{S_c}},\ M_3 = E_8 - \frac{E_1}{\sqrt{S_c}},\ M_4 = E_7 - \frac{E_2}{\sqrt{S_c}},$$

$$M_5 = E_9 - \frac{E_3}{\sqrt{S_c}},\ M_6 = P_5 - P_{12},\ M_7 = P_6 - P_{13},\ M_8 = P_7 - P_{14},$$

$$M_9 = P_1 - P_8,\ M_{10} = P_3 - P_{10},\ M_{11} = P_2 - P_9,\ M_{12} = P_4 - P_{11},$$

$$E_1 = -\frac{1}{2ib^{3/2}\left(i\sqrt{b}+\sqrt{a}\right)},\ E_2 = -\frac{1}{2ib^{3/2}\left(i\sqrt{b}-\sqrt{a}\right)},\ E_3 = \frac{1}{a(a+b)},\ E_4 = -\frac{1}{b\sqrt{a}},$$

$$E_5 = -E_1 - E_2 - E_3,\ E_6 = -\frac{\sqrt{K}}{b\sqrt{a}},\ E_7 = -\frac{\sqrt{K-b}}{2ib^{3/2}\left(i\sqrt{b}-\sqrt{a}\right)},$$

$$E_8 = -\frac{\sqrt{K-b}}{2ib^{3/2}\left(i\sqrt{b}+\sqrt{a}\right)},\ E_9 = \frac{\sqrt{K+a}}{a(a+b)},\ E_{10} = -E_8 - E_7 - E_9,$$

$$P_1 = -\frac{\sqrt{KS_c}}{b\sqrt{a}},\ P_2 = \frac{\sqrt{(K-b)S_c}}{2ib^2\sqrt{b}\left(i\sqrt{b}-\sqrt{a}\right)},\ P_3 = \frac{\sqrt{(K-b)S_c}}{2ib^2\sqrt{b}\left(i\sqrt{b}+\sqrt{a}\right)},$$

$$P_4 = \frac{\sqrt{(K+a)S_c}}{a^2(a+b)},\ P_5 = -P_2 - P_3 - P_4,\ P_6 = \sqrt{a}P_5 + \sqrt{a}P_3 + i\sqrt{b}P_3 + \sqrt{a}P_2 - i\sqrt{b}P_2,$$

$$P_7 = -bP_5 + \sqrt{a}P_6 - i\sqrt{ab}P_3 + i\sqrt{ab}P_2 - bP_4,\ P_8 = -\frac{1}{b\sqrt{a}},\ P_9 = \frac{1}{2ib^2\sqrt{b}\left(i\sqrt{b}-\sqrt{a}\right)},$$

$$P_{10} = \frac{1}{2ib^2\sqrt{b}\left(i\sqrt{b}+\sqrt{a}\right)},\ P_{11} = \frac{1}{a^2(a+b)},\ P_{12} = -P_{10} - P_9 - P_{11},$$

$$P_{13} = \sqrt{a}P_{12} + \sqrt{a}P_{10} + i\sqrt{b}P_{10} + \sqrt{a}P_9 - i\sqrt{b}P_9,\ P_{14} = -bP_{12} + \sqrt{a}P_{13} - i\sqrt{ab}P_{10} + i\sqrt{ab}P_9 - bP_{11},$$

$$C_1 = \frac{G_r Q_1 bd}{aK^2S_c^2},\ C_2 = \frac{G_r}{1-a},\ C_3 = \frac{G_r Q_1 ab}{KS_c(1-a)\sqrt{a}},\ C_4 = \frac{G_m}{KS_c}.$$

# 14

# Heat and Mass Transfer in Visco-Elastic Fluid through Rotating Porous Channel with Hall Effect

**Pradip Kumar Gaur, Abhay Kumar Jha**

Department of Mathematics, JECRC University, Jaipur, India

Email: p79gaur@gmail.com, itsabhay@rediffmail.com

## Abstract

**This paper examined the hydromagnetic boundary layer flow of viscoelastic fluid with heat and mass transfer in a vertical channel with rotation and Hall current. A constant suction and injection is applied to the plates. A strong magnetic field is applied in the direction normal to the plates. The entire system rotates with uniform angular velocity ($\Omega$), about the axis perpendicular to the plates. The governing equations are solved by perturbation technique to obtain an analytical result for velocity, temperature, concentration distributions and results are presented graphically for various values of viscoelastic parameter ($K_2$), Prandtl number (Pr), Schmidt number (Sc), radiation parameter (R), heat generation parameter (Qh) and Hall parameter (m).**

## Keywords

**Visco-Elastic Fluid, MHD, Hall Effect, Porous Medium**

## 1. Introduction

Hydromagnetic convection with heat transfer in a rotating medium has important applications in geophysics, nuclear power reactors and in underground water and energy storage system. When the strength of the magnetic field is strong, one cannot neglect the effects of Hall currents. A comprehensive discussion of Hall current is given by Cowling [1], Soundalgekar [2], Soundalgekar and Uplekar [3]. Hossain and Rashid [4] analyzed Hall effect of MHD free convective flow along porous plate with mass transfer. Attia [5] studied Hall current on the velocity and temperature fields on unsteady Hartmann flow. Effects of Hall current on free convective flow past on accelerated vertical plate in a rotating system with heat source/sink is analyzed by Singh and Garg [6]. Saha *et al.* [7] studied Hall current effect on MHD natural convection from a vertical plate. Aboeldahad and Elbarbary

[8] examined heat and mass transfer over a vertical plate in the presence of magnetic field and Hall effect. Abo-Eldahab and El Aziz [9] investigated the Hall current and Joule heating effects on electrically conducting fluid past a semi-infinite plate with strong magnetic field and heat generation/absorption. Radiation effects on free convection flow have become very important due to its applications in space technology, processes having high temperature and design of pertinent equipments. Moreover, heat and mass transfer with thermal radiation on convective flows is very important due to its significant role in the surface heat transfer. Recent developments in gas cooled nuclear reactors, nuclear power plants, gas turbines, space vehicles, and hypersonic flights have attracted research in this field. The unsteady convective flow in a moving plate with thermal radiation was examined by Cogley *et al.* [10] and Mansour [11]. The combined effects of radiation and buoyancy force past a vertical plate were analyzed by Hossain and Takhar [12]. Hossain *et al.* [13] analyzed the influence of thermal radiation on convective flows over a porous vertical plate. Seddeek [14] explained the importance of thermal radiation and variable viscosity on unsteady forced convection with an align magnetic field. Muthucumaraswamy and Senthil [15] studied the effects of thermal radiation on heat and mass transfer over a moving vertical plate. Pal [16] investigated convective heat and mass transfer in a stagnation-point flow towards a stretching sheet with thermal radiation. Aydin and Kaya [17] justified the effects of thermal radiation on mixed convection flow over a permeable vertical plate with magnetic field. Mohamed [18] studied unsteady MHD flow over a vertical moving porous plate with heat generation and Soret effect. Chauhan and Rastogi [19] analyzed the effects of thermal radiation, porosity and suction on unsteady convective hydromagnetic vertical rotating channel. Ibrahim and Makinde [20] investigated radiation effect on chemical reaction MHD boundary layer flow of heat and mass transfer past a porous vertical flat plate. Pal and Mondal [21] studied the effects of thermal radiation on MHD Darcy-Forchheimer convective flow past a stretching sheet in a porous medium. The study of heat and mass transfer due to chemical reaction is also very important because of its occurrence in most of the branches of science and technology. The processes involving mass transfer effects are important in chemical processing equipments which are designed to draw high value products from cheaper raw materials with the involvement of chemical reaction. In many industrial processes, the species undergo some kind of chemical reaction with the ambient fluid which may affect the flow behavior and the production quality of final products. Kandasamy *et al.* [22] discussed the effects of chemical reaction and magnetic field on heat and mass transfer over a vertical plate stretching surface. Muthu-cumaraswamy and Janakiraman [23] analyzed the effects of mass transfer over a vertical oscillating plate with chemical reaction. Sharma and Singh [24] have analyzed the unsteady MHD free convection flow and heat transfer over a vertical porous plate in the presence of internal heat generation and variable suction. Sudheer Babu and Satya Narayan [25] examined chemical reaction and thermal radiation effects on MHD convective flow in a porous medium in the presence of suction. Makinde and Chinyoka [26] studied the effects of magnetic field on MHD Couette flow of a third-grade fluid with chemical reaction. Recently, Pal and Talukdar [27] investigated the influence of chemical reaction and Joule heating on unsteady convective viscous dissipating fluid over a vertical plate in porous media with thermal radiation and magnetic field. Despite the above studies, attention has hardly been focused to study the effects of the Hall current on unsteady hydromagnetic non-Newtonian fluid flows. Such work seems to be important and useful partly for gaining a basic understanding of such flows and partly possible applications of these fluids in chemical process industries, food and construction engineering, movement of biological fluids. Another important field of application is the electromagnetic propulsion. The study of such system, which is closely associated with magneto-chemistry, requires a complete understanding of the equation of state shear stress-shear rate relationship, thermal conductivity and radiation. Some of these properties are undoubtedly influenced by the presence of an external magnetic field. Sarpkaya (1961) discussed the steady flow of a uniformly conducting non-Newtonian incompressible fluid between two parallel plates. The fluid considered is under the influence of constant pressure gradient. Aldos *et al.* [28] studied MHD mixed convection flow from a vertical plate embedded in porous medium. Rajgopal *et al.* [29] analyzed an oscillatory mixed convection flow of a viscoelastic electrically conducting fluid in an infinite vertical channel filed with porous medium. Considering the Hall effects Attia [30] discussed unsteady Hartmann flow of a viscoelastic fluid. Chaudhary and Jha [31] analyzed heat and mass transfer in elastic-viscous fluid past an impulsively started infinite vertical plate with Hall current. Singh [32] investigated MHD mixed convection visco-elastic slip-flow through a porous medium in a vertical porous channel with thermal radiation. The objective of the present study is to analyze the effects of Hall current, thermal radiation, and first-order chemical reaction on the oscillatory convective flow of viscoelastic fluid with suction/injection in a rotating vertical porous channel.

## 2. Mathematical Formulation

The constitutive equations for the rheological equation of state for an elastico-viscous fluid (Walter's liquid $B'$) are

$$p_{ik} = -pg_{ik} + p'_{ik} \,. \quad (1)$$

$$p'_{ik} = 2\int_{-\infty}^{t} \psi(t-t') e_{ik}^{(1)}(t') \mathrm{d}t' \,. \quad (2)$$

in which

$$\psi(t-t') = \int_{0}^{\infty} \frac{N(\tau)}{\tau} \mathrm{e}^{-(t-t')\tau} \mathrm{d}\tau \,. \quad (3)$$

$N(\tau)$ is the distribution function of relaxation times. In the above equations $p_{ik}$ is the stress tensor, p an arbitrary isotropic pressure, $g_{ik}$ is the metric tensor of a fixed co-ordinate system $x_i$ and $e_{ik}^{(1)}$ is the rate of strain tensor. It was shown by Walter's [33] that Equation (2) can be put in the following generalized form which is valid for all types of motion and stress

$$p'^{ik}(x,t) = 2\int_{-\infty}^{t} \psi(t-t') \frac{\partial x^i}{\partial x'^m} \frac{\partial x^k}{\partial x'^r} e^{(1)mr}(x't') \mathrm{d}t' \quad (4)$$

where $x''$ is the position at time $t'$ of the element that is instantaneously at the print $x^i$ at time "$t$". The fluid with equation of state (1) and (4) has been designated as liquid $B'$. In the case of liquids with short memories, *i.e.* short relaxation times, the above equation of state can be written in the following simplified form

$$p'^{ik}(x,t) = 2\eta_0 e^{(1)ik} - 2k_0 \frac{\partial e^{(1)ik}}{\partial t} \,. \quad (5)$$

In which $\eta_0 = \int_0^{\infty} N(\tau)\mathrm{d}\tau$ is the limiting viscosity at small rates of shear, $K_0 = \int_0^{\infty} N(\tau)\mathrm{d}\tau$ and $\frac{\partial}{\partial t}$ denotes the convected time derivative. We consider Oscillatory free convective flow of a viscous incompressible and electrically conducting fluid between two insulating infinite vertical permeable plates. A strong transverse magnetic field of uniform strength $H_0$ is applied along the axis of rotation by neglecting induced electric and magnetic fields. The fluid is assumed to be a gray, emitting, and absorbing, but non scattering medium. The radiative heat flux term can be simplified by using the Rosseland approximation. It is also assumed that the chemically reactive species undergo first-order irreversible chemical reaction.

The equations governing the flow of fluid together with Maxwell's electromagnetic equations are as follows.

Equation of Continuity

$$\nabla \cdot V = 0 \,. \quad (6)$$

Momentum Equation

$$\frac{\partial V}{\partial t} + (V\cdot\nabla)V = -\frac{1}{\rho}\nabla P + \nabla\cdot p_{ij} + g\beta(T-T_\infty) + g\beta^*(C-C_\infty) + \frac{1}{\rho}(J\times B) \,. \quad (7)$$

Energy Equation

$$\rho C_p\left[\frac{\partial T}{\partial t} + (V\cdot\nabla)T\right] = k\nabla^2 T - \nabla q \,. \quad (8)$$

Concentration Equation

$$\rho C_p\left[\frac{\partial C}{\partial t} + (V\cdot\nabla)C\right] = k\nabla^2 C - K_c C \,. \quad (9)$$

The generalized Ohm's law, in the absence of the electric field [34], is of the form.

$$\boldsymbol{J} + \frac{\omega_e \tau_e}{H_0}(\boldsymbol{J}\times\boldsymbol{H}) = \sigma\left(\mu_e \boldsymbol{V}\times\boldsymbol{H} + \frac{1}{en_e}\nabla p_e\right) \quad (10)$$

where $\bar{V}, \omega_e, \tau_e, \mu_e, n_e, e, \sigma$ and $p_e$ are velocity, the electrical conductivity, the magnetic permeability, the cyclotron frequency, the electron collision time, the electric charge, the number density of the electron and the electron pressure, respectively. Under the usual assumption, the electron pressure (for a weakly ionized gas), the thermoelectric pressure, and ion slip are negligible, so we have from the Ohm's law.

$$J_x^* + \omega_e \tau_e J_y^* = \sigma \mu_e H_0 v^*, \; J_y^* + \omega_e \tau_e J_x^* = \sigma \mu_e H_0 v^* . \tag{11}$$

From which we obtain that

$$j_x^* = \frac{\sigma \mu_e H_0 \left( mu^* + v^* \right)}{1+m^2}, \; J_y^* = \frac{\sigma \mu_e H_0 \left( mv^* - u^* \right)}{1+m^2} . \tag{12}$$

The solenoidal relation for the magnetic field $\nabla \cdot \boldsymbol{H} = 0$ where $\boldsymbol{H} = \left( \boldsymbol{H}_x, \boldsymbol{H}_y, \boldsymbol{H}_z \right)$ gives $\boldsymbol{H}_z = \boldsymbol{H}_0$ (constant) everywhere in the flow, which gives $\boldsymbol{H} = \left(0, 0, H_0\right)$. If $\left( \boldsymbol{J}_x, \boldsymbol{J}_y, \boldsymbol{J}_z \right)$ are the component of electric current density $\boldsymbol{J}$, then the equation of conservation of electric charge $\nabla \cdot \boldsymbol{J} = 0$ gives $\boldsymbol{J}_z = \text{constant}$.

Since the plates are infinite in extent, all the physical quantities except the pressure depend only on $z^*$ and $t^*$. A cartesian coordinate system is assumed and $z^*$-axis is taken normal to the plates, while $x^*$-axis and $y^*$-axis are in the upward and perpendicular directions on the plate $z^* = 0$ (origin), respectively .The velocity components $u^*$, $v^*$, $w^*$ are in the $x^*$-, $y^*$-, $z^*$-directions respectively. The governing equations in the rotating system in presence of Hall current, thermal radiation and chemical reaction are given by the following equations.

$$\frac{\partial w^*}{\partial z^*} = 0 \Rightarrow w^* = w_0 \tag{13}$$

$$\frac{\partial u^*}{\partial t^*} + w_0 \frac{\partial u^*}{\partial z^*} - 2\Omega^* v^* = -\frac{1}{\rho} \frac{\partial p^*}{\partial x^*} + v \frac{\partial^2 u^*}{\partial z^{*2}} + g_0 \beta \left( T^* - T_d \right) + g_0 \beta^* \left( C^* - C_d \right) \frac{H_0}{\rho} J_y^* - K_0 \frac{\partial^3 u^*}{\partial z^{*2} \partial t^*} \tag{14}$$

$$\frac{\partial v^*}{\partial t^*} + w_0 \frac{\partial v^*}{\partial z^*} + 2\Omega^* u^* = -\frac{1}{\rho} \frac{\partial p^*}{\partial y^*} + v \frac{\partial^2 v^*}{\partial z^{*2}} - \frac{H_0}{\rho} J_y^* - K_0 \frac{\partial^3 u^*}{\partial z^{*2} \partial t^*} \tag{15}$$

$$\frac{\partial T^*}{\partial t^*} + w_0 \frac{\partial T^*}{\partial z^*} = \frac{k}{\rho c_p} \frac{\partial^2 T^*}{\partial z^{*2}} - \frac{Q_0}{\rho c_p} \left( T^* - T_d \right) - \frac{1}{\rho c_p} \frac{\partial q_r^*}{\partial z^*} \tag{16}$$

$$\frac{\partial C^*}{\partial t^*} + w_0 \frac{\partial C^*}{\partial z^*} = D_m \frac{\partial^2 C^*}{\partial z^{*2}} - K_1 \left( C^* - C_d \right) \tag{17}$$

where $m \left( = \omega_e \tau_e \right)$ is the Hall parameter, $\beta$ and $\beta^*$ are coefficients of thermal and solutal expansion, $c_p$ is the specific heat at constant pressure, $\rho$ is the density of the fluid, v is the kinematics viscosity $k$ is the fluid thermal conductivity, $g_0$ is the acceleration of gravity, $Q_0$ is the additional heat source, $q_r^*$ is the radiative heat flux, $D_m$ is the molecular diffusivity, $K_1$ is the chemical reaction rate constant .The radiative heat flux is given by $q_r^* = -\frac{4\sigma^*}{3K^*} \frac{\partial T^{*4}}{\partial z^*}$, in which $\sigma^*$ and $K^*$ are the Stefan-Boltzmann constant and the mean absorption coefficient, respectively.

The initial and boundary conditions as suggested by the physics of the problem are

$$\begin{aligned} & u^* = v^* = 0, T^* = T_0 + \varepsilon \left( T_0 - T_d \right) \cos \omega^* t^* \\ & C^* = T^* = C_0 + \varepsilon \left( C_0 - C_d \right) \cos \omega^* t^* \text{ at } z^* = 0 \\ & u^* = U^* \left( t^* \right) = U_0 \left( 1 + \varepsilon \cos \omega^* t^* \right), v^* = 0 \text{ at } z^* = 0 \\ & T^* = T_d, C^* = C_d \text{ at } z^* = d \end{aligned} \tag{18}$$

where $\varepsilon$ is a small constant.

We now introduce the dimensionless variables and parameter as follows:

$$\eta = \frac{z^*}{d},\ u = \frac{u^*}{U_0},\ v = \frac{v^*}{U_0},\ t = t^*\omega^*,\ \omega = \frac{\omega^* d^2}{v},\ \Omega = \frac{\Omega^* d^2}{v},$$
$$\lambda = \frac{w_0 d}{v},\ \theta = \frac{T^* - T_d}{T_0 - T_d},\ \phi = \frac{C^* - C_d}{C_0 - C_d}. \tag{19}$$

After combining (14) and (15) and taking $q = u + iv$, then (14)-(17) reduce to

$$\omega\frac{\partial q}{\partial t} + \lambda\frac{\partial q}{\partial \eta} = \frac{\partial^2 q}{\partial \eta^2} + \omega\frac{\partial U}{\partial t} - 2i\Omega(q - U) - \frac{M^2(1+im)}{1+m^2}(q-U) + \lambda^2(Gr\theta + Gm\phi) - K_0\frac{\partial^3 q}{\partial \eta^2 \partial t}$$
$$\omega\frac{\partial \theta}{\partial t} + \lambda\frac{\partial \theta}{\partial \eta} = \frac{1}{\Pr}\left[1 + \frac{4}{3R}\right]\frac{\partial^2 q}{\partial \eta^2} - \frac{Q_H}{\Pr}\theta \tag{20}$$
$$\omega\frac{\partial \varphi}{\partial t} + \lambda\frac{\partial \varphi}{\partial \eta} = \frac{1}{Sc}\frac{\partial^2 \phi}{\partial \eta^2} - \xi\phi$$

where $Gr = g_0\beta v(T_0 - T_d)/U_0\,\omega_0^2$ is the modified Grashof number, $\Pr = v\rho c_p/k$ is the Prandtl number, $Gm = g_0\beta^* v(C_0 - C_d)/U_0\,\omega_0^2$ is the modified solutal Grashof number, $M = H_0 d\sqrt{\sigma/\mu}$ is the Hartmann number, $R = KK^*/4\sigma^* T_d$ is the radiation parameter, $Sc = v/Dm$ is the Schmidt number, and $\xi = K_1 d^2/v$ is the reaction parameter.

The boundary conditions (9) can be expressed in complex form as:

$$q = 0, \theta = 1 + \frac{\varepsilon}{2}\left(e^{it} + e^{-it}\right),\ \phi = 1 + \frac{\varepsilon}{2}\left(e^{it} + e^{-it}\right)$$
$$q = U(t) = 1 + \frac{\varepsilon}{2}\left(e^{it} + e^{-it}\right),\ \phi = 0, \theta = 0 \text{ at } \eta = 1 \tag{21}$$

## 3. Method of Solution

The set of partial differential Equations (20) cannot be solved in closed form. So it is solved analytically after these equations are reduced to a set of ordinary differential equations in dimensionless form. We assume that

$$R(\eta, t) = R_0(\eta) + \frac{\varepsilon}{2}\left(R_1(\eta)e^{it} + R_2(\eta)e^{-it}\right) \tag{22}$$

where R stands for q or $\theta$ or $\phi$ and $\varepsilon \ll 1$ which is applicable for small perturbation.

Substituting (22) into (20) and comparing the harmonic and non harmonic terms, we obtain the following ordinary differential equations:

$$q_{0'}'' - \lambda q_0' - Sq_0 = -S - \lambda^2(Gr\theta_0 + Gm\phi_0);$$
$$q_1'' - \lambda q_1' - (S + iw)q_1 = -(S + iw) - \lambda^2(Gr\theta_1 + Gm\phi_1)$$
$$q_2'' - \lambda q_2' - (S - iw)q_1 = -(S - iw) - \lambda^2(Gr\theta_2 + Gm\phi_2)$$
$$\theta_{0'}'' - \frac{3\lambda\Pr R}{3R+4}\theta_0' - \frac{3RQ_H}{3R+4}\theta_0 = 0$$
$$\theta_1'' - \frac{3\lambda\Pr R}{3R+4}\theta_1' - \frac{3R}{3R+4}(iw\Pr + Q_H)\theta_1 = 0 \tag{23}$$
$$\theta_2'' - \frac{3\lambda\Pr R}{3R+4}\theta_2' + \frac{3R}{3R+4}(iw\Pr - Q_H)\theta_2 = 0$$
$$\phi_0'' - Sc\lambda\phi_0' - Sc\xi\phi_0 = 0$$
$$\phi_1'' - Sc\lambda\phi_1' - Sc(iw + \xi)\phi_1 = 0$$
$$\phi_2'' - Sc\lambda\phi_2' + Sc(iw - \xi)\phi_2 = 0$$

where $S = \left(M^2(1+im)/\left(1+m^2\right)\right) + 2i\Omega$ and dashes denote the derivatives w.r.t $\eta$.

The Transformed boundary conditions are

$$\begin{aligned} &q_0 = 0, q_1 = 0, q_2 = 0; \theta_0 = 1, \theta_1 = 1, \theta_2 = 1; \phi_0 = 1, \phi_1 = 1, \phi_2 = 1 \text{ at } \eta = 0 \\ &q_0 = 1, q_1 = 1, q_2 = 1; \theta_0 = 0, \theta_1 = 0, \theta_2 = 0; \phi_0 = 0, \phi_1 = 0, \phi_2 = 0 \text{ at } \eta = 1 \end{aligned} . \tag{24}$$

The solutions of (23) under the boundary conditions are

$$\begin{aligned} q_0 = {} & 1 - e^{h_{14}\eta} + A_1\left(e^{h_{14}\eta} - e^{h_8\eta}\right) - A_2\left(e^{h_{14}\eta} - e^{h_7\eta}\right) + A_3\left(e^{h_{14}\eta} - e^{h_2\eta}\right) - A_4\left(e^{h_{14}\eta} - e^{h_1\eta}\right) \\ & + \frac{\left(e^{h_{14}\eta} - e^{h_{13}\eta}\right)}{\left(e^{h_{13}} - e^{h_{14}}\right)} \times \left[A_1\left(e^{h_{14}} - e^{h_8}\right) - A_2\left(e^{h_{14}} - e^{h_7}\right) + A_3\left(e^{h_{14}} - e^{h_2}\right) - A_4\left(e^{h_{14}} - e^{h_1}\right)\right] \end{aligned} \tag{25}$$

$$\begin{aligned} q_1 = {} & 1 - e^{h_{16}\eta} + A_5\left(e^{h_{16}\eta} - e^{h_{10}\eta}\right) - A_6\left(e^{h_{16}\eta} - e^{h_9\eta}\right) + A_7\left(e^{h_{16}\eta} - e^{h_4\eta}\right) - A_8\left(e^{h_{16}\eta} - e^{h_3\eta}\right) \\ & + \frac{\left(e^{h_{15}\eta} - e^{h_{16}\eta}\right)}{\left(e^{h_{15}} - e^{h_{16}}\right)} \times \left[A_5\left(e^{h_{16}} - e^{h_{10}}\right) - A_6\left(e^{h_{16}} - e^{h_9}\right) + A_7\left(e^{h_{16}} - e^{h_4}\right) - A_8\left(e^{h_{16}} - e^{h_3}\right)\right] \end{aligned} \tag{26}$$

$$\begin{aligned} q_2 = {} & 1 - e^{h_{18}\eta} + A_9\left(e^{h_{18}\eta} - e^{h_{12}\eta}\right) - A_{10}\left(e^{h_{18}\eta} - e^{h_{11}\eta}\right) + A_{11}\left(e^{h_{18}\eta} - e^{h_6\eta}\right) - A_{12}\left(e^{h_{18}\eta} - e^{h_5\eta}\right) \\ & + \frac{\left(e^{h_{17}\eta} - e^{h_{18}\eta}\right)}{\left(e^{h_{17}} - e^{h_{18}}\right)} \times \left[A_9\left(e^{h_{18}} - e^{h_{12}}\right) - A_{10}\left(e^{h_{18}} - e^{h_{11}}\right) + A_{11}\left(e^{h_{18}} - e^{h_6}\right) - A_{12}\left(e^{h_{18}} - e^{h_5}\right)\right] \end{aligned} \tag{27}$$

$$\theta_0 = \frac{e^{h_7 + h_8\eta} - e^{h_8 + h_7\eta}}{e^{h_7} - e^{h_8}} \tag{28}$$

$$\theta_1 = \frac{e^{h_9 + h_{10}\eta} - e^{h_{10} + h_9\eta}}{e^{h_9} - e^{h_{10}}} \tag{29}$$

$$\theta_2 = \frac{e^{h_{11} + h_{12}\eta} - e^{h_{12} + h_{11}\eta}}{e^{h_{11}} - e^{h_{12}}} \tag{30}$$

$$\phi_0 = \frac{e^{h_1 + h_2\eta} - e^{h_2 + h_1\eta}}{e^{h_1} - e^{h_2}} \tag{31}$$

$$\phi_1 = \frac{e^{h_3 + h_4\eta} - e^{h_4 + h_3\eta}}{e^{h_3} - e^{h_4}} \tag{32}$$

$$\phi_2 = \frac{e^{h_5 + h_6\eta} - e^{h_6 + h_5\eta}}{e^{h_5} - e^{h_6}} . \tag{33}$$

### 3.1. Amplitude and Phase Difference Due to Steady and Unsteady Flow

Equation (25) corresponds to the steady part, which gives $_0$ as the primary and $V_0$ as secondary velocity components. The amplitude (resultant velocity) and phase difference due to these primary and secondary velocities for the steady flow are given by

$$R_0 = \sqrt{\left(u_0^2 + v_0^2\right)}, \quad \alpha = \tan^{-1}\left(\frac{v_0}{u_0}\right) \tag{34}$$

where $q_0(\eta) = u_0(\eta) + iv_0(\eta)$.

Equation (26) and (27) together give the unsteady part of the flow. Thus unsteady primary and secondary velocity components $u_1(\eta)$ and $v_1(\eta)$, respectively, for the fluctuating flow can be obtained from the following.

$$\begin{aligned} u_1(\eta,t) &= \left[\text{Real}q_1(\eta) + \text{Real}q_2(\eta)\right]\cos t - \left[\text{IM}q_1(\eta) + \text{Real}q_2(\eta)\right]\sin t, \\ v_1(\eta,t) &= \left[\text{Real}q_1(\eta) - \text{Real}q_2(\eta)\right]\sin t + \left[\text{IM}q_1(\eta) + \text{Real}q_2(\eta)\right]\cos t \end{aligned} . \tag{35}$$

The amplitude (resultant velocity) and the phase difference of the unsteady flow are given by

$$R_0 = \sqrt{\left(u_1^2 + v_1^2\right)}, \quad \alpha = \tan^{-1}\left(\frac{v_1}{u_1}\right) \tag{36}$$

where $q_1(\eta)e^{it} + q_2(\eta)e^{-it} = u_1(\eta) + iv_1(\eta)$.

The amplitude (resultant velocity) and the phase difference are given by

$$R_n = \sqrt{\left(u^2 + v^2\right)}, \quad \alpha = \tan^{-1}\left(\frac{v}{u}\right) \tag{37}$$

where $u$ = Real part of $q$ and $v$ = Imaginary part of $q$.

## 3.2. Amplitude and Phase Difference of Shear Stresses Due to Steady and Unsteady Flow at the Plate

The amplitude and phase difference of shear stresses at the stationary plate ($\eta$ = 0), the steady flow can be obtained as

$$\tau_{0r} = \sqrt{\left(\tau_{0x}^2 + \tau_{0y}^2\right)}, \quad \beta = \tan^{-1}\left(\frac{\tau_{0y}}{\tau_{0x}}\right). \tag{38}$$

For the unsteady part of flow, the amplitude and phase difference of shear stresses at the stationary plate ($\eta = 0$) can be obtained as.
where

$$\begin{aligned}\left(\frac{\partial q_0}{\partial \eta}\right)_{\eta=0} &= \tau_{0x} + i\tau_{0y} = -h_{14} + A_1\left(h_{14} - h_8\right) - A_2\left(h_{14} - h_7\right) + A_3\left(h_{14} - h_2\right) - A_4\left(h_{14} - h_1\right) \\ &\quad + \frac{h_{13} - h_{14}}{\left(e^{h_{13}} - e^{h_{14}}\right)} \times \left[e^{h_{14}} - A_1\left(e^{h_{14}} - e^{h_8}\right) + A_2\left(e^{h_{14}} - e^{h_7}\right) - A_3\left(e^{h_{14}} - e^{h_2}\right) + A_4\left(e^{h_{14}} - e^{h_1}\right)\right]\end{aligned} \tag{39}$$

$$\tau_{1r} = \sqrt{\left(\tau_{1x}^2 + \tau_{1y}^2\right)}, \ \beta_1 = \tan^{-1}\left(\frac{\tau_{1y}}{\tau_{1x}}\right) \text{ where } \tau_{1x} + i\tau_{1y} = \left(\frac{\partial q_1}{\partial \eta}\right)_{\eta=0} e^{it} + \left(\frac{\partial q_2}{\partial \eta}\right)_{\eta=0} e^{-it} \tag{40}$$

where

$$\begin{aligned}\left(\frac{\partial q_1}{\partial \eta}\right)_{\eta=0} &= -h_{16} + A_5\left(h_{16} - h_{10}\right) - A_6\left(h_{16} - h_9\right) + A_7\left(h_{16} - h_4\right) - A_8\left(h_{16} - h_3\right) \\ &\quad + \frac{\left(h_{15} - h_{16}\right)}{\left(e^{h_{15}} - e^{h_{16}}\right)} \times \left[e^{h_{16}} - A_5\left(e^{h_{16}} - e^{h_{10}}\right) + A_6\left(e^{h_{16}} - e^{h_9}\right) - A_7\left(e^{h_{14}} - e^{h_4}\right) + A_8\left(e^{h_{16}} - e^{h_3}\right)\right] \\ \left(\frac{\partial q_2}{\partial \eta}\right)_{\eta=0} &= -h_{18} + A_9\left(h_{18} - h_{12}\right) - A_{10}\left(h_{18} - h_{11}\right) + A_{11}\left(h_{18} - h_6\right) - A_{12}\left(h_{18} - h_5\right) \\ &\quad + \frac{\left(h_{17} - h_{18}\right)}{\left(e^{h_{17}} - e^{h_{18}}\right)} \times \left[e^{h_{16}} - A_9\left(e^{h_{18}} - e^{h_{12}}\right) + A_{10}\left(e^{h_{18}} - e^{h_{11}}\right) - A_{11}\left(e^{h_{18}} - e^{h_6}\right) + A_{12}\left(e^{h_{18}} - e^{h_5}\right)\right].\end{aligned} \tag{41}$$

The amplitude and phase difference of shear stresses at the stationary plate ($\eta$ = 0) can be as

$$\tau = \left(\frac{\partial q}{\partial \eta}\right)_{\eta=0} = \sqrt{\left(\tau_x^2 + \tau_y^2\right)}, \quad \beta_2 = \tan^{-1}\left(\frac{\tau_y}{\tau_x}\right) \tag{42}$$

where $\tau_x =$ Real part of $\left(\frac{\partial q}{\partial \eta}\right)_{\eta=0}$ and Where $\tau_y =$ Imaginary part of $\left(\frac{\partial q}{\partial \eta}\right)_{\eta=0}$.

The Nusselt number

$$Nu=-\left(1+\frac{4}{3R}\right)\left(\frac{\partial q}{\partial \eta}\right)_{\eta=0}=N_x+iN_y. \tag{43}$$

The rate of heat transfer (*i.e.* heat flux) at the plate in terms of amplitude and phase difference is given by

$$\Theta=\sqrt{\left(N_x^2+N_y^2\right)},\ \gamma=\tan^{-1}\left(\frac{N_y}{N_x}\right). \tag{44}$$

The Sherwood number

$$Sh=\left(\frac{\partial \phi}{\partial \eta}\right)_{\eta=0}=M_x+iM_y. \tag{45}$$

The rate of mass transfer (*i.e.* mass flux) at the plate in terms of amplitude and phase difference is given by.

$$\Phi=\sqrt{\left(M_x^2+M_y^2\right)},\ \gamma=\tan^{-1}\left(\frac{M_y}{M_x}\right). \tag{46}$$

## 4. Results and Discussion

The system of ordinary differential Equation (23) with boundary condition (24) is solved analytically using perturbation technique. The solutions are obtained for the steady and unsteady velocity fields from (25)-(27), temperature fields from (28)-(30) and concentration fields are given by (31)-(33). The influences of each of the parameters on the thermal mass and hydrodynamic behaviors of buoyancy-induced flow in a rotating vertical channel are studied. The results are presented graphically. Temperature of the heated wall (left wall) $z^*=0$ is a function of time as given in the boundary conditions and the cooled wall at $z^*=d$ is maintained at a constant temperature. Further it is assumed that the temperature difference is small enough so that the density changes of the fluid in the system will be small. When the injection/suction parameter $\lambda$ is positive, fluid is injected through the hot wall into the channel and sucked out through the cold wall. The effect of various physical parameters on flow, heat, concentration fields, skin-friction Nusselt number, and Sherwood number are presented graphically in **Figures 1-14**. The profiles for resultant velocity Rn for the flow are in **Figures 1-4** for suction/injection parameter $\lambda$, rotation parameter $\Omega$, viscoelastic parameter $K_2$, and $\varepsilon$ respectively. **Figure 1** shows that increase in the suction parameter $\lambda$ leads to an increase of Rn within the stationary plates. Similarly the

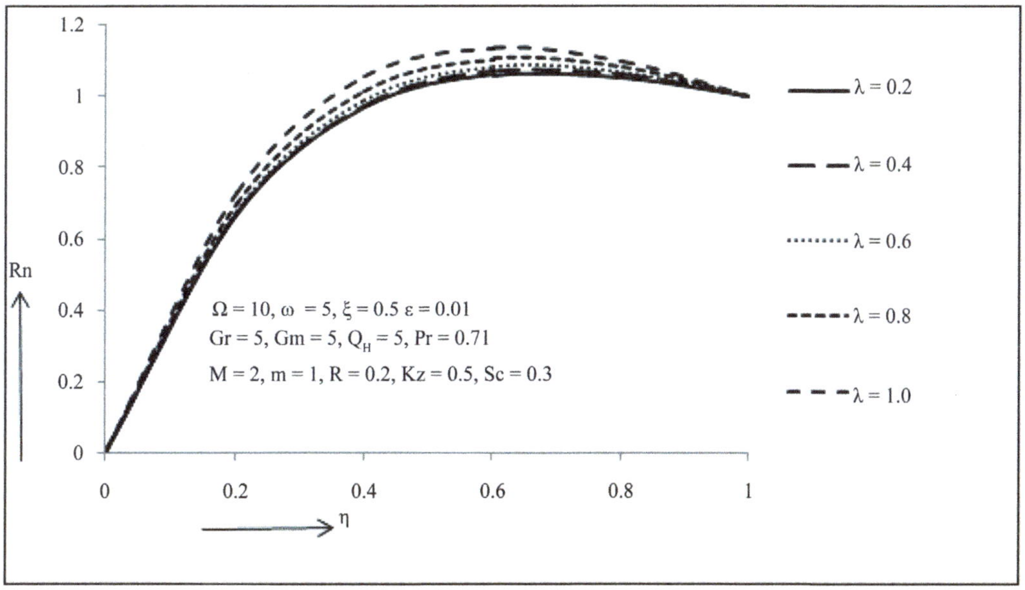


**Figure 1.** Resultant velocity Rn due to $u$ and $v$ versus $\lambda$ for different values of $\eta$ at $t=\pi/4$.

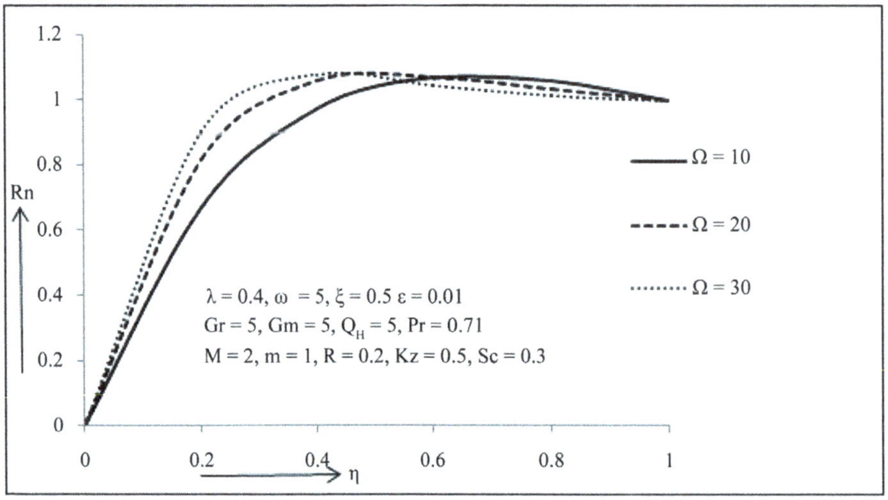


**Figure 2.** Resultant velocity Rn due to *u* and *v* versus $\eta$ for different values of $\Omega$ at $t = \pi/4$.

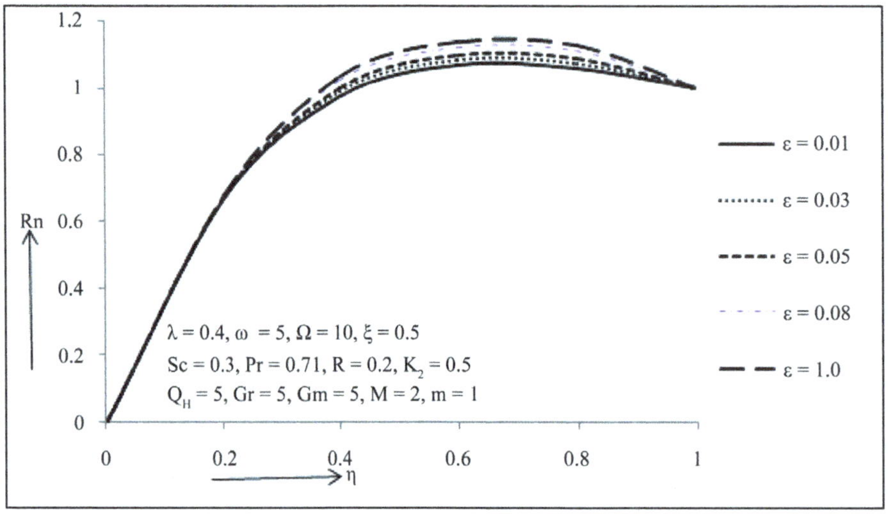


**Figure 3.** Resultant velocity Rn due to *u* and *v* versus $\eta$ for different values of $\varepsilon$ at $t = \pi/4$.

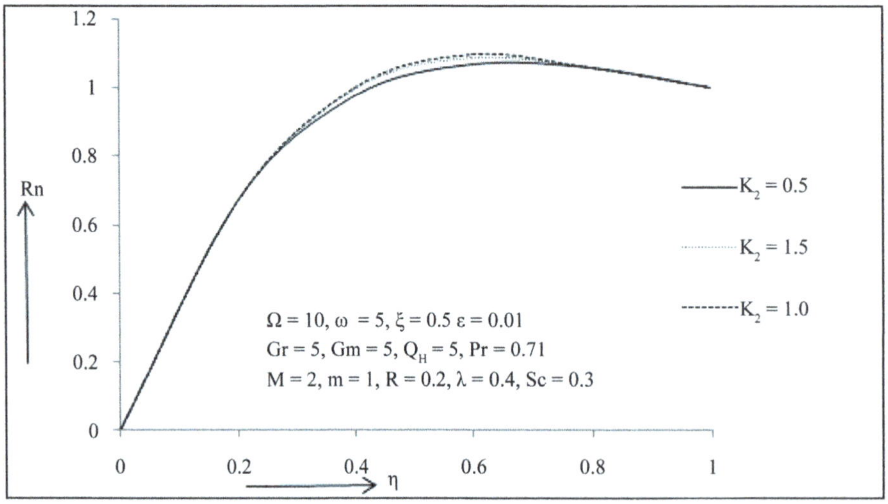


**Figure 4.** Resultant velocity Rn due to *u* and *v* versus $\eta$ for different values of $K_2$ at $t = \pi/4$.

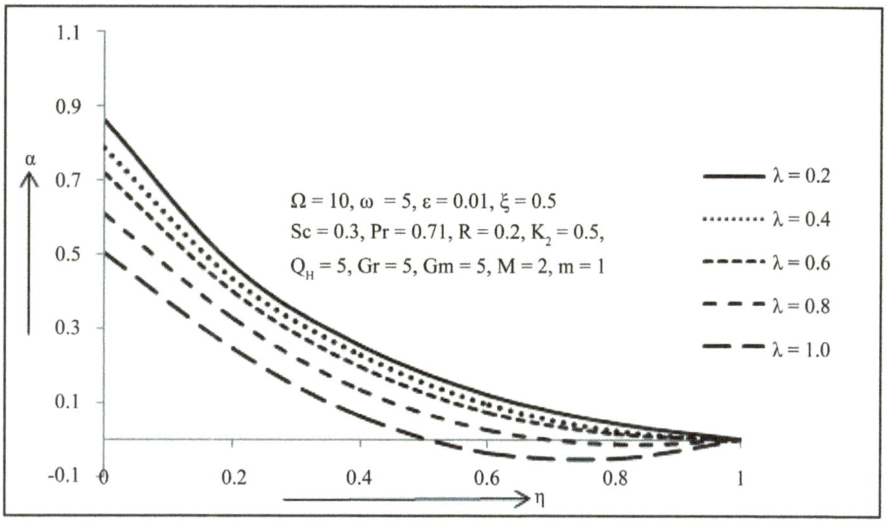


**Figure 5.** Phase difference $\alpha$ due to $u$ and $v$ versus $\eta$ for different values of $\lambda$ at $t = \pi/4$.

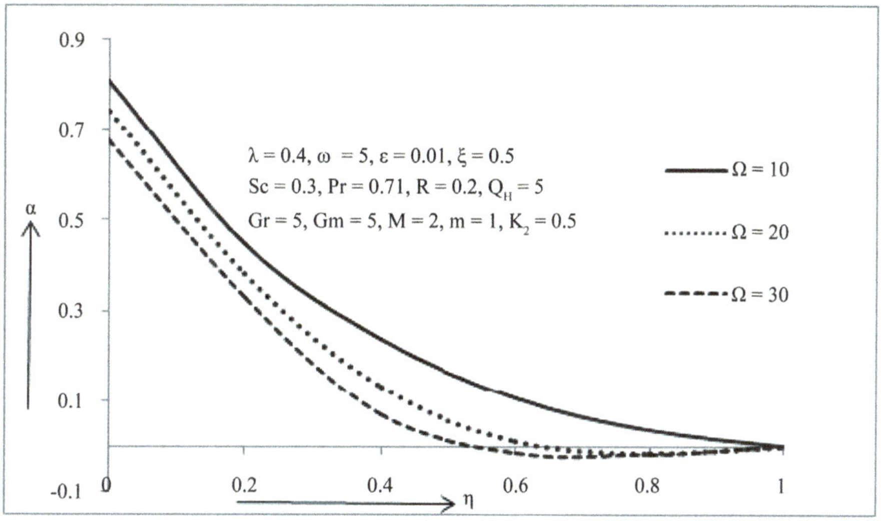


**Figure 6.** Phase difference $\alpha$ due to $u$ and $v$ versus $\eta$ for different values of $\Omega$ at $t = \pi/4$.

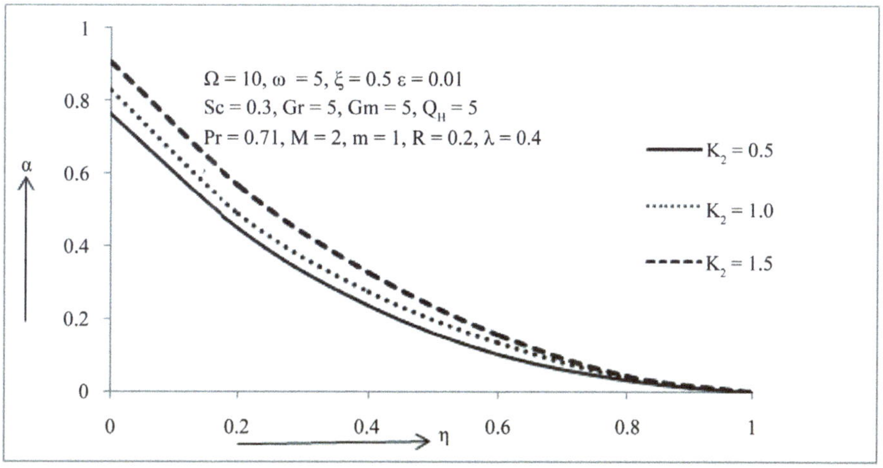


**Figure 7.** Phase difference $\alpha$ due to $u$ and $v$ versus $\eta$ for different values of $K_2$ at $t = \pi/4$.

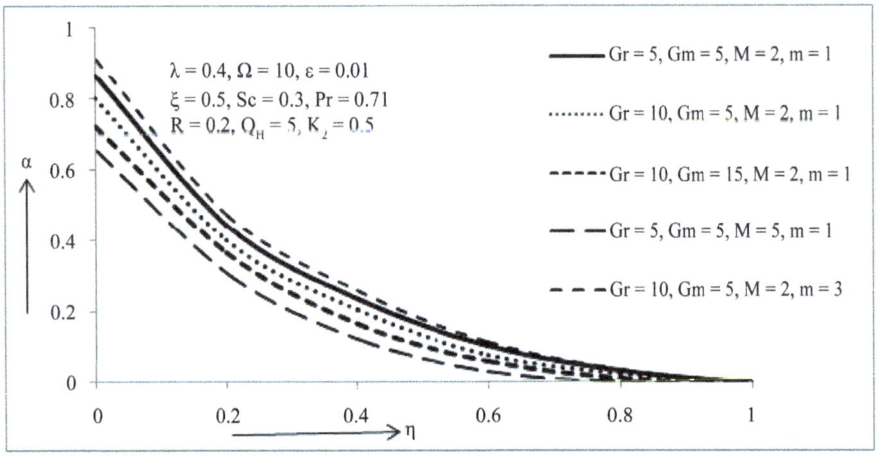


**Figure 8.** Phase difference $\alpha$ due to $u$ and $v$ versus $\eta$ for different values of Gr, Gm, M and m at $\pi/4$.

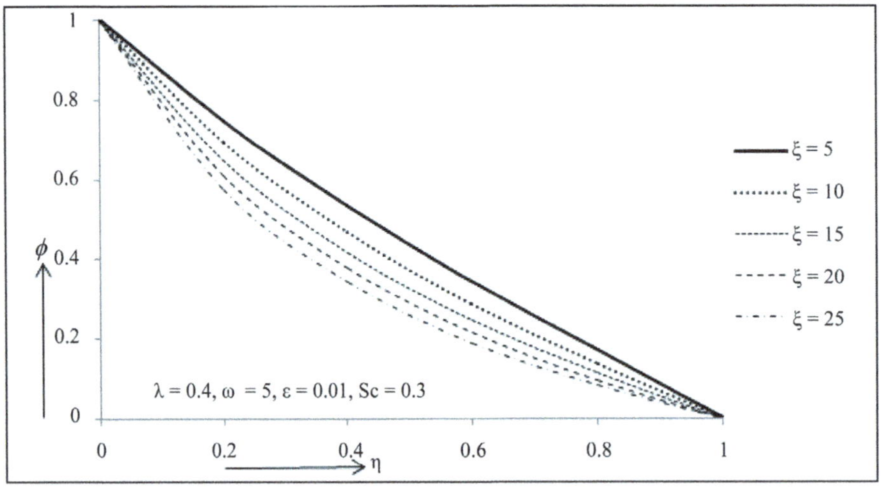


**Figure 9.** Concentration profiles against $\phi$ for different values of $\xi$ at $t = \pi/4$.

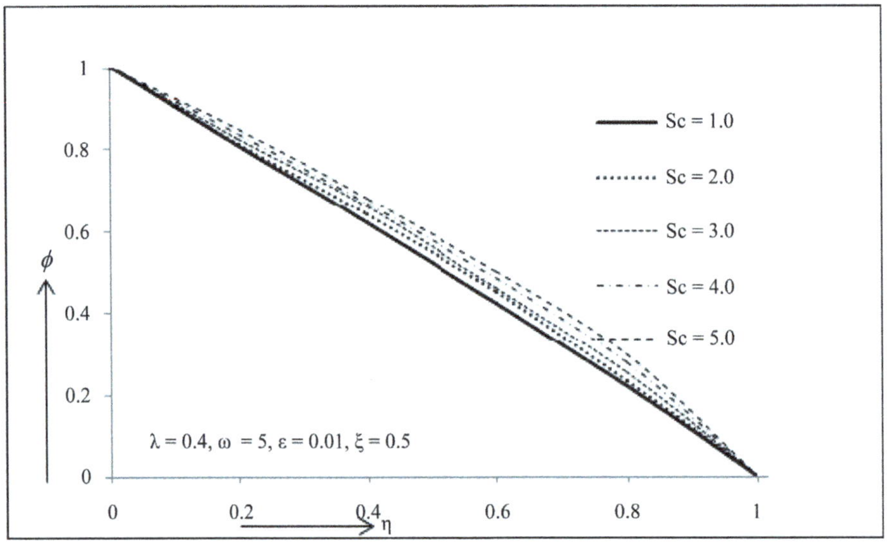


**Figure 10.** Concentration profiles against $\phi$ for different values of Sc at $t = \pi/4$.

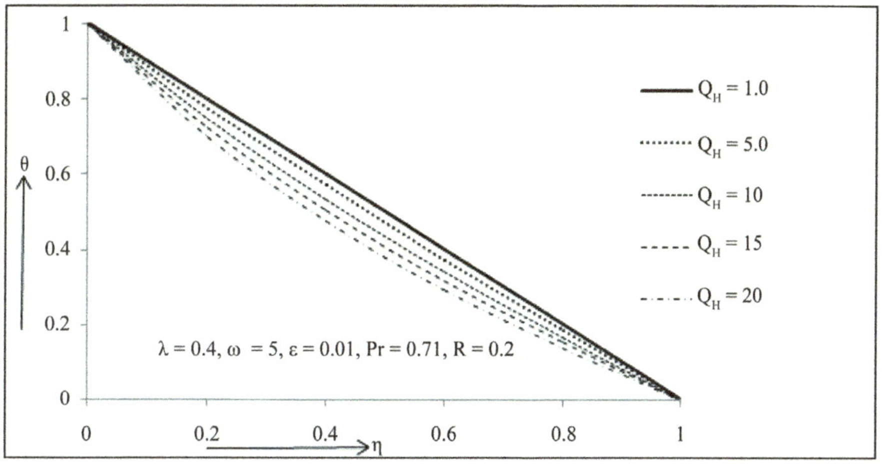


**Figure 11.** Temperature profiles $\theta$ against $\eta$ for different values of QH at $t = \pi/4$.

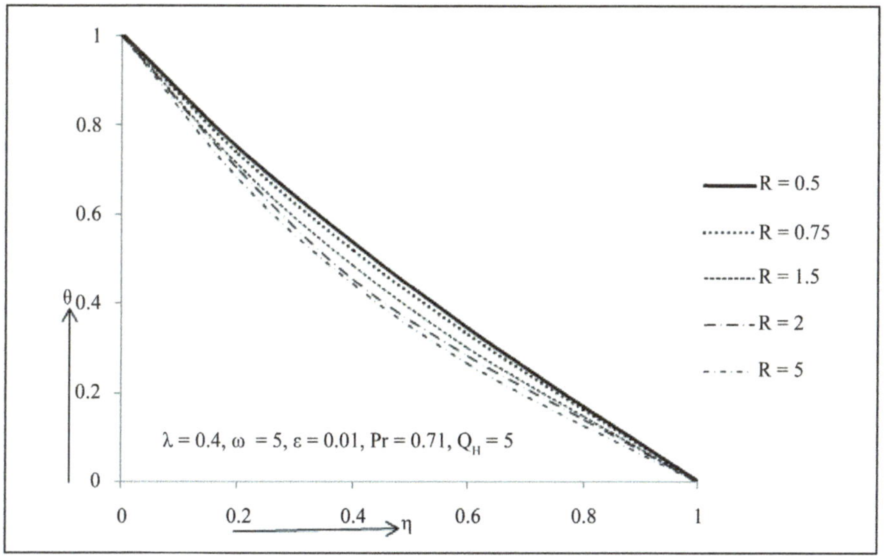


**Figure 12.** Temperature profiles $\theta$ against $\eta$ for different values of R at $t = \pi/4$

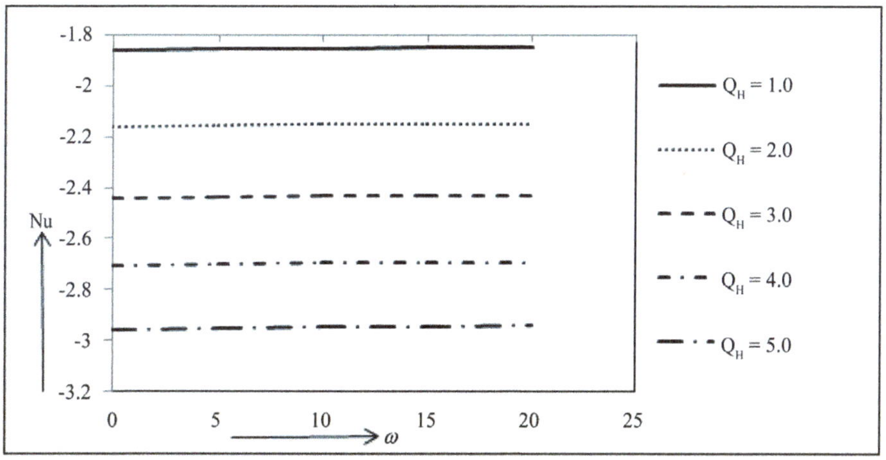


**Figure 13.** Nusselt number against for different values of $Q_H$ with $\lambda = 0.5$, Pr = 0.71, R = 2, $\varepsilon = 0.01$ at $t = \pi/4$.

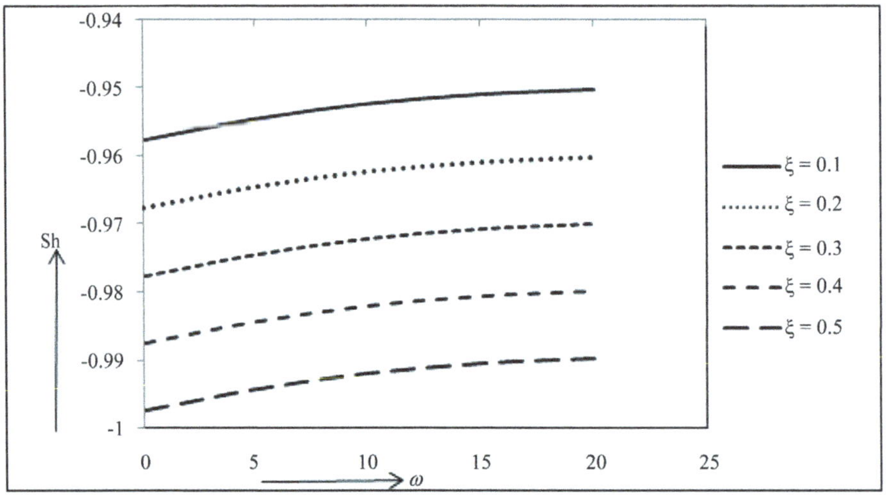


**Figure 14.** Sherwood number against $\omega$ for different values of $\xi$ with $\lambda = 0.5$, $\varepsilon = 0.01$, Sc = 0.3 at $t = \pi/4$.

resultant velocity increases with increasing values of rotation parameter $\Omega$. This is due to the fact that the rotation effects being more dominant near the walls, so when $\Omega$ reaches high values secondary velocity component $v$ decreases with increases in $\Omega$ as shown in **Figure 2**. From **Figure 3**, it is observed that the increase in the $\varepsilon$ leads to an increase of Rn within the stationary plates. From **Figure 4**, it inferred that resultant velocity Rn goes on increasing with increasing value of viscoelastic parameter $K_2$.

The phase difference $\alpha$ for the flow is shown graphically in **Figures 5-8**. **Figure 5** shows phase angle for various positive values of suction/injection parameter $\lambda$. The figure shows that the phase angle $\alpha$ decreases with the increases of suction parameter. **Figure 6** is the phase angle for various values of rotation parameter $\Omega$. From this figure, it is observed that the phase angle $\alpha$ decreases with an increase in rotation parameter. From **Figure** 7, it is observed that the phase angle $\alpha$ increases with an increase in visco elastic parameter. **Figure 8** shows the variation of $\alpha$ against $\eta$ for different values of thermal Grashof number Gr, Solutal Grashof number Gm, and Hartmaan number. From this figure, it is found that the values of phase $\alpha$ decreases with increasing values of Gr, Gm and M while reversed effect is observed for the Hall parameter m.

The concentration profile $\phi$ for the flow is shown graphically in **Figure 9**, **Figure 10**. From **Figure 9**, it is observed that with the increasing the value of the chemical reaction parameter $\xi$ decreases the concentration of species in the boundary layer; this due to the fact that destructive chemical reduces the solutal boundary thickness and increases the mass transfer. Opposite trend is seen in the case when Schmidt number is increased as noted in **Figure 10**. It may also be observed from this figure that the effect of Schmit number Sc is to be increases the concentration distribution in the solutal boundary layer.

The temperature profiles $\theta$ are shown graphically in **Figure 11** and **Figure 12**. **Figure 11** has been plotted to depict the variation of temperature profiles against $\eta$ for different values of heat absorption parameter $Q_H$ by fixing other physical parameters. From this figure, we observe that temperature $\theta$ decreases with increase in the heat absorption parameter $Q_H$ because when heat is absorbed, the buoyancy force decreases the temperature profile. **Figure 12** represents graph of temperature distribution with $\eta$ for different values of radiation parameter. From this figure, we note that initial temperature $\theta = 1$ decreases zero satisfying boundary condition at $\eta = 1.0$ Further, it is observed from this figure that increase in the radiation parameter decreases the temperature distribution in the thermal boundary layer due to decreases in the thickness of the thermal boundary layer with thermal radiation parameter R. This is because large values of radiation parameter correspond to an increase in dominance of conduction over radiation, thereby decreasing the buoyancy force and temperature in the thermal boundary layer.

**Figure 13** and **Figure 14** show the amplitude of skin-friction, Nusselt number, and Sherwood number against frequency parameter w for different values of $Q_H$, and $\xi$, respectively. The amplitude of Nusselt number decreases with increasing the value of heat source parameter $Q_H$ which is shown in **Figure 13**. **Figure 14** shows the variation of Sherwood number with $\xi$ and $\omega$. From this figure, it is observed that the Sherwood number decreases with increasing the value of chemical reaction parameter $\xi$, and opposite trend is seen with increasing the values of $\omega$.

## References


[1] Cowling, T. (1957) Magnetohydrodynamics. Interscience Publications Inc., New York.

[2] Soundalgekar, V.M. (1979) Hal and Ion-Slip Effects in MHD Coutte Flow with Heat Transfer. *IEEE Transactions on Plasma Science*, **PS-7**, 178-186. http://dx.doi.org/10.1109/TPS.1979.4317226

[3] Soundalgekar, V.M. and Uplekar, A.G. (1986) Hal Effects in MHD Coutte Flow with Heat Transfer. *IEEE Transactions on Plasma Science*, **PS-7**, 579-587. http://dx.doi.org/10.1109/TPS.1986.4316600

[4] Hossain, M.A. and Rashid, R.I.M.I. (1987) Hall Effect on Hydromagnetic Free Convection Flow along a Porous Flat Plate with Mass Transfer. *Journal of the Physical Society of Japan*, **56**, 97-104. http://dx.doi.org/10.1143/JPSJ.56.97

[5] Attia, H.A. (1998) Hall Current Effects on Velocity and Temperature Fields of an Unsteady Hartmann Flow. *Canadian Journal of Physics*, **76**, 73-78. http://dx.doi.org/10.1139/cjp-76-9-739

[6] Singh, K.D. and Garg, B.P. (2010) Radiative Heat Transfer in MHD Oscillatory Flow through Porous Medium Bounded by Two Vertical Porous Plates. *Bulletin of Calcutta Mathematical Society*, **102**, 129-139.

[7] Saha, L.K., Siddqia, S. and Hossain, M. A. (2011) Effect of Hall Current on MHD Natural Convection Flow from Vertical Permeable Flat Plate with Uniform Surface Heat Flux. *Applied Mathematics and Mechanics*, **32**, 1127-1146. http://dx.doi.org/10.1007/s10483-011-1487-9

[8] Aboeldahab, E.M. and Elbarby, E.M.E. (2001) Hall Current Effect on Magnetohydrodynamic Free-Convection Flow Past a Semi-Infinite Vertical Plate with Mass Transfer. *International Journal of Engineering Science*, **39**, 1641-1652. http://dx.doi.org/10.1016/S0020-7225(01)00020-9

[9] Abo-Eldahab, M. and ElAziz, M.A. (2005) Viscous Dissipation and Joule Heating Effects on MHD Free Convection from a Vertical Plate with Power-Law Variation in Surface Temperature in the Presence of Hall and Ion-Slip Currents. *Apllied Mathematical Modeling*, **29**, 579-595. http://dx.doi.org/10.1016/j.apm.2004.10.005

[10] Cogley, A.C., Vinceti, W.C. and Gilles, S.E. (1968) Differential Approximation for Radiation Transfer in a Nongray Gas near Equilibrium. *American Institute of Aeronautics and Astronautics Journal*, **6**, 551-555. http://dx.doi.org/10.2514/3.4538

[11] Mansour, M.A. (1990) Radiative and Free Convection Effects on the Oscillatory Flow past a Vertical Plate. *Astrophysics and Space Science*, **166**, 269-275. http://dx.doi.org/10.1007/BF01094898

[12] Hossain, M.A. and Thakar, H.S. (1996) Radiation Effect on Mixed Convection along a Vertical Plate with Uniform Surface Temperature. *Heat and Mass Transfer*, **314**, 243-248. http://dx.doi.org/10.1007/BF02328616

[13] Hossain, M.A., Alim, M.A. and Rees, S. (1999) The Effect of Radiation on Free Convection from a Porous Vertical Plate. *International Journal of Heat and Mass Transfer*, **42**, 181-191. http://dx.doi.org/10.1016/S0017-9310(98)00097-0

[14] Seddek, M.A. (2002) Effect of Radiation and Variable Viscosity on a MHD Free Convection Flow past a Semi Infinite Flat Plate with an Aligned Magnetic Field in the Case of Unsteady Flow. *International Journal of Heat and Mass Transfer*, **45**, 931-935. http://dx.doi.org/10.1016/S0017-9310(01)00189-2

[15] Muthucumarswamy, R. and Senthil, G.K. (2004) Studied the Effect of Heat and Mass Transfer on Moving Vertical Plate in the Presence of Thermal Radiation. *Journal of Theoretical and Applied Mechanics*, **31**, 35-46. http://dx.doi.org/10.2298/TAM0401035M

[16] Pal, D. (2009) Heat and Mass Transfer in Stagnation-Point Flow towards a Stretching Surface in the Presence of Buoyancy Force and Thermal Radiation. *Meccanica*, **44**, 145-158. http://dx.doi.org/10.1007/s11012-008-9155-1

[17] Aydin, A. and Kaya, A. (2008) Radiation Effect on MHD Mixed Convection Flow about a Permeable Vertical Plate. *Heat and Mass Transfer*, **45**, 239-246. http://dx.doi.org/10.1007/s00231-008-0428-y

[18] Mohamed, R.A. (2009) Double-Diffusive Convection Radiation Interaction on Unsteady MHD Flow over a Vertical Moving Porous Plate with Heat Generation and Soret Effects. *Applied Mathematical Sciences*, **3**, 629-651.

[19] Chauhan, D.S. and Rastog, P. (2010) Radiation Effects on Natural Convection MHD Flow in a Rotating Vertical Porous Channel Partially Filled with a Porous Medium. *Applied Mathematical Sciences*, **4**, 643-655.

[20] Ibrahim, S.Y. and Makinde, O.D. (2011) Radiation Effect on Chemically Reacting MHD Boundary Layer Flow of Heat and Mass Transfer through a Porous Vertical Flat Plate. *International Journal of Physical Sciences*, **6**, 1508-1516.

[21] Pal, D. and Mondal, H. (2011) The Influence of Thermal Radiation on Hydromagnetic Darcy-Forchheimer Mixed Covection Flow past a Stretching Sheet Embedded in a Porous Medium. *Mecccanica*, **46**, 739-753. http://dx.doi.org/10.1007/s11012-010-9334-8

[22] Kandasamy, R., Periasamy, K. and Sivagnana Prabhu, K.K. (2005) Chemical Reaction, Heat and Mass Transfer on MHD Flow over a Vertical Stretching Surface with Heat Source and Thermal Stratification Effects. *International Journal of Heat and Mass Transfer*, **48**, 4457-4561. http://dx.doi.org/10.1016/j.ijheatmasstransfer.2005.05.006

[23] Muthucumaraswamy, R. and Janakiraman, B. (2008) Mass Transfer Effects on Isothermal Vertical Oscillating Plate in the Presence of Chemical Reaction. *International Journal of Applied Mathematics and Mechanics*, **4**, 66-74.

[24] Sharma, P.R. and Singh, K.D. (2009) Unsteady MHD Free Convective Flow and Heat Transfer along a Vertical Porous Plate with Variable Suction and Internal Heat Generation. *International Journal of Applied Mathematics and Mechanics*, **4**, 1-8.

[25] Sudheer Babu, M. and Satya Narayan, P.V. (2009) Effects of the Chemical Reaction and Radiation Absorption on Free Convection Flow through Porous Medium with Variable Suction in the Presence of Uniform Magnetic Field. *Journal of Heat and Mass Transfer*, **3**, 219-234.

[26] Makinde, D. and Chinyoka, T. (2011) Numerical Study of Unsteady Hydromagnetic Generalized Coutte Flow of a Reactive Third-Grade Fluid with Asymmetric Convective Cooling. *Computers and Mathematics with Applications*, **61**, 1167-1179. http://dx.doi.org/10.1016/j.camwa.2010.12.066

[27] Pal, D. and Talukdar, B. (2011) Combined Effects of Joule Heating and Chemical Reaction on Unsteady Magnetohydrodynamic Mixed Convection of a Viscous Dissipating Fluid over a Vertical Plate in Porous Media with Thermal Radiation. *Mathematical and Computer Modeling*, **54**, 3016-3036. http://dx.doi.org/10.1016/j.mcm.2011.07.030

[28] Aldoss, T.K., Nimar, A.L., Jarrah, M.A. and Shaer, A.L. (1995) Magnetohydrodynamic Mixed Convection from a Vertical Plate Embedded in a Porous Medium. *Numerical Heat Transfer*, **28**, 635-642. http://dx.doi.org/10.1080/10407789508913766

[29] Rajgopal, K., Veena, P.H. and Parvin, V.K. (2006) Oscillatory Motion of an Electrically Conducting Viscoelastic Fluid over a Stretching Sheet in Saturated Porous Medium with Suction/Blowing. *Mathematical Problems in Engineering*, **1**, 1-14. http://dx.doi.org/10.1155/MPE/2006/60560

[30] Attia, H.A. (2004) Unsteady MHD Coutte Flow of a Viscoelastic Fluid Considering the Hall Effect. *Canadian Journal of Physics*, **82**, 127-135. http://dx.doi.org/10.1139/p03-117

[31] Chaudhary, R.C. and Jha, A.K. (2012) Heat and Mass Transfer in Elastic-Viscous Fluid past an Impulsively Started Infinite Vertical Plate with Hall Effect. *Latin American Applied Research*, **38**, 17-26.

[32] Singh, K.D. (2012) Viscoelastic Mixed Convection MHD Oscillatory Flow through a Porous Medium Filled in a Vertical Channel. *International Journal of Physics and Mathematical Sciences*, **3**, 194-205.

[33] Walters, K. (1962) The Solution of Flow Problems in the Case of Materials with Memories. *Journal de Mécanique*, **1**, 474-478.

[34] Meyer, R.C. (1958) On Reducing Aerodynamics Heat Transfer Rates by Magnetohydrodynamic Techniques. *Journal of the Aerospace Science*, **25**, 561-566. http://dx.doi.org/10.2514/8.7781

## Appendix

$$R_1 = \frac{R}{3R+4}, \quad R_2 = R_1 \Pr, \quad K_2 = \frac{K_0}{vd^2}$$

$$h_1 = \frac{Sc\lambda + \sqrt{(Sc\lambda)^2 + 4\xi Sc}}{2},$$

$$h_2 = \frac{Sc\lambda - \sqrt{(Sc\lambda)^2 + 4\xi Sc}}{2}$$

$$h_3 = \frac{Sc\lambda + \sqrt{(Sc\lambda)^2 + 4(i\omega + \xi)Sc}}{2}$$

$$h_4 = \frac{Sc\lambda - \sqrt{(Sc\lambda)^2 + 4(i\omega + \xi)Sc}}{2}$$

$$h_5 = \frac{Sc\lambda + \sqrt{(Sc\lambda)^2 - 4(i\omega + \xi)Sc}}{2}$$

$$h_6 = \frac{Sc\lambda - \sqrt{(Sc\lambda)^2 - 4(i\omega + \xi)Sc}}{2}$$

$$h_7 = \frac{R_2\lambda + \sqrt{(R_2\lambda)^2 + 4Q_H R_1}}{2}$$

$$h_8 = \frac{R_2\lambda - \sqrt{(R_2\lambda)^2 + 4Q_H R_1}}{2}$$

$$h_9 = \frac{R_2\lambda + \sqrt{(R_2\lambda)^2 + 4R_1(i\omega \Pr + Q_H)}}{2}$$

$$h_{10} = \frac{R_2\lambda - \sqrt{(R_2\lambda)^2 + 4R_1(i\omega \Pr + Q_H)}}{2}$$

$$h_{11} = \frac{R_2\lambda + \sqrt{(R_2\lambda)^2 - 4R_1(i\omega \Pr - Q_H)}}{2}$$

$$h_{12} = \frac{R_2\lambda - \sqrt{(R_2\lambda)^2 - 4R_1(i\omega \Pr - Q_H)}}{2}$$

$$h_{13} = \frac{\lambda + \sqrt{\lambda^2 + 4s}}{2}, \quad h_{14} = \frac{\lambda - \sqrt{\lambda^2 + 4s}}{2}$$

$$h_{15} = \frac{\lambda + \sqrt{\lambda^2 + 4(1 - iK_2)(S + i\omega)}}{2(1 - iK_2)}$$

$$h_{16} = \frac{\lambda - \sqrt{\lambda^2 + 4(1 - iK_2)(S + i\omega)}}{2(1 - iK_2)}$$

$$h_{17} = \frac{\lambda + \sqrt{\lambda^2 + 4(1 + iK_2)(S - i\omega)}}{2(1 + iK_2)}$$

$$h_{18} = \frac{\lambda - \sqrt{\lambda^2 + 4(1+iK_2)(S-i\omega)}}{2(1+iK_2)}$$

$$A_1 = \frac{\lambda^2 Gr e^{h_7}}{\left(e^{h_7} - e^{h_8}\right)\left[h_8^2 - \lambda h_8 - S\right]}$$

$$A_2 = \frac{\lambda^2 Gr e^{h8}}{\left(e^{h_7} - e^{h_8}\right)\left[h_7^2 - \lambda h_7 - S\right]}$$

$$A_3 = \frac{\lambda^2 Gm e^{h_1}}{\left(e^{h_1} - e^{h_2}\right)\left[h_2^2 - \lambda h_2 - S\right]}$$

$$A_4 = \frac{\lambda^2 Gm\, e^{h_2}}{\left(e^{h_1} - e^{h_2}\right)\left[h_1^2 - \lambda h_1 - S\right]}$$

$$A_5 = \frac{\lambda^2 Gr\, e^{h_9}}{\left(e^{h_9} - e^{h_{10}}\right)\left[h_{10}^2(1-iK_2) - \lambda h_{10} - (i\omega + S)\right]}$$

$$A_6 = \frac{\lambda^2 Gr\, e^{h_{10}}}{\left(e^{h_9} - e^{h_{10}}\right)\left[h_9^2(1-iK_2) - \lambda h_9 - (i\omega + S)\right]}$$

$$A_7 = \frac{\lambda^2 Gm\, e^{h_3}}{\left(e^{h_3} - e^{h_4}\right)\left[h_4^2(1-iK_2) - \lambda h_4 - (i\omega + S)\right]}$$

$$A_8 = \frac{\lambda^2 Gm\, e^{h_4}}{\left(e^{h_3} - e^{h_4}\right)\left[h_3^2(1-iK_2) - \lambda h_3 - (i\omega + S)\right]}$$

$$A_9 = \frac{\lambda^2 Gr e^{h_{11}}}{\left(e^{h_{11}} - e^{h_{12}}\right)\left[h_{12}^2(1+iK_2) - \lambda h_{12} - (S - i\omega)\right]}$$

$$A_{10} = \frac{\lambda^2 Gr e^{h_{12}}}{\left(e^{h_{11}} - e^{h_{12}}\right)\left[h_{11}^2(1+iK_2) - \lambda h_{11} - (S - i\omega)\right]}$$

$$A_{11} = \frac{\lambda^2 Gm e^{h_5}}{\left(e^{h_5} - e^{h_6}\right)\left[h_6^2(1+iK_2) - \lambda h_6 - (S - i\omega)\right]}$$

$$A_{12} = \frac{\lambda^2 Gm e^{h_6}}{\left(e^{h_5} - e^{h_6}\right)\left[h_5^2(1+iK_2) - \lambda h_5 - (S - i\omega)\right]}$$

## Nomenclature

$C^*$ : Dimensional concentration
$J_x$ : $x$-component of current density
$C_0$ : Concentration at the left plate
$K^*$ : Mean absorption coefficient
$C_d$ : Specific heat at constant pressure
$K_1$ : Chemical reaction rate constant
$C_p$ : Specific heat at constant pressure
$m$ : Hall parameter
$d$ : Distance of the plate
$M$ : Hartmann number
$D_m$ : Chemical molecular diffusivity

$Nu$ : Nusselt number
$e$ : Electric charge
$n_e$ : Number density of the electron
$g_0$ : Acceleration due to gravity
$P^*$ : Dimensional pressure
$G_m$ : Modified Grashof number for mass transfer
$P_e$ : Electron pressure
$G_r$ : Modified Grashof number for heat transfer
$P_r$ : Prandtl number
$\boldsymbol{H}$ : Magnetic field
$q_r^*$ : Radiative heat flux
$H_0$ : Magnetic field of uniform strength
$Q_0$ : Dimensional heat source
$H_x$ : $\boldsymbol{x}$-Component of magnetic field
$Q_H$ : Heat source parameter
$\boldsymbol{J}$ : Current density
$R$ : Radition parameter
$R_0$ : Amplitude for steady flow
$T_d$ : Temperature at the right wall
$R_n$ : Resultant velocity
$t^*$ : Dimensional time
$R_v$ : Amplitude for unsteady flow
$Sc$ : Schmidt number
$U_0$ : Non zero constant mean velocity
$Sh$ : Sherwood number
$T^*$ : Dimensional temperature
$\boldsymbol{V}$ : Electron velocity
$T_0$ : Temperature at the left wall
$K_2$ : Visco-elastic parameter
$w_0$ : Dimensional injection /suction velocity
$u_0$ : Primary velocity component for unsteady flow
$u_1$ : Primary velocity component for steady flow
$v_0$ : Secondary velocity component for steady flow
$v_1$ : Secondary velocity component for unsteady flow
$u^*, v^*, w^*$ : Velocity components are in the $x^*-, y^*-, z^*-$ directions respectively.

## Greek Symbols

$\alpha_0$ : Phase difference for steady flow
$\lambda$ : Injection/suction parameter
$\alpha_1$ : Phase difference for unsteady flow
$\mu$ : Dynamic viscosity
$\alpha$ : Phase difference for the flow
$\mu_e$ : Magnetic permeability
$\delta$ : Phase difference of mass flux
$\nu$ : Kinematic viscosity
$\varepsilon$ : Small positive constant
$\omega$ : Oscillation parameter
$\eta$ : Dimensionless distance
$\Omega^*$ : Dimensional parameter
$\gamma$ : Phase difference of heat flux
$\Omega$ : Angular velocity
$K$ : Fluid thermal conductivity

$\omega_e$ : Cyclotron frequency
$\sigma$ : Electric conductivity
$\Phi$ : Amplitude of mass flux
$\sigma^*$ : Stefan-Boltzmann constant
$\Theta$ : Amplitude of heat flux
$\phi$ : Non dimensional concentration
$\rho$ : Density
$\theta$ : Non-dimensional temperature
$\sigma$ : Coefficient of thermal expansion
$\beta$ : Coefficient of thermal expansion
$\beta^*$ : Coefficient of solutal expansion
$\beta_0$ : Phase difference of shear stresses for the steady flow
$\beta_1$ : Phase difference of shear stresses for the unsteady flow
$\beta_2$ : Phase difference of shear stresses for the flow
$\tau$ : Amplitude of shear stresses for the flow
$\tau_{0r}$ : Amplitude of shear stresses for the steady flow
$\tau_{1r}$ : Amplitude of shear stresses for the unsteady flow

# 15

# Combined Effect of Magnetic Field and Compressibility on Rayleigh Taylor Instability

**Aniruddha Mitra[1], Labakanta Mandal[2], Rajkumar Roychoudhury[3], Manoranjan Khan[1]**

[1]Dept. of Instrumentation Science & Centre for Plasma Studies, Jadavpur University, Kolkata, India
[2]Dept. of Physics, Budge Budge Institute of Technology, Kolkata, India
[3]Advanced Centre for Nonlinear & Complex Phenomena, Kolkata, India
Email: anibabun@gmail.com, labakanta@gmail.com, rajdaju@rediffmail.com, mkhan.ju@yahoo.com

## Abstract

**The nonlinear analysis of the combined effect of magnetic field and compressibility on the growth rate of Rayleigh-Taylor (RT) instability has been investigated for inviscid two fluid interface. We have considered an interface-parallel density dependent magnetic field and used Layzer's approach to analyze the problem. We have also investigated the relative effect of magnetic pressure and hydrodynamic pressure on RT instability through the variation of the ratio of hydromagnetic pressure to magnetic pressure ($\beta$). Dynamics of bubble and spike has been studied analytically and numerically. Finally, we have obtained the stability conditions of our result through linear stability analysis**

## Keywords

**Rayleigh-Taylor Instability, Bubbles, Spikes, Self Generated Magnetic Field, Compressibility, Density Dependent Magnetic Field, ICF**

## 1. Introduction

Under the gravitational force, when a denser fluid overlies a lighter fluid, the interfacial instability occurs between those two fluids, and it is called Rayleigh-Taylor instability (RTI). The development of nonlinear structures at the two fluid interface consequent to RTI is of much current interest both from theoretical and experimental points of view. The evolution of the structure is called a bubble if the lighter fluid pushes across the unperturbed interface into the heavier fluid and it is termed as a spike if the opposite event takes place. The importance of such instabilities arises in connection with a wide range of applications ranging from astrophysical phenomena such as Supernova remnant to Inertial Confinement Fusion (ICF) problems. A core collapse Super Nova (SN) is driven by an external powerful shock, and strong shocks are the breeding ground of hydrodynamic instabilities like RTI. The interfacial RT instabilities have been investigated by several authors [1]-[11]. In an ICF situation, ablation front of an imploding capsule is subject to the RT instability because dense core is compressed and accelerated by low density ablating plasma. Recent experiments [12] [13] in LLNL shows that RTI at the ablation front of a composite ablator (CH, Si dopped CH) is induced, which requires mitigation for efficient ICF burn. Ponderomotive force due to spatially inhomogeneous laser beam is an effective mechanism for generation of large amplitude self generated magnetic field [14]-[16] in laser produced plasmas. Srivastava *et al.* [17] analyzed theoretically the spatially inhomogeneous laser beam could generate an axial magnetic field via ponderomotive force. The analytical results were compared with the experimental observation by [18]. The effect of self-generated magnetic field on RTI has been studied by Evans [19]. When the explosion of a Ia type supernova (SNIa) starts in a white dwarf as a laminar deflagration at the center of the star, RTI begins to act [20] [21]. In a white dwarf, magnetic field with strength up to $10^8$ - $10^9$ G exists at the surface and the field may be ~10 times greater near its center. RTI arising during type Ia Supernova explosion is also associated with strong magnetic field.

The effect of magnetic field on RTI was initially studied in details by [22]. In his linear analysis, he has shown that when the magnetic field is normal to the surface of separation of the two fluids, the RTI is shown to be unaffected by the magnetic field for small wave number "$k$" of the perturbation. However, as $k$ increases, the RTI growth rate does not increase indefinitely with "$k$" but tends to be a saturation value as $k \to \infty$ [22]. This situation may happen for sufficiently intense magnetic field.

If the external magnetic field is parallel to the interface, it is found in the linear theory that there exists no effect of the magnetic field on the instability [22]. However, in compressible and magnetized fluids, RTI has been shown to be reduced [23] even in linear theory. The RTI growth rate is found to be lowered for continuously accelerated two fluid interface when $k$ has component parallel to the magnetic field [24]. The nature of the depression has close resemblance to that of the effect of surface tension [22] and also has useful application in the astrophysical context [20] [25].

In all the above studies, the RTI was studied in linear approximation. However, in nonlinear theory, it has been studied separately, the evolution of the interfacial fluid structure due to RTI in compressible fluids [26] and that due to presence of magnetic field [27]. The present paper is addressed to the problem of the evolution of the nonlinear interfacial structure caused by RTI in presence of a magnetic field parallel to the surface of separation of the two compressible fluids. With such a geometry, there is no effect of the magnetic field in the classical linear approximation [22]. However, it is no longer the case for nonlinear theory. This may be understood from the following consideration.

In presence of only magnetic field [27], there exists a magnetic pressure in addition to the usual hydrodynamic pressure. As a result, the RTI driving pressure difference $g(\rho_h - \rho_l)y$ is changed by the inclusion of the magnetic pressure difference $(1/2\mu)(B_h^2 - B_l^2)$ [the suffix h(l) corresponds to the dynamical variable associated with the heavier (lighter) fluid]. In our earlier study [27] it has been shown that, the growth rate of RTI may be enhanced or depressed according to whether the extra contribution is positive or negative. Moreover, stabilization may occur in the sense that the surface of separation executes periodic undulation resulting from time lag in the temporal variation of $B_h$ and $B_l$. These are entirely nonlinear effects and disappear in the linear approximation.

In compressible fluids [26], there have been incorporated the effect of compressibility by introducing an extra work function proportional to surface pressure $p_0$ and density gradient scale length $(\delta\rho/\rho_0)$. The degree of stabilization or destabilization (depending on algebric sign of the density gradient at the surface) was shown to be proportional to $p_0$.

In this study, we have also included the compressibility in addition to the magnetic field on RTI. We have developed a model up to second order of nonlinearity in the transverse coordinate system to investigate the combined effect of compressibility and magnetic field by using Layzer's approach [28]. We have shown that, the growth rate of bubble and spikes is substantially reduced and shows oscillatory stabilization due to the combined effect which is conducive in ICF burn. We are mainly concentrated on a single mode bubble (spike) and our analysis is restricted in it (near the tip). The growth rate of instability is defined by the growth rate of the tip of the bubble (spike). The maximum spatial range of applicability of our solution has also been determined. We have tested the stability criteria of the solution.

The paper is organized as follows: Sec. 2 has been devoted for a brief discussion of Layzer's model. In Sec. 3 we have discussed the geometry of the problem and the structures of bubble and spike. Basic mathematical model and assumptions regarding the problem have been formulated in Sec. 4. The stability analysis of the model has been studied in Sec. 5. The numerical results regarding the dynamics of the tip of the bubble and spike have been discussed in Sec. 6. Sec. 7 is kept for conclusion.

## 2. Layzer's Model: A Brief Discussion

Potential flow model was first introduced in the nonlinear regime by Layzer in 1955 to study the stability of superposed fluid in a gravitation field in a three dimensional axially symmetric cylindrical or in two dimensional parallel walls from initially slight perturbed position. He obtained an approximate analytical solution considering a perfect incompressible fluid occupying the upper half of a vertical cylinder, being supported by a diaphragm against gravity. The lower half of the tube is empty *i.e.* $A_T = 1$. The diaphragm is removed initially and applied a small disturbance at the interface then consequent motion has been investigated in 3D tubular and 2D periodic flows. He expanded the flow potentials using the Bessel (3D) and Fourier (2D) series, and retaining the lowest order terms, re-expanded the conservation laws in the vicinity of a regular point of the free surface (bubble tip). In this way, the nonlinear boundary value problem for the Euler equations was reduced to a system of ordinary differential equations.The asymptotic solutions for this dynamical system at short and long times have agreed with the linear theory of Rayleigh and with the results of Taylor, respectively. Though Layzers solution was the local one, as it described the temporal evolution of the flow in the vicinity of the bubble tip only, yet it accounted for the dynamics in the entire domain via the use of the Bessel and Fourier series .

## 3. Geometrical Structure of the Problem

We consider such a system in which heavier fluid lies above the lighter one and their interface is in the X-Z plane (**Figure 1**). Heavier fluid (density $\rho_h$) above this interface is extended to infinity towards positive Y axis and lighter fluid (density $\rho_l$) is below the interface and extended to infinity along negative Y axis. An external magnetic field has been applied across the interface towards positive Z axis. The gravity acts along negative Y axis. Since the force due to magnetic field is only in the perpendicular direction of the magnetic field, our study is essentially two dimensional.

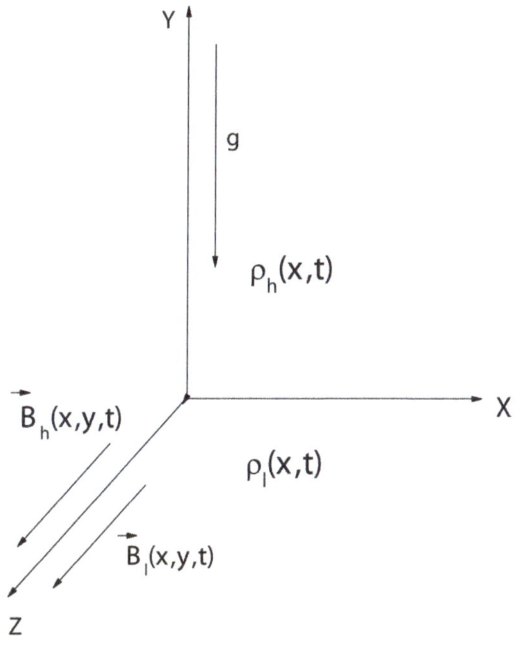


**Figure 1.** Geometry of the system.

We assume the perturbed interface to be of parabolic form which is given by

$$y(x,t)=\eta(x,t)=\eta_0(t)+\eta_2(t)x^2 \tag{1}$$

For a bubble (here the lower fluid is pushing across the interface into the upper fluid with $\rho_h > \rho_l$) we have,

$$\eta_0 > 0 \text{ and } \eta_2 < 0 \tag{2}$$

and for spike:

$$\eta_0 < 0 \text{ and } \eta_2 > 0 \tag{3}$$

here $\eta_0(t), \eta_2(t)$ are respectively, the height and curvature of the bubble.

## 4. Basic Equations and Mathematical Model of Hydrodynamics

We assume that the fluids are compressible, inviscid and the flow is irrotational. So there exists a velocity potential for both heavier and lighter fluids. Both the fluids obey the law of barotropic equation of state *i.e.* $p=c_s^2\rho^\gamma$, where $c_s$ is speed of sound through the fluid and $\gamma$ is adiabatic index of the fluid and $\rho$ is the density. The fluids are infinitely conducting in nature. We consider both high $\beta$ $\left(p \gg B^2/2\mu\right)$ and low $\beta$ $\left(p \ll B^2/2\mu\right)$, where p is the fluid pressure and B is the magnetic field. We work in a frame in which the interface is at rest,so the zeroth order velocity $\boldsymbol{u}_0=0$ and we take the first order velocity to be $\boldsymbol{u}(x,y,t)$. We assume that the flow is irrotational. Moreover, in this situation the fluctuations are assumed to be incompressible, which may not require the fluid itself to be incompressible [29]. Thus we may consider the compressibility of the fluid in this condition. This assumption may be expressed as $\nabla\cdot\boldsymbol{u}=0$ *i.e.* $\nabla^2\phi=0$.

Following [1] velocity potential flow model, we get velocity potentials for heavier fluid as

$$\phi_h(x,y,t)=a_1(t)\cos(kx)\mathrm{e}^{-k(y-\eta_0(t))};\ y>0 \tag{4}$$

Velocity potential for lighter fluid is given by

$$\phi_l(x,y,t)=b_0(t)+b_1(t)\cos(kx)\mathrm{e}^{k(y-\eta_0(t))};\ y<0 \tag{5}$$

### 4.1. Kinematic Boundary Condition

In our study, we like to focus into the dynamics of the tip of the bubble and spike. Our whole analysis and results are concentrated on this particular dynamics of instability.

The kinematical boundary conditions satisfied at the interface $y=\eta(x,t)$ are

$$\frac{\partial\eta}{\partial t}-\frac{\partial\phi_h}{\partial x}\frac{\partial\eta}{\partial x}=-\frac{\partial\phi_h}{\partial y} \tag{6}$$

$$-\frac{\partial\phi_h}{\partial x}\frac{\partial\eta}{\partial x}+\frac{\partial\phi_h}{\partial y}=-\frac{\partial\phi_l}{\partial x}\frac{\partial\eta}{\partial x}+\frac{\partial\phi_l}{\partial y} \tag{7}$$

Now substituting the values of $\phi_h,\phi_l$, and $\eta$ into above equations and expanding in powers of the transverse coordinate x and neglecting terms $O(x^n)$ $(n\geq 3)$ we get,

$$\frac{\mathrm{d}\xi_1}{\mathrm{d}\tau}=\xi_3 \tag{8}$$

$$\frac{\mathrm{d}\xi_2}{\mathrm{d}\tau}=-3\xi_3\left(\xi_2+\frac{1}{6}\right) \tag{9}$$

$$b_0=\left(\frac{-6\xi_2}{3\xi_2-0.5}\right)ka_1,\ b_1=\left(\frac{3\xi_2+0.5}{3\xi_2-0.5}\right)a_1 \tag{10}$$

where $\xi_1=k\eta_0;\xi_2=\eta_2/k;\xi_3=k^2a_1/\sqrt{kg}$ are dimensionless displacement, curvature and velocity of the tip of the bubble, respectively.

## 4.2. Governing Equations

Basic equations to study our problem are : continuity equation, momentum equation, barotropic equation of state, and magnetic induction equation. The general forms of the equation are as follows:

$$\frac{\partial \rho}{\partial t}+\nabla\cdot(\rho \boldsymbol{V})=0 \tag{11}$$

$$\rho\left[\frac{\partial \boldsymbol{V}}{\partial t}+(\boldsymbol{V}\cdot\nabla)\boldsymbol{V}\right]=-\nabla p-\rho \boldsymbol{g}+\frac{1}{\mu}(\nabla\times \boldsymbol{B})\times \boldsymbol{B} \tag{12}$$

Neglecting the magnetic tension term as the magnetic field to be planer, we get

$$\rho\left[\frac{\partial \boldsymbol{V}}{\partial t}+(\boldsymbol{V}\cdot\nabla)\boldsymbol{V}\right]=-\nabla p-\rho \boldsymbol{g}-\nabla\left(\frac{B^2}{2\mu}\right) \tag{13}$$

and for infinitely conducting fluid

$$\frac{\partial \boldsymbol{B}}{\partial t}=\nabla\times(\boldsymbol{V}\times \boldsymbol{B}) \tag{14}$$

Consequence of this infinite conductance is flux conservation (Alfven theorem), which tells that the the magnetic lines of force are frozen-in into the fluid.

## 4.3. Time Evolution of Densities and Magnetic Fields

Substituting the value of velocity potential for heavier fluid from Equation (4) into Equation (11), we have the following equation

$$\frac{\partial \rho_h}{\partial t}-\nabla\phi_h\cdot\nabla\rho_h=0 \tag{15}$$

To solve this equation with the given form of velocity potential, we set,

$$\rho_h(x,y,t)=\rho_{h0}(t)+\left[\rho_{h1}(t)\cos(k(x))\right]e^{-k(y-\eta_0(t))} \tag{16}$$

Putting this form of density into Equation (15), and equating the coefficients of $x^0$ we get

$$\rho_{h0}(t)+\rho_{h1}(t)=\tilde{\rho}_{h0}\ (\text{constant})$$

Expanding Equation (16) up to $x^2$ order, we get the expression for density of the heavier fluid as follows

$$\rho_h(x,y,t)=\tilde{\rho}_{h0}-\rho_{h1}(t)k^2(\xi_2+0.5)x^2 \tag{17}$$

where $\xi_2=\eta_2/k$.

Now equating the coefficients of $x^2$ in the Equation (15) we have,

$$\frac{\dot{\rho}_{h1}}{\rho_{h1}}=\left[\frac{\xi_2-0.5}{\xi_2+0.5}\right]\xi_3 \tag{18}$$

where $\xi_3=a_1k^2$

In our model, we assume the magnetic field is not uniform throughout the system but depends on the density of the fluids. This is very much realistic because density dependent magnetic field may be present in Hybrid stars and Neutron stars [30]. To incorporate such a concept, we use the $\beta$ parameter ( ratio of hydrodynamic pressure and magnetic pressure) which allows us to extend our results from laboratory experiment to solar corona by just changing the value of $\beta$. So using Equation (17) and the expression for **B**, viz $\boldsymbol{B}^2=\rho^\gamma\left[\left(2\mu C_s^2\right)/\beta\right]$, we find the expression for magnetic field and its evolution for heavier density fluid as follows

$$\boldsymbol{B}_h(t) = \tilde{\boldsymbol{B}}_{h0}\left\{1 - \frac{\gamma_h k^2}{2}\left(\frac{\boldsymbol{B}_{h1}(t)}{\tilde{\boldsymbol{B}}_{h0}}\right)^{\frac{2}{\gamma_h}}(\xi_2 + 0.5)x^2\right\} \tag{19}$$

where

$$\tilde{B}_{h0}^2 = \left(\frac{2\mu C_{sh}^2}{\beta}\right)_h \tilde{\rho}_{h0}^{\gamma_h}$$

is magnetic field at equilibrium for heavier fluid. We also assume that the magnetic curvature force is too small in comparison with the magnetic pressure force. Time evolution of the magnetic field for heavier fluid is (using Equation (18)

$$\frac{\dot{B}_{h1}(t)}{B_h(t)} = \frac{\gamma_h}{2}\left(\frac{\xi_2 - 0.5}{\xi_2 + 0.5}\right)\xi_3 \tag{20}$$

Similarly for lighter fluid using Equation (5), Equation (15) and setting the density of lighter fluid as

$$\rho_l(x,y,t) = \rho_{l0}(t) + \rho_{l1}(t)\cos(kx)\exp\left[k\left(y - \eta_0(t)\right)\right]$$

we get the following form of density for lighter fluid (upto $x^2$ term)

$$\rho_l(x,y,t) = \tilde{\rho}_{l0} + \rho_{l1}(t)k^2(\xi_2 - 0.5)x^2 \tag{21}$$

and the equation of time evolution of density is

$$\frac{\dot{\rho}_{l1}(t)}{\rho_{l1}(t)} = \left(\frac{\xi_2 + 0.5}{\xi_2 - 0.5}\right)\left(\frac{\xi_2 + \frac{1}{6}}{\xi_2 - \frac{1}{6}}\right)\xi_3 \tag{22}$$

Similarly the value of magnetic field in lighter fluid is as follows

$$\boldsymbol{B}_l(t) = \tilde{\boldsymbol{B}}_{l0}\left\{1 + \frac{\gamma_l k^2}{2}\left(\frac{B_{l1}(t)}{\tilde{B}_{l0}}\right)^{\frac{2}{\gamma_l}}(\xi_2 - 0.5)x^2\right\} \tag{23}$$

and time evolution of this magnetic field is given by

$$\frac{\dot{B}_l(t)}{B_l(t)} = \frac{\gamma}{2}\left(\frac{\xi_2 + 0.5}{\xi_2 - 0.5}\right)\left(\frac{\xi_2 + \frac{1}{6}}{\xi_2 - \frac{1}{6}}\right)\xi_3 \tag{24}$$

Now substituting $V = -\nabla\phi$ into force equation Equation (13) and integrating once, we have,

$$\frac{-\partial\phi}{\partial t} + \frac{1}{2}(\nabla\phi)^2 = -\int\frac{\mathrm{d}p}{\rho} - \int\frac{\mathrm{d}\left(B^2\right)}{2\mu\rho} + f(t)$$

$f(t)$ is a time dependent integrating constant (independent of space variable). We denote the values of $f(t)$ as $f_l(t)$ and $f_h(t)$ for lighter and heavier fluid, respectively. However, $f_l(t)$ and $f_h(t)$ do not contribute anything in the dynamics of the interface both in linear and nonlinear theory [31]. Substituting the values of magnetic field, pressure, density for both fluids, we get two separate equations and subtracting them we get a combined equation as follows:

$$\rho_{h0}^{*}\left[-\frac{\partial\phi_h}{\partial t}+\frac{1}{2}(\nabla\phi_h)^2+gy\right]-\rho_{l0}^{*}\left[-\frac{\partial\phi_l}{\partial t}+\frac{1}{2}(\nabla\phi_l)^2+gy\right]$$

$$=-\left(p_{h0}+\frac{\boldsymbol{B}_{h0}^2}{\beta_h}\right)\left[1-(\gamma_h-1)\times\frac{\rho_{h1}}{\tilde{\rho}_{h0}}k^2(\xi_2+0.5)x^2\right]$$

$$+\left(p_{l0}+\frac{\boldsymbol{B}_{l0}^2}{\beta_l}\right)\left[1+(\gamma_l-1)\frac{\rho_{l1}}{\tilde{\rho}_{l0}}k^2(\xi_2-0.5)x^2\right]+\rho_{h0}^{*}f_h(t)-\rho_{l0}^{*}f_l(t)$$

where $\rho_0^{*}=\tilde{\rho}_0(\gamma-1)/\gamma$. Now into the above equation, putting the values of

$$(\nabla\phi_h)^2,(\nabla\phi_l)^2,\partial\phi_h/\partial t,\partial\phi_l/\partial t,$$

and calculating only the coefficients of $x^2$ terms, we get

$$\dot{\xi}_3=\chi_1+\chi_2+\chi_3 \tag{25}$$

where

$$\chi_1=-\frac{N(\xi_2,r^*)}{D(\xi_2,r^*)}\frac{\xi_3^2}{6\xi_2-1}$$

$$\chi_2=2kg(r^*-1)\frac{\xi_2(6\xi_2-1)}{D(\xi_2,r^*)}$$

$$\chi_3=-\frac{6\xi_2-1}{D(\xi_2,r^*)}\pi_1-\frac{6\xi_2-1}{D(\xi_2,r^*)}\pi_2$$

$$\pi_1=\left(\frac{r^*\gamma_h k^2}{\tilde{\rho}_{h0}}\left(p_{h0}\delta\rho_h(t)+\frac{\boldsymbol{B}_{h0}^2}{\beta_h}(\delta B_h(t))^{\frac{2}{\gamma_h}}\right)(\xi_2+0.5)\right)$$

$$\pi_2=\left(\frac{\gamma_l k^2}{\tilde{\rho}_{l0}}\left(p_{l0}\delta\rho_l(t)+\frac{\boldsymbol{B}_{l0}^2}{\beta_l}(\delta B_l(t))^{\frac{2}{\gamma_l}}\right)(\xi_2-0.5)\right)$$

$$\frac{\rho_{h1}(t)}{\rho_{h0}}=\delta\rho_h(t);\ \frac{\rho_{l1}(t)}{\rho_{l0}}=\delta\rho_l(t);\ \frac{B_{h1}(t)}{B_{h0}}=\delta B_h(t);$$

$$\frac{B_{l1}(t)}{B_{l0}}=\delta B_l(t);\ r^*=\frac{\rho_{h0}^{*}}{\rho_{l0}^{*}}$$

$$N(\xi_2,r^*)=36(1-r^*)\xi_2^2+12(4+r^*)\xi_2+(7-r^*)$$

$$D(\xi_2,r^*)=12(1-r^*)\xi_2^2+4(1-r^*)\xi_2+(1+r^*)$$

Finally we get the full set of equations as follows

$$\dot{\xi}_1=\xi_3$$

$$\dot{\xi}_2=-\frac{1}{2}(6\xi_2+1)\xi_3$$

$$\dot{\xi}_3=\chi_1+\chi_2+\chi_3$$

$$\frac{\dot{\rho}_{l1}(t)}{\rho_{l1}(t)} = \left(\frac{\xi_2 + 0.5}{\xi_2 - 0.5}\right)\left(\frac{\xi_2 + \frac{1}{6}}{\xi_2 - \frac{1}{6}}\right)\xi_3 \tag{26}$$

$$\frac{\dot{\rho}_{h1}}{\rho_{h1}} = \left(\frac{\xi_2 - 0.5}{\xi_2 + 0.5}\right)\xi_3$$

$$\frac{\dot{B}_{h1}(t)}{B_h(t)} = \frac{\gamma_h}{2}\left(\frac{\xi_2 - 0.5}{\xi_2 + 0.5}\right)\xi_3$$

$$\frac{\dot{B}_{l1}(t)}{B_l(t)} = \frac{\gamma}{2}\left(\frac{\xi_2 + 0.5}{\xi_2 - 0.5}\right)\left(\frac{\xi_2 + \frac{1}{6}}{\xi_2 - \frac{1}{6}}\right)\xi_3$$

The above set of equations (26) describe the time evolution of bubble.

The time evolution equation for spike can be obtained by making the following transformations [7]: $\xi_1 \to -\xi_1$, $\xi_2 \to -\xi_2$, $\xi_3 \to \xi_3$ and $r^* \to 1/r^*$.

$$\dot{\xi}_1 = -\xi_3$$

$$\dot{\xi}_2 = \frac{1}{2}(1 - 6\xi_2)\xi_3$$

$$\dot{\xi}_3 = \tilde{\chi}_1 + \tilde{\chi}_2 + \tilde{\chi}_3$$

$$\frac{\dot{\rho}_{l1}(t)}{\rho_{l1}(t)} = \left(\frac{\xi_2 - 0.5}{\xi_2 + 0.5}\right)\left(\frac{\xi_2 - \frac{1}{6}}{\xi_2 + \frac{1}{6}}\right)\xi_3 \tag{27}$$

$$\frac{\dot{\rho}_{h1}}{\rho_{h1}} = \left(\frac{\xi_2 + 0.5}{\xi_2 - 0.5}\right)\xi_3$$

$$\frac{\dot{B}_{h1}(t)}{B_h(t)} = \frac{\gamma_h}{2}\left(\frac{\xi_2 + 0.5}{\xi_2 - 0.5}\right)\xi_3$$

$$\frac{\dot{B}_{l1}(t)}{B_l(t)} = \frac{\gamma}{2}\left(\frac{\xi_2 - 0.5}{\xi_2 + 0.5}\right)\left(\frac{\xi_2 - \frac{1}{6}}{\xi_2 + \frac{1}{6}}\right)\xi_3$$

where

$$\chi_1 = \frac{N(\xi_2, 1/r^*)}{D(\xi_2, 1/r^*)}\frac{\xi_3^2}{6\xi_2 + 1}$$

$$\chi_2 = 2kg\left(1 - 1/r^*\right)\frac{\xi_2(6\xi_2 + 1)}{D(\xi_2, 1/r^*)}$$

$$\chi_3 = \frac{6\xi_2 + 1}{D(\xi_2, 1/r^*)}\tilde{\pi}_1 + \frac{6\xi_2 + 1}{D(\xi_2, 1/r^*)}\tilde{\pi}_2$$

$$\tilde{\pi}_1 = -\frac{\gamma_h k^2}{r^* \tilde{\rho}_{h0}}\left( p_{h0}\delta\rho_h(t) + \frac{\boldsymbol{B}_{h0}^2}{\beta_h}(\delta B_h(t))^{\frac{2}{\gamma_h}} \right)(\xi_2 - 0.5)$$

$$\tilde{\pi}_2 = -\frac{\gamma_l k^2}{\tilde{\rho}_{l0}}\left( p_{l0}\delta\rho_l(t) + \frac{\boldsymbol{B}_{l0}^2}{\beta_l}(\delta B_l(t))^{\frac{2}{\gamma_l}} \right)(\xi_2 + 0.5)$$

$$N(\xi_2, 1/r^*) = 36(1 - 1/r^*)\xi_2^2 + 12(4 + 1/r^*)\xi_2 + (7 - 1/r^*)$$

$$D(\xi_2, 1/r^*) = 12(1 - 1/r^*)\xi_2^2 + 4(1 - 1/r^*)\xi_2 + (1 + 1/r^*)$$

## 5. Stability Analysis

We check the stability of the system given by the set of equations Equation (26), assuming the initial values $\rho_{h0}/\tilde{\rho}_{h0}$ and $\rho_{l0}/\tilde{\rho}_{l0}$ to be unity. Let us consider the following coupled equations,

$$\dot{\xi}_3 = \tilde{\chi}_1 + \tilde{\chi}_2 + \tilde{\chi}_3$$

and

$$\dot{\xi}_2 = -\frac{1}{2}(6\xi_2 + 1)\xi_3 = g(\xi_2, \xi_3)\ (\text{say}) \tag{28}$$

There are two sets of fixed points of the system given by the above equations, are

$$\left\{ \xi_2 = \frac{1}{6}, \xi_3 = 0 \right\}$$

and

$$\left\{ \xi_2 = \frac{1}{2}\left[ \frac{r^*\alpha_h - \alpha_l}{\frac{2g}{k}(r^* - 1) - r\alpha_h - \alpha_l} \right], \xi_3 = 0 \right\}$$

where

$$\alpha_h = C_{sh}^2 + \frac{\gamma_h \mu_h}{\beta_h} V_{ah}^2;$$

$$\alpha_l = C_{sl}^2 + \frac{\gamma_l \mu_l}{\beta_l} V_{al}^2$$

$$C_s = \sqrt{\frac{\gamma p}{\rho}}\quad (\text{sound velocity})$$

$$V_a = \sqrt{\frac{B^2}{\mu\rho}}\quad (\text{Alfven velocity})$$

Now to check the stability of the solutions we calculate the eigen values of the following matrix, called Jacobian ($J$), at those fixed points

$$J = \begin{pmatrix} \frac{\partial g}{\partial \xi_2} & \frac{\partial g}{\partial \xi_3} \\ \frac{\partial h}{\partial \xi_2} & \frac{\partial h}{\partial \xi_3} \end{pmatrix}$$

If $\lambda$ be the eigen value of the Jacobian, then

$$\begin{pmatrix} \dfrac{\partial g}{\partial \xi_2} - \lambda & \dfrac{\partial g}{\partial \xi_3} \\ \dfrac{\partial h}{\partial \xi_2} & \dfrac{\partial h}{\partial \xi_3} - \lambda \end{pmatrix} = 0 \tag{29}$$

The eigen values at the fixed point $\xi_2 = 1/6, \xi_3 = 0$ are

$$\left\{ \frac{1}{2}\varpi, -\frac{1}{2}\varpi \right\},$$

where

$$\varpi = \sqrt{6\left[ r^* \alpha_h + \alpha_l - kg\left(r^* - 1\right)\right]}$$

So, the system will be stable if

$$r^* \alpha_h < kg\left(r^* - 1\right) - \alpha_l$$

The above mentioned condition also holds for the stability of the fixed point

$$\left( \xi_2 = \frac{1}{2}\left\{ \frac{r^* \alpha_h - \alpha_l}{\frac{2g}{k}\left(r^* - 1\right) - r^* \alpha_h - \alpha_l} \right\}, \xi_3 = 0 \right)$$

This has been checked thoroughly by using numerical method. The phase diagram (**Figure 2**) also ensures the stability of the system under that above mentioned condition.

However, the eigen value of the second set of fixed point is of very lengthy expression and not presented in this paper.

## 6. Numerical Results and Discussion

According to the linear analysis,when the externally applied magnetic field is parallel to the interface of separation of the two fluid system (heavier fluid lies over the lighter fluid under the action of gravity), the RTI remain unaffected [22]. However, nonlinear analysis shows a decreasing nature of RTI growth rate for the above configuration [27]. It was shown that, compressibility also decreases the growth rate of RTI [26] [32]. In our

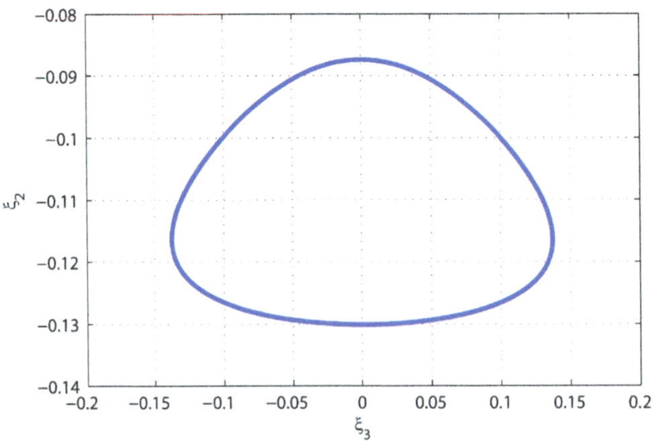


**Figure 2.** Phase diagram with initial values $\xi_1 = -\xi_2 = \xi_3 = -0.1$ and $r = 1.5$.

nonlinear study, we consider the compressibility and interface-parallel magnetic field simultaneously, which has not been studied previously.

To verify our model numerically, we have used Runge-Kutta method and set the values of constants in our equations as follows: $kg = 1, \beta_h = \beta_l$,

The results of our studies are discussed as follows.

## 6.1. Bubble

1) From **Figure 3**, it is clear that, both compressibility and magnetic field decreases the RTI growth rate separately. When they act together the growth rate decreases and system shows oscillatory stabilization. The reason behind such oscillatory stabilization is as follows : When both fluids are conducting, the hydrodynamic pressure difference and magnetic pressure difference for both fluids acts in opposite direction on two sides of the interface. The resultant force due to those two pressure differences induces an oscillatory motion at the interface. The amplitude and the velocity of that oscillatory motion are finite but change rapidly in periodic manner. Moreover, the magnetic pressure difference brings all the periodicity in the motion. The force due to hydrodynamic pressure difference can not lead the system to the oscillatory stabilization, alone.

2) **Figure 4** shows the relative effect of magnetic field strength and compressibility on the growth rate of RTI. The advantage of our model is that, we can observe the domination of magnetic field or hydrodynamic pressure force on the growth rate of instability through the changing value of $\beta$. When $\beta$ is small, the magnetic field dominates the dynamics and when $\beta$ is large (>1) the hydrodynamic pressure dominates the dynamics of instability. As we mentioned earlier the dynamics of the interface is governed by the resultant force due to compressibility (hydrodynamic force) and magnetic pressure (magnetic force). We can see from **Figure 4** that, as $\beta$ increases, the oscillatory nature of the motion of the interface decreases. The reason behind this changes is as follows: when $\beta$ increases the effect of magnetic field decreases and hydrodynamic pressure force increases. As we stated earlier, the periodicity of the motion occurs due to the magnetic field solely. So, as $\beta$ increases *i.e.* the magnetic force decreases the periodicity of the motion decreases and the motion becomes hyd-

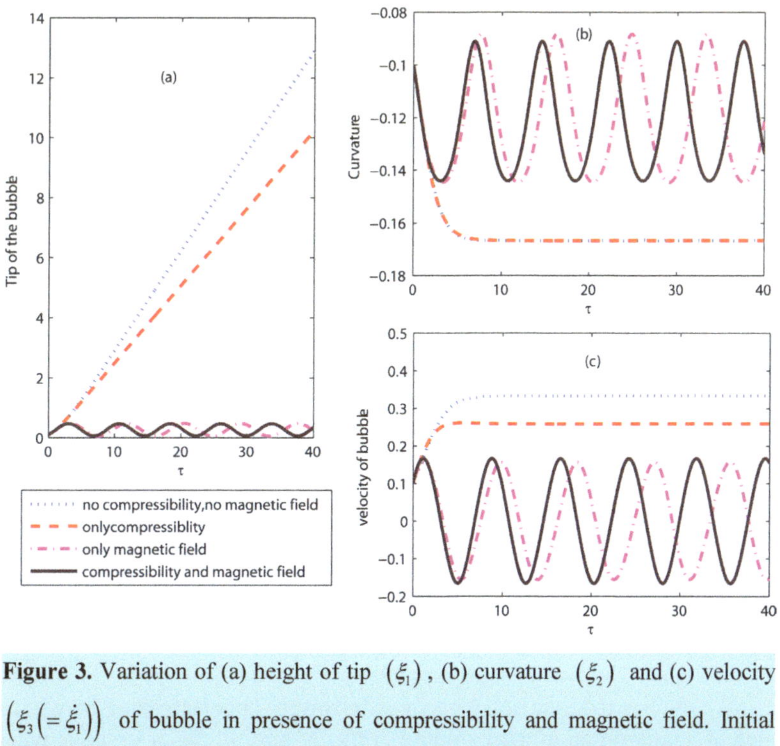


**Figure 3.** Variation of (a) height of tip $(\xi_1)$, (b) curvature $(\xi_2)$ and (c) velocity $(\xi_3(=\dot{\xi}_1))$ of bubble in presence of compressibility and magnetic field. Initial values $\xi_1 = -\xi_2 = \xi_3 = -0.1$ and $r = 1.5$, $\beta_h = \beta_l = 1.0$, $\gamma_h = \gamma_l = 1.0$ (for incompressible case) and $\gamma_h = \frac{5}{3}; \gamma_l = \frac{3}{2}$ (for compressible case), $A_T = 0.20$.

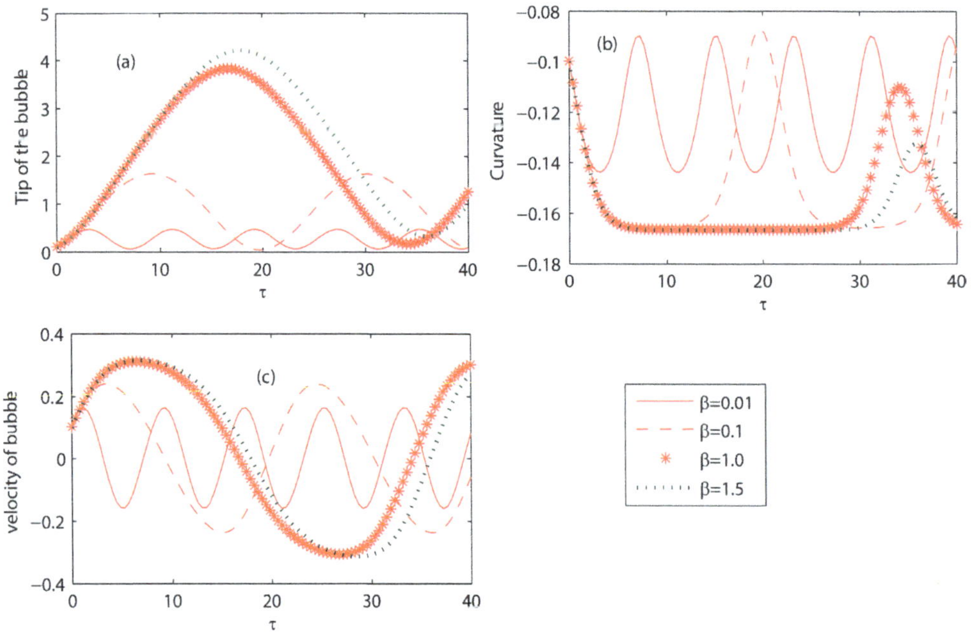


**Figure 4.** Variation of (a) height of tip $(\xi_1)$, (b) curvature $(\xi_2)$ and (c) velocity $\left(\xi_3\left(=\dot{\xi}_1\right)\right)$ of bubble in presence of compressibility and magnetic field for different values of $\beta(\beta_h=\beta_l)$. Initial values $\xi_1=-\xi_2=\xi_3=-0.1$ and $r=1.5$, $A_T=0.20$.

rodynamic force dominant. According to our analysis, for a fixed value of Atwood number $(A_T)$, their is a certain critical value of $\beta(=\beta_c)$, above which the motion of the interface loses its periodic nature and become completely unstable. We show this for a critical beta value $\beta_c=500$ at $A_T=0.20$ (**Figure 5**).

3) The effect of variation of Atwood number on the dynamics of instability has been shown in the **Figure 6**. For a fixed value of $\beta$, the growth rate of instability increases with increasing value of Atwood number. As energy is conserved in the system, the height of the bubble decreases as velocity increases with increasing value of Atwood number. This result is obvious because, as Atwood number increases, the density difference increases, which ensure a lower buoyancy force on the upper (heavier) fluid due to the lower (lighter) fluid. A lower buoyancy force means a lower resistance for the penetration of heavier fluid into the lighter one. As the total volume of the system is conserved, the lighter fluid penetrates into the heavier fluid in the form of bubble. So, as Atwood number increases, the buoyancy force decreases, the growth rate of bubble increases; our numerical analysis (**Figure 6**) ensures this conclusion.

## 6.2. Spike

When heavier fluid penetrates into the lighter fluid under the action of some acceleration (here gravity), spikes form. We will discuss here the results of our analysis about spike.

1) From **Figure 7** it is clear that, under the action of an oscillatory magnetic field, the height, curvature and velocity of the spike changes in periodic manner with finite amplitude. At the first stage, when magnetic pressure increases, the height increases but curvature decreases almost exponentially. The reason behind this is the following. As the magnetic field is parallel to the interface and depends on the densities of the fluids, total pressure on the interface increases in the opposite direction of density gradient. As a result, the heavier fluid falls rapidly into the lighter fluid with large amplitude in the form of spike. The decreasing character of the curvature of the spike implies that, the spike become narrower. This increasing height and decreasing curvature is the consequence of conservation of total volume of the system. In the next stage, where the magnetic field is almost constant, the height, curvature and velocity also remain unchanged. At the last stage of a complete magnetic cycle, the height decreases and curvature increases. This is again due to the conservation of total volume of the system. However, at this stage, the velocity of the spike changes significantly. For first few moments of this

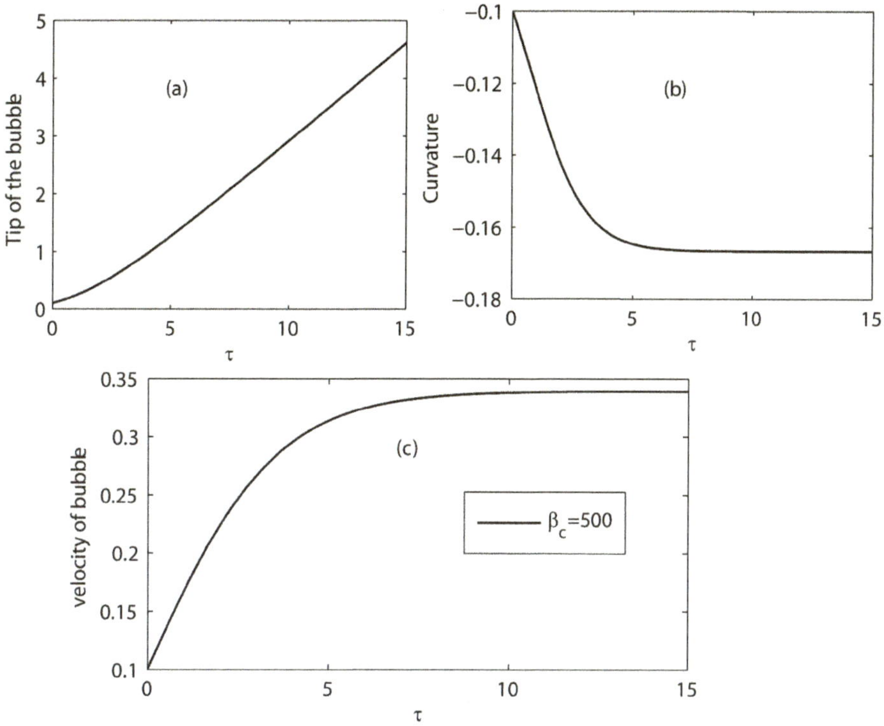


**Figure 5.** Variation of (a) height of tip $(\xi_1)$, (b) curvature $(\xi_2)$ and (c) velocity $(\xi_3(=\dot{\xi}_1))$ of bubble at $\beta(\beta_h=\beta_l=\beta_{hc}=500)$. Initial values $\xi_1=-\xi_2=\xi_3=-0.1$ and $A_T=0.20$, $r=1.5$.

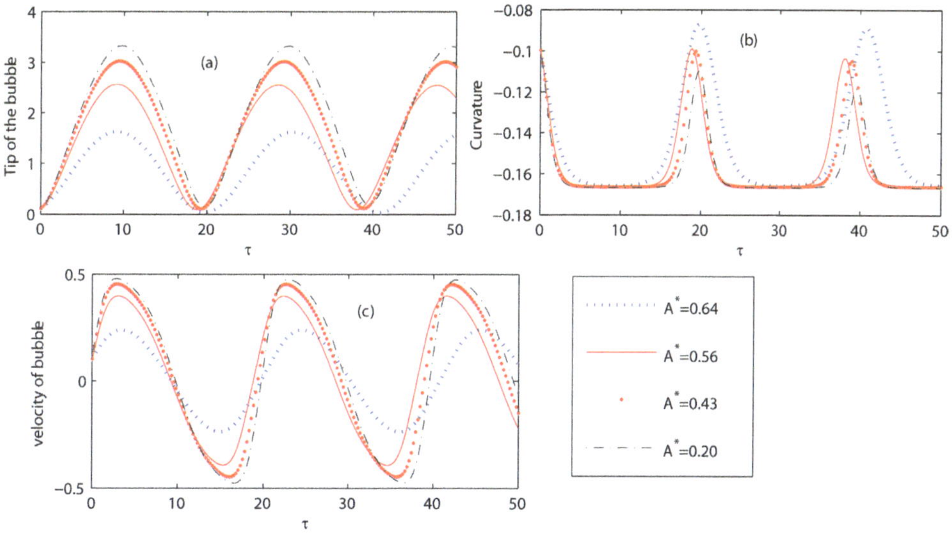


**Figure 6.** Variation of (a) height of tip $(\xi_1)$, (b) curvature $(\xi_2)$ and (c) velocity $(\xi_3(=\dot{\xi}_1))$ of bubble for different values of Atwood number. Initial values $\xi_1=-\xi_2=\xi_3=-0.1$.

phase, velocity increases almost linearly. We believe that this sudden rise of velocity is due to a rapid compression of density of the fluids. The final stage of this phase is characterized by rapid expansion of density and subsequent fall of spike velocity.

This whole cycle of changes in height, curvature and velocity of the spike, repeats itself throughout the evolution of instability. Moreover we can see from the same figure (**Figure 7**) that, as magnetic field decreases

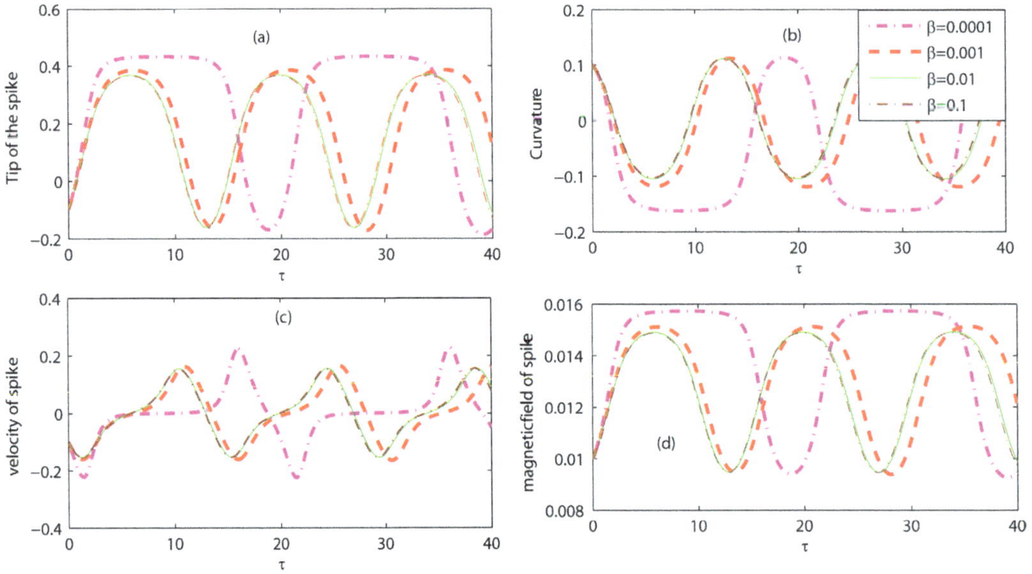


**Figure 7.** Variation of (a) height of tip $(\xi_1)$, (b) curvature $(\xi_2)$ and (c) velocity $\left(\xi_3\left(=\dot{\xi}_1\right)\right)$ and (d) magnetic field of spike with the value of different $\beta(\beta_h=\beta_l)$. Initial values $-\xi_1=-\xi_3=-0.1;-\xi_2=0.1$ and $r=\frac{1}{1.5}$, $A_T=0.20$, $\gamma_h=\frac{5}{3};\gamma_l=\frac{3}{2}$.

*i.e.* $\beta$ increases, the stable region of spike velocity decreases and height of the spike decreases.

2) **Figure 8** shows the dependence of the height, curvature and velocity of spike on Atwood number $(A_T)$. The velocity of the spike decreases as $A_T$ decreases in oscillatory manner. For a fixed value of $A_T$, velocity decreases however, height and curvature of the spike increases for a short period of time. After that, velocity increases very slowly but height of the spike remain almost unchanged over a long time period. After this long time range, the velocity and height of the spike decreases very rapidly over a small time scale. The highest value of spike velocity depends on $A_T$. The maximum attainable value of spike velocity increases with increasing $A_T$. This is very natural because, the hydrodynamic pressure on the interface increases with increasing value of $A_T$, so spike falls more rapidly against gravity.

3) In case of spike, the critical value of magnetic field (at which the spike become unstable) depends largely on Atwood number. As we can see from **Figure 9** that, $\beta=0.0001$ becomes the critical value for all Atwood number $(A_T)$ greater than 0.5, however, $\beta=0.0001$ is not the critical value for $A_T=0.20$.

## 7. Conclusions

We found in earlier studies that interface parallel magnetic field decreases the growth rate nonlinearly, whereas, compressibility effect also linearly decreases the RTI growth rate for bubble and spikes. When they act together, the resultant decreasing character of growth rate depends on their relative strength. This relative strength depends on the value of $\beta$. If $\beta$ is small, *i.e.* $\beta<1$, the magnetic field dominates the growth rate when $\beta>1$ compressibility has the dominating effect on the growth rate.

In our calculation, we assume the variation of $\beta$ for both fluids are negligible and their values are almost constant. We also see that, Atwood number $(A_T)$ has very profound effect on the dynamics of instability. For fixed value of magnetic field, $A_T$ governs the dynamics of the interface. Moreover, in presence of magnetic field, the evolution of the bubble is quite different from that of the spike, near the critical $\beta$ value, for a fixed $A_T$.

In this paper we have considered single mode bubble (spike) and the whole analysis is valid for the spatial scale lengths $0<x\le\frac{1}{k}\sqrt{\frac{2\tilde{\rho}_{h(l)0}}{\rho_{h(l)1}(t)(2\xi_2+1)}}$. This particular limit in length scale arises due to our assumption

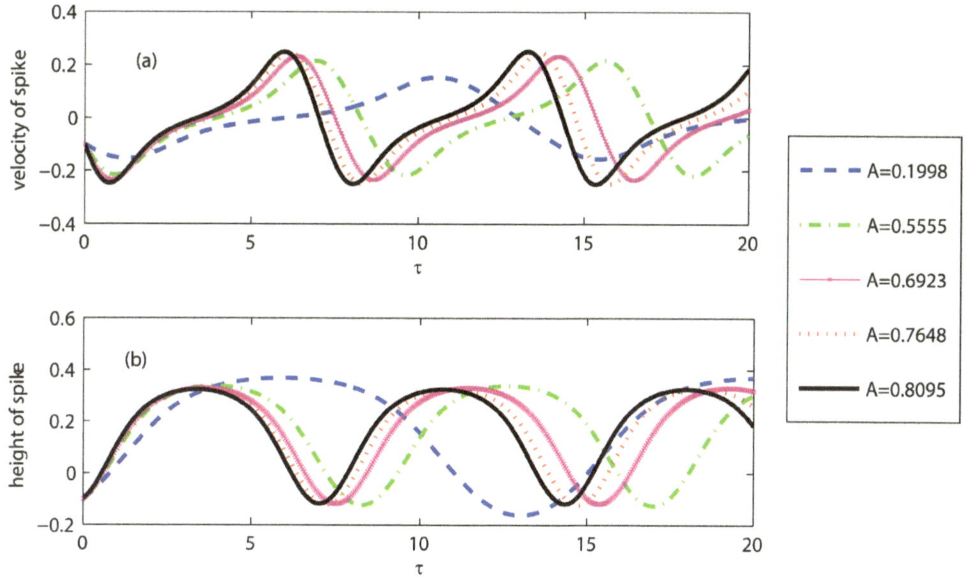


**Figure 8.** Ariation of (a) velocity $\left(\xi_3\left(=\dot{\xi}_1\right)\right)$ and (b) height of tip $(\xi_1)$ of spike for different value of Atwood number (*A*). Initial values $-\xi_1=-\xi_3=-0.1;-\xi_2=0.1$ and $r=\frac{1}{1.5}$, $\beta_h=\beta_l=1.0$, $\gamma_h=\frac{5}{3};\gamma_l=\frac{3}{2}$.

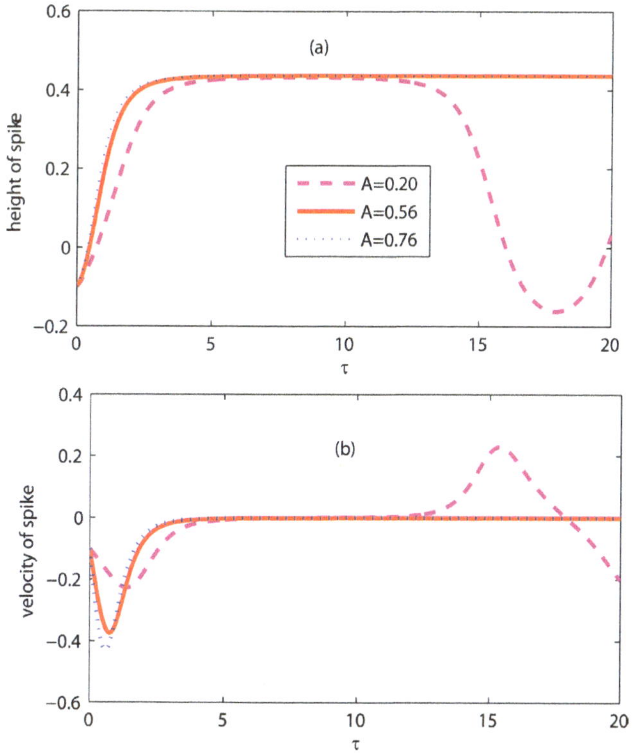


**Figure 9.** Variation of (a) height of tip $(\xi_1)$ and (b) velocity $\left(\xi_3\left(=\dot{\xi}_1\right)\right)$ of at a fixed value of $\beta(\beta_h=\beta_l=0.0001)$. Initial $-\xi_1=-\xi_3=-0.1;-\xi_2=0.1$, $\gamma_h=\frac{5}{3};\gamma_l=\frac{3}{2}$, $r=\frac{1}{1.5}$.

about the shape of the bubble and spike as a parabola and also the fact that we ignored the higher order terms in the expansion of velocity potential. However, this length scale gives an overview of the effect of magnetic field and compressibility on the growth rate of RT instability. From our general model of RT instability, all previous results concerning effect of a) only compressibility [26] [32]; b) only magnetic field [27] can be derived, shown in the numerical graphs (**Figure 3**, **Figure 4**, **Figures 6-8**).

Recent experiments [12] [13] on hydrodynamic instability and mix experiments at Nation Ignition Facilities (NIF) reveal that such RTI at the ablation front in composite ablator is primarily responsible for the hot spot mix, which requires mitigation. In our analysis, it is evident that the compressibility in the composite ablator and magnetic field may play a key role in mitigating RTI leading to ignition at the hot spot.

## References

[1] Goncharov, V.N. (2002) Analytical Model of Nonlinear, Single-Mode, Classical Rayleigh-Taylor Instability at Arbitrary Atwood Numbers. *Physical Review Letters*, **88**, Article ID: 134502. http://dx.doi.org/10.1103/PhysRevLett.88.134502

[2] Gupta, M.R., Banerjee, B., Mandal, L.K., Bhar, R., Pant, H.C., Khan, M. and Srivastava, M.K. (2012) Effect of Viscosity and Surface Tension on the Growth of Rayleigh-Taylor Instability and Richtmyer-Meshkov Instability Induced Two Fluid Interfacial Nonlinear Structure. *Indian Journal of Physics*, **86**, 471-479. http://dx.doi.org/10.1007/s12648-012-0077-3

[3] Gupta, N.K. and Lawande, S.V. (1986) Rayleigh-Taylor Instability in Multi-Structured Spherical Targets. *Plasma Physics and Controlled Fusion*, **28**, 925-941. http://dx.doi.org/10.1088/0741-3335/28/6/008

[4] Hazak, G. (1996) Lagrangian Formalism for the Rayleigh-Taylor Instability. *Physical Review Letters*, **76**, 4167-4170. http://dx.doi.org/10.1103/PhysRevLett.76.4167

[5] Hecht, J., Alon, U. and Shvarts, D. (1994) Potential Flow Models of Rayleigh-Taylor and Richtmyer-Meshkov Bubble Fronts. *Physics of Fluids*, **6**, 4019-4030. http://dx.doi.org/10.1063/1.868391

[6] Piriz, A.R., Lpez Cela, J.J. and Cortzar, O.D. (2005) Rayleigh-Taylor Instability in Elastic Solids. *Physical Review E*, **72**, Article ID: 056313. http://dx.doi.org/10.1103/PhysRevE.72.056313

[7] Ramaprabhu, P. and Dimonte, G. (2005) Single-Mode Dynamics of the Rayleigh-Taylor Instability at Any Density Ratio. *Physical Review E*, **71**, Article ID: 036314. http://dx.doi.org/10.1103/PhysRevE.71.036314

[8] Sohn, S.-I. (2003) Simple Potential-Flow Model of Rayleigh-Taylor and Richtmyer-Meshkov Instabilities for All Density Ratios. *Physical Review E*, **67**, Article ID: 026301.

[9] Velikovich, A.L. and Dimonte, G. (1996) Nonlinear Perturbation Theory of the Incompressible Richtmyer-Meshkov Instability. *Physical Review Letters*, **76**, 3112. http://dx.doi.org/10.1103/PhysRevLett.76.3112

[10] Zhang, Q. (1998) Analytical Solutions of Layzer-Type Approach to Unstable Interfacial Fluid Mixing. *Physical Review Letters*, **81**, 3391-3394. http://dx.doi.org/10.1103/PhysRevLett.81.3391

[11] Banerjee, R., Mandal, L.K., Khan, M. and Gupta, M.R. (2013) Bubble and Spike Growth Rate of Rayleigh Taylor and Richtmeyer Meshkov Instability in Finite Layers. *Indian Journal of Physics*, **87**, 929-937. http://dx.doi.org/10.1007/s12648-013-0300-x

[12] Smalyuk, V.A., Barrios, M., Caggiano, J.A., *et al.* (2014) Hydrodynamic Instability Growth and Mix Experiments at the National Ignition Facilitya. *Physics of Plasmas*, **21**, Article ID: 056301. http://dx.doi.org/10.1063/1.4872026

[13] Edwards, M.J., *et al.* (2013) Progress towards Ignition on the National Ignition Facility. *Physics of Plasmas*, **20**, Article ID: 070501.

[14] Mason, R.J. and Tabak, M. (1998) Magnetic Field Generation in High-Intensity-Laser-Matter Interactions. *Physical Review Letters*, **80**, 524-527. http://dx.doi.org/10.1103/PhysRevLett.80.524

[15] Sudan, R.N. (1993) Mechanism for the Generation of 10 9 G Magnetic Fields in the Interaction of Ultraintense Short Laser Pulse with an Overdense Plasma Target. *Physical Review Letters*, **70**, 3075-3078. http://dx.doi.org/10.1103/PhysRevLett.70.3075

[16] Wilks, S.C., Kruer, W.L., Tabak, M. and Langdon, A.B. (1992) Absorption of Ultra-Intense Laser Pulses. *Physical Review Letters*, **69**, 1383-1386. http://dx.doi.org/10.1103/PhysRevLett.69.1383

[17] Srivastava, M.K., Lawande, S.V., Khan, M., Das, C. and Chakraborty, B. (1992) Axial Magnetic Field Generation by Ponderomotive Force in a Laser-Produced Plasma. *Physics of Fluids B*, **4**, 4086. http://dx.doi.org/10.1063/1.860315

[18] Khan, M., Das, C., Chakraborty, B., *et al.* (1998) Self-Generated Magnetic Field and Faraday Rotation in a Laser-Produced Plasma. *Physical Review E*, **58**, 925-930. http://dx.doi.org/10.1103/PhysRevE.58.925

[19] Evans, R.G. (1986) The Influence of Self-Generated Magnetic Fields on the Rayleigh-Taylor Instability. *Plasma Physics and Controlled Fusion*, **28**, 1021. http://dx.doi.org/10.1088/0741-3335/28/7/006

[20] Ghezzi, C.R., de Gouveia Dal Pino, E.M. and Horvath, J.E. (2001) Magnetic Field Effects on the Thermonuclear Combustion Front of Chandrasekhar Mass White Dwarfs. *The Astrophysical Journal Letters*, **548**, L193. http://dx.doi.org/10.1086/319091

[21] de Gouveia Dal Pino, E.M. and Benz, W. (1993) Three-Dimensional Simulations of Protostellar Jets. *Astrophysical Journal*, **410**, 686-695. http://dx.doi.org/10.1086/172785

[22] Chandrasekhar, S. (1961) Hydrodynamic and Hydromagnetic Stability. Dover Publications, Inc., New York.

[23] Liberatore, S. and Bouquet, S. (2008) Analytical Modeling of Magnetic Rayleigh-Taylor Instabilities in Compressible Fluids. *Physics of Fluids*, **20**, Article ID: 116101. http://dx.doi.org/10.1063/1.3025832

[24] Samtaney, R. (2003) Suppression of the Richtmyer-Meshkov Instability in the Presence of a Magnetic Field. *Physics of Fluids*, **15**, L53. http://dx.doi.org/10.1063/1.1591188

[25] Jun, B.I., Norman, M.L. and Stone, J.M. (1995) A Numerical Study of Rayleigh-Taylor Instability in Magnetic Fluids. *The Astrophysical Journal*, **453**, 322. http://dx.doi.org/10.1086/176393

[26] Gupta, M.R., Roy, S., Khan, M., *et al.* (2009) Effect of Compressibility on the Rayleigh-Taylor and Richtmyer-Meshkov Instability Induced Nonlinear Structure at Two Fluid Interface. *Physics of Plasmas*, **16**, Article ID: 032303. http://dx.doi.org/10.1063/1.3074789

[27] Gupta, M.R., Mandal, L.K., Roy, S. and Khan, M. (2010) Effect of Magnetic Field on Temporal Development of Rayleigh-Taylor Instability Induced Interfacial Nonlinear Structure. *Physics of Plasmas*, **17**, Article ID: 012306. http://dx.doi.org/10.1063/1.3293120

[28] Layzer, D. (1955) On the Instability of Superposed Fluids in a Gravitational Field. *The Astrophysical Journal*, **122**, 1. http://dx.doi.org/10.1086/146048

[29] Drake, R.P. (2006) High Energy-Density Physics. Springer, New York.

[30] Casali, R., Castro, L.B. and Menezes, D.P. (2014) Hadronic and Hybrid Stars Subject to Density-Dependent Magnetic Fields. *Physical Review C*, **89**, Article ID: 015805. http://dx.doi.org/10.1103/PhysRevC.89.015805

[31] Lamb, H. (1932) Hydrodynamics. Cambridge University Press, Cambridge.

[32] Bernstein, I.B. and Book, D.L. (1983) Effect of Compressibility on the Rayleigh-Taylor Instability. *Physics of Fluids*, **26**, 453. http://dx.doi.org/10.1063/1.864158

# 16

# Numerical Investigation of Effective Heat Conductivity of Fluid in Charging Process of Thermal Storage Tank

**Hadi Taheri, Ferdinand P. Schmidt, Martin Gabi**

Institute of Fluid Machinery, Department of Mechanical Engineering, Karlsruhe Institute of Technology, Karlsruhe, Germany
Email: hadi.taheri@kit.edu, ferdinand.schmidt@kit.edu, martin.gabi@kit.edu

## Abstract

**This paper presents a numerical case study of heat transfer mechanisms during the charging process of a stratified thermal storage tank applied in a specific adsorption heat pump cycle. The effective thermal conductivity of the heat transfer fluid during the charging process is analyzed through CFD simulations using Unsteady Reynolds-averaged Navier-Stokes equations (URANS). The aim of the study is to provide an equivalent thermal conductivity for a one-dimensional storage tank model to be used in a system simulation of the complete adsorption heat pump cycle. The influence of the turbulent mixing and also the advection effect due to fluid bulk motion are investigated. The results show that in the case considered here, the turbulence effect on the effective thermal conductivity is more considerable than the advection effect.**

## Keywords

**Thermal Storage Tank, Stratification Pipe, Effective Thermal Conductivity**

## 1. Introduction

The achievable coefficient of performance (COP) of adsorption heat pump systems can be increased by coupling the adsorber in a specific way to a stratified thermal storage tank [1]. In this cycle concept (cf. **Figure 1**), the COP depends on the degree of mixing in the thermal storage tank. Rational design of a system utilizing such an adsorption cycle with an integrated stratified storage tank requires a transient analysis of the coupled system of adsorber, storage tank, evaporator and condenser as well as external heater and cooler. In order to limit the computational effort involved in system simulations, it makes sense to use a simplified 1-D model of the storage

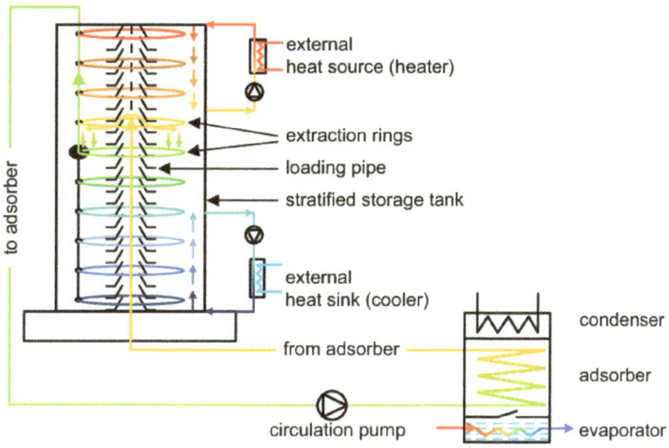


**Figure 1.** Schematic model of the storage tank coupled with adsorption module [3].

tank. Such models are commonly employed in system simulations of solar thermal heating and cooling systems, see e.g. [2].

In order to be able to describe the thermal behavior of the storage tank in a system model, the influences of turbulent mixing and advection on the effective thermal conductivity have been investigated by means of a CFD model of thermal storage tank. Effective thermal conductivity is the determining parameter for the description of stratification and mixing in the thermal storage tank. Regarding operation of the thermal storage, the main aim is to maintain thermal stratification by minimizing any mixing effects that may be caused by inertial effects of fluid insertion, turbulence effects or buoyancy effects in case of a temperature mismatch of inserted fluid. For this reason, the convection heat transfer plays an important role in the thermal storage tank. Convection heat transfer includes both diffusion (individual motion of fluid particle in small scale) and advection (bulk fluid motion in large scale). Thermal analysis of the storage tank has been widely studied both experimentally and numerically by a number of researchers. Zurigat *et al.* [4] studied the influence of different inlet geometries on thermal stratification in a thermocline storage tank. Alizadeh [5] studied the thermal behavior of the storage tank experimentally and numerically. He analyzed specially the discharging process of the storage tank. Nelson *et al.* [6] presented a one-dimensional transient heat conduction model to describe the decay of the thermocline in a stratified water tank. Bouhdjar *et al.* [7] studied the influence of the fluid properties and aspect ratios on stratification. Furbo *et al.* [8] analyzed the entrance effects in solar storage tanks. They performed an experimental and theoretical analysis of water jets entering a solar storage tank. Consul *et al.* [9] performed the transient numerical simulation of thermal behavior in a storage tank forming part of a thermo-syphon solar heating system. In this study, the influence of the inlet mass flow rate on the degree of thermal stratification during a discharging process has been analyzed. Shah *et al.* [10] investigated the charging of the storage tank with inlet stratifiers by CFD and experimental work. Altuntop *et al.* [11] analyzed the effect of using different obstacles on thermal stratification in a cylindrical hot water tank. Andersen *et al.* [12] analyzed the thermal stratification devices for hot water heat stores by means of CFD and experimental study. They investigated the inlet stratification pipe effect with a limited number of operating conditions appearing in solar collector loops. Different charging systems for maintaining the stratification in thermal storage tanks are described and analyzed in the previous works. One of the prevalent stratifier systems consists of pipes with vertically arranged outlets. This patented charging system is marketed by German company Sailer GmbH [13]. According to the fluid density, incoming fluid can flow to the proper tank layer matching its temperature. In the present work, the same construction of stratification pipe with vertical openings has been implemented for the charging of the storage tank. There are different reasons that diminish the thermal stratification in the storage tank. These are as follows,

1) Plume entrainment

Near the inlet region, momentum effects of the fluid entering the storage are commonly more important than buoyancy effects. In such situations, fluid plume can be defined as fluid jet. Because of the velocity gradient and shear effect, the surrounding fluid will be entrained, resulting in an increased mixing effect [14]. In the present work, the entrainment of the fluid from the tank to stratification pipe through the openings has been described as suction effect.

2) Inlet jet mixing

Because of the high momentum of the entering fluid, turbulent mixing in the near of the inlet leads to a locally increased mixing. One of the advantages of the stratification pipe for the charging process is that turbulent kinetic energy can be partly bounded inside of the stratification pipe. Therefore, the mixing effect influences have been more restricted.

3) Conduction heat transfer within the storage medium and storage tank wall

Conductive heat transfer tends to reduce the temperature gradient and consequently more destratification. In the current study the storage tank for application in the adsorption heat pump cycle has been simulated and its charging process has been investigated. The charging of the storage tank has been carried out through a rigid inlet stratification pipe with a similar structure to the stratification pipe marketed by Sailer GmbH [13]. The main aim of using the stratification devices is the suppression of mixing effects due to high inflow velocities. This stratification pipe also enables an inflow at different heights of the storage due to the influence of buoyancy. The inlet stratification pipe has an inner diameter of 34 mm and an outer diameter of 89.4 mm and has been mounted vertically in the storage tank. Fluid arises through the stratification pipe in the vertical direction and can leave the pipe through the openings at different levels. Vertical distance between each two openings is 9.5 mm.

## 2. Modeling and Simulation

**Figure 1** shows the schematic model of storage tank in adsorption heat pump cycle. The geometry of the cylindrical tank used for the current investigation is described as follows:

- Height: 1.9 m;
- Diameter: 19 cm;
- Width of the heater and cooler ports: 0.41 cm (as ring around tank perimeter);
- Heater outlet position: 57% of the storage height from the bottom (1.08 m);
- Heater inlet position: at the top of the storage tank;
- Cooler outlet position: 25% of the storage height from the bottom (0.47 m);
- Cooler inlet position: at the bottom of the storage tank.

The development potential of this adsorption heat pump cycle bases upon the internal heat recovery between adsorption and desorption processes [1]. The thermal storage tank functions as intermediate heat storage between two half cycles of adsorption and desorption. In **Figure 2**, the storage tank CFD model with the implemented stratification pipe and boundary conditions in an axisymmetric domain has been depicted. The very high aspect ratio (height/diameter) of 10 has been chosen for this tank to decrease the temperature gradient for a certain temperature difference inside of the storage tank and therefore to reduce the heat transfer process due to conduction. Reduced heat transfer maintains the stratification level and causes less mixing effect. It should be noted, however, that such a large aspect ratio may increase the thermal losses of the tank to its environment.

The function of the heater cycle is to provide that part of the desorption heat, which is not covered from the released heat during the adsorption half cycle. The heat transfer fluid is extracted at the level of the heater position (heater outlet in CFD model) and after heating up, will be inserted at the top of the tank. The function of cooler cycle in this application is to provide a heat sink. The cooler cycle absorbs that part of the released heat during the adsorption that will not be used for the desorption half cycle.

### 2.1. Heat Transfer Fluid

The storage tank is considered here for a high-temperature adsorption cycle utilizing zeolite 13X/water as the

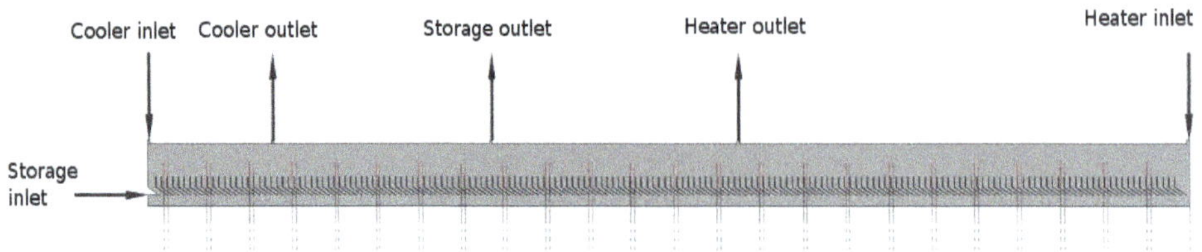


**Figure 2.** Storage tank 2-D axisymmetric domain with implemented stratification pipe and boundary conditions.

adsorption pair [1]. Therefore, the heat transfer fluid used in the storage is a synthetic thermal oil, Marlotherm LH [15]. Its properties such as density, specific heat capacity, thermal conductivity and viscosity have been defined in a piecewise linear manner in Ansys Fluent CFD model. The thermal conductivity of Marlotherm LH in the applied temperature range varies between 0.111 and 0.129 $(\text{W/m}\cdot\text{K})$). The viscosity ranges from 0.0004 to 0.00256 $(\text{kg/m}\cdot\text{s})$.

## 2.2. Numerical Method

Commercial CFD code Ansys Fluent 14.0 based on the finite volume method has been used for the simulation. The fluid cylindrical domain in the storage tank has been simulated as a 2-D axisymmetric CFD model. The pressure-based coupled algorithm has been applied in order to solve the flow governing equations. Second order upwind discretization scheme for momentum, turbulent kinetic energy and dissipation rate has been employed in the computation. The fluid domain has been discretized with about 400,000 cells, which is a combination of block-structured and unstructured grid. Second order implicit time discretization has been applied for the transient formulation of the charging process with a time step of 0.2 second.

## 2.3. Initial and Boundary Condition

Since there is no single state of the storage that could be considered as representative of the whole adsorption/desorption cycle, an arbitrary state that might occur during the adsorption half cycle has been selected for this case study. The fluid inside of the storage tank has been initialized with linear function of temperature with 38˚C at the bottom of the tank to 200˚C at the top of the tank (as shown in **Figure 3**). The storage has been charged from the bottom through the stratification pipe with charging fluid at 119˚C and 0.2 kg/s mass flow rate. Simultaneously, the heater and cooler extract and reinsert the fluid with 20 percent of the storage charging mass flow rate. Therefore, the solution domain contains six different boundary conditions for inflow and outflow, and also wall and symmetry axis. In the cooler cycle, fluid will be extracted from the storage and will be cooled down to the fluid temperature at the bottom of the tank (38˚C) and enters again to the tank at the bottom. In the heater cycle, similarly after fluid extraction, it will be heated up to fluid temperature at the top of the storage (200˚C) and enters again to the tank at the top.

## 2.4. Turbulence Modeling

The modified version of $k-\varepsilon$ RANS (Reynolds Averaged Navier-Stokes) model suggests alternative eddy viscosity formulation and dissipation rate transport equation, which shows better prediction of the flow structures with round jet and recirculation. The realizable $k-\varepsilon$ model has been used in order to compute the turbulent viscosity. Transport equations for turbulent kinetic energy and turbulent energy dissipation rate have been represented in Equations (1) and (2), respectively [16].

$$\frac{\partial}{\partial t}(\rho k)+\frac{\partial}{\partial x_j}(\rho k u_j)=\frac{\partial}{\partial x_j}\left[\left(\mu+\frac{\mu_t}{\text{Pr}_{t,k}}\right)\frac{\partial k}{\partial x_j}\right]+G_k+G_b-\rho\varepsilon-Y+S_k \tag{1}$$

$$\frac{\partial}{\partial t}(\rho \varepsilon)+\frac{\partial}{\partial x_j}(\rho \varepsilon u_j)=\frac{\partial}{\partial x_j}\left[\left(\mu+\frac{\mu_t}{\text{Pr}_{t,\varepsilon}}\right)\frac{\partial \varepsilon}{\partial x_j}\right]+\rho C_1 S_\varepsilon-\rho C_2\frac{\varepsilon^2}{k+\sqrt{v\varepsilon}}+C_{1\varepsilon}\frac{\varepsilon}{k}C_{3\varepsilon}G_b+S_\varepsilon \tag{2}$$

where

$$\eta=S\frac{k}{\varepsilon}\quad C_1=\max\left[0.43,\frac{\eta}{\eta+5}\right]\quad S=\sqrt{2S_{ij}S_{ij}}$$

In transport equations of turbulent kinetic energy and turbulent energy dissipation rate, $\text{Pr}_{t,k}$ and $\text{Pr}_{t,\varepsilon}$ are the turbulent Prandtl numbers for $k$ and $\varepsilon$.

In the $k-\varepsilon$ turbulence model, turbulent eddy viscosity is computed by means of turbulent kinetic energy and turbulent energy dissipation rate via the Equation (3):

$$\mu_t=\rho C_\mu\frac{k^2}{\varepsilon} \tag{3}$$

The value of $k$ and $\varepsilon$ is determined directly from differential transport equations for the turbulence kinetic energy and turbulent energy dissipation rate. The turbulent kinetic energy contour in Thermal Energy Storage has been shown in **Figure 4**.

The effective thermal conductivity due to turbulent mixing in the fluid domain is calculated from the Equation (4),

$$k_{\text{eff}} = k + \frac{C_p \mu_t}{\text{Pr}_t} \tag{4}$$

In Equation (4) $k$ is fluid thermal conductivity, $\mu_t$ turbulent eddy viscosity and $\text{Pr}_t$ turbulent Prandtl number.

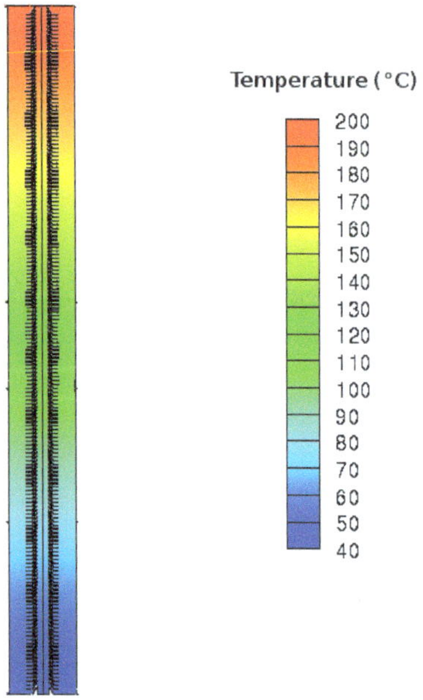


**Figure 3.** Initial temperature distribution in thermal storage tank.

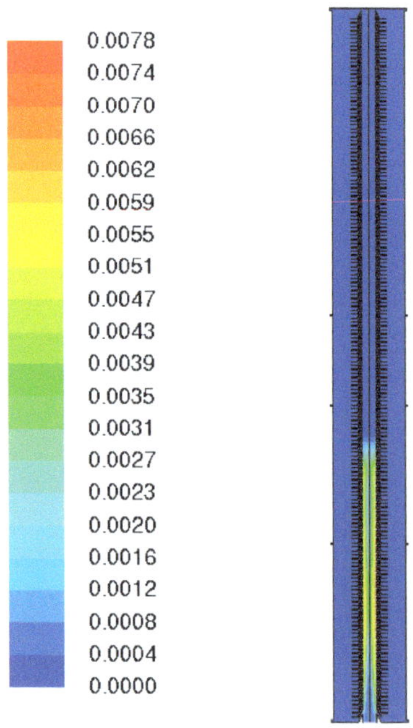


**Figure 4.** Turbulent kinetic energy contour after 4 seconds charging.

### 2.5. Buoyancy Modeling

Due to the temperature difference and, therefore, density difference, the natural buoyancy results in a buoyancy-driven flow in thermal energy storage.

In simulation of TES, the buoyancy term has been modeled by means of a body-force term in momentum equation (Boussinesq Model). This body-force term has been shown in Equation (5) [16].

$$(\rho - \rho_{\text{ref}})g \tag{5}$$

where $\rho_{\text{ref}}$ is the reference density which is usually considered as averaged density value in simulation domain and $g$ the gravitational acceleration.

### 2.6. Calculation of Advection Effect in Charging Process of the Storage Tank

Heat advection caused by the bulk fluid motion contributes to effective thermal conductivity with an additional part. This part of effective thermal conductivity has been calculated for the current selected case of charging process in adsorption heat pump cycle. The equations used for the calculation of advection part in effective thermal conductivity have been represented in Equation (6).

$$\begin{aligned}
\dot{q}_{\text{conv}} &= \rho \cdot C_p (\boldsymbol{U} \cdot \nabla T) \mathrm{d}V \\
\dot{Q}_{\text{conv}} &= \int_{V_\text{cs}} \rho C_p (\boldsymbol{U} \cdot \nabla T) \mathrm{d}V \\
\dot{Q}_{\text{conv}} &= k_{\text{eff_conv}} \cdot A_{\text{hor_cs}} \cdot \overline{\frac{\partial T}{\partial x}} \\
k_{\text{eff_conv}} &= \frac{\int_{V_\text{cs}} \rho C_p (\boldsymbol{U} \cdot \nabla T) \mathrm{d}V}{A_{\text{hor_cs}} \cdot \overline{\frac{\partial T}{\partial x}}}
\end{aligned} \tag{6}$$

where $A_{\text{hor\_cs}}$ is the horizontal cross section of thermal storage tank between outer diameter of the stratification pipe and storage tank wall. $\overline{\frac{\partial T}{\partial x}}$ is the average value of the temperature gradient over the horizontal cross section of the storage tank.

## 3. Results and Discussion

The main mechanisms influencing the transient development of the temperature profile in a Thermal Energy Storage are heat transport process, mixing of the charging fluid with stored fluid, and local turbulence because of stratification [17]. From the temperature distribution inside of the storage it can be seen that the region above to the heater outlet doesn't show a significant mixing. The fluid at the extraction level of heater cycle has been extracted (heater outlet in CFD model) and the fluid with the same temperature as the upper layer in tank has been introduced at the upper part. Because of the insertion at the heater inlet and extraction at the heater outlet, temperature profile has been shifted to the left and with the time progress, this displacement increases. This behavior can also be observed by ideal plug flow charging model. The average values of the temperature between outer diameter of the stratification pipe and storage tank wall have been demonstrated in **Figure 5**. It shows that the fluid insertion level from stratification pipe to storage tank is about 1.1 m from the bottom of the tank which is close to the heater outlet position (as shown in **Figure 6**). Because of high momentum of entering fluid in stratification pipe an overrun of the fluid jet and higher insertion height from stratification pipe to the storage is observed. At this height, the initial, linearly stratified temperature profile starts to break. At lower levels especially until cooler outlet, mixing leads to destroy the linear stratification with higher intensity.

### 3.1. Analysis of Effective Thermal Conductivity Due to Advection

In the 1-D plug flow model of thermal storage tank in adsorption heat pump system simulation, the effect of all heat transfer processes can only be taken into account as an effective heat conduction between adjacent fluid

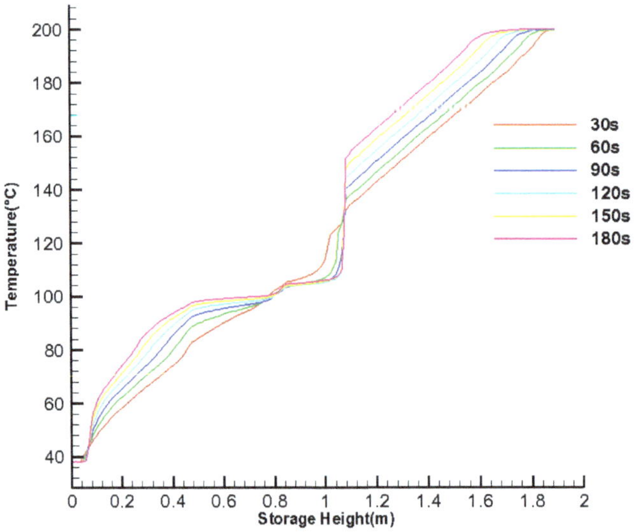


**Figure 5.** Average temperature distribution in storage tank for different charging times in transient simulation.

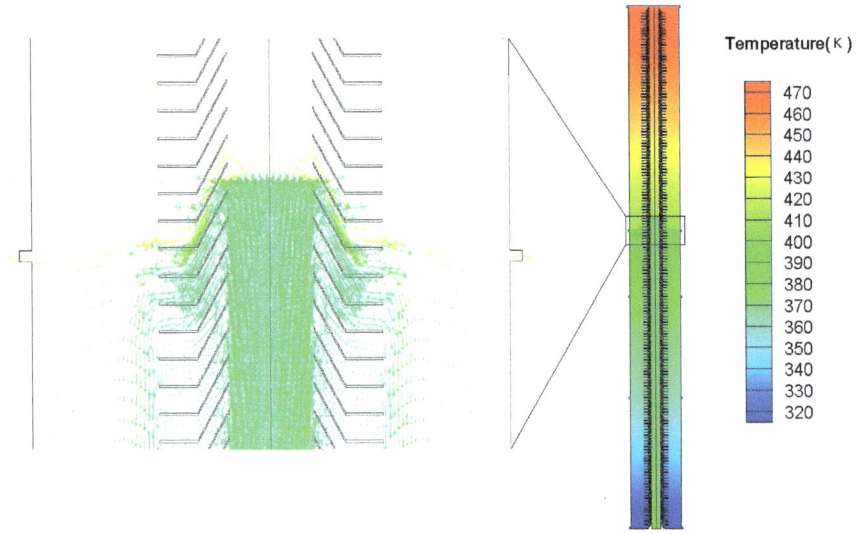


**Figure 6.** Temperature contour and velocity vector with the magnification of the insertion region from stratification pipe to the storage tank after 60 s charging.

layers. In the current numerical investigation of the charging process, advection due to the bulk fluid flow has been considered as another heat transfer mechanism, which leads to stronger mixing in the tank. In order to calculate the advection part of the effective thermal conductivity according to Equation (6), a User Defined Function (UDF) has been applied to the CFD model. The UDF has been interpreted and executed to calculate the desired variables in Equation (6). The advective heat transfer part has been integrated over horizontal surfaces at different heights of the tank. In addition, the production of horizontal surface and average value of temperature gradient in axial direction has been exported over different horizontal surfaces at different heights of the tank. By means of these parameters, the effective thermal conductivity due to advection has been calculated at different charging times.

## 3.2. Analysis of Effective Thermal Conductivity Due to Turbulence

Turbulent diffusion in fluid flow in TES enhances the mixing due to increasing the transport rate of momentum, which leads to homogenization. On the other hand, enhanced energy transport rate due to turbulence causes an intensified heat transfer process. This enhancement can be considered on the basis of additional thermal conductivity called effective thermal conductivity due to turbulent motions.

In transient simulation, effective thermal conductivity in the storage with the realizable $k-\varepsilon$ turbulence model has been calculated and the average values over the whole horizontal area at different storage heights have been plotted (**Figure 7**). The results demonstrate also the maximum value in the region of main flow insertion from stratification pipe into the storage. Turbulent viscosity contour depicted in **Figure 8** demonstrates the highest value of turbulent viscosity, and consequently, effective thermal conductivity due to turbulence in this region. The comparison of two parts of the effective thermal conductivity reveals that the effective thermal conductivity part due to turbulence is considerably higher than effective thermal conductivity part due to advection except for a small region at the lower part of the storage tank (as shown in **Figure 9**). Computation results with SST $k-\omega$ turbulence model show the similar trend of effective thermal conductivity inside of the storage tank. Although this model predicts relatively smaller values in regions of main flow insertion from stratification pipe to the storage and heater cycle, however, the turbulent part of effective thermal conductivity still plays a key role in comparison with advection part.

## 4. Conclusions

The effective thermal conductivity due to turbulence and advection in the thermal storage tank is an important parameter influencing the performance of the whole adsorption heat pump system under consideration. It accounts for convection effect and turbulent mixing in the storage tank. Therefore, it can only be determined for a specific geometry of tank, inlet and outlet ports, initial and boundary conditions. For any given geometry, it depends on the fluid charging volume flow rate, heat transfer fluid, and the temperature profile in the storage tank. The results reveal that in this CFD model and under the conditions chosen here, the influence of the turbulence part on the effective thermal conductivity is much more important compared to the advection part. Turbulent mixing leads to more intense mixing effect and destratification and shows stronger influence on the effective thermal conductivity. Under these circumstances, the results can also be expected to strongly depend on the details of the turbulence model used, and therefore should be treated with caution. A reassessment of these results with more fine-grained turbulent flow computation methods such as Large Eddy Simulation (LES) or Direct Numerical Simulation (DNS) would be welcome.

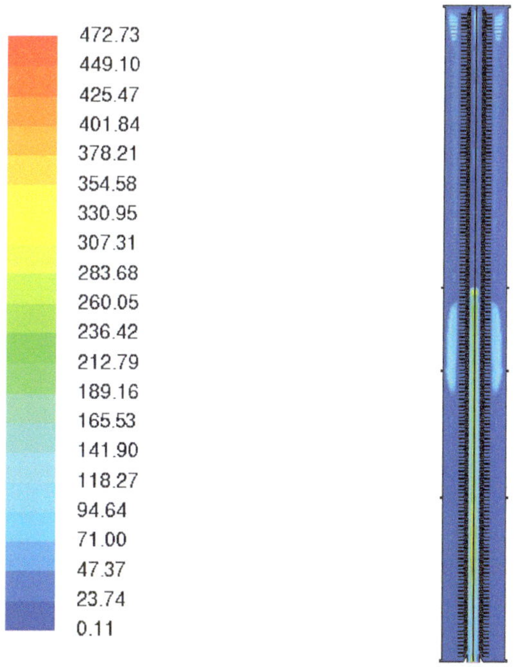


**Figure 7.** Contour of effective thermal conductivity (W/mK) due to turbulence after 60 s of charging.

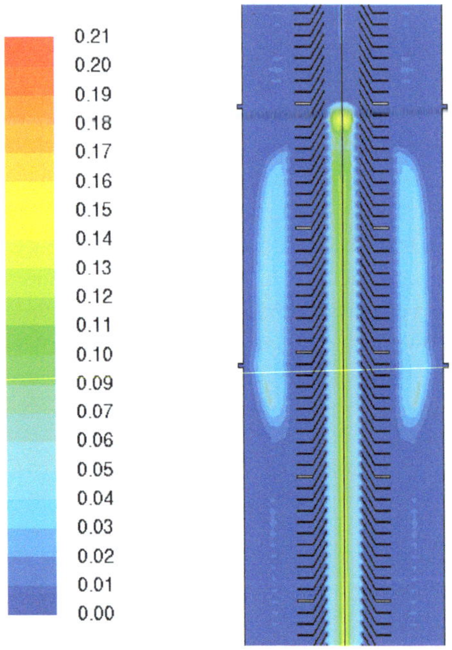


**Figure 8.** Turbulent viscosity contour (kg/m·s) in the storage outlet region after 60 s of charging.

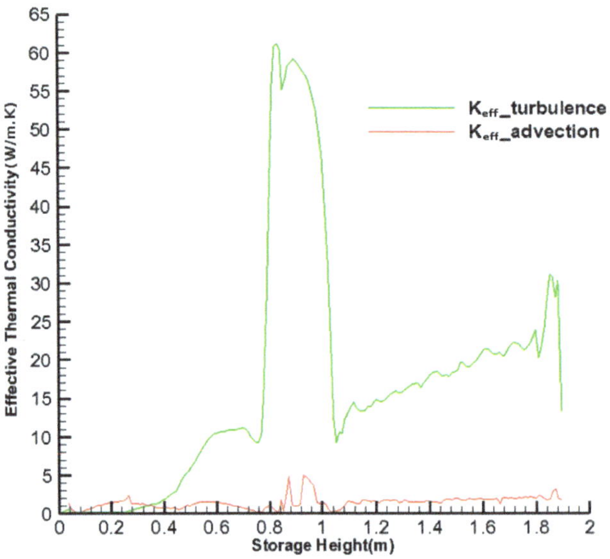


**Figure 9.** Comparison of effective thermal conductivity due to turbulence and advection after 60 s of charging.

Application of an inlet stratification pipe restricts the impact of the turbulent mixing partly to the inner region of that pipe and therefore would be advantageous in order to avoid higher level of turbulent mixing and effective thermal conductivity inside of the storage tank. One of the demonstrative disadvantage of the applied geometry for the stratification pipe is the entrainment effect which leads to suction of the fluid from storage to the stratification pipe.

## Acknowledgements

H. Taheri would like to sincerely thank Karlsruhe Institute of Technology for providing a scholarship (Rektorstipendium), and also Balazs Pritz for his constructive comments. F. P. Schmidt gratefully acknowledges project funding by the state of Baden-Württemberg (ZO-III Initiative) through the Project Management Agency Karlsruhe (PTKA-BWP).

## References

[1] Schwamberger, V., Joshi, C. and Schmidt, F.P. (2011) Second Law Analysis of a Novel Cycle Concept for Adsorption Heat Pumps. *International Sorption Heat Pump Conference* (*ISHPC*11), Padua, 28-29 April 2011, 991-998.

[2] Eicker, U. and Pietruschka, D. (2009) Design and Performance of Solar Powered Absorption Cooling Systems in Office Buildings. *Energy and Buildings*, **41**, 81-91. http://dx.doi.org/10.1016/j.enbuild.2008.07.015

[3] Schmidt, F. (2012) Entwicklungspotenzial thermisch angetriebener Wärmepumpen. Jahrestagung KIT-Zentrum Energie.

[4] Zurigat, Y.H., Liche, P.R. and Ghajar, A.J. (1991) Influence of Inlet Geometry on Mixing in Thermocline Thermal Energy Storage. *International Journal of Heat and Mass Transfer*, **34**, 115-125. http://dx.doi.org/10.1016/0017-9310(91)90179-I

[5] Alizadeh, S. (1999) An Experimental and Numerical Study of Thermal Stratification in a Horizontal Cylindrical Solar Storage Tank. *Solar Energy*, **66**, 409-421. http://dx.doi.org/10.1016/S0038-092X(99)00036-5

[6] Nelson, J.E.B., Balakrishnan, A.R. and Murthy, S.S. (1998) Transient Analysis of Energy Storage in a Thermally Stratified Water Tank. *International Journal of Energy Research*, **22**, 867-883. http://dx.doi.org/10.1002/(SICI)1099-114X(199808)22:10<867::AID-ER410>3.0.CO;2-L

[7] Bouhdjar, A. and Harhad, A. (2002) Numerical Analysis of Transient Mixed Convection Flow in Storage Tank: Influence of Fluid Properties and Aspect Ratios on Stratification. *Renewable Energy*, **25**, 555-567. http://dx.doi.org/10.1016/S0960-1481(01)00090-8

[8] Shah, L.J. and Furbo, S. (2003) Entrance Effects in Solar Storage Tanks. *Solar Energy*, **75**, 337-348. http://dx.doi.org/10.1016/j.solener.2003.04.002

[9] Cònsul, R., *et al.* (2004) Virtual Prototyping of Storage Tanks by Means of Three-Dimensional CFD and Heat Transfer Numerical Simulations. *Solar Energy*, **77**, 179-191. http://dx.doi.org/10.1016/j.solener.2004.04.009

[10] Shah, L.J., Andersen, E. and Furbo, S. (2005) Theoretical and Experimental Investigations of Inlet Stratifiers for Solar Storage Tanks. *Applied Thermal Engineering*, **25**, 2086-2099. http://dx.doi.org/10.1016/j.applthermaleng.2005.01.011

[11] Altuntop, N., *et al.* (2005) Effect of Obstacles on Thermal Stratification in Hot Water Storage Tanks. *Applied Thermal Engineering*, **25**, 2285-2298. http://dx.doi.org/10.1016/j.applthermaleng.2004.12.013

[12] Andersen, E., *et al.* (2008) Investigations on Stratification Devices for Hot Water Heat Stores. *International Journal of Energy Research*, **32**, 255-263. http://dx.doi.org/10.1002/er.1345

[13] (2010) Firma Sailer, Sailer Einschichtvorrichtung.

[14] Kaminski, E., Tait, S. and Carazzo, G. (2005) Turbulent Entrainment in Jets with Arbitrary Buoyancy. *Journal of Fluid Mechanics*, **526**, 361-376. http://dx.doi.org/10.1017/S0022112004003209

[15] (2004) Produktinformationen zu Marlotherm SH, Marlotherm LH (Datenblatt). Firma Sasol. http://www.marlotherm.com/pdf/MARLOTHERM_SH_GB.pdf

[16] (2011) Ansys Fluent 14.0 User's Guide. Ansys, Inc.

[17] Steinert, P., Göppert, S. and Platzer, B. (2013) Transient Calculation of Charge and Discharge Cycles in Thermally Stratified Energy Storages. *Solar Energy*, **97**, 505-516. http://dx.doi.org/10.1016/j.solener.2013.08.039

## Nomenclature

### Symbols

$A$ $\left[\mathrm{m}^2\right]$ Surface area
$C$ $\left[\frac{\mathrm{J}}{\mathrm{kg}\cdot\mathrm{K}}\right]$ Specific heat
$C$ $[-]$ Constant coefficient of $k-\varepsilon$ turbulence model
$C_p$ $\left[\frac{\mathrm{J}}{\mathrm{kg}\cdot\mathrm{K}}\right]$ Specific heat at constant pressure
$C_\mu$ $[-]$ Turbulent viscosity coefficient
$G$ $[-]$ Generation term
$g$ $\left[\mathrm{m/s}^2\right]$ Gravitational acceleration
$H$ $[\mathrm{m}]$ Storage tank height
$k$ $\left[\frac{\mathrm{W}}{\mathrm{m}\cdot\mathrm{K}}\right]$ Thermal conductivity
$k$ $\left[\mathrm{m}^2/\mathrm{s}^2\right]$ Turbulent kinetic energy
$P$ $[\mathrm{Pa}]$ Pressure
$\mathrm{Pr}$ $[-]$ Prandtl number
$Q$ $[\mathrm{J}]$ Heat energy
$\dot{q}$ $[\mathrm{W}]$ Heat transfer rate
$S$ $[-]$ Source term
$S_{ij}$ $\left[\mathrm{s}^{-1}\right]$ Mean strain rate tensor
$T$ $[\mathrm{K}]$ Temperature
$t$ $[\mathrm{s}]$ Time
$u$ $[\mathrm{m/s}]$ Velocity
$u, v, w$ $[\mathrm{m/s}]$ Velocity components
$V$ $\left[\mathrm{m}^3\right]$ Volume
$Y$ $[-]$ Dissipation term
$x, y, z$ $[\mathrm{m}]$ Coordinates

### Greek Symbols

$\rho$ $\left[\mathrm{kg/m}^3\right]$ Density
$\mu$ $\left[\frac{\mathrm{kg}}{\mathrm{m}\cdot\mathrm{s}}\right]$ Dynamic viscosity
$v$ $\left[\frac{\mathrm{m}^2}{\mathrm{s}}\right]$ Kinematic viscosity
$\varepsilon$ $\left[\frac{\mathrm{m}^2}{\mathrm{s}^3}\right]$ Turbulent kinetic energy dissipation rate

### Subscripts

cs Cross section
conv Convection
eff Effective
hor Horizontal

| | |
|---|---|
| ref | Reference |
| $t$ | Turbulent |
| $k$ | Turbulent kinetic energy |
| $\varepsilon$ | Turbulent kinetic energy dissipation rate; |
| $b$ | Buoyancy. |

## Superscripts

| | |
|---|---|
| $\overline{\phi}$ | Mean value; time averaged component |
| $\boldsymbol{\phi}$ | Vector |
| $\dot{\phi}$ | Time derivative |

## List of Abbreviations

| | |
|---|---|
| COP | Coefficient of Performance |
| CFD | Computational Fluid Dynamics |
| RANS | Reynolds Averaged Navier-Stokes Equation |
| UDF | User defined function |
| 1-D | One Dimensional |
| 2-D | Two Dimensional |

# 17

# Numerical Study of the Influence of Elements inside the Wheelhouse on the Passenger Vehicle Aerodynamic

**Edvaldo Angelo[1], Gabriel Angelo[2,3], Pedro Henrique Di Giovanni Santos[3], Delvonei Alves de Andrade[3]**

[1]Grupo de Simulação Numérica (GSN), Universidade Presbiteriana Mackenzie, São Paulo, Brazil
[2]Department of Mechanical Engineering, Centro Universitário da Fundação Educacional Inaciana Padre Sabóia de Medeiros (FEI), São Bernardo do Campo, Brazil
[3]Instituto de Pesquisas Energéticas e Nucleares, IPEN CNEN/SP, Universidade de São Paulo, São Paulo, Brazil
Email: eangelo@mackenzie.br, gangelo@fei.edu.br, pedro_digiovanni_s@hotmail.com, delvonei@ipen.br

## Abstract

**Models for the study of computational fluid dynamics in vehicles to determine aerodynamic loads usually take into account only the geometry of the body. Several constructive elements such as the wheel geometry or suspension components are disregarded in the computational models. This work presents the study of the aerodynamics of a one-fourth model passenger vehicle, which contains the wheelhouse interior elements. The goal is to identify the aerodynamic loads produced by these components and their effect on the flow dynamics. Wheel and tire set, brake components, suspension and drive shaft are contemplated. Computer simulations were performed to the vehicle speed varying from 0 to 120 km/h and included the rotation of the tire and wheel assembly, considering the tire geometry in dynamic conditions. The computational model is solved by the finite volume method, wherein the computational domain is divided into tetrahedral and hexahedral elements. The turbulence model used is the standard $k - \varepsilon$.**

## Keywords

**Computational Fluid Dynamics, Wheelhouse, Aerodynamics, Vehicle Aerodynamics**

## 1. Introduction

The numerical study for the calculation of aerodynamic forces arising in a vehicle has become more conventional. The methods usually employed for the analysis of such forces in vehicles are designed after the devise of prototypes in wind tunnels, which are very expensive demanding a large amount of equipment, facilities and skilled labor.

Currently companies have performed three-dimensional construction of their models, thus facilitating processes of tools construction, assembly interference checking and also allowing the numerical analysis.

Most of the articles related to numerical studies of vehicle aerodynamics aim at the vehicle body [1] or phenomena associated with lateral wind [2] generally disregarding components as the wheel geometries, tire, brake assembly (disc and caliper), suspension, shock absorber, stub axle and wheelhouse.

This article aims to determine the flow influence on a passenger vehicle using a one-fourth model of a complete vehicle as seen in **Figure 1**.

Many researchers have been trying to understand the complex flow patterns around vehicles, developing measurement techniques in wind tunnels [3] and computational efforts [4]. Few of the studies [5], however, are concentrated in the region of wheelhouse. Generally, the studies include the complete vehicle and indirectly study the region of the wheelhouse.

The aim of this study is to identify, through numerical simulations, the effects of internal elements of the wheelhouse, the wheelhouse itself, the wheel and tire assembly in drag force of a passenger vehicle, through the modelling of a quarter car.

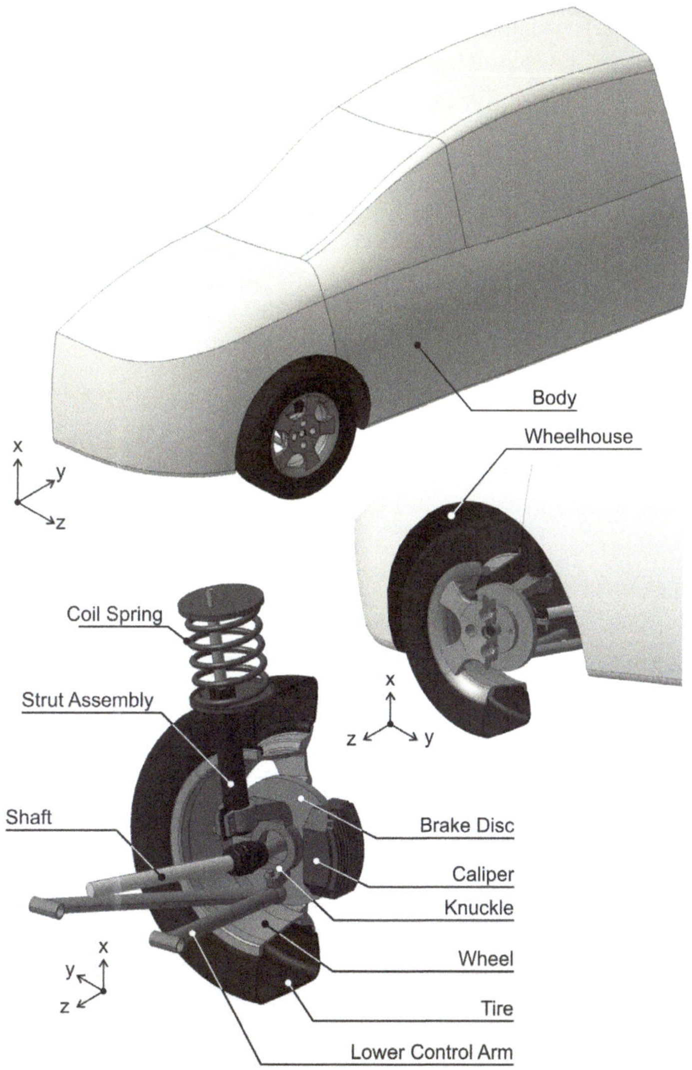


**Figure 1.** One-fourth model and components to be evaluated.

## 2. Mathematical Model and Boundary Conditions

The finite volume method was used to solve the conservation equations applied to an unstructured mesh. The equations solved for the model are respectively, the mass and the momentum conservation equations. It is used the standard $k$ $\varepsilon$ model [6] for the turbulence treatment and the fluid is air at a temperature ($T$) of 25°C. The flow regime is transient, varying the inlet velocity ($v_{in}$) of the fluid and the soil ($v_g$) from zero to 33.5 m/s ($v_{in}$ = $v_g$). Mach number ($M$) for the proposed condition in this analysis is less than 0.3 so the flow can also be considered as incompressible [7] and thermal effects are neglected, *i.e.* isothermal flow. The wheel angular velocity ($w$) can be expressed as a function of ground speed and the dynamic tire radius ($r_d$) [8]. The boundary conditions are: non-slip flow for the vehicle surfaces, free-slip flow on the top and side walls and symmetry for the $zx$ plane as presented in **Figure 2**.

A simplification of some elements was adopted for the proposed model given that other geometries exert little influence on the flow. These elements are the hood of the constant velocity joint, wheel bolts case, braking set details, etc. The respective changes can be observed in **Figure 3**.

The approach presented by Stern and Wilson [9] [10] to define and verify the mesh is used in this paper. They discuss the mesh dependency on the results focusing the element size definition in order to validate the CFD models. The methodology considers an increase of the mesh density for the same boundary condition using predefined ratios. This procedure must be performed in such a way that property variation or small variations are not present. When this condition is satisfied the solution is considered independent of the mesh. The computational domain for the last refinement iteration process has 2.8 million elements. In **Figure 4(a)** it is possible to observe the surface elements applied to the vehicle body. A greater density of elements has been applied inside the wheelhouse and the tire. **Figure 4(b)** and **Figure 4(d)**, and an auxiliary view is presented for the volume density elements verification, **Figure 4(c)**.

## 3. Results and Discussion

The numerical study was performed in transient regime. In order to demonstrate the highest gradients of the quantities involved, air input speed was chosen to 33.5 m/s (120 km/h). The pressure variation as a function of the model surface is indicated on the color map of **Figure 5(a)**. Sections in the computational domain were constructed and can be observed in **Figure 5(c)**. **Figure 5(b)** shows a view (cut) in the $xy$ plane and height ($z$) equal to $r_d$. This section region is protected by the vehicle body which contributes to a small pressure variation against the other variations. The sections in the $xz$ plane, **Figure 5(d)** and **Figure 5**, detail A, show respectively, the

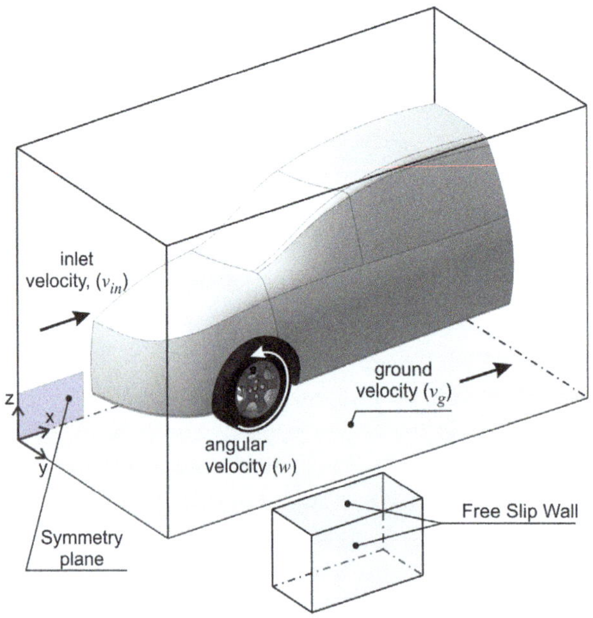


**Figure 2.** Boundary conditions for the numerical model.

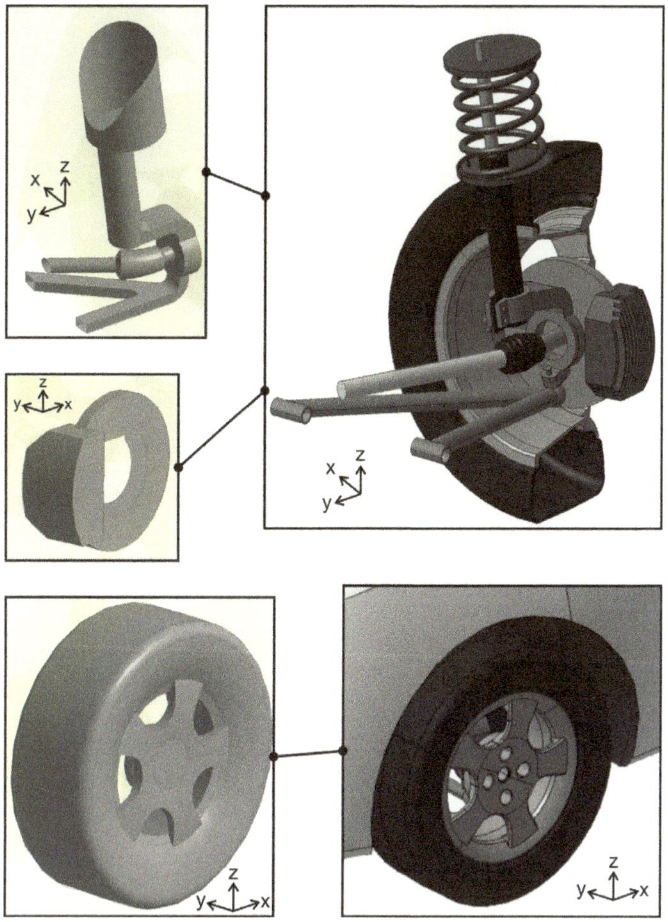


Figure 3. Geometrical simplifications.

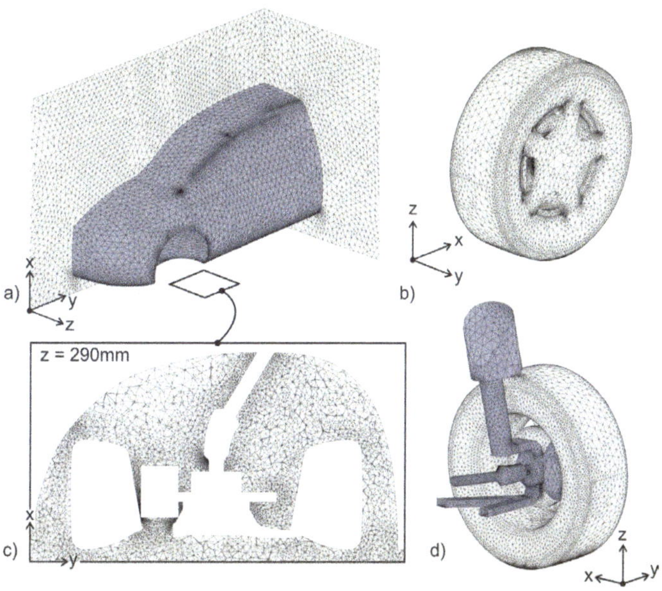


**Figure 4.** (a) Triangular elements in the region of the vehicle body; (b) Surface elements applied on the wheel geometry; (c) Detail view in the xy plane of the wheelhouse region; (d) Superficial elements inside the wheelhouse.

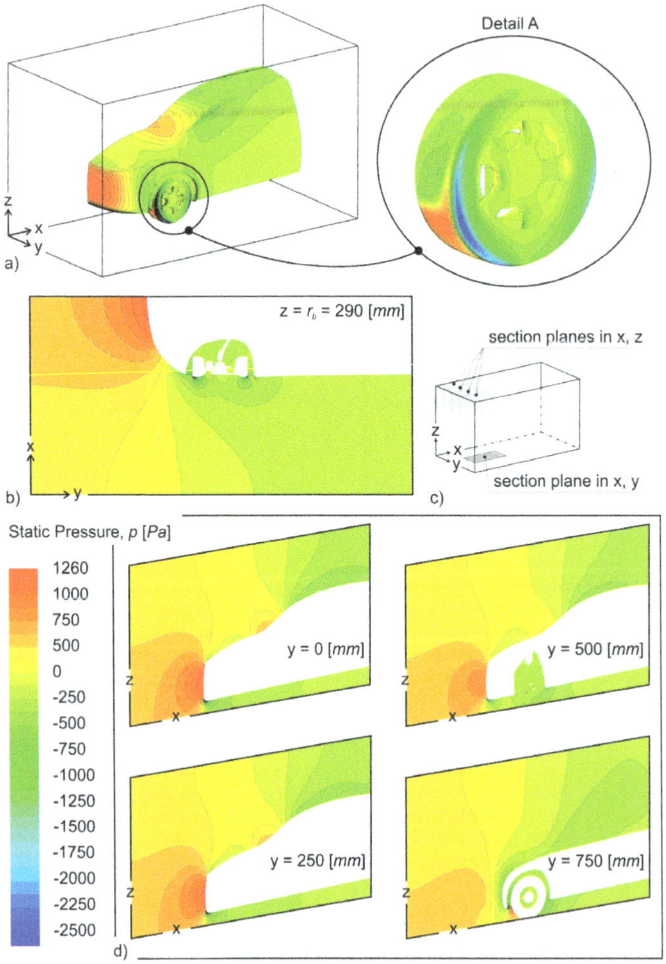


**Figure 5.** (a) Pressure gradient applied to the vehicle surface; (b) Pressure gradient on *xy* plane at axis *z* equal to $r_d$; (c) Illustration showing the positions of sections; and (d) Pressure gradient applied to *xz* plane at various positions.

pressure gradient in the section and on the surface geometry tire/wheel. Also in **Figure 5** the frontal location of the vehicle shows stagnation regions, this behavior indicates consistency in the numerical results [11].

In **Figure 6** streamlines colored by velocity gradient were constructed in order to make evident the flow behavior. **Figure 6** also shows that the vehicle external geometry does not originates vortex structures. This vortex structures occur only in the lower region of the wheel, tire and wheelhouse.

Vortex structures appeared depending on the geometry of the wheelhouse as shown in **Figure 7**. This characteristic in some cases depending on the vortex frequency emission can excite surface so as to promote noise. However, this simulation does not take into consideration such factors. Another problem associated with the formation of these vortex inside the wheelhouse is the decrease of the air flow that cools the brake system. Thus in the wheelhouse design these recirculation structures should be avoided [12].

The aerodynamic drag force ($F_a$) according to Fox and McDonald [7] is defined by Equation (1):

$$F_a = C_d \cdot \rho \cdot A_f \cdot \frac{v^2}{2} \quad (1)$$

where: $C_d$—aerodynamic drag coefficient, $\rho$—density and $A_f$—frontal area.

The aerodynamic drag coefficient is not constant [13]. It depends directly on the flow characteristics. Thus the Reynolds number is defined by Equation (2):

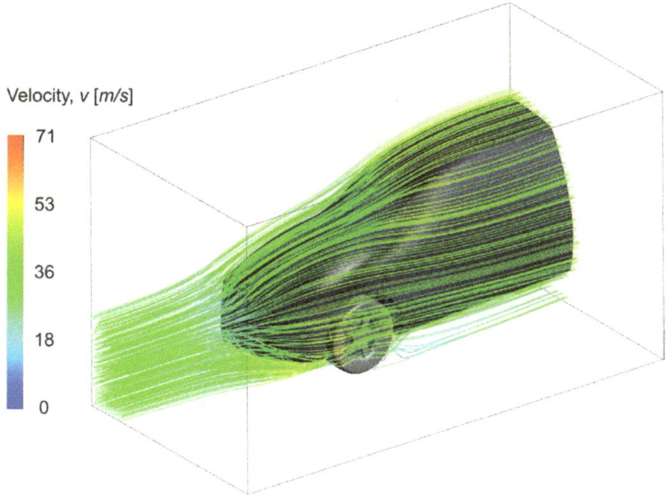


**Figure 6.** Streamlines in the proximity of the vehicle surface.

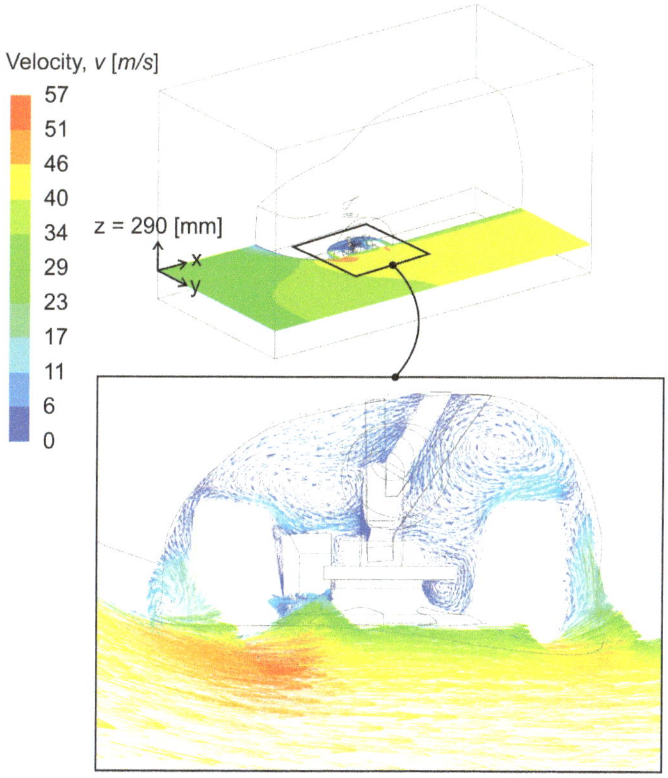


**Figure 7.** Cross section of the gradient and velocity vectors on the *xy* plane and *z* = 290 mm.

$$\mathrm{Re} = \frac{\rho \cdot v \cdot L}{\mu} \tag{2}$$

where: $L$ is the is the characteristic length chosen as $L = \frac{A_f}{P_f}$ and $P_f$ perimeter as function of the frontal area.

The respective areas and perimeters for this model can be observed in **Figure 8**.

The numerical simulations and experimental results reported in the literature indicate that the drag coefficient values vary slightly from a given speed [14]. In such cases, the value stabilizes, even with the increased speed and allows it to be defined as aerodynamic characteristics of the body. Gillespie, 1992 [8] presents average values tables for the drag coefficient as a function of the vehicle category which is proper only for estimates.

For the verification of such feature a graph of velocity versus drag coefficient was built. The vehicle body and the whole vehicle body have different drag coefficients, but show the same trend as shown in **Figure 9**.

**Figure 10** presents the drag force projected in the $x$ direction as a function of velocity. As expected the greatest strength comes from the vehicle body, nevertheless the geometry effect of the tire, wheel and wheelhouse are significant and together come to 16.3% of the total drag force in this direction at a speed of 33.5 (m/s) and is directly proportional to the velocity.

The internal elements of the wheelhouse exert a small influence on the total drag force of the vehicle. It is about 0.4%, so that adopted simplification compared with the order of magnitude of other elements is very much reasonable.

## 4. Conclusions

The main conclusions of the study are:

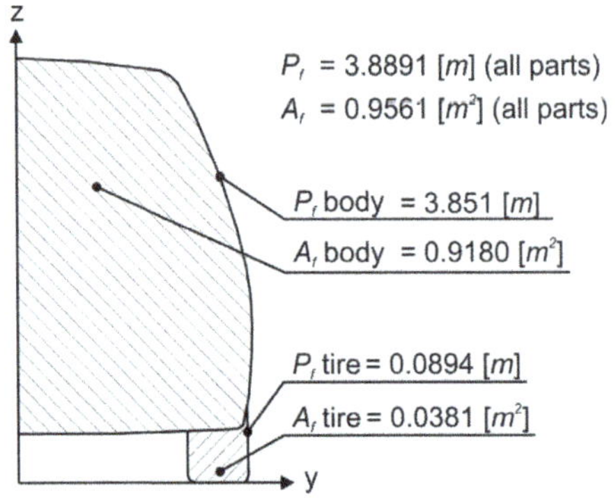


**Figure 8.** Vehicle frontal projection.

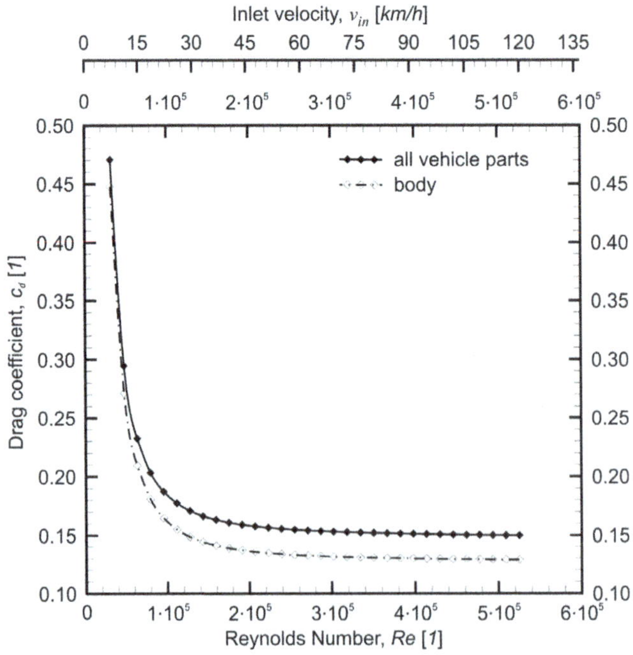


**Figure 9.** Inlet velocity and Reynolds number as a function of the drag coefficient.

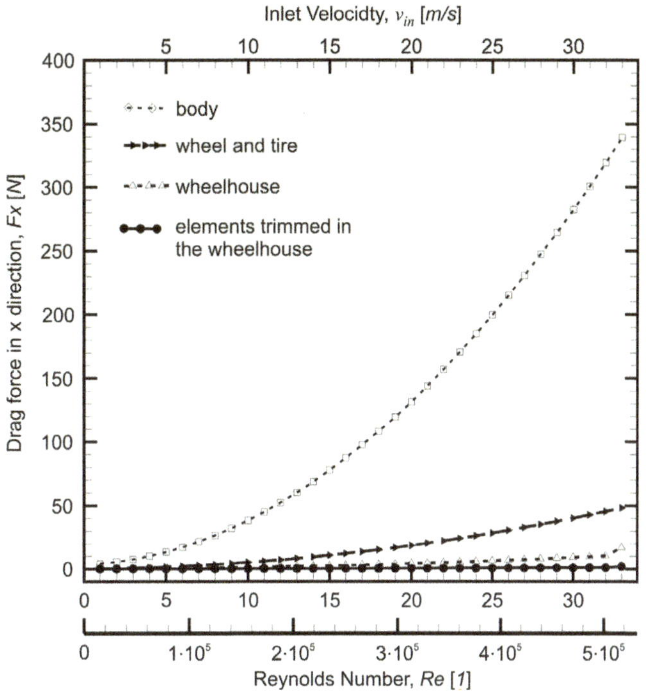


**Figure 10.** Inlet speed as a function of the strength's projection in the *x* direction.

1) It is possible the use of mathematical models for predicting aerodynamic loads;

2) Flow inside the wheelhouse is complex and presents recirculation regions;

3) Mechanical elements within the wheelhouse had simplified geometry and were incorporated into the model; however, the contribution to the drag force of such elements is quantitatively less than that of the wheelhouse and assembly wheel and tire;

4) According to the simulations, presented in paper, neglecting the effects of the box on wheels and wheel tire set is not suitable in aerodynamic vehicle research, except at low speeds where the aerodynamic loads are low for any vehicle.

## References


[1] Beaudoin, J.-F. and Aider, J.-L. (2008) Drag and Lift Reduction of a 3D Bluff Body Using Flaps. *Experiments in Fluids*, **44**, 491-501. http://dx.doi.org/10.1007/s00348-007-0392-1

[2] Corin, R.J., He, L. and Dominy, R.G. (2008) A CFD Investigation into the Transient Aerodynamic Forces on Overtaking Road Vehicle Models. *Journal of Wind Engineering and Industrial Aerodynamics*, **96**, 1390-1411. http://dx.doi.org/10.1016/j.jweia.2008.03.006

[3] Cogotti, A. (2008) Evolution of Performance of an Automotive Wind Tunnel. *Journal of Wind Engineering and Industrial Aerodynamics*, **96**, 667-700. http://dx.doi.org/10.1016/j.jweia.2007.06.007

[4] Guilmineau, E. (2008) Computational Study of Flow around a Simplified Car Body. *Journal of Wind Engineering and Industrial Aerodynamics*, **96**, 1207-1217. http://dx.doi.org/10.1016/j.jweia.2007.06.041

[5] Regert, T. and Lajos, T. (2007) Description of Flow Field in the Wheelhouses of Cars. *International Journal of Heat and Fluid Flow*, **28**, 616-629. http://dx.doi.org/10.1016/j.ijheatfluidflow.2007.04.017

[6] Launder, B.E. and Spalding, D.B. (1974) The Numerical Computation of Turbulent Flows. *Computer Methods in Applied Mechanics and Engineering*, **3**, 269-289. http://dx.doi.org/10.1016/0045-7825(74)90029-2

[7] Fox, R.W., McDonald, A.T. and Pritchard, P.J. (1985) Introduction to Fluid Mechanics. Vol. 7, John Wiley & Sons, New York.

[8] Gillespie, T.D. (1992) Fundamentals of Vehicle Dynamics. Vol. 400, Society of Automotive Engineers Warrendale. http://dx.doi.org/10.4271/R-114

[9] Stern, F., Wilson, R.V., Coleman, H.W. and Paterson, E.G. (2001) Comprehensive Approach to Verification and Vali-

dation of CFD Simulations—Part 1: Methodology and Procedures. *Journal of Fluids Engineering*, **123**, 793-802. http://dx.doi.org/10.1115/1.1412235

[10] Wilson, R.V., Stern, F., Coleman, H.W. and Paterson, E.G. (2001) Comprehensive Approach to Verification and Validation of CFD Simulations—Part 2: Application for Rans Simulation of a Cargo/Container Ship. *Journal of Fluids Engineering*, **123**, 803-810. http://dx.doi.org/10.1115/1.1412236

[11] Duell, E.G. and George, A.R. (1999) Experimental Study of a Ground Vehicle Body Unsteady near Wake. Technical report, SAE Technical Paper.

[12] Landström, C., Löfdahl, L. and Walker, T. (2009) Detailed Flow Studies in Close Proximity of Rotating Wheels on a Passenger Car. Technical Report, SAE Technical Paper.

[13] Cengel, Y.A. and Cimbala, J.M. (2006) Fluid Mechanics. Vol. 1, Tata McGraw-Hill Education, New York.

[14] White, F.M. and Corfield, I. (2006) Viscous Fluid Flow. Vol. 3, McGraw-Hill, New York.

# 18

# Dynamics of Unsteady MHD Convective Flow with Thermophoresis of Particles and Variable Thermo-Physical Properties past a Vertical Surface Moving through Binary Mixture

**Isaac Lare Animasaun**

Department of Mathematical Sciences, Federal University of Technology, Akure, Nigeria
Email: anizakph2007@gmail.com

## Abstract

**The dynamics of unsteady magnetohydrodynamic convective fluid flow with radiation and thermophoresis of particles past a vertical porous plate moving through a binary mixture in an optically thin environment is investigated. The approximate form of the radiative heat flux is considered as the fourth power of temperature. Chemical reaction that occurs as the chemically reacting fluid flow through binary mixture is accounted for in energy and species concentration equations. Exponential space dependent heat source is introduced to generate additional heat energy across the fluid domain. The corresponding influence of heat energy is properly accounted for. It is assumed that viscosity and thermal conductivity vary as a linear function of temperature. The governing boundary layer equations are converted to nonlinear ordinary differential equations using similarity variables. A novel method of obtaining root finding starting with three guesses in shooting techniques is presented. The corresponding nonlinear coupled ordinary differential equations is solved numerically by shooting technique along with quadratic interpolation scheme. Graphical results of the dimensionless velocity, temperature and concentration distributions are shown for certain pertinent parameters controlling the fluid flow. The quadratic interpolation method is found to produce better estimated values of $f'(0)$, $-\theta'(0)$, $-\phi'(0)$ which satisfy the degree of accuracy and proportional to the physical quantities.**

## Keywords

**Variable Viscosity, Variable Thermal Conductivity, Binary Mixture, Quadratic Interpolation,**

**Unsteady Flow, Thermophoresis, Exponential Heat Source, Shooting Technique**

## 1. Introduction

In August 1904, Ludwig Prandtl presented a comprehensive report on aerodynamic boundary layer at International Congress of Mathematicians, Heidelberg, Germany. Since then, the study of thin layer formed on a vertical surface as fluid flows over a surface has received great attention because of its application in industry and engineering processes. Boundary layer flow past a surface moving through binary mixture is of important type of flow occurring in a number of engineering processes (*i.e.* chemical and petroleum). In industry, a familiar example of a binary mixture of fluids is an emulsion. Typical example of emulsion is oil dispersed within water or water within oil. A familiar example shows up when mixture of oil and water after shaken together is subjected to flow along a vertical heated surface due to buoyancy and pressure gradient. An emulsion is a dispersion (droplets) of one liquid in another immiscible liquid. Extensive reviews of this theory and its application can be found in Al-Sharif *et al.* [1] and Wang *et al.* [2]. By means of this model, Makinde *et al.* [3] carried out a research on unsteady convection with chemical reaction and radiative heat transfer past a flat porous plate moving through a binary mixture using the classical Boussinesq approximation. They assumed constant fluid viscosity within the boundary layer formed on vertical surface. The effects of increasing magnitude of suction/injection, wall temperature, Damkohler number and radiation parameter over velocity, temperature and concentration profiles are reported extensively. Their idea was then followed and extended by Makinde and Olanrewaju [4]; hence, they introduced thermo-diffusion (Soret), Diffuso-thermal (Dufour) effects and reported the corresponding effects while Sastry and Murti [5] followed another direction by assuming that the fluid in question is electrically conducting and reported the effect of magnetic field parameter.

According to Somasundaran [6], thermophoresis in liquid mixtures was first observed and reported by Carl Ludwig in 1856 and that of gas mixtures was first observed and reported by John Tyndall in 1870. Thermophoresis has many applications most especially as a method used to separate different polymer particles in field flow fractionation as reported by Giddings *et al.* [7]. The term thermophoresis often applies to aerosol mixtures but may also refer to the phenomenon in all phases of matter like liquid and gas (fluid). This phenomenon can be observed in mixtures of mobile particles where different particles exhibit different responses to the force of temperature gradient. A common example of thermophoresis is the blackening of the glass globe of a kerosene lantern; the temperature gradient established between the flame and the globe drives the carbon particles produced in the combustion process towards the globe where they deposit [8]. This contribution to the body of knowledge attracted Chamkha and Isaa [9] to study effects of heat generation/absorption and thermophoresis on hydromagnetic flow with heat and mass transfer over a flat surface. Animasaun [10] presented the effects of thermophoretic parameter on non-darcian MHD dissipative Casson fluid flow along linearly stretching vertical surface with migration of colloidal particles in response to macroscopic temperature. Recently, Fagbade *et al.* [11] employed SHAM to solve dimensionless governing equations which corresponds to a problem of Darcy-Forcheimer mixed convection flow in the presence of magnetic field and thermophoresis. Extensive studies on dynamics of MHD fluid flow past continuously moving semi-infinite surface with large suction, classical heat generation; radiation effects on MHD flow, unsteady MHD convection flow of polar fluids and effects of space and temperature internal heat source can be found in [12]-[19].

Numerical solutions are of particular interest due to the fact that exact solutions do not exist in closed form for most engineering and scientific problems. A commonly used numerical method for the solution of two point boundary value problems is the shooting method. Shooting method is a classical and special methodology for solving boundary value problems of differential equations. Numerous strong nonlinear boundary-value ordinary differential equations arise in engineering and science. Shooting method has been adopted to solve several kinds of B.V.P. in published articles. The basic idea behind the proposed numerical method can be introduced as theoretical approach to solve System of Ordinary Differential Equations (Boundary Value Problem) based on conversion of B.V.P. to I.V.P. by using method of superposition introduced by Na [20]; to obtain better estimate of $f'(0)$, $\theta'(0)$ and $\phi'(0)$ which satisfy the boundary conditions at infinity by starting with three guesses, and finally to solve the I.V.P. using the initial conditions together with the better estimate. Through literature review, it is noticed that in all the studies on chemically reacting fluid flow past a surface moving through binary mixture, thermo-physical properties of the fluid are treated as constant function of temperature. However, it is known that these properties may change with temperature, most especially when Engineer introduces heat energy to enhance the transport phenomena. To accurately predict the dynamics of the flow and heat transfer rates, it is necessary to account for the variation in viscosity and thermal conductivity. This present paper is an extension of [3]-[5] by considering the Arrhenius kinetics, thermophoresis and thermal radiation in an unsteady MHD convective flow with temperature dependent viscosity and thermal conductivity over a moving plate through a binary mixture with suction at the plate surface. The governing equations are converted to ordinary differential equations by applying the similarity transformation. Numerical solutions of the reduced nonlinear similarity equations are then obtained by adopting Runge-Kutta Gill scheme along with the shooting method and the quadratic interpolation. The results of the numerical solution are then presented graphically for difference values of the various parameters.

## 2. Governing Equations of the Flow

An unsteady one-dimensional hydrodynamic convective flow of a viscous incompressible, electrically conducting and chemically reacting fluid flow past a porous vertical heated surface moving through a binary mixture is considered. The flow is assumed to be in the $x$-direction, which is taken along the semi-infinite plate and $y$-axis is normal to it. Since the surface is of infinite length in $x$-direction, therefore all the physical quantities (*i.e.* velocity, temperature and species concentration) are assumed to be independent of $x$. A uniform magnetic field $B_o$ is applied in the direction perpendicular to the surface. The fluid is assumed to be slightly conducting, and the magnetic Reynolds number $M_{re} \ll 1$; hence, the induced magnetic field is negligible in comparison with the applied magnetic field. It is further assumed that there is no applied voltage, so that electric field is absent. The fluid is assumed to be optically thin with absorption coefficient $\alpha \ll 1$. Following Cheng [21], the approximate form of the radiative heat flux equation $(\partial q_r/\partial y)$ is taken as the fourth power of temperature in the energy balance equation. It is assumed that heat and mass is transferred by free convection in the boundary layer over a vertical flat plate of constant temperature ($T_w$) and concentration ($C_w$) which is embedded in a fluid-saturated porous medium of ambient temperature ($T_\infty$) and concentration ($C_\infty$) where $T_w > T_\infty$ and $C_w > C_\infty$ respectively. This formulation is based on the notion that just as temperature gradient constitutes the driving potential for heat transfer, species concentration gradient in a mixture provides the driving potential for mass transfer [22]. The effect of thermophoresis is usually prescribed by means of the average velocity, which a particle will acquire when exposed to a temperature gradient [10] [23]. In boundary layer flow, the temperature gradient in the horizontal $y$-direction is very much larger than in the vertical $x$-direction, and therefore only the thermophoretic velocity in $y$-direction is considered [3]. In view of this, the component of thermophoretic velocity along the plate is negligible compared to the component of its velocity normal to the surface. Considering the mathematical model introduced by [4] [5] which is an extension of [3] together with all assumptions stated above along with classical Boussinesq's approximation; the governing equations for continuity, momentum, energy, and species diffusion in laminar incompressible boundary layer flow can be written as follows:

$$\frac{\partial v}{\partial y} = 0, \tag{1}$$

$$\frac{\partial u}{\partial t} + v\frac{\partial u}{\partial y} = \frac{1}{\rho}\frac{\partial}{\partial y}\left(\mu\frac{\partial u}{\partial y}\right) + g\beta\left(T - T_\infty\right) + g\beta^*\left(C - C_\infty\right) - \frac{\sigma^* B_o^2}{\rho}u, \tag{2}$$

$$\frac{\partial T}{\partial t} + v\frac{\partial T}{\partial y} = \frac{1}{\rho C_P}\frac{\partial}{\partial y}\left(\kappa\frac{\partial T}{\partial y}\right) + \frac{(-\Delta H)}{\rho C_P}R_A - \frac{4\sigma\alpha^2}{\rho C_P}T^4 + \frac{Q_0\left(T - T_\infty\right)}{\rho C_P}\mathrm{e}^{-g\left(\frac{y}{2\sqrt{\vartheta t}}\right)}, \tag{3}$$

$$\frac{\partial C}{\partial t} + v\frac{\partial C}{\partial y} + \frac{\partial}{\partial y}\left[V_T\left(C - C_\infty\right)\right] = D_m\frac{\partial^2 C}{\partial y^2} - R_A. \tag{4}$$

Equations (1)-(4) are subject to boundary conditions

$$t \le 0: u(y,0) = 0,\ T(y,0) = T_w,\ C(y,0) = C_w, \tag{5}$$

$$t > 0: u(0,t) = U_o,\ T(0,t) = T_w,\ C(0,t) = C_w, \tag{6}$$

$$t>0: u(\infty,t)\to 0, T(\infty,t)\to T_\infty, C(\infty,t)\to C_\infty \tag{7}$$

where $U_o$ is the plate characteristic velocity. From the continuity Equation (1), it is worth noticing that velocity component ($v$) is either constant or a function of time. Following [24], velocity component along $y$-axis is considered as

$$v=-c\sqrt{\vartheta/t}, \tag{8}$$

where $c>0$ and $c<0$ are known as suction and injection respectively. In the energy Equation (3), the relationship between the activation energy and the rate at which a reaction proceeds is accounted for by using

$$Q=(-\Delta H)R_A,\ R_A=K_r\text{Exp}(-E_A/R_G T)C^n. \tag{9}$$

In Equations (3), (4) and (9), it is very important to state that the frequency factor for the chemical reaction is $K_r C^n$. In this present study, it is worth mentioning that exponential space dependent internal heat source is introduced to increase the temperature of the fluid and to positively influence buoyancy. This concept can be traced to the idea of Crepeau and Clarksean [25], El-Aziz and Salem [26] [27] and Animasaun *et al.* [28]. The increase of temperature may also leads to a local increase in the transport phenomena by reducing the viscosity across the momentum boundary layer and so the heat transfer rate at the wall may also be affected greatly. In [3]-[5], fluid viscosity and thermal conductivity have been assumed to be constant function of temperature within the boundary layer. However, it is known that physical properties of the fluid may change significantly when expose to space dependent internal generated temperature or thermal stratification [29]. For lubricating fluids, heat generated by the internal friction and the corresponding rise in temperature affect the viscosity of the fluid and so the fluid viscosity can no longer be assumed constant. In order to account for the variation in thermo-physical properties of the fluid as it flows past a vertical heated surface moving through a binary mixture; classical Boussinesq's approximation is adopted such that temperature of the wall/surface is greater than temperature of the fluid at free stream. In order to invoke this assumption, it is valid to consider the mathematical model of temperature dependent viscosity model used in [31] [32] which was developed using the experimental data of Batchlor [30] together with the mathematical model of temperature dependent thermal conductivity model of Charraudeau [33] as

$$\mu(T)=\mu^*\left[a+b(T_w-T)\right] \text{ and } \kappa(T)=\kappa^*\left[a+\delta(T-T_\infty)\right] \tag{9}$$

where $\mu^*$ and $\kappa^*$ are the constant value of the coefficient of viscosity and thermal conductivity at the free stream respectively. A case where $a=1$ is considered in this work and $(b,\delta)>0$. The thermophoretic velocity parameter in Equation (4) was given in [8] and later by Tsai [34] as

$$V_T=-\frac{k^{Th}}{T_{ref}}\frac{\partial T}{\partial y},\ \text{and}\ \tau=-\frac{k^{Th}T_w}{\vartheta T_{ref}} \tag{10}$$

$k^{Th}$ is the thermophoretic coefficient which ranges in values from, 0.2 to 1.2 as indicated by Batchelor and Shen [35] and properly defined using the theory [8] as,

$$k^{Th}=\frac{2C_s\left(\dfrac{\kappa_{fluid}}{\kappa_{diff.P}}+C_tK_n\right)C_u}{(1+3C_mK_n)\left(1+\dfrac{2\kappa_{fluid}}{\kappa_{diff.P}}+2C_tK_n\right)}, \tag{11}$$

$C_s, C_m, C_t$ are constants, $\kappa_{fluid}$ is the thermal conductivity of the fluids, $\kappa_{diff.P}$ is the thermal conductivity of the diffused particles, $C_u$ is the Cunningham correction factor, $K_n$ is the Knudsen number. Substituting all into dimensional governing Equations (1)-(4) we have

$$\frac{\partial u}{\partial t}+v\frac{\partial u}{\partial y}=\vartheta^*\left[1+\xi-\frac{\theta}{\theta_w}\xi\right]\frac{\partial^2 u}{\partial y^2}-\vartheta^*\frac{\xi}{\theta_w}\frac{\partial\theta}{\partial y}\frac{\partial u}{\partial y}+g\beta(T-T_\infty)+g\beta^*(C-C_\infty)-\frac{\sigma^* B_o^2}{\rho}u, \tag{12}$$

$$\rho C_P\left(\frac{\partial T}{\partial t}+v\frac{\partial T}{\partial y}\right)=\kappa^*\left[1+\frac{\lambda\theta}{\theta_w}-\frac{\lambda}{\theta_w}\right]T_\infty\frac{\partial^2\theta}{\partial\eta^2}\frac{1}{4\vartheta t}+\frac{\lambda}{\theta_w}\left(\frac{\partial\theta}{\partial\eta}\right)^2\frac{\kappa^* T_\infty}{4\vartheta t}-4\sigma\alpha^2\theta^4T_\infty^4$$
$$+(-\Delta H)K_r\text{Exp}\left(-\frac{E_A}{R_GT_\infty}\right)\text{Exp}\left(\frac{E_A}{T_\infty R_G}\left(1-\frac{1}{\theta}\right)\right)\phi^nC_\infty^n+Q_0T_\infty(\theta-1)\text{Exp}\left(-g\left(\frac{y}{2(\vartheta t)^{1/2}}\right)\right), \quad (13)$$

$$\frac{\partial C}{\partial t}+v\frac{\partial C}{\partial y}+C_\infty(\phi-1)\frac{\tau}{\theta_w}\frac{\partial^2\theta}{\partial\eta^2}\frac{1}{4t}+\frac{C_\infty}{\theta_w}\frac{\partial\theta}{\partial\eta}\frac{\partial\phi}{\partial\eta}\frac{1}{4t}$$
$$=D_m\frac{\partial^2C}{\partial y^2}-K_r\text{Exp}\left(-\frac{E_A}{R_GT_\infty}\right)\text{Exp}\left(\frac{E_A}{T_\infty R_G}\left(1-\frac{1}{\theta}\right)\right)\phi^nC_\infty^n. \quad (14)$$

Considering the following dimensionless variables

$$f(\eta)=\frac{u}{U_o},\ \eta=\frac{y}{2\sqrt{\vartheta t}},\ \theta=\frac{T}{T_\infty},\ \theta_w=\frac{T_w}{T_\infty},\ \phi=\frac{C}{C_\infty},\ \phi_w=\frac{C_w}{C_\infty},\ \chi=\frac{tQ_0}{\rho C_P}$$
$$G_r=\frac{4\text{tg}\beta}{U_ob},\ G_c=\frac{4\text{tg}\beta^*}{U_ob},\ M=\frac{4t\sigma^*B_o^2}{\rho},\ \gamma=\kappa^*/\rho C_P,\ \lambda=\delta T_w,\ \xi=bT_w, \quad (15)$$
$$h=\frac{(-\Delta H)C_\infty}{\rho C_PT_\infty},\ P_r=\frac{\vartheta}{\gamma}=\frac{\mu C_P}{\kappa^*},\ R_a=4t\frac{4\sigma\alpha^2T_\infty^4}{\rho C_PT_\infty},\ \omega=\frac{E_A}{T_\infty R_G},\ D_a=4tK_re^{-\frac{E_A}{R_GT_\infty}}C_\infty^{n-1}.$$

Using (15), the following dimensionless non-linear ordinary differential equations are obtain

$$\left[1+\xi-\frac{\theta\xi}{\theta_w}\right]\frac{d^2f}{d\eta^2}-\frac{\xi}{\theta_w}\frac{d\theta}{d\eta}\frac{df}{d\eta}+2(\eta+c)\frac{df}{d\eta}+G_r\frac{\xi}{\theta_w}(\theta-1)+G_c\frac{\xi}{\phi_w}(\phi-1)-Mf=0 \quad (16)$$

$$\left[1+\frac{\lambda\theta}{\theta_w}-\frac{\lambda}{\theta_w}\right]\frac{d^2\theta}{d\eta^2}+\frac{\lambda}{\theta_w}\frac{d\theta}{d\eta}\frac{d\theta}{d\eta}+2P_r(\eta+c)\frac{d\theta}{d\eta}+P_rhD_a\text{Exp}\left[\omega\left(1-\frac{1}{\theta}\right)\right]\phi^n$$
$$-P_r\theta^4R_a+4\chi(\theta-1)\text{Exp}[-g\eta]=0 \quad (17)$$

$$\frac{d^2\phi}{d\eta^2}+2S_c(\eta+c)\frac{d\phi}{d\eta}-S_cD_a\text{Exp}\left[\omega\left(1-\frac{1}{\theta}\right)\right]\phi^n-S_c(\phi-1)\frac{\tau}{\theta_w}\frac{d^2\theta}{d\eta^2}-S_c\frac{\tau}{\theta_w}\frac{d\theta}{d\eta}\frac{d\phi}{d\eta}=0. \quad (18)$$

The dimensionless boundary conditions of the above problem are

$$f(0)=1,\ \theta(0)=\theta_w(>1),\ \phi(0)=\phi_w(>1), \quad (19)$$

$$f(\infty)\to 0,\ \theta(\infty)\to 1,\ \phi(\infty)\to 1 \quad (20)$$

where $D_a$ is Damköhler number, $R_a$ is the radiation parameter, $\omega$ is the activation energy parameter, $G_r$ is the modified thermal Grashof number, $G_c$ is the modified solutal Grashof number, $\chi$ is the heat source parameter, $K_r$ is the chemical reaction rate, $P_r$ is the Prandtl number, $S_c$ is known as Schmidt number, $\tau$ is known as thermophoretic parameter and $M$ is the magnetic field parameter. Also, other quantities of physical interest in this problem are the local skin friction $C_f$, local Nusselt number $N_u$ and Sherwood numbers $S_h$ which are defined by Makinde [3] as

$$C_f=\frac{2\vartheta\sqrt{t}}{U_o\sqrt{\vartheta}}\frac{\partial u}{\partial y}\bigg|_{y=0}=\frac{df}{d\eta}\bigg|_{\eta=0},\quad N_u=-\frac{2\sqrt{\vartheta t}}{T_\infty}\frac{\partial T}{\partial y}\bigg|_{y=0}=-\frac{d\theta}{d\eta}\bigg|_{\eta=0},$$
$$S_h=-\frac{2\sqrt{\vartheta t}}{C_\infty}\frac{\partial C}{\partial y}\bigg|_{y=0}=-\frac{d\phi}{d\eta}\bigg|_{\eta=0}. \quad (21)$$

And from that we can easily compute the results of the local skin friction, local Nusselt number and local Sherwood number.

## 3. Numerical Procedure

Following process for the step-by-step integration of differential equations in Gill [36], the MATLAB Code (*m-file*) for solving system of O.D.E. using Runge-Kutta Gill can be found under Appendix I. In this code, $y(i,:)$ denotes $f(i,1)$, $\theta(i,2)$ and $\phi(i,2)$. The initial guess values are *a*, *b* and *c*. Runge Kutta Gill is also a fourth order Runge Kutta method with Local truncation error of $O(h^5)$ and global truncation error $O(h^4)$. Delin and Zheng [37], formulated a general formula of fourth Order Runge Kutta Method. Using a Lemma, they obtained the general method which is dependent on a parameter "*t*". Five different fourth order Runge Kutta was obtained using the formula with $t = 2$ (the general scheme yielded classical Runge Kutta), $t = 4$ (the general scheme yielded England Runge Kutta Method) and when $t = 1$, 3 and 5 three other fourth Order Runge Kutta methods was also obtained. It was reported that numerical calculations using all these formulas have almost the same accuracy as the classical RK4. When we used $t = 2+\sqrt{2}$, Runge Kutta Gill was obtained. There are two types of error involved in a Runge Kutta as an approximation method of Ordinary differential equations. They are Round off error and Truncation error. Runge-Kutta Gill method is selected because it reduces (minimize) round off error and this numerical scheme is one of the most widely used fourth order methods because it reduces the amount of storage required in the solution of a large number of simultaneous first order differential equations. According to Finlayson [38], Order analysis, Consistency analysis and Stability analysis shows that Runge Kutta Gill is also of order four, stable and consistent. Considering system of second order ordinary differential equations (boundary value problem) as in (16)-(20), shooting method can be explained as follows. Let

$$\begin{aligned}
f''(0) &= F_1\left[\eta, f(\eta), \theta(\eta), \phi(\eta), f'(\eta), \theta'(\eta), \phi'(\eta)\right], \text{ subject to } f(\eta_o) = f_o \text{ and } f(\eta_\infty) = f_1, \\
\theta''(0) &= F_2\left[\eta, f(\eta), \theta(\eta), \phi(\eta), f'(\eta), \theta'(\eta), \phi'(\eta)\right], \text{ subject to } \theta(\eta_o) = \theta_o \text{ and } \theta(\eta_\infty) = \theta_1, \\
\phi''(0) &= F_3\left[\eta, f(\eta), \theta(\eta), \phi(\eta), f'(\eta), \theta'(\eta), \phi'(\eta)\right], \text{ subject to } \phi(\eta_o) = \phi_o \text{ and } \phi(\eta_\infty) = \phi_1
\end{aligned} \tag{22}$$

represent the boundary value problem. Let $f(\eta;a)$, $\theta(\eta;b)$ and $\phi(\eta;c)$ denote the solution of the corresponding initial value problem after using method of superposition to reduce B.V.P. to I.V.P. as

$$\begin{aligned}
f''(0) &= F_1\left[\eta, f(\eta), \theta(\eta), \phi(\eta), f'(\eta), \theta'(\eta), \phi'(\eta)\right], \text{ subject to } f(\eta_o) = f_o \text{ and } f'(\eta_o) = a, \\
\theta''(0) &= F_2\left[\eta, f(\eta), \theta(\eta), \phi(\eta), f'(\eta), \theta'(\eta), \phi'(\eta)\right], \text{ subject to } \theta(\eta_o) = \theta_o \text{ and } \theta'(\eta_o) = b, \\
\phi''(0) &= F_3\left[\eta, f(\eta), \theta(\eta), \phi(\eta), f'(\eta), \theta'(\eta), \phi'(\eta)\right], \text{ subject to } \phi(\eta_o) = \phi_o \text{ and } \phi'(\eta_o) = c.
\end{aligned} \tag{23}$$

Next step is to find *a*, *b*, *c* which must satisfy tolerance and conditions at infinity $f(\eta_\infty) = f_1$, $\theta(\eta_\infty) = \theta_1$ and $\phi(\eta_\infty) = \phi_1$. We may define the function $F_1(a)$, $F_2(b)$ and $F_3(c)$ as

$$F_1(a) = f(\eta_\infty;a) - f_1, \; F_2(b) = \theta(\eta_\infty;b) - \theta_1, \; F_3(c) = \phi(\eta_\infty;c) - \phi_1. \tag{24}$$

If $F_1(a)$ in Equation (24) has a root "*a*"; certainly, the solution $f(\eta;a)$ of the corresponding I.V.P. is also a solution of the B.V.P. in Equation (22). Also applicable to $F_2(b)$ and $F_3(c)$. In addition, if the boundary value problem in Equation (22) has a solution $f(\eta)$, $\theta(\eta)$ and $\phi(\eta)$, then $f(\eta)$, $\theta(\eta)$ and $\phi(\eta)$ are also the unique solution $f(\eta;a)$, $\theta(\eta;b)$ and $\phi(\eta;c)$ of the initial value problem where

$$a = f'(\eta_o), \; b = \theta'(\eta_o), \; c = \phi'(\eta_o). \tag{25}$$

Thus, "*a*" is a root of $F_1$, "*b*" is a root of $F_2$ and "*c*" is a root of $F_3$. In this study, the *m-file* ("Animasaun OJFD") that contains the system of first order can be found under Appendix II. In the code, $f = y(1)$, $f' = y(2)$, $\theta = y(3)$, $\theta' = y(4)$, $\phi = y(5)$ and $\phi' = y(6)$. Using shooting method to solve nonlinear ordinary differential equation(s), zero finding problem for the function (Root finding) is a critical problem. It is very important to state that the two *m-files* must be in the same directory (folder) and patched to the command window of MATLAB. This problem is related to how to improve the solution (*i.e.* to get better estimate(s) of *a*, *b* and *c* that satisfy the B.V.P. and also meet up to our desirable degree of accuracy when compare to the condition(s) at infinity). Most researchers prefer to use Newton's method, or Secant method namely linear interpolation. In this research, quadratic interpolation which resembles Muller's method is adopted. This new approach is based on locally approximating the nonlinear function $f(x)$ by a quadratic function $g(x)$ and the root of the quadratic function $g(x)$ is taken as an improved approximation to the root of the nonlinear function $f(x)$. The procedure is applied repetitively to convergence. Animasaun [39] stated that three initial approximations *W*, *X*, and $Y$ which may or may not bracket the root are required to start the algorithm (see **Table 1**). The only difference between this new approach and the Secant method is that $g(x)$ is a quadratic function while $g(x)$ is a linear function in the Secant method. General procedure for fitting a polynomial to a set of equally spaced or unequally spaced data. The *nth* polynomial is

$$P_n(x) = a_0 + a_1x^1 + a_2x^2 + a_3x^3 + \cdots + a_nx^n. \tag{22}$$

When $n = 1$, we shall obtain linear polynomial of the form $P_1(x) = a_0 + a_1x^1$.
When $n = 2$, we shall obtain Quadratic Polynomial of the form

$$P_2(x) = a_0 + a_1x^1 + a_2x^2. \tag{23}$$

Using $P_2(x) = a_0 + a_1x^1 + a_2x^2$, when $W_i$ is guessed, set $x = R_i$

$$W_i = a_0 + a_1R_i + a_2R_i^2.$$

When $X_i$ is guessed, set $x = S_i$

$$X_i = a_0 + a_1S_i + a_2S_i^2.$$

Lastly, when $X_i$ is guessed, set $x = T_i$

$$Y_i = a_0 + a_1T_i + a_2T_i^2.$$

Solving for $a_0$, $a_1$ and $a_2$ simultaneously and then substitute into (23) to obtain

$$\begin{aligned} P_2(x) = &\left( \frac{S_iY_iR_i^2 - T_iR_i^2X_i - S_i^2R_iY_i + R_iT^2X_i + TS_i^2W_i - W_iT^2S_i}{S_i^2T_i - ST_i^2 + R_iT_i^2 - R_iS_i^2 - R_i^2T_i + R_i^2S_i} \right) \\ &+ \left( \frac{Y_iS_i^2 - S_i^2W_i + R_i^2X_i - Y_iR_i^2 + W_iT_i^2 - XT_i^2}{(S_i - T_i)(T_iS_i - S_iR_i + R^2 - T_iR)} \right) x \\ &+ \left( \frac{-T_iX_i + Y_iS_i - R_iY_i + R_iX + W_iT_i - W_iS_i}{S_i^2T_i - S_iT_i^2 + R_iT_i^2 - R_iS_i^2 - R_i^2T_i + R_i^2S_i} \right) x^2. \end{aligned}$$

Finally leads to

$$u^n(p) = P_2(p) \text{ for } i = 1:N. \tag{24}$$

The procedure is continued till the result of $u^n(p)$ at $N$ satisfy the desired degree of accuracy. Details can be found in [39]. It is worth mentioning that the convergence rate of the new approach of finding roots while shooting is faster than the rate of the secant method. Generally speaking, the secant method is specified because of its simplicity, even though its convergence rate is slightly smaller than the convergence rate of the new designed method which can be called Muller. The BVP cannot be solved on an infinite interval, and it would be impractical to solve it on a very large finite interval. In this research, the author imposed the infinite boundary conditions at a finite point of $\eta = 3$. To integrate the corresponding IVP, $f'(0)$, $\theta'(0)$ and $\phi'(0)$ are require but no such values exist after the non-dimensionalization of the boundary conditions (19) and (20). The suitable guess values for $W = f'(0)$, $X = \theta'(0)$ and $Y = \phi'(0)$ are chosen and integration is carried out using the command under Appendix III. The calculated values for $f(\eta = 3)$, $\theta(\eta = 3)$ and $\phi(\eta = 3)$ are compared with that of boundary condition (20). Adjustment of $W = f'(0)$, $X = \theta'(0)$ and $Y = \phi'(0)$ were carried out for better approximation for the solution of Equation (24). Care has been taken to shoot in steps; shoots are improved in stages and round off error is avoided by computing with 15 decimal places. Values of $W = f'(0)$, $X = \theta'(0)$ and $Y = \phi'(0)$ together with conditions in (19) are solved using Runge Kutta Gill

**Table 1.** Guess values for quadratic interpolation.

| | | | | |
|---|---|---|---|---|
| Corresponding values at the boundary $= x$ | $f(\eta = 3) = R_i$ | $\theta(\eta = 3) = S_i$ | $\phi(\eta = 3) = T_i$ | $p$ |
| Guess values $= f(x)$ | $a = W_i$ | $b = X_i$ | $c = Y_i$ | $u^n(p)$ |

method with $h = 0.01$. To improve the solutions, quadratic interpolation (Muller's method) is adopted. The above procedure is repeated until we get the results up to the desired degree of accuracy $10^{-5}$. The numerical solution described above is first written in 27 steps of Algorithm and implemented in a MATLAB environment together with Microsoft Excel software for proper analysis and extraction of data.

## 4. Results and Discussion

Numerical computations are carried out for the present problem by employing the similarity solution for $f(\eta)$, $\theta(\eta)$ and $\phi(\eta)$ against $\eta$ with variations of parameters controlling the fluid dynamics in the flow regime. The values of Schmidt number for hydrogen $S_c = 0.22$ at 25°C and one atmospheric pressure is considered in the analysis of the solution. It should be mentioned here that $D_a > 0$ indicates an increase in the destructive chemical reaction rate, while $D_a < 0$ corresponds to an increase in generative chemical reaction rate. The value of the Prandtl number is chosen to be $P_r = 0.71$, which represents air at 25°C and one atmospheric pressure. When modified local thermal Grashof number $G_r > 0$ this corresponds to cooling of the surface and when $G_r < 0$ this corresponds to heating of the surface. In addition, when modified local solutal Grashof number $G_c > 0$ this indicates that the chemical species concentration in the free stream region is less than the concentration at the surface/wall and when $G_c < 0$ this indicates that the chemical species concentration in the free stream region is greater than the concentration at the surface/wall. From **Table 2**, it is observed that the convergence rate of linear interpolation (secant method) is slightly smaller than the convergence rate of the new technique (quadratic interpolation) method which can be referred to as Muller. This result is in good agreement with a related result in Hoffman [40]. According to the result, both linear interpolation (Secant Method) and quadratic interpolation (Muller's method) are adopted to solve nonlinear problem (Four-bar linkage problem), it was reported that the convergence rate of Muller's method is 1.84, which is faster than the 1.62 rate of the secant method and slower than the 2.0 rate of Newton's method. Generally speaking, the secant method is preferred because of its simplicity even though its convergence rate is 1.62. In order to check the computational method for numerical accuracy, the author compares a special case of this present study with the numerical data published in Table 8 of Sastry and Murti [5]. As shown in **Table 2** below, the results obtained are in good agreement. It is important to remark that due to the kind of fluid under consideration, using $\xi > 1$, $\lambda > 1$ the state of the fluid can not be changed from liquid to gas.

**Figure 1** illustrates the influence of temperature dependent variable fluid viscosity and thermal conductivity on the fluid as it flows vertically upward past a moving surface through binary mixture in the presence of space dependent heat source. When modified buoyancy parameters $(G_r, G_c) < 0$, it is observed that as ($\xi$ and $\lambda$) increases, the velocity asymptotically decreases near the vertical wall to a point near free stream and satisfies boundary condition. It is also seen that the velocity profile when $(\xi = \lambda = 5)$ greatly over shoot downward and tends to have negative values of velocity within $0.5 \le \eta < 2$. This result implies that, at a constant value of ($b$ and $\delta$) and $G_r = G_c = -1$, increase in ($\xi = bT_w$ and $\lambda = \delta T_w$) generates heat energy which can never subdue

**Table 2.** Comparison of the values of $C_f$ with Sastry and Murti [2] using ( $\xi = \lambda = \tau = \chi = 0$, $G_r\left(\frac{\xi}{\theta_w}\right) = G_r = 0.1$, $G_c\left(\frac{\xi}{\phi_w}\right) = G_c = 0.1$, $D_a = 0.1$, $\omega = 0.1$, $\theta_w = \phi_w = 0.1$, $R_a = 0.1$, $P_r = 0.71$, $S_c = 0.22$, $h = n = 1$.

| | $f'(0)\,c = 0.1$ in Sastry and Murti [5] | Using shooting method along with quadratic interpolation present $f'(0)\,c = 0.1$ | Using shooting method along with linear interpolation present $f'(0)\,c = 0.1$ |
|---|---|---|---|
| $M = 5$ | −2.635097137 | −2.635097136618 | −2.6314758242887 |
| $M = 10$ | −3.475594470 | −3.4755944595667 | −3.4738132412511 |
| | $f'(0)\,c = 0.1$ | $f'(0)$ when $c = -0.1$ | $f'(0)$ when $c = -0.1$ |
| $M = 5$ | −2.418368678 | −2.418368675767 | −2.418388801774 |
| $M = 10$ | −3.265944783 | −3.265944783052 | −3.265943456084 |

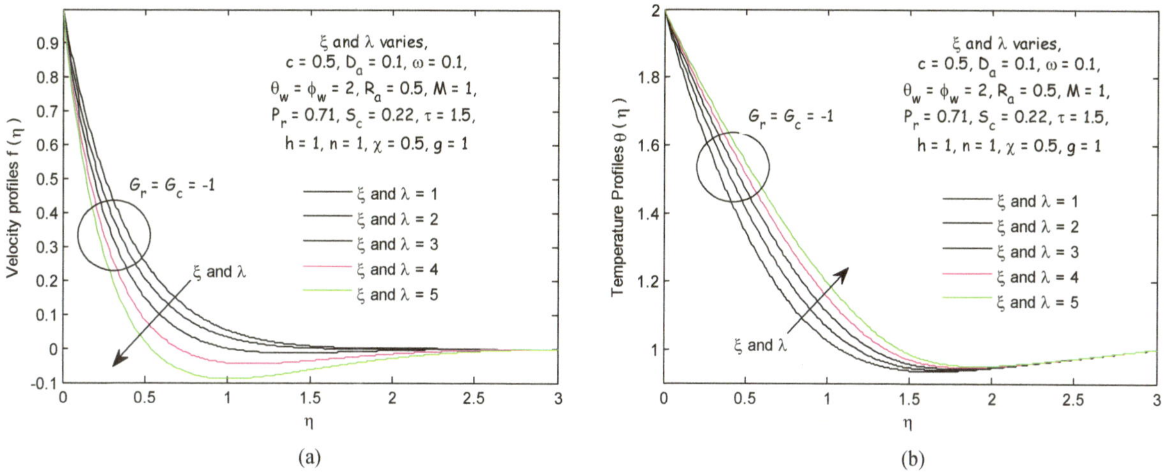


**Figure 1.** Effects of variable thermo-physical properties on (a) velocity profiles and (b) temperature profiles when $G_r = G_c = -1$.

corresponding influence of cooling of the fluid at the surface. This account for a drastic increase in the viscosity of the fluid as it flows (see **Figure 1(a)**). It is worth mentioning that at a constant value of space heat source parameters ( $\chi = 0.5$ , $g = 1$ ), as ($\xi$ and $\lambda$) increases, the corresponding increase in heat energy tends to increase the temperature within the fluid domain. This account for the increase in temperature in the fluid domain as it flows (see **Figure 1(b)**). This also account for the decrease in the local skin friction coefficient $f'(0)$ as magnitude of ($\xi$ and $\lambda$) increases (see **Figure 4**). In order to further unravel the dynamic of fluid as it flows past a moving surface through binary mixture when variable thermo-physical properties is considered, a case where $G_r = G_c = 0.1$ is investigated.

It is observed that as ($\xi$ and $\lambda$) increases, the velocity profiles decreases near the vertical wall $(0 \le \eta \le 0.2)$ and increases significantly thereafter as $\eta \to 3$ (see **Figure 2(a)**). This result is true since the rate of cooling of the surface has been reduced. This result also account for negligible increase of local skin friction coefficient $f'(0)$ with an increase in ($\xi$ and $\lambda$) (see **Figure 4**). It is further observed that the temperature profiles increases with an increase in ($\xi$ and $\lambda$). As the fluid rises along the vertical surface, it becomes cooled by the surface in the process. Due to this, influence of increasing ($\xi$ and $\lambda$) is subdued near the wall. Few distances away from the wall, influence of increasing variable thermo-physical parameters (*i.e.* temperature dependent viscosity and thermal conductivity) enhances the transport phenomena. The variations of $f'(\eta)$ known as velocity profiles along $\eta$ with different values of ($\xi$ and $\lambda$) are plotted in **Figure 3** when $G_r = G_c = 1$. It is seen that the increase of both ($\xi$ and $\lambda$) leads to significant increase of velocity profiles at all points within $0 \le \eta < 2.5$. When the fluid heats up, it's molecules become excited and begin to move. The energy of this movement is enough to overcome the forces that bind the molecules together, allowing the fluid to move faster and decreasing its viscosity; hence, the velocity increases significantly from the wall $\eta = 0$ to a point near the free stream with an increase in the magnitude of ($\xi$ and $\lambda$). This account for the reason why local skin friction coefficient $f'(0)$ is an increasing function of ($\xi$ and $\lambda$) as shown in **Figure 4**.

**Figure 5(a)** represents the velocity profiles for different values of magnetic field parameter ($M$). It is observed that velocity profile is a decreasing function of $M$. The magnetic field parameter represents the importance of magnetic field on the fluid as it flows vertically upward past a surface moving through binary mixture. The presence of transverse magnetic field sets in Lorentz force, which results in retarding force on the velocity field and therefore as magnetic field parameter increases, so does the retarding force. This account for the decrease in the velocity profiles. When variable thermo-physical properties are properly accounted for (*i.e.* $\xi = \lambda = 2$), it is observed that $f'(0)$ which is related to skin friction coefficient decreases with magnetic field parameter ($M$) and increases with buoyancy parameters ($G_r$ and $G_c$) (see **Figure 5(b)**). **Figure 6(a)** depicts the velocity profiles for different values of suction parameter. It is observed that the momentum boundary layer thickness and velocity distribution decreases with increase in suction ($c > 0$). This is in agreement with the usual fact that suction stabilize the boundary layer growth, as reported by (Makinde and Olanrewaju [4]; Alam and Rahman [41]). Wall suction removes the low-momentum fluid and delays both transition and separation. The effect of suction

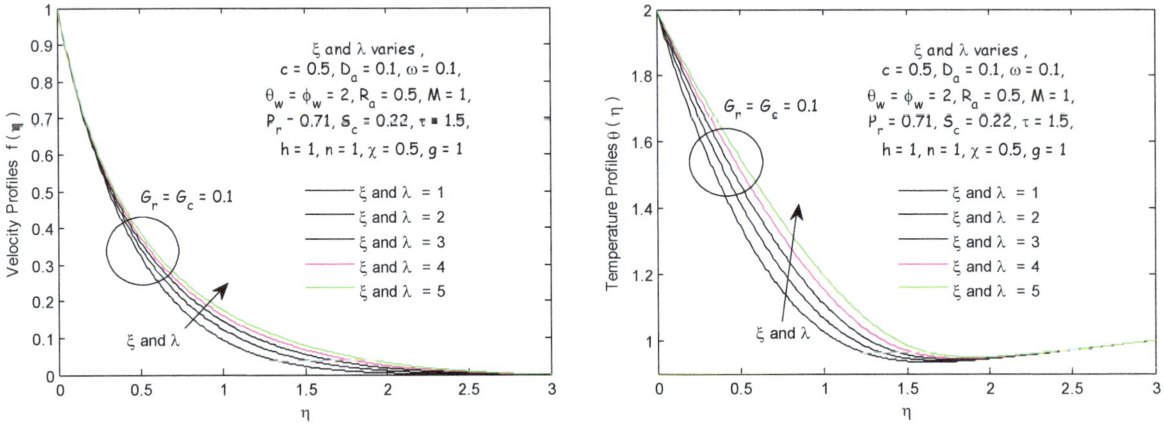


(a) (b)

**Figure 2.** Effects of variable thermo-physical properties on (a) velocity profiles and (b) temperature profiles when $G_r = G_c = 0.1$.

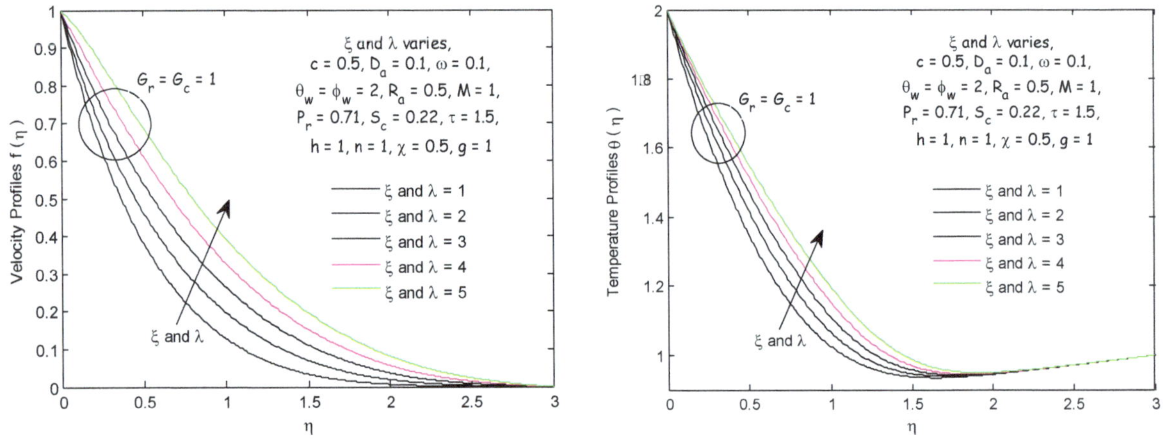


(a) (b)

**Figure 3.** Effects of variable thermo-physical properties on (a) velocity profiles and (b) temperature profiles when $G_r = G_c = 1$.

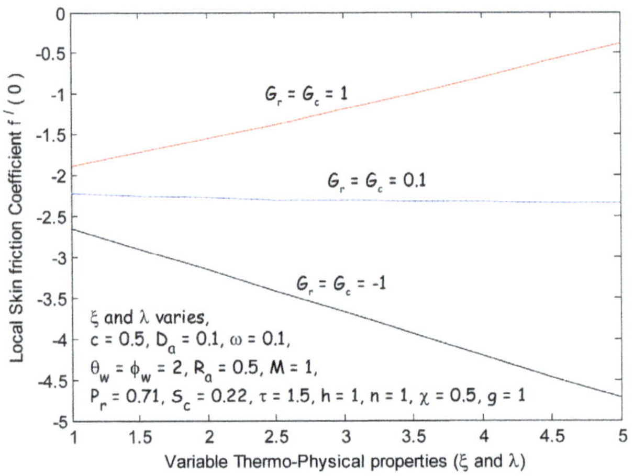


**Figure 4.** Variation of Skin friction coefficient for different values of variable thermo-physical properties (viscosity and thermal conductivity) together with $G_r$ and $G_c$.

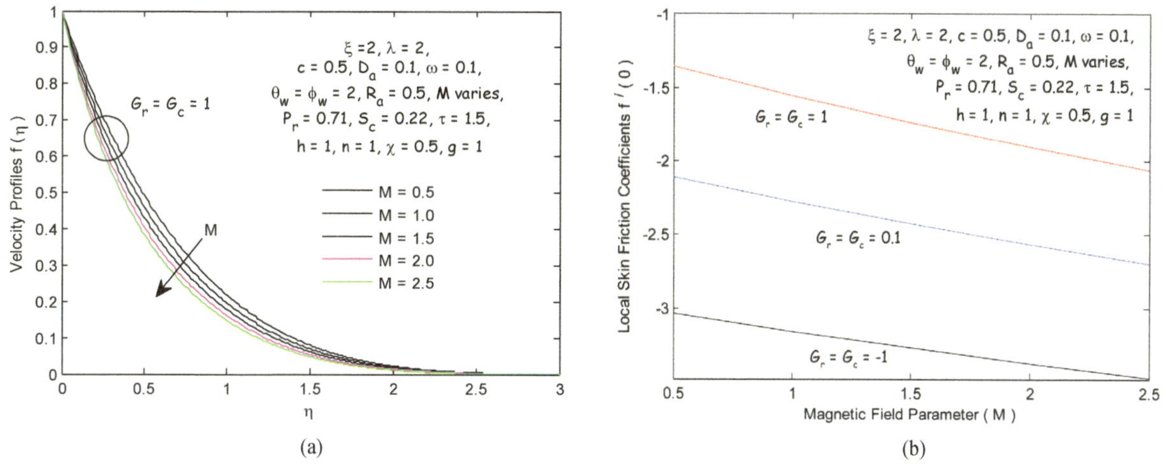


**Figure 5.** (a) Effects of magnetic field parameter on velocity; (b) Variation of skin friction coefficient for different values of magnetic field parameter against buoyancy parameters ($G_r$ and $G_c$).

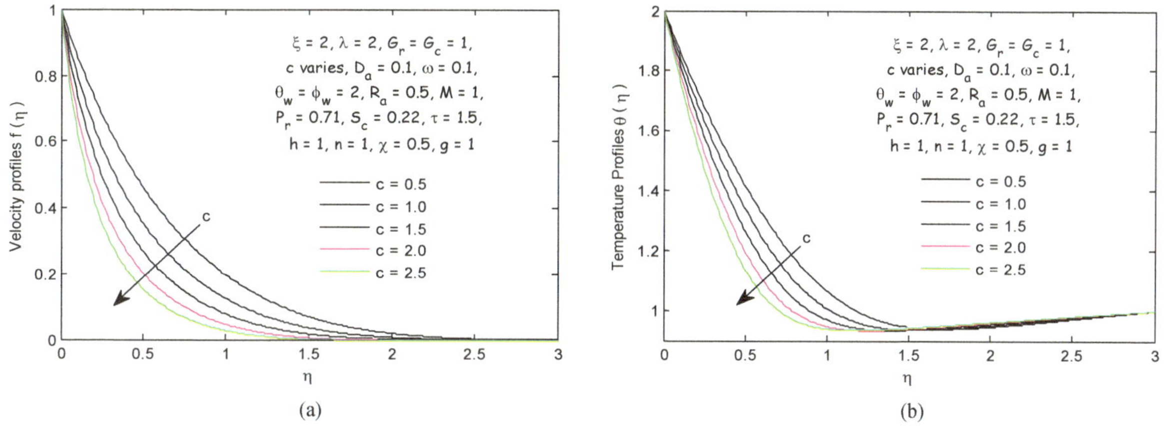


**Figure 6.** Effects of Suction parameter on (a) velocity profiles and (b) temperature profiles when variable thermo-physical properties are accounted for (*i.e.* $\xi=\lambda=2$).

parameter on the fluid temperature is highlighted in **Figure 6(b)**. It is observed that the fluid temperature decreases near the wall and increases negligibly far from the wall. As the flow develops along a vertical surface moving through binary mixture, one way of dealing with boundary layer transition (flow separation) is to suck the thin boundary layer through the vertical porous surface. As this method reduces drag, heat energy escaped away from the flow regime; hence the temperature reduces as magnitude of suction increases.

As we all know in thermal science that heat energy migrates from region of high temperature to region of low temperature, this account for the reason why temperature profiles over shoot downward towards the region of low heat energy within the fluid domain as shown in **Figure 6(b)**. The effect of suction parameter on the chemical species concentration in the boundary is depicted in **Figure 7(a)**. From this figure, it is seen that the species concentration within the boundary layer decreases with an increase in the magnitude of $c$. **Figure 7(b)** displays the effect of $\tau$ over concentration profile when $\xi=1.5$, $\lambda=1.5$, $S_c=0.22$; it is observed that solutal boundary layer decreases with an increasing values of parameter$\tau$. This can be interpreted as follows; when all the small particles are suspended in the fluid domain and temperature gradient exists, all the particles experienced a force which pushes the particles towards the free stream (area of low heat energy). In particular, the effect of increasing the thermophoretic parameter $\tau$ is limited to decreasing the concentration profiles. This is true only for small values of Schmidt number for which the Brownian diffusion effect is large compared to the convection effect. The thermophoretic parameter is expected to alter the concentration boundary layer significantly. The effect of increasing $\tau$ in this research where $\phi(0)=2$, $\phi(3)=1$ is in good agreement with that of [31] where $\phi(0)=1$, $\phi'(4)=0$ in which concentration profiles decreases with an increase in thermophoretic parameter.

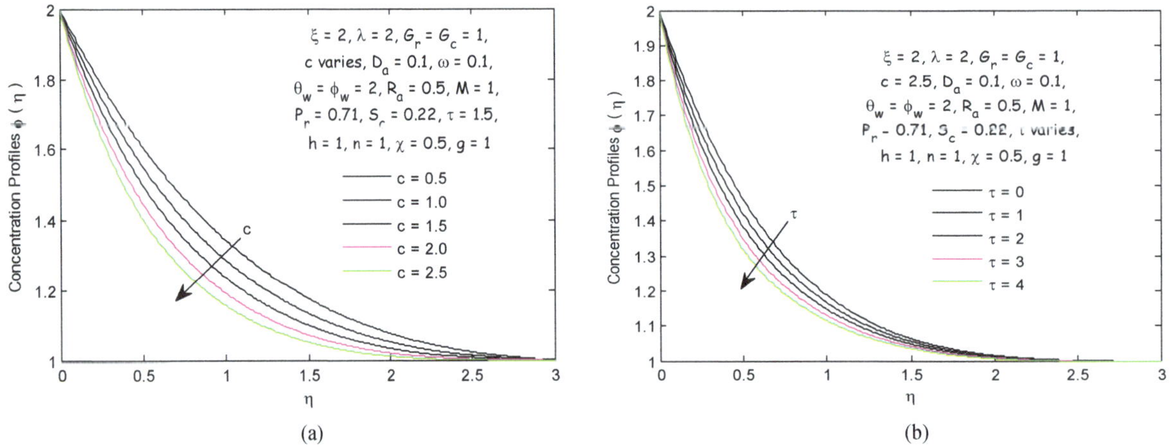


**Figure 7.** (a) Effects of Suction parameter "$c$" and (b) Effects of thermophoretic parameter "$\tau$" on concentration profiles when variable thermo-physical properties are accounted for (*i.e.* $\xi = \lambda = 2$).

## 5. Concluding Remarks

In this paper, effects of thermo-physical properties on unsteady convective flow with thermophoresis of particles, chemical reaction and radiative heat transfer past a porous vertical plate moving through a binary mixture are investigated. Quadratic interpolation technique is introduced with the fourth order Runge-Kutta Gill scheme to form shooting method. The following conclusions can be drawn as a result of the computation and analysis:

1) Velocity increases with increasing value of $\xi$ when $(G_r, G_c) \gg 0$; velocity decreases with increasing values of $\xi$ when $(G_r, G_c) \ll 0$. The two effects exist when $G_r = G_c = 0.1$.

2) Within the boundary layer of the problem considered, the effect of increasing suction ($c > 0$) leads to decrease of velocity, temperature (near the wall only) and concentration profiles.

3) Increase in thermophoretic parameter ($\tau$) leads to significant decrease in concentration profiles.

4) Local skin friction coefficient is a decreasing function of magnetic field parameter at all values of buoyancy parameter within $-1 \le (G_r, G_c) \le 1$. When variable thermo-physical properties are accounted for in transport phenomena of fluid flow along a vertical surface moving through binary mixture in the presence of exponential heat source, Grashof number related to thermal and solutal is an important yard stick to control skin friction drag which arises from the friction of the fluid against the skin of the vertical surface that is moving through binary mixture. The convergence rate of the new approach (Quadratic Interpolation) of finding roots while shooting is faster than the rate of the secant method (Linear Interpolation).

## References

[1] Al-Sharif, A., Chamniprasart, K., Rajagopal, K.R. and Szeri, A.Z. (1993) Lubrication with Binary Mixtures: Liquid-Liquid Emulsion. *Journal of Tribology*, **115**, 46-55. http://dx.doi.org/10.1115/1.2920985

[2] Wang, S.H., Al-Sharif, A., Rajagopal, K.R. and Szeri, A.Z. (1993) Lubrication with Binary Mixtures: Liquid-Liquid Emulsion in an EHL Conjunction. *Journal of Tribology*, **115**, 515-522. http://dx.doi.org/10.1115/1.2921668

[3] Makinde, O.D., Olanrewaju, P.O. and Charles, W.M. (2011) Unsteady Convection with Chemical Reaction and Radiative Heat Transfer past a Flat Porous Plate Moving through a Binary Mixture. *Afrika Matematika*, **21**, 1-17.

[4] Makinde, O.D. and Olanrewaju, P.O. (2011) Unsteady Mixed Convection with Soret and Dufour Effects past a Porous Plate Moving through a Binary Mixture of Chemically Reacting Fluid. *Chemical Engineering Communications*, **22**, 65-78. http://dx.doi.org/10.1080/00986445.2011.545296

[5] Sastry, D.R.V.S.R.K. and Murti, A.S.N. (2013) A Double Diffusive Unsteady MHD Convective Flow past a Flat Porous Plate Moving through a Binary Mixture with Suction or Injection. *Journal of Fluids*, **2013**, Article ID: 935156.

[6] Somasundaran, P. (2006) Encyclopaedia of Surface and Colloid Science. CRC Press, Boca Raton, 711.

[7] Giddings, J.C., Yang, F.J. and Myers, M.N. (1976) Flow Field-Flow Fractionation: A Versatile New Separation Method. *Science*, **193**, 1244-1245. http://dx.doi.org/10.1126/science.959835

[8] Talbot, L., Cheng, R.K., Scheffer, R.W. and Wills, D.P. (1980) Thermophoresis of Particles in a Heated Boundary Layer. *Journal Fluid Mechanics*, **101**, 737-758. http://dx.doi.org/10.1017/S0022112080001905

[9] Chamkha, A.J. and Isaa, C. (2000) Effects of Heat Generation/Absorption and Thermophoresis on Hydromagnetic Flow with Heat and Mass Transfer over a Flat Surface. *International Journal of Numerical Methods for Heat and Fluid flow*, **4**, 432-448. http://dx.doi.org/10.1108/09615530010327404

[10] Animasaun, I.L (2015) Effects of Thermophoresis, Variable Viscosity and Thermal Conductivity on Free Convective Heat and Mass Transfer of Non-Darcian MHD Dissipative Casson Fluid Flow with Suction and nth Order of Chemical Reaction. *Journal of the Nigerian Mathematical Society*, **34**, 11-31. http://dx.doi.org/10.1016/j.jnnms.2014.10.008

[11] Fagbade, A.I., Falodun, B.O. and Boneze, C.U. (2015) Influence of Magnetic Field, Viscous Dissipation and Thermophoresis on Darcy-Forcheimer Mixed Convection Flow in Fluid Saturated Porous Media. *American Journal of Computational Mathematics*, **5**, 18-40. http://dx.doi.org/10.4236/ajcm.2015.51002

[12] Singh, A.K. and Dikshit, C.K. (1988) Hydromagnetic Flow past a Continuously Moving Semi-Infinite Plate for Large Suction. *Astrophysics and Space Science*, **148**, 249-256. http://dx.doi.org/10.1007/BF00645964

[13] Vajravelu, K. and Hadjinicolaou, A. (1993) Heat Transfer in a Viscous Fluid over a Stretching Sheet with Viscous Dissipation and Internal Heat Generation. *International Communications in Heat and Mass Transfer*, **20**, 417-430. http://dx.doi.org/10.1016/0735-1933(93)90026-R

[14] Takhar, H.S., Gorla, R.S.R. and Soundalgekar, V.M. (1996) Radiation Effects on MHD Free Convection Flow of a Gas past a Semi-Infinite Vertical Plate. *International Journal of Numerical Methods for Heat and Fluid Flow*, **6**, 77-83. http://dx.doi.org/10.1108/09615539610113118

[15] Kim, Y.J. (2001) Unsteady MHD Convection Flow of Polar Fluids past a Vertical Moving Porous Plate in a Porous Medium. *International Journal of Heat and Mass Transfer*, **44**, 2791-2799. http://dx.doi.org/10.1016/S0017-9310(00)00332-X

[16] Raptis, A., Perdikis, C. and Leontitsis, A. (2003) Effects of Radiation in an Optically Thin Gray Gas Flowing past a Vertical Infinite Plate in the Presence of a Magnetic Field. *Heat and Mass Transfer*, **39**, 771-773. http://dx.doi.org/10.1007/s00231-002-0317-8

[17] Israel-Cookey, C., Ogulu, A. and Omubo-Pepple, V.B. (2003) Influence of Viscous Dissipation and Radiation on Unsteady MHD Free-Convection Flow past an Infinite Heated Vertical Plate in a Porous Medium with Time-Dependent Suction. *International Journal of Heat and Mass Transfer*, **46**, 2305-2311. http://dx.doi.org/10.1016/S0017-9310(02)00544-6

[18] Chamkha, A.J. (2004) Unsteady MHD Convective Heat and Mass Transfer past a Semi-Infinite Vertical Permeable Moving Plate with Heat Absorption. *International Journal of Engineering Science*, **42**, 217-230. http://dx.doi.org/10.1016/S0020-7225(03)00285-4

[19] Koriko, O.K., Omowaye, A.J. and Animasaun, I.L. (2015) Effects of Some Thermo-Physical Parameters on Free Convective Heat and Transfer over Vertical Stretching Surface at Absolute Zero. *Journal of Heat and Mass Transfer Research*, **1**, in Press.

[20] Na, T.Y. (1979) Computational Methods in Engineering Boundary Value Problems. Academic Press, New York.

[21] Cheng, P. (1964) Two Dimensional Radiation Gas Flow by Moment Method. *American Institute of Aeronautics and Astronautics Journal*, **2**, 1662-1664. http://dx.doi.org/10.2514/3.2645

[22] Incropera, F. (1985) Fundamentals of Heat and Mass Transfer. John Wiley and Sons, New York.

[23] Chamkha, A.J., Al-Mudhaf, A.F. and Pop, I. (2006) Effect of Heat Generation or Absorption on Thermophoretic Free Convection Boundary Layer from a Vertical Flat Plate Embedded in a Porous Medium. *International Communications in Heat and Mass Transfer*, **33**, 1096-1102. http://dx.doi.org/10.1016/j.icheatmasstransfer.2006.04.009

[24] Makinde, O.D. (2005) Free Convection Flow with Thermal Radiation and Mass Transfer past a Moving Vertical Porous Plate. *International Communications in Heat and Mass Transfer*, **32**, 1411-1419. http://dx.doi.org/10.1016/j.icheatmasstransfer.2005.07.005

[25] Crepeau, J.C. and Clarksean, R. (1997) Similarity Solutions of Natural Convection with Internal Heat Generation. *Transactions of ASME—Journal of Heat Transfer*, **119**, 184-185. http://dx.doi.org/10.1115/1.2824086

[26] Salem, A.M. and El-Aziz, M.A. (2007) MHD-Mixed Convection and Mass Transfer from a Vertical Stretching Sheet with Diffusion of Chemically Reactive Species and Space- or Temperature-Dependent Heat Source. *Canadian Journal of Physics*, **85**, 359-373. http://dx.doi.org/10.1139/P07-048

[27] Salem, A.M. and El-Aziz, M.A. (2008) Effect of Hall Currents and Chemical Reaction on Hydromagnetic Flow of a Stretching Vertical Surface with Internal Heat Generation/Absorption. *Applied Mathematical Modelling*, **32**, 1236-1254. http://dx.doi.org/10.1016/j.apm.2007.03.008

[28] Animasaun, I.L., Adebile, E.A. and Fagbade, A.I. (2015) Casson Fluid Flow with Variable Thermo-Physical Property along Exponentially Stretching Sheet with Suction and Exponentially Decaying Internal Heat Generation Using the Homotopy Analysis Method. *Journal of the Nigerian Mathematical Society*, in Press. http://dx.doi.org/10.1016/j.jnnms.2015.02.001

[29] Animasaun, I.L. (2015) Casson Fluid Flow of Variable Viscosity and Thermal Conductivity along Exponentially Stretching Sheet Embedded in a Thermally Stratified Medium with Exponentially Heat Generation. *Journal of Heat and Mass Transfer Research*, **1**, in Press.

[30] Batchelor, G.K. (1987) An Introduction to Fluid Dynamics. Cambridge University Press, London.

[31] Sivagnana-Prabhu, K.K., Kandasamy, R. and Saravanan, R. (2009) Lie Group Analysis for the Effect of Viscosity and Thermophoresis Particle Deposition on Free Convective Heat and Mass Transfer in the Presence of Suction/Injection. *Theoretical and Applied Mechanics*, **36**, 275-298. http://dx.doi.org/10.2298/TAM0904275S

[32] Animasaun, I.L. and Oyem, A.O. (2014) Effect of Variable Viscosity, Dufour, Soret and Thermal Conductivity on Free Convective Heat and Mass Transfer of Non-Darcian Flow past Porous Flat Surface. *American Journal of Computational Mathematics*, **4**, 357-365. http://dx.doi.org/10.4236/ajcm.2014.44030

[33] Charraudeau, J. (1975) Influence de gradients de properties physiques en convection force application au cas du tube. *International Journal of Heat and Mass Transfer*, **18**, 87-95. http://dx.doi.org/10.1016/0017-9310(75)90011-3

[34] Tsai, R. (1999) A Simple Approach for Evaluating the Effect of Wall Suction and Thermophoresis on Aerosol Particle Deposition from a Laminar Flow over a Flat Plate. *International Communications in Heat and Mass Transfer*, **26**, 249-257. http://dx.doi.org/10.1016/S0735-1933(99)00011-1

[35] Batchelor, G.K. and Shen, C. (1985) Thermophoretic Deposition of Particles in Gas Flowing over Cold Surfaces. *Journal of Colloid and Interface Science*, **107**, 21-37. http://dx.doi.org/10.1016/0021-9797(85)90145-6

[36] Gill, S. (1951) A Process for the Step-by-Step Integration of Differential Equations in an Automatic Digital Computing Machine. *Mathematical Proceedings of the Cambridge Philosophical Society*, **47**, 96-108. http://dx.doi.org/10.1017/S0305004100026414

[37] Delin, T. and Zheng, C. (2012) On a General Formula of Fourth Order Runge Kutta Method. *Journal of Mathematics Sciences and Mathematics Education*, **7**, 1-10.

[38] Finlayson, B.A. (1980) Nonlinear Analysis in Chemical Engineering. McGrawHill, New York.

[39] Animasaun, I.L. (2014) Effects of Some Thermo-Physical Parameters on Free Convective Heat and Mass Transfer over Vertical Surface. Master of Technology Thesis, Federal University of Technology, Akure.

[40] Hoffman, J.D. (1992) Numerical Method's for Engineers and Scientists. McGraw Hill, New York.

[41] Alam, M.S. and Rahman, M.M. (2006) Dufour and Soret Effects on Mixed Convection Flow past a Vertical Porous Flat Plate with Variable Suction. *Nonlinear Analysis: Modeling and Control*, **11**, 3-12.

## Appendix I

```
function [eta,y] = Rungekgill(t0,tf,y0,f,n)
m = sqrt(2);
h = (t0 – tf)/n;          % In this study, t0 = 0 and tf = 3
t = linspace(t0,tf,n+1);  % To discretize the independent variable "η"
y(1,:) = y0;              % y0 is known as initial conditions in column vector
for i = 1: n;             % here n = 300, implies that stepsize h = 0.01
        k1 = h*feval(f,t(i),y(i,:));
        k2 = h*feval(f,t(i)+(h/2),y(i,:)+(k1/2));
        k3 = h*feval(f,t(i)+(h/2),y(i,:)+(m-1)*(k1/2)+(m-1)*(k2/m));
        k4 = h*feval(f,t(i)+h,y(i,:)-(k2/m)+(m+1)*k3/m);
y(i+1,:) = y(i,:)+(k1)/6+((k2)*(2-m))/6+((k3)*(2+m))/6+(k4)/6;
end
```

## Appendix II

```
function dy = AnimasaunOJFD(eta,y)
xi = 2; lamda = 2; c = 2.5; Gr = 1; Gc = 1; Da = 0.1; omega = 0.1; thetaw = 2; phiw = 2;
Ra = 0.5; M = 1; Pr = 0.71; Sc = 0.22; tau = 4; h = 1; n = 1; chi = 0.5; g = 1;
dy(1) = y(2);
dy(2) = ((xi/thetaw)*y(4)*y(2)-2*(eta+c)*y(2)-Gr*(xi/thetaw)*(y(3)-1)-Gc*(xi/phiw)*(y(5)-1)+M*y(1))/
(1+ xi - (y(3)*xi)/thetaw);
dy(3) = y(4);
dy(4) =(-(lamda/thetaw)*y(4)*y(4) - 2*Pr*(eta+c)*y(4) - (exp(omega - (omega/y(3))))*Da*Pr*h*(y(5))^n +
Ra*Pr*y(3)*y(3)*y(3)*y(3)- 4*chi*(y(3)-1 )*exp(-g*eta))/(1+((lamda*y(3))/thetaw)-(lamda/thetaw)) ;
dy(5) = y(6);
dy(6) = - Sc*2* (eta+c)*y(6)+(exp(omega-(omega/y(3))))*Da*Sc*(y(5))^n+Sc*(tau/thetaw)*y(4)*y(6) +
Sc*(tau/thetaw)*(y(5)-1)*((-(lamda/thetaw)*y(4)*y(4)-2*Pr*(eta+c)*y(4)-(exp(omega-(omega/
y(3))))*Da*Pr*h*(y(5))^n+Ra*Pr*y(3)*y(3)*y(3)*y(3)- 4*chi*(y(3)-1 )*exp(-g*eta))/(1 +
((lamda*y(3))/thetaw)-(lamda/thetaw)));
```

## Appendix III

The*m-files* "Rungekgill"and "AnimasaunOJFD" must be in the same folder, and then patch with MATLAB command window.Lastly, paste the command below on Command window.

```
>>[eta,y]=Rungekgill(0,3,[1;a;2;b;2;c],'AnimasaunOJFD',300) % Assign values for a, b and c.
```

# 19

# Effect of Curvature of Tip and Convexity of Electrode on Localization of Particles

**Sudarshan Ghonge[1], D. Nagendra Prasad[1], Swarnim Narayan[1], Hains Francis[1], Astha Sethi[2], Subimal Deb[3], Souri Banerjee[1]**

[1]Department of Physics, Birla Institute Technology and Science-Pilani, Hyderabad Campus, Hyderabad, India
[2]Department of Physics and Frederick Seitz Materials Research Laboratory, University of Illinois, Urbana-Champaign, IL, USA
[3]Department of Physics, GIT, GITAM University, Visakhapatnam, India
Email: souri@hyderabad.bits-pilani.ac.in

## Abstract

**We investigate the effect of curvature of the tip and the convexity of an electrode on the localization of suspended particles under the combined effect of dielectrophoresis and AC electroosmosis through simulations using COMSOL Multiphysics. A systematic analysis of the parameters defining the convexity of the electrode—the radius of the tip and the apex angle shows that suspended particles can be trapped closely to the electrode edges for comparatively larger tip radii and apex angles. This in turn should favour the trapping of polarizable molecules between the electrodes only if the fluid velocities at the vortices are not very strong.**

## Keywords

**Finite Element Method, Transport Dynamics, AC Electroosmosis, Dielectrophoresis, Vortex**

## 1. Introduction

The manipulation of polarizable biomolecules like DNA by electrical forces in microdevices has become a topic of increasing interest as it finds wide-spread applications in lab-on-a-chip technology or in DNA diagnostics [1]. In a suitable condition, a DNA molecule in an electrolytic solution will develop a strong electric dipole moment [2]. This induced dipole allows trapping and manipulation of the molecule when placed within micro-electrodes owing to a process known as dielectrophoresis (DEP) [3] [4]. In this context it is worth mentioning that the dielectrophoretic motion of DNA molecules requires alternate current (AC) electric field to suppress the electrophoric effect of the molecules' net charge [5]. However, the motion of the DNA molecules and hence their trapping/manipulation possibilities within a set of microelectrodes will also be countered by the flow of the electrolyte known as AC electro-osmosis (ACEO) [6]. Thus balancing DEP and ACEO is found to be a promising technique to study non-contact manipulation/trapping of DNA molecules suspended in an electrolyte solution and placed within electrically fed microelectrodes [7]. Washizu *et al.* [8] used a set of interdigitated sinusoidally corrugated aluminum microelectrodes to experimentally observe site-specific immobilization of double-stranded DNA molecules under AC electric field and concluded that the orientation of DNA molecules depended strongly on the electric field strength. Similar interdigitated rectangular metal microelectrodes of different dimensions and gap lengths were later investigated, both experimentally and numerically, by several researchers [9] in order to understand movement of DNA molecules within microelectrodes under electrokinetic effects. From these studies it appears that one might further exploit state-of-art lithographic technique to fabricate smaller electrodes with sharper bends which would intensify the non-uniformity of electric field and significantly influence DNA manipulation. To this end, use of triangular shaped electrodes has recently been very popular [10] [11]. Porath *et al.* [10] reported the first experimental evidence of trapping of a single, double stranded DNA molecule between the two vertices of a set of triangular electrodes where a few nanometers were apart. It clearly revealed that electric field strength which was extremely large at the vertices of the triangular electrode, coupled with polarization properties, played an important role in guiding the flow of DNA molecules in and around the tips. However, the most modern lithographic technique has its limitation to reproduce the bends or sharpness of the electrodes [11]. Thus it is interesting to study how small variations in electrode configurations would affect the nature of immobilization of the suspended DNA molecules confined between them.

In this report, we present a systematic analysis of the effect of curvature of the electrode tips on the localization of insoluble particles. We have also tried to assess how the convexity of the electrodes could influence the localization process.

## 2. Theory

The system under consideration consisting of two triangular electrodes is schematically represented in **Figure 1**. The vertices of the triangular electrodes are rounded as would be in the case of an actual fabrication. The rounded vertex results in a lower gradient of electric field around the tip in contrast to sharp vertices. We shall call the radius $r$ of the rounded electrode corner (represented by a dashed circle) the fillet radius. The vertex angle $\theta$ (acute angle between the electrode edges) defines the convexity of the electrodes. An increase in $r$ keeping $\theta$ constant would result in electrodes with blunter tips with the same convexity. This is to match a practical scenario where lithographic tools are used to fabricate microlectrodes [11]. Such a pair of electrodes placed at the base of a cubical region filled with an electrolyte, is studied numerically for the possibility of localization of suspended particles.

The properties of the electrolyte used for this configuration correspond to that of KCl at infinite dilution with electrical conductivity of 0.0021 S/m, relative permittivity of 78.49, density of 1000 kg/m$^3$ and a dynamic viscosity (at 293 K) of 0.001 Pa-s [12].

We consider the variation of the apex angle with a fixed separation ($d$) between the tips of the electrode. The

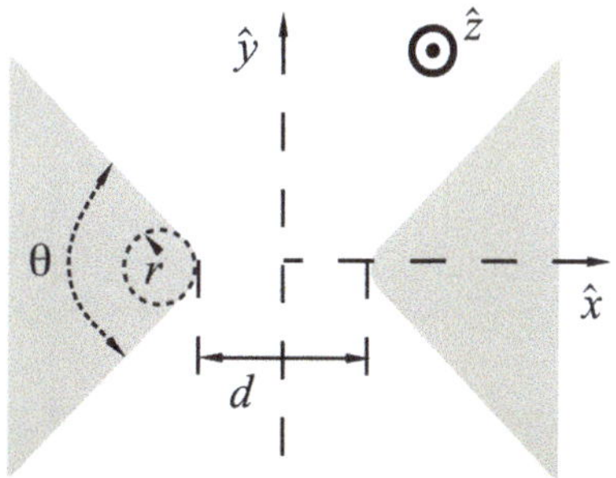


**Figure 1.** Schematic representation of the top view of the electrode setup. The dashed lines along $\hat{x}$ and $\hat{y}$ represent the planes of mirror symmetry transverse to the plane of the electrodes.

$y$ - $z$ plane, a plane of mirror symmetry between the electrodes, reduces the domain of the problem by half owing to the symmetry of the configuration and opposite polarities of the electrodes. Clearly an obtuse apex angle is a concave configuration of the electrode edges facing each other. Convex electrodes produce the strongest electric fields near their apices owing to the proximity of the tips. This produces a region of strong electric field at the center of the configuration that gradually decreases away from the origin along the y-axis on either side. A concave electrode, on the other hand, has a gradually decreasing spacing between the edges (measured along the x-axis) which produces a field of least strength at the origin, gradually increasing along the positive and negative y-axis. Changing the apex angle from acute to obtuse is a cross-over from the first regime of the electric field decreasing away from the origin to the one where the field density change is reversed. We cover both the regimes by changing the apex angle.

The special case of plane parallel electrodes ($\theta = \pi$) exhibits two symmetries. The first is a translational symmetry along the edge of the electrodes. The second is a mirror symmetry about the *y-z* plane through the origin. The mirror symmetry warrants the use of the method of images under an applied electric potential and reduces the problem space by half. The translational symmetry reduces the configuration to two dimensions confined to the plane of translational symmetry. This configuration was studied in detail by Green *et al.* [13].

The arrangement of triangular electrodes has two planes of mirror symmetry in the arrangement: the first being the *x-z* plane passing through the apex of both the electrodes; the second being the y-z plane through the origin. This space, however, must be studied as a three-dimensional system since the convexity of the electrodes breaks the translational symmetry that existed for $\theta = \pi$. The three parameters for our study could be $r$, $\theta$ and $d$, where one could systematically vary a parameter holding the other two fixed. Here we present the simulation studies with varying $\theta$ keeping $r$ constant and vice-versa for a fixed value of $d$. The simulations were performed using COMSOL Multiphysics version 4.3a [14]. The method of images is used for the analysis of the system owing to the antisymmetry of the applied potential across the *y-z* plane which will be at a potential of 0 V. To study the effect of AC electroosmosis, the double layer was modelled as distributed impedance over the electrode surface. The potential applied on the surface of the electrodes is

$$V(t) = V_{ep} \sin \omega t \tag{1}$$

We use $V_{ep} = 0.25\ \text{V}$ with the applied potential having an angular frequency $\omega = 5\ \text{kHz}$. The distributed impedance model defines the spatial variation of the potential ($\phi_{ep}$) at the edges of the double layer formed

$$\sigma \nabla \phi_{ep} = \frac{\phi_{ep} - V_{ep}}{Z_{DL}};\ Z_{DL} = \frac{1}{i\omega C_{DL}} \tag{2}$$

where $\sigma$ is the electrical conductivity, $C_{DL}$ is the capacitance of the double layer. The components of the flow velocities are computed using Navier-Stokes equations to identify regions capable of localizing or trapping particles around the electrodes. The time averaged DEP force for a spherical particle of radius $R$ is given by the expression [15]

$$F_{dep} = 2\pi\epsilon_m R^3 \Re\left\{\left[K(\omega)\right]\right\} \nabla \left|E_{rms}\right|^2 \tag{3}$$

where the Clausius-Mossotti factor

$$K(\omega) = \frac{\tilde{\epsilon}_p - \tilde{\epsilon}_m}{\tilde{\epsilon}_p + 2\tilde{\epsilon}_m};\ \tilde{\epsilon}_{p,m} = \epsilon_{p,m} - \frac{i\sigma_{p,m}}{\omega} \tag{4}$$

Here $\epsilon$ and $\sigma$ are the complex permittivities and conductivities respectively with subscripts $p$ and $m$ referring to the particle and medium respectively. We choose $\epsilon_p = 4\epsilon_0$, $\sigma_p = 2.4\ \text{S}\cdot\text{cm}^{-1}$ at 25°C, $K$ (5 kHz) = $0.9999 - 1.3345 \times 10^{-7} i$.

The radius of the Lambda DNA is known to be in the range 2 to 4 μm. A similar study by Bown and Meinhart [16] on the DNA of Lambda phage bacteria was reported in a frequency range of 100 Hz to 1.6 MHz with a rotationally symmetric electrode configuration. The expression for the DEP force on a particle, also given by an alternate expression [3], provides a measure of the effective polarizability of the particle and the magnitude of the electric field, which has been used in earlier simulations [17] and the formula as given by Equation 3 is found suitable to calculate DEP forces for spherical particles [15]. Washizu and Kurosawa [2] have shown that the DNA would stretch under an electric field. Stretching the DNA can result in a change of the effective polarizability of the molecule resulting in a shift of the trapping points. Regtmeier *et al.* [18] have reported that the polarizability of the DNA in linear and supercoiled forms is of similar magnitude within error margins. We therefore use the same effective polarizability for free and trapped particles assuming that the stretched DNA (under the applied field) will not affect the trapping points.

## 3. Results and Discussion

We use a particle size of $R = 4\ \mu\text{m}$ to study the feasibility of polarized and stretched DNA molecules forming a bridge across electrodes. The forces on the stretched DNA molecule are the DEP force (calculated from the computed electric fields) and the drag force (calculated from Stoke's law). **Figure 2** shows the variation of the $x$-component of the force on a stretched DNA molecule along $x$-axis for an electrode system having typical parameters $d = 25\ \mu\text{m}$, $r = 0.5\ \mu\text{m}$, and $\theta = 102^\circ$. It is important to point out that at any point of time the electrodes under the application of an alternating voltage have opposite polarities. The symmetry of the electrode configuration and the simultaneous antisymmetry of the electric potential exist for all ranges of parameters $r$, $d$ and $\theta$. The range of parameters examined show the same nature of the x-component of the force. Here we present only a typical plot of the force for $d = 25$ m, $r = 0.5$ m, $\theta = 102^\circ$. Therefore **Figure 2** is representative of any electrode configuration defined by the parameters $d$, $r$ and $\theta$. It is worth noting from the plot that the force has a positive and a negative value on either side of a point where the force vanishes. Such a force can represent only a vortex. The suspended particles in the electrolyte upon entering the vortex get trapped in it due to the torque set up by the vortex. The results show that the vortex is located near the edge of the electrode and therefore the insoluble particles in the vortex will have a higher probability of bridging the gap between the electrodes. This also requires that the torque on the trapped particles should not be too great to prohibit it from exiting the vortex.

### 3.1 Variation of Apex Angle

A variation of $\theta$ for a fixed $r$ and electrode separation leads to increasing electric field density with decreasing $\theta$. Quickly varying electric field densities indicate the existence of strong vortices with a short span along the $x$-axis, as envisaged from **Figure 2** as well. We test it for four different angles keeping $r = 0.5\ \mu\text{m}$ and $d = 25\ \mu\text{m}$.

**Figure 3** shows the variation of the slope (stiffness) of the $x$ component of the force as a function of $\theta$ for the convex (dashed curve) and the concave (solid curve) electrodes where the magnitude of the slope is taken as a measure of the stiffness. The plot clearly shows a systematic increase of the stiffness of the force with increasing $\theta$ for the set of concave electrodes. It should be noted that in **Figure 3**. We have plotted the slope of the x-component of the force for the two cases namely, convex and concave electrodes. It is worthwhile to mention that angle in the plot is the acute angle between the edges of the electrode. Hence, the angle measured for the concave electrode, if measured on the same side of the electrode edge as for the convex electrode, would be a

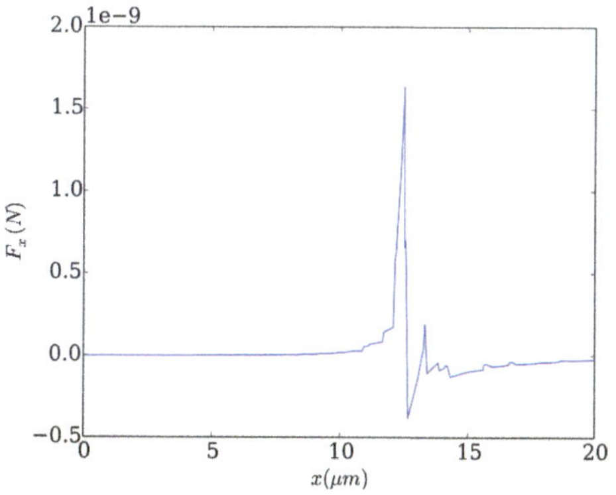


**Figure 2.** Variation of the $x$-component of the force on a stretched DNA molecule along the x-axis. The specific parameters used are $d = 25\ \mu\text{m}$, $r = 0.5\ \mu\text{m}$, $\theta = 102^\circ$.

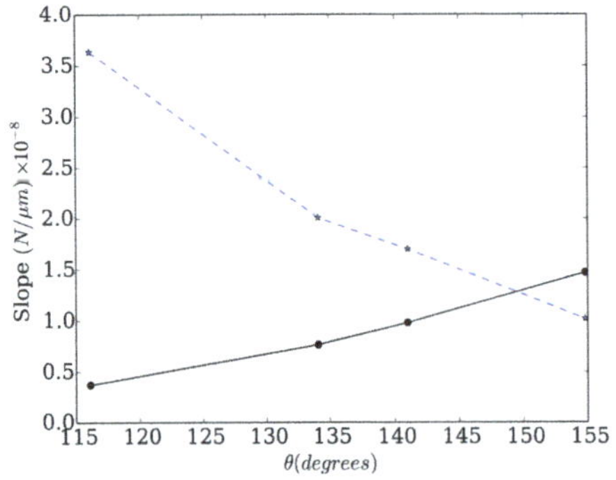


**Figure 3.** Variation of the slope of the $x$-component of the force as a function of the apex angle $\theta$ for the concave (solid curve) and convex (dashed curve) electrodes

reflex angle, and therefore, the two plots in **Figure 3** are consistent with each other. Smaller slope (stiffness) at larger $\theta$ indicates that electrodes with less convexity is likely to generate vortex sites of weaker trapping capacity. This prompts us to further investigate the scenario with electrodes having moderate convexity but with varying apex radius as it would address a more practical issue. It is well accepted that lithographically fabricated micro/nano electrodes would yield rounded tips of different curvature due to the resolution limit of the respective lithographic tool used [10] [11].

### 3.2 Variation of Apex Radius

A variation of the fillet radius $r$ changes the electric field density around the tip of the electrodes. A sharper electrode tip produces a stronger electric field and strengthens the DEP force. Thus, the net force across the vortex is expected to change faster for sharper electrodes. We test this by plotting the variation of $x$-component of the force $F_x$ as a function of the fillet radius $r$ at the center of the vortex keeping electrode separation $d = 25\ \mu\text{m}$ and apex angle $\theta = 102^\circ$. **Figure 4** shows the variation of the slope of the $x$-component of the force as a function of $r$ for the convex (dashed curve) and the concave (solid curve) electrodes. Specifically, the slope was calculated by taking a linear approximation of the curve between the two points where the force reaches its extremal values. $F_x$ varied at a rate of $0.48 \times 10^{-10}\ \text{N}/\mu\text{m}$ for flat electrodes ($r \to \infty$). The decrease of the stiffness of the force with the fillet radius occurs for all angles checked in the range of $116.1^\circ$ to $154.8^\circ$. The plot (**Figure 4(a)**) suggests that electrodes with smaller $r$ (*i.e.* having sharper tip) would produce stronger vortices within the electrolyte. Keeping in mind real life fabrication issues, the strength and the span of the vortex are of interest to us. We calculate the field strength by taking the average of the magnitude of the extremal values of $F_x$ and the span is calculated by finding the spacing between the locations where $F_x$ takes the extremal values. **Figure 4(b)** and **Figure 4(c)** represent the variation of the strength of the field and span respectively as a function of fillet radius. From the plots it is clear that the strength decreases and the span increases as the electrode tips become more and more blunt.

## 4. Conclusion

In conclusion, we have systematically analyzed two of the three parameters (the fillet radius and the apex angle) that may be tuned to have a localizing region for suspended DNA molecules in electrolytes. It is found that stiffness of the force decreases with increase of the apex angle, indicating weaker trapping sites in the electrolyte as the convexity of the electrode increases. In this context we tried to address a practical situation where electrodes of fixed and moderate convexity but having different curvature of the tips were considered. Our study reveals that the strength of the vortex gets weaker and its span becomes wider as the tip gets more and more rounded. It also provides an intuitive insight that, if polarization of DNA molecules has any role to facilitate the DNA attachment process, it is desirable to have vortex of moderate strength and span around the edge of the electrode such that the molecules are localized (rather than being strongly trapped) in the vortex facilitating polarization to complete the bridging process. Therefore, it is suggested that a set of electrodes with fixed convexity but having rounded tips may better serve the purpose of bridging electrodes by DNA molecules. It would be

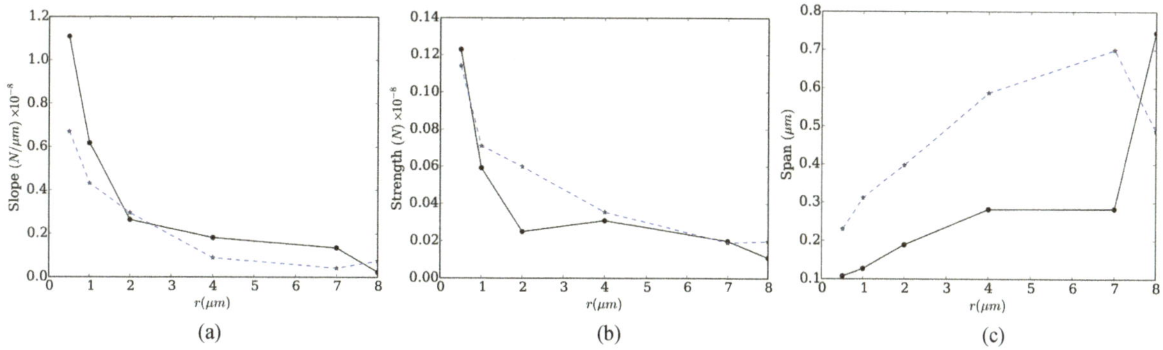


**Figure 4.** Variation of the (a) slope, (b) strength and (c) span of the $x$-component of the force as a function of the fillet radius $r$ for the concave (solid curve) and convex (dashed curve) electrodes.

interesting to further study the strength, distribution and the span of vortex points within the electrolyte as a function of electrode separation as well (keeping $r$ and $\theta$ constant), which is already underway.

## Acknowledgements

One of the authors (SB) acknowledges the financial support of the Board of Research in Nuclear Sciences, Department of Atomic Energy, Govt. of India for financial assistance.

## References

[1] Abramowitz, S. (1996) Towards Inexpensive DNA Diagnostics. *Trends in Biotechnology*, **14**, 397-401. http://dx.doi.org/10.1016/0167-7799(96)10051-2

[2] Washizu, M. and Kurosawa, O. (1990) Electrostatic Manipulation of DNA in Microfabricated Structures. *IEEE Transactions on Industry Applications*, **26**, 1166-1172. http://dx.doi.org/10.1109/28.62403

[3] Pohl, H.A. (1978) Dielectrophoresis: The Behavior of Neutral Matter in Nonuniform Electric Fields. Cambridge University Press, Cambridge.

[4] Asbury, C.L. and van den Engh, G. (1998) Trapping of DNA in Nonuniform Oscillating Electric Fields. *Biophysical Journal*, **74**, 1024-1030. http://dx.doi.org/10.1016/s0006-3495(98)74027-5

[5] Bakewell, D.J. and Morgan, H. (2006) Dielectrophoresis of DNA: Time- and Frequency-Dependent Collections on Microelectrodes. *IEEE Transactions on Nano Bioscience*, **5**, 1-8.

[6] Loucaides, N.G., Ramos, A. and Georghiou, G.E. (2008) Trapping and Manipulation of Nanoparticles by Using Jointly Dielectrophoresis and AC Electroosmosis. *Journal of Physics*: *Conference Series*, **100**, Article ID: 052015. http://dx.doi.org/10.1088/1742-6596/100/5/052015

[7] Du, J.-R., Juang, Y.-J., Wu, J.-T. and Wei, H.-H. (2008) Long-Range and Superfast Trapping of DNA Molecules in an AC Electrokinetic Funnel. *Biomicrofluidics*, **2**, Article ID: 044103. http://dx.doi.org/10.1063/1.3037326

[8] Washizu, M., Kurosawa, O., Arai, I., Suzuki, S. and Shimamoto, N. (1995) Applications of Electrostatic Stretch-and-Positioning of DNA. *IEEE Transactions on Industry Applications*, **31**, 447-456. http://dx.doi.org/10.1109/28.382102

[9] Asbury, C.L., Diercks, A.H. and van den Engh, G. (2002) Trapping of DNA by Dielectrophoresis. *Electrophoresis*, **23**, 2658-2666. http://dx.doi.org/10.1002/1522-2683(200208)23:16<2658::aid-elps2658>3.0.co;2-o

[10] Porath, D., Bezryadin, A., De Vries, S. and Dekker, C. (2000) Direct Measurement of Electrical Transport through DNA Molecules. *Nature*, **403**, 635-638.

[11] Zheng, L.F., Brody, J.P. and Burke, P.J. (2004) Electronic Manipulation of DNA, Proteins, and Nanoparticles for Potential Circuit Assembly. *Biosensors and Bioelectronics*, **20**, 606-619. http://dx.doi.org/10.1016/j.bios.2004.03.029

[12] Ramos, A., Morgan, H., Green, N.G. and Castellanos, A. (1998) AC Electrokinetics: A Review of Forces in Microelectrode Structures. *Journal of Physics D*: *Applied Physics*, **31**, 2338. http://dx.doi.org/10.1088/0022-3727/31/18/021

[13] Green, N.G., Ramos, A., Gonzalez, A., Morgan, H. and Castellanos, A. (2002) Fluid Flow Induced by Nonuniform AC Electric Fields in Electrolytes on Microelectrodes. III. Observation of Streamlines and Numerical Simulation. *Physical Review E*, **66**, Article ID: 026305. http://dx.doi.org/10.1103/PhysRevE.66.026305

[14] Comsol 4.3a. http://www.comsol.com

[15] Green, N.G. and Morgan, H. (1997) Dielectrophoretic Investigations of Sub-Micrometre Latex Spheres. *Journal of Physics D: Applied Physics*, **30**, 2626. http://dx.doi.org/10.1088/0022-3727/30/18/018

[16] Bown, M.R. and Meinhart, C.D. (2006) AC Electroosmotic Flow in a DNA Concentrator. *Microfluidics and Nanofluidics*, **2**, 513-523. http://dx.doi.org/10.1007/s10404-006-0097-4

[17] Tuukkanen, S., Kuzyk, A., Toppari, J.J., Häkkinen, H., Hytönen, V.P., Niskanen, E., Rinkiö, M. and Törmä, P. (2007) Trapping of 27 bp - 8 kbp DNA and Immobilization of Thiol-Modified DNA Using Dielectrophoresis. *Nanotechnology*, **18**, Article ID: 295204. http://dx.doi.org/10.1088/0957-4484/18/29/295204

[18] Regtmeier, J., Eichhorn, R., Bogunovic, L., Ros, A. and Anselmetti, D. (2010) Dielectrophoretic Trapping and Polarizability of DNA: The Role of Spatial Conformation. *Analytical Chemistry*, **82**, 7141-7149. http://dx.doi.org/10.1021/ac1005475

20

# Heat Transfer Characteristics of Work Fluid Including Phase Change Material That Flow into Heating Surface from Narrow Path

**Shin-Ichi Morita[1], Yasutaka Hayamizu[1], Takanobu Yamada[2], Akihiko Horibe[3], Naoto Haruki[3], Toshiaki Setoguchi[4], Kazuma Adachi[5]**

[1]Department of Mechanical Engineering, National Institute of Technology, Yonago College, Yonago, Japan
[2]Department of Mechanical Engineering, Kitami Institute of Technology, Kitami, Japan
[3]Graduate School of Natural Science and Technology, Okayama University, Okayama, Japan
[4]Department of Advanced Technology Fusion, Saga University, Saga, Japan
[5]Graduate School of Machine Creative Engineering, Nagaoka University of Technology, Nagaoka, Japan
Email: s-morita@yonago-k.ac.jp

## Abstract

**Use of the low temperature (less than 100°C) energy contributes to effective use of heat resources. The cost recovery by power generation is difficult by using an existing system (the binary cycle or the thermoelectric conversion element), because the initial investment is large. The final purpose of this research is development of the low temperature difference drive engine supposing use in a hot-springs resort as a power source for electric power generation. In order that a traveler may look at and delight a motion of an engine, it is made to drive at low-speed number of rotations. An engine cycle of this study is aimed at the development of Stirling cycle engine which can maintain high efficiency in small size. This kind of engine has simple structure; it brings low cost, and it is easy to perform maintenance. However, it is difficult to obtain enough output by this type of engine, because of its low temperature difference. This paper deals with the heat transfer characteristic that the working fluid including a phase change material flows into the heating surface from the narrow path. In order to increase the amount of the heat transmission, Diethylether is added to the working fluid. Diethylether is selected as a phase change material (PCM) that has the boiling point which exists between the heat source of high temperature and low temperature. The parameters of the experiment are additive amount of PCM, rotational speed of the displacer piston and temperature of heat transfer surface. It is shown that it is possible to make exchange of heat amount increase by adding phase change material. The result of this research shows the optimal condition of the difference in temperature in heat processing, number of revolutions, and addition concentration of PCM.**

## Keywords

**Phase Change Material, Heat Transfer, Low Temperature Difference, Working Fluid**


# 1. Introduction

Many attentions have been paid to the researches which use the low-temperature heat source for the purpose of effective use of energy. Using of the low temperature (less than 100˚C) energy contributes to effective use of heat resources. Japan has more than 3000 [1] places of hot-springs resorts with accommodations. That number is the highest in the world. The number of hot-spring resource in Japan is about 27,000 that are including the resource which has 12,000 places of more than 42˚C hot-spring temperature and about 900 places of the water vapor gas. Although the electric power generation by using the binary cycle [2] or the thermoelectric conversion element [3] is beginning to be tried at some places, since initial investment is large, a very long period is needed for cost recovery by power generation.

The final purpose of this research is development of the low temperature difference drive engine supposing use in a hot-springs resort as a motor for electric power generation. In order that a traveler may look at and delight a motion of an engine, it is made to drive at low-speed number of rotations. An engine cycle of this study is aimed at the Stirling cycle which can maintain high efficiency also in small size. This kind of engine has simple structure, it brings low cost, and it is easy to perform maintenance. However, it is difficult to obtain enough output by this type of engine, because of its low temperature difference. Authors found out the increasing of engine output by adding a little phase change material which has the boiling point between the source of high temperature and the source of low temperature in a working fluid. This paper deals with the heat transfer characteristic that the work fluid including phase change material flows into the heating surface from the narrow path. Some researches [4]-[6] on heat transfer evaluation of a working fluid have been published, but these studies were examined by using a single phase gas as working fluid. In order to increase the amount of the heat transmission, a little amount of Diethylether ($C_4H_{10}O$, molecular weight 74.12 g/mol, density 0.7134 g/cm$^3$ on liquid phase, boiling point 34.6˚C, colorless liquid), as a phase change material, is added to the working fluid. Diethylether is selected as a phase change material (PCM) that has the boiling point between the source of high temperature and low temperature.

# 2. Experimental Procedure

## 2.1. Experimental Apparatus

**Figure 1** shows the schematic diagram of the experimental apparatus of this study. The experimental apparatus consists of a test section, an actuator, two sets of constant temperature bath and data logger. The examination part is made of the clear acrylic resin, and thermal insulation panels 10 mm in thickness have been stuck on the internal surface. The inner size of the test section is 120 mm in height, 300 mm in width and 150 mm in length. The upper surface and the bottom surface of the test section are made of copper plate 3 mm in thickness, and are a cooling area and a heating side, respectively. The cooling and heating surface temperature are measured by 6 points of K type thermo couples that have 0.18 mm diameter. The heating and cooling surface temperature are controlled within ±0.2 K of the experimental condition. The pressure change on the experiment is measured by the pressure sensor (Keyence AP-43, range 0 - +1.000 MPa, resolution 0.1 kPa, response time 1 ms) that is attaching to the acrylic test sections wall. Putting the phase change material into the test section is performed using digital buret (Continuous E 2.5 ml/rev., minimum scale 0.01 ml, accuracy ±0.2%). A displacer piston that is set in the test section is connected to an actuator (Misumi single axis robot RSD2 position accuracy ±0.02 mm, maximum load 25 kg, stroke 50 - 300 mm) through a SUS seal rod (3 mm in diameter), and is moved up and down by computer control. The increasing of pressure in the test section is caused by moving of displacer piston to upper side. The process distance of a displacer piston is 60 mm. Working fluid flows through 3 mm gaps (length of 150 mm) of side wall by up-and-down motion of a displacer piston. The working fluid is flowed from upper side to bottom side by moving of displacer piston, so that the pressure of the test section is increased by heat exchange between working fluid and hot heat transfer surface.

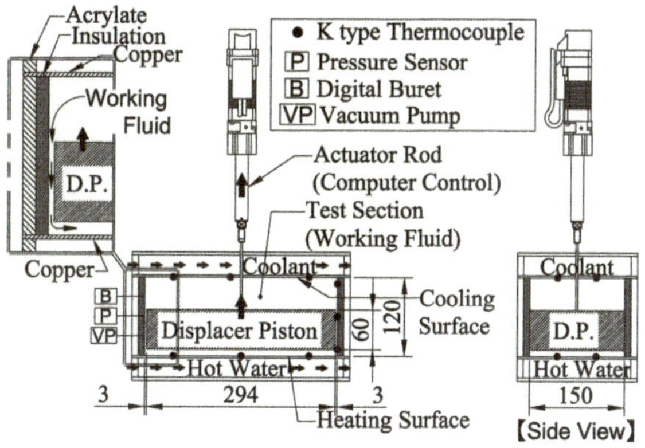


**Figure 1.** Experimental apparatus and measuring system.

## 2.2. Thermophysical Properties of Test Samples

The working fluid of this study research is air or air-Diethylether mixture gas. Diethylether ($C_4H_{10}O$) is selected as the phase change material which has the boiling point in temperature range between cooling surface and heating surface temperature. **Table 1** shows the thermophysical properties of air and Diethylether. For example, the quantity of 1.0 mass% addition at 24°C of Diethylether is 0.1114 ml or 0.07953 g.

## 2.3. Experimental Procedures

The following experimental procedures are conducted.

1) Exhaust the working fluid which remains to test section by vacuum pump.

2) Air as a working fluid of the temperature of 24°C and pressure 101.3 kPa (absolute humidity at this time is $0.0112 \pm 0.0007$ kg/kg$^*$) is introduced into test section. Diethylether of specified quantity is thrown into test section by using digital buret.

3) A cooling surface and a heating surface are set to the experimental condition temperature.

4) A displacer piston is moved to upper side from the bottom side by actuator which is dominated by computer. At this time, the pressure, heating surface temperature, and working-fluid temperature of test section are measured, and it collects in a data logger.

## 2.4. Experimental Conditions

Temperature difference of heat source is set as 20 - 70 K containing the boiling point of phase change material. The high temperature heat source is set 80°C (or less) that aims as source temperature of hot-spring. The low temperature heat source is maintained 35°C (and below). The speed of a displacer piston is set as low-speed rotation 1 - 40 rpm ($u_{mc}$ = 0.04 - 1.8 m/s). The mixed rate to the working fluid (air) of PCM (Diethylether) is evaluated in not more than 4 mass%.

**Figure 2** indicates the time history of displacer piston position and velocity at this experimental condition. The position of a displacer piston is decided by the rotation angle regardless of number of revolutions. Sine wave operation is performed in the rise process of a displacer piston, because a displacer piston on the actual engine is controlled by rotation angle. The displacer piston operates by the computer controlled actuator. The heating process times are 0.75 seconds at 40 rpm, 1.5 seconds at 20 rpm, 3.0 seconds at 10 rpm and 6.0 seconds at 5 rpm.

**Figure 3** shows the variation of mean velocity on heating surface with crank angle. The flow velocities in the graph indicate that the values change in a process on each rotation. The heat transfer characteristic in a process evaluates by the flow velocity in each time.

**Figure 4** indicates the variation of the cycle mean velocity on the heat transfer surface of the heating process. The heat transfer characteristic of the heating process in a cycle is examined using the cycle mean velocity on each number of rotations. The cycle mean velocity is calculated by next equation using process time.

$$H = St \times \sin\theta \tag{1}$$

**Table 1.** Physical properties of air & Diethylether [7].

| Working fluid | | Air | Diethylether $C_4H_{10}O$ |
|---|---|---|---|
| Boiling point °C | | −193 | 34.6 |
| Density kg/m$^3$ | *L*: Liquid | - | 713.8 |
| | *V*: Vapor | 1.181 | 3.308 |
| Heat of evaporation J/g | | - | 392.1 |
| *Cv*·J/(g·K) | | 0.7171 | 2.33:*L* |
| *λ*·W/(m·K) | | 0.00257 | 0.129:*L* |

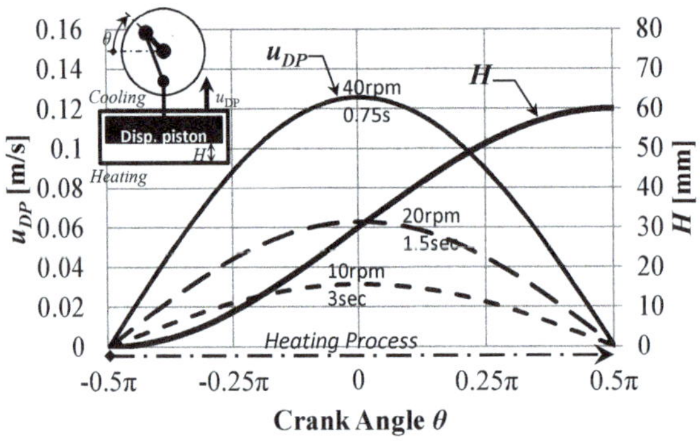


**Figure 2.** Position and velocity of displacer piston on process.

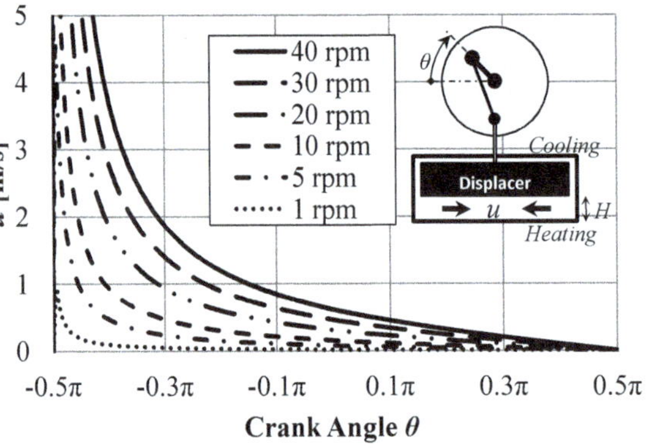


**Figure 3.** Variation of mean velocity with rotation angle.

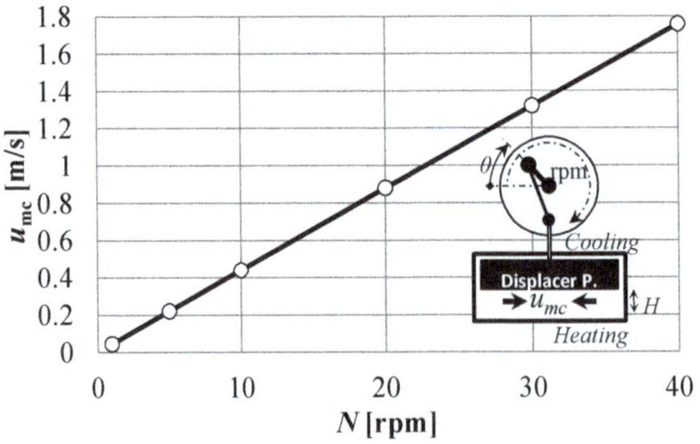


**Figure 4.** Variation of mean velocity on heating process.

$$u = \frac{N}{60}\frac{\Delta H \times W_{DP} \times D_{DP}}{2 \times H \times D_{DP}} \tag{2}$$

$$u_{mc} = \frac{1}{t_{pr}}\int_{-0.5\pi}^{0.5\pi} u\mathrm{d}\theta \quad \left(\text{Integration range } t_{pr} \geq 0.1 \text{ sec}\right) \tag{3}$$

The flow velocity on a heat transfer (heating) side is estimated by the following equations as Reynolds number. Two kinds of Reynolds-number are defined as $Re$ (using a flow velocity in a heating process) and $Re_{mc}$ (using an averaged flow velocity of the heating process on each number of rotations), respectively. The characteristic length is set to 1/2 ($L$ = 150 mm) of heating surface width, because the working fluid flows into center from both ends. The amount of additional phase change material is very little, so that the air value of dynamic viscosity is used for calculation.

$$Re = \frac{uL}{\nu_{\text{air}}} \tag{4}$$

$$Re_{mc} = \frac{u_{mc}L}{\nu_{\text{air}}} \tag{5}$$

The quantity of heat which a working fluid receives in heating process is calculated by the following equation with the application of a constant-volume change.

$$q = G_w C_w \left(\frac{P}{P_0}T_i - T_{i-1}\right) \tag{6}$$

In which, $C_w$ and $G_w$ are the specific heat at constant volume and the mass of working fluid. These values are calculated by using the thermophysical properties of composition.

$$C_w = Cv_{\text{air}}G_{\text{air}} + Cv_d G_d \tag{7}$$

$$G_w = (1 - Add) \times G_{\text{air}} + Add \times G_d \tag{8}$$

The quantity of heat in the heating process which a working fluid receives is calculated by next integral equation.

$$Q = \int_{-0.5\pi}^{0.5\pi} q\mathrm{d}\theta \tag{9}$$

$$q_{mc} = \frac{Q}{t_{pr}} \tag{10}$$

A heat transfer coefficient and Nusselt number are defined as follows.

$$\alpha = \frac{q}{S(T_h - T_{w0})} \tag{11}$$

$$Nu = \frac{\alpha L}{\lambda_{\text{air}}} \tag{12}$$

The averaged heat transfer coefficient and averaged Nusselt number in the heating process are defined by next equations.

$$\alpha_{mc} = \frac{1}{t_{pr}}\int_{-0.5\pi}^{0.5\pi} \alpha\mathrm{d}\theta \tag{13}$$

$$Nu_{mc} = \frac{\alpha_{mc}L}{\lambda_{\text{air}}} \tag{14}$$

The data of this study will be compared with the calculated values of next equation that is given Nusselt number of average laminar flow convective-heat-transfer on the plate [8] [9]. *Pr* is Plandtle number of air.

$$Nu_{th} = 0.664 Re^{\frac{1}{2}} Pr^{\frac{1}{3}} \tag{15}$$

# 3. Experimental Results and Discussions

## 3.1. Heat Exchange in Process

**Figure 5** shows the time history of the pressure increasing at 60 K in temperature difference of heat source, at 5 rpm in displacer piston speed. The pressure of test section is rising gently in response to the movement speed of displacer-piston. Then, pressure indicates the rapid increase that is caused by the high moving speed of a displacer-piston which shows the maximum velocity at the middle (3 seconds) of the process. At the last of process, it reaches to the maximum pressure by the small pressure increasing rate. The maximum pressures at the condition of using mixed-gas as working-fluid are larger than the condition of using air. The PCM (Diethylether) in a working fluid is changed from a liquid phase to gas phase by heating. It is certain that the maximum pressure enhancement is brought by the volume increasing by phase change.

**Figure 6** shows the variation of exchange heat $q$ and exchange heat amount of heating process $Q$ with time at 55 K in temperature difference of heat source, at 10 rpm in displacer piston speed. The exchange heat $q$ shows the maximum value at around the middle (1.5 seconds) of heating process. It is caused by the large movement speed of displacer-piston (the large flow rate of the working fluid to the heating side). The maximum of exchange heat $q$ is shown in little previous side from middle of a process (1.3 seconds). It is because the flow velocity on a heating surface is large in the first half of a process as above-mentioned. The exchange heat amount $Q$ is the integrated value of exchange heat $q$. The exchange heat is increased by integration of exchange heat, and the maximum exchange heat amount at finishing time of a process (3 seconds) is the value which is obtained heat amount of working fluid in heating process.

**Figure 7** indicates the variation of heat transfer coefficient with time at the condition of 55 K and 5 rpm. The heat transfer coefficient indicates the peak at around the middle (3.0 seconds) of heating process. It is caused by the large flow rate of the working fluid to the heating side (the large movement speed of displacer-piston). The maximum of heat transfer coefficient is indicated in little previous side from middle of a process (2.7 seconds). It is because the flow velocity on a heating surface is large in the first half of a process as above-mentioned.

**Figure 8** shows the relationship between averaged Nusselt number with averaged Reynolds number of heating process. The theoretical figure of the laminar flow *Nu* is shown in the same figure. The averaged *Nu* increases with increase of averaged Reynolds number. The flow space height keeps increase at the process, so that the flow on the heating surface of this study includes the forced convection flow. It is caused that the averaged *Nu* is larger than the theoretical figure of the laminar flow *Nu*.

## 3.2. Effect of Temperature Difference of Heat Source

**Figure 9** indicates the variation of the maximum exchange heat amount $Q_e$ with the temperature difference of heat source at the condition of 1 mass% Diethylether. A large difference in temperature brings about increase of the amount of evaporation. The maximum exchange heat amount $Q_e$ increases with increasing of the temperature difference of heat source.

## 3.3. Effect of Revolution Speed of Displacer Piston

**Figure 10** shows the variation of exchange heat amount ratio with number of revolution $N$ (process time). The exchange heat amount ratio decreases with increase of revolution number (or with decrease of process time). One or more values are measured in less than 20 rpm because of its long process time. In the range of this study, the effect of phase-change-material addition is observed at less than (or more than 1.5 seconds process time) 20 rpm of revolution number. Effect of additional ratio of PCM.

**Figure 11** indicates the variation of exchange heat amount ratio with the mass-additional-ratio. It is implied that the value of larger exchange heat amount ratio at the range with exceeding 1 is larger influence of PCM (Diethylether) addition. The exchange heat amount ratio shows the over one value at the range of less than 4 mass% on 55 - 60 K and 5 - 10 rpm.

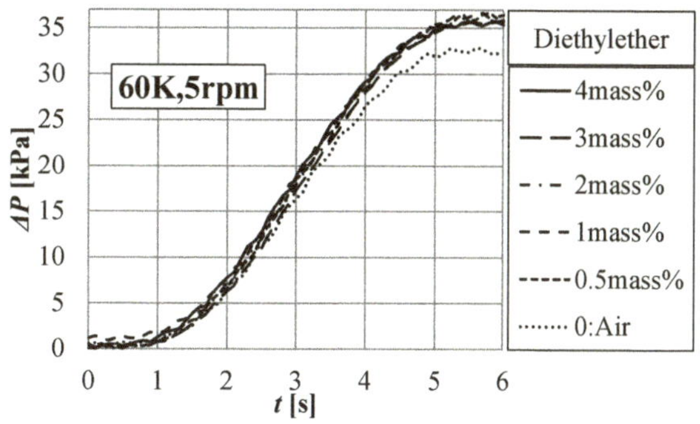


**Figure 5.** Time history of pressure increase at 60 K, 5 rpm.

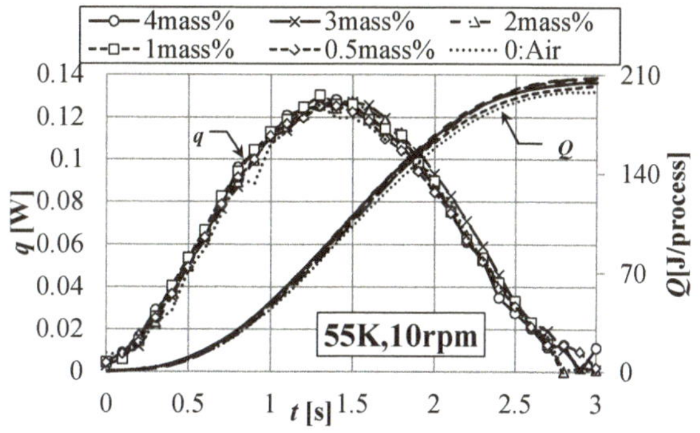


**Figure 6.** Variation of exchange heat and heat amount.

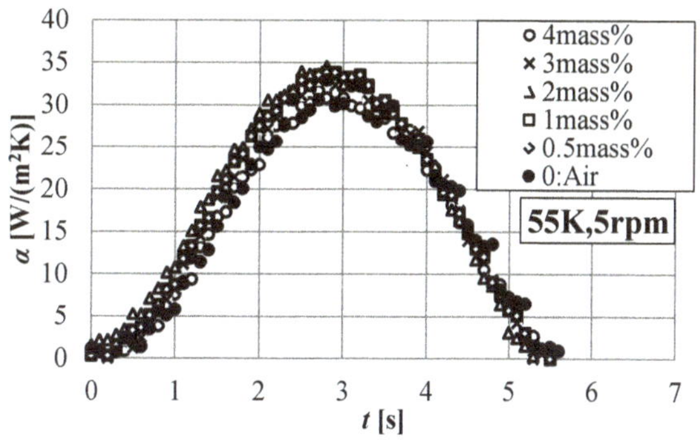


**Figure 7.** Time history of heat transfer coefficient.

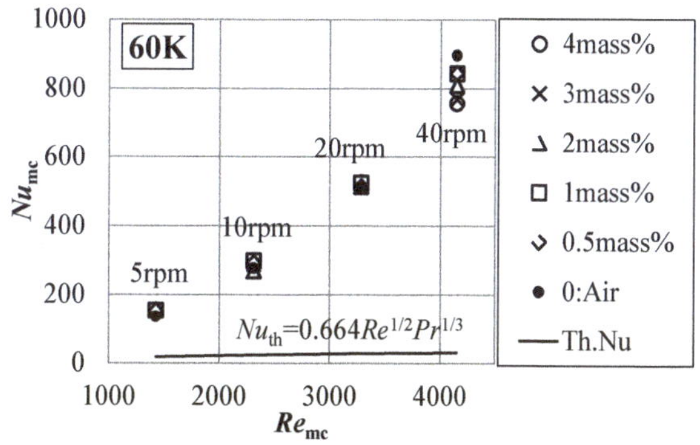


**Figure 8.** Variation of mean $Nu$ with mean $Re$ of process.

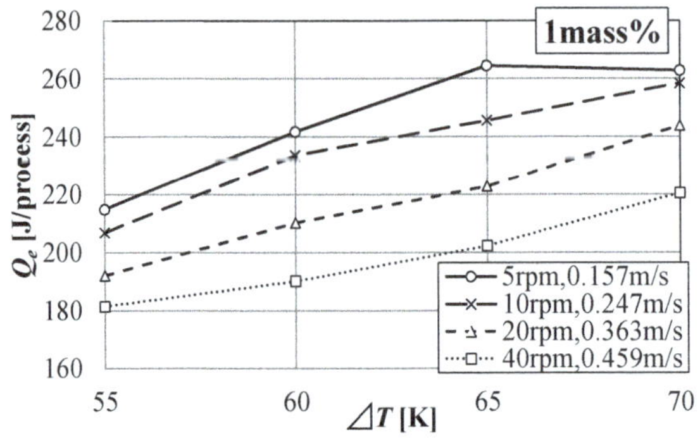


**Figure 9.** Relationship between heat $Q_e$ and $\Delta T$.

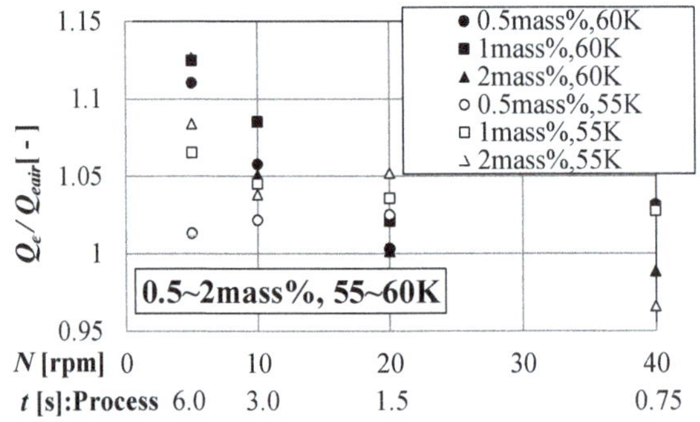


**Figure 10.** Exchange heat amount ratio with $N$ or $t$.

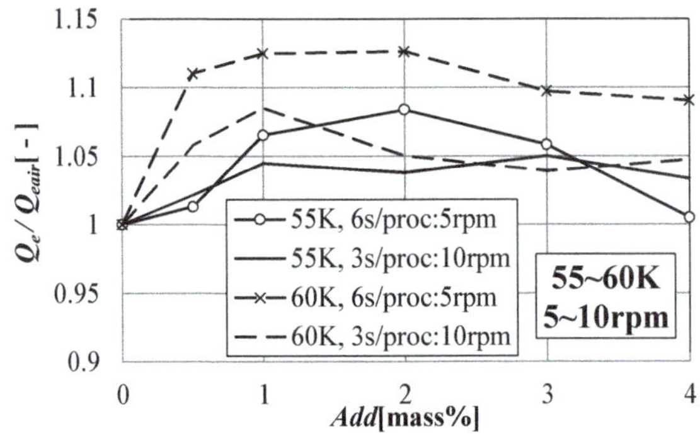


**Figure 11.** Variation of exchange heat amount ratio with *Add*.

## 4. Conclusions

Experimental study of the heat transfer by using the working fluid including PCM (Diethylether) is carried out, and the following conclusions are obtained.

1) It is suggested that the maximum of heat transfer coefficient is indicated in little previous side from middle of a process.

2) It is indicated that the exchange heat amount increases with increase of temperature difference of heat source.

3) It is clarified that the optimum ranges are 20 rpm (the number of rotations) and less than 1 mass% (the mass-additional-ratio of PCM).

## References

[1] Ministry of the Environment, Government of Japan (2012) Hot Spring Use Situation Data 2012.

[2] Osato, K., Muraoka, H. (2008) Hot Spring Binary Cycle Power Generation, the Latest Research and Development of Geothermal Energy. *Journal of Japan Institute of Energy*, **87**, 812-818.

[3] Ogawa, Y., Watamabe, H., Sakai, M. and Tunou, K. (1994) Analysis of Thermoelectric Power Generation Using Thermoelectric Element. *The Transactions of the Institute of Electronics, Information and Communication Engineers*, **J77-C-2(1)**, 34-43.

[4] Shiraki, T., Toda, F., Yamashita, Y., Ohkubo, T. and Kuribara, T. (2010) Development of Exhaust-Heat-Recovering Multiple Cylinder Type Stirling Engine. *Proceeding of 13th Stirling Cycle Symposium*, **1**, 57-58.

[5] Tahara, T. and Akazawa, T. (2010) Development of 5 kW Class Waste Heat Recovery Stirling Engine: 1st Report, Design and Trial Running of Prototype Engine. *Proceeding of 13th Stirling Cycle Symposium*, **1**, 19-20.

[6] Haramura, Y. and Nakamura, K. (2010) Heat Transfer Due to Annular Jet Induced by Displacer Motion (the Effects of Annular Channel Position and Cross Sectional Area). *Proceeding of 12th Stirling Cycle Symposium*, **1**, 81-84.

[7] Japan Society of Thermophysical Properties (2008) Thermophysical Properties Handbook. Yokendo Co. Ltd., Tokyo, 469.

[8] The Japan Society of Mechanical Engineering (2008) JSME Heat Transfer Handbook, Maruzen Co. Ltd., Tokyo, 35.

[9] The Japan Society of Mechanical Engineering (2008) JSME Data Book of Heat Transfer. 5th Edition, Maruzen Co. Ltd., Tokyo, 40.

## Nomenclature

| | | | |
|---|---|---|---|
| *Add* | Mass additional ratio to Air (mass%) | **Greek letters** | |
| *C* | Specific heat (J/(kg·K)) | $\alpha$ | Heat transfer coefficient (W/(m$^2$·K) ) |
| *D* | Depth (m) | $\nu$ | Kinematic viscosity (m$^2$/s) |
| *G* | Mass (kg) | $\theta$ | Rotation angle (rad) |
| *H* | Height between displacer piston and heat transfer surface (m, mm) | $\lambda$ | Thermal conductivity (W/(m·K) ) |
| *L* | Length (m) | **Subscripts** | |
| *N* | Number of revolutions (rpm) | 0 | Initial |
| *Nu* | Nusselt number (-) | *air* | Air |
| *P* | Pressure (kPa) | *c* | Cycle process |
| $\Delta P$ | Pressure difference (kPa) | *d* | Diethylether |
| *Pr* | Plandtle number (-) | *DP* | Displacer pinton |
| *q* | Exchange heat per time (J/s) | *e* | End of process |
| *Q* | Exchange heat on heating process (J/process) | *h* | Heat transfer surface |
| $S (=W_h \times D)$ | Area of heat transfer surface (m$^2$) | *i* | Data step |
| *St* | Stroke (60 mm) | *m* | Mean |
| *t* | Time (s) | *max* | Maximum |
| *T* | Temperature (°C) | *pr* | Process |
| $\Delta T$ | Temperature difference of heat source (K) | *th* | The oritical |
| *u* | Mean velocity on heat transfer surface (m/s) | *v* | Constant volume |
| *W* | Width (m) | *w* | Working fluid (Air, Air + PCM) |

# 21

# Numerical Study of RBC Motion and Deformation through Microcapillary in Alcohol Plasma Solution

**Aleksey Ni[1], Taqi Ahmad Cheema[2], Cheol Woo Park[1*]**

[1]School of Mechanical Engineering, Kyungpook National University, Daegu, Korea
[2]Department of Mechanical Engineering, GIK Institute of Engineering Sciences and Technology, Topi, Pakistan
Email: [*]chwoopark@knu.ac.kr

## Abstract

**Alcohol influences human health condition by starving red blood cells (RBCs) of oxygen, which results in poor blood circulation. Starved RBCs clump together and restrict blood flow, especially in capillaries. In this study, a finite element method-based moving mesh technique was applied to simulate the motion and deformation of a single RBC under different flow conditions. A 2-D model of a single RBC floating in plasma-alcohol solution was created using Arbitrary Lagrangian-Eulerian (ALE) method with moving mesh for a fluid structure interaction problem. Cell deformability and stability were studied in an alcoholic plasma solution at different fluid flow conditions. Poor blood circulation was observed with RBC tending to rotate and oscillate at low flow rates. Moreover, RBC exhibited a parachute shape while moving without oscillation, which indicated improved micro-circulation at increased flow rates. In both cases, RBC exhibited a parachute shape while moving through micro-channel at increased flow rates. The simulation also showed the significant increase of RBC deformability with the increasing viscosity of plasma as a result of alcohol presence in blood.**

## Keywords

**Moving Mesh, Fluid Structure Interaction, Alcohol, RBC, Microvessel, Deformability, Micro-Circulation**

[*]Corresponding author.

## 1. Introduction

A healthy and long life is a prime concern for every human being. Among the major threats to a long lifespan are circulatory diseases, which are related to the circulation of blood and its components (*i.e.*, blood cells and plasma) in the body. Hematocrit in human arteries is about 40%, and this value is even lower in micro-vessels, such as capillaries. An abnormality in the morphology of healthy red blood cells (RBCs) and blood flow velocity may lead to a circulatory disease, which may increase in severity at the micro-scale. Many experimental and numerical studies have been conducted to improve the current knowledge and understanding of the interaction between RBCs and plasma in micro-vessels [1]-[3].

Under normal conditions, RBCs are biconcave-shaped discs of about 8 μm in diameter. The membrane of the cell is highly deformable to external forces and returns to its original shape after the removal of forces [4]. Blood flow rheology in micro vessels is affected considerably by RBC deformability, which may be determined by the mechanical properties of the membrane and its internal fluid. In general, the degree of RBC deformability is controlled by the flexibility of the cell membrane. Therefore, numerous measurements and simulations of cell membrane rheology have been performed on populations of RBCs and at the single-cell level [5] [6].

Alcohol is a factor that influences human health condition; its presence in the body starves RBCs of oxygen, thickens the blood, and kills cells, resulting in poor blood circulation [7] [8]. Starved RBCs clump together and restrict the blood flow in small blood vessels and capillaries. Over time, alcohol intake reduces the production of RBCs, which are essential in maintaining oxygen supply to the skin, muscles, liver, and pancreas. Previous studies have focused on the influence of alcohol on blood flow, RBCs, and plasma [9] [10]. Experimental studies that have been conducted have focused on the effects of the mechanical properties of RBCs under alcoholic influence; however, these earlier works restricted their analysis only to RBC deformability under different alcoholic concentrations without tracking RBCs in a micro-vessel [11]. Although numerical simulation tools are fully capable of tracking RBC motion and deformation in a micro-vessel, no numerical study has ever considered the alcoholic influence on RBC motion and deformation in a microcapillary using the fluid structure interaction (FSI). Therefore, the current work aims to fill the research gap on the mechanics of RBCs under alcoholic influence in a microcapillary using FSI technique to investigate motion and deformation.

In this study, a finite element method-based moving mesh technique was applied to simulate the motion and deformation of a single RBC under different flow conditions. The effects of blood plasma viscosity and velocity due to alcohol consumption on RBC behavior and deformability while moving through a microcapillary channel were investigated. This study could provide theoretical knowledge on RBC behavior at the micro-scale level.

## 2. Mathematical Modeling and Governing Equations

A 2-D simulation model was created for RBC motion in plasma fluid between two parallel plates. The length of flow channel L and the distance between the plates D are 100 and 10 μm, respectively. RBC was placed 15 μm from the inlet of the channel, as shown in **Figure 1**.

### Governing Equations

The interaction of an incompressible fluid and rigid structure belongs to the class of fluid-structure interaction problems. The equation for the motion of an incompressible Newtonian liquid is described by the Navier-Stokes equations:

$$\rho\frac{\partial \mathbf{u}_{\text{fluid}}}{\partial t}+\rho\left(\mathbf{u}_{\text{fluid}}\cdot\nabla\right)\mathbf{u}_{\text{fluid}}=\nabla\cdot\left[-p\mathbf{I}+\mu\nabla\mathbf{u}_{\text{fluid}}+\left(\nabla\mathbf{u}_{\text{fluid}}\right)^{\mathrm{T}}\right]+\mathbf{F} \tag{1}$$

$$\rho\left(\nabla\cdot\mathbf{u}_{\text{fluid}}\right)=0 \tag{2}$$

where $\mathbf{u}_{\text{fluid}}$ is fluid velocity, $\rho$ is fluid density, $\mathbf{I}$ is identity tensor, $\mu$ is viscosity, $p$ is pressure, and $\mathbf{F}$ is body force or volumetric force.

The behavior of the cell and membrane interaction with fluid is described by the following elastodynamic equation:

$$\rho\frac{\partial^2 \mathbf{u}_{\text{solid}}}{\partial t^2}-\nabla\cdot\sigma=\mathbf{Fv} \tag{3}$$

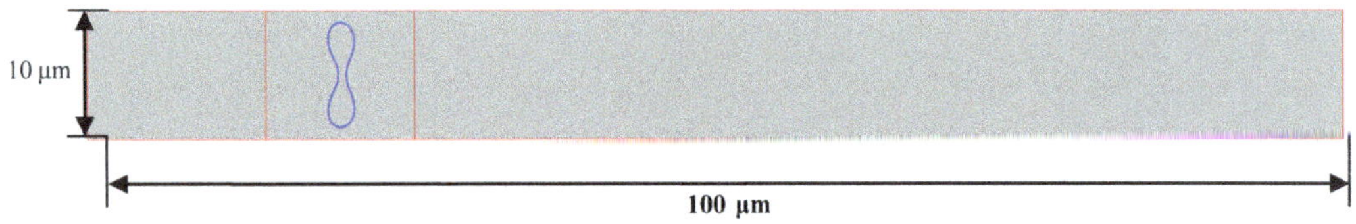


**Figure 1.** Schematic view of 2-D model for RBC in a microvessel (units: μm).

where $\sigma$ is the Cauchy stress tensor, and Fv represents the force per unit volume.

The hyperelastic smoothing method was used to simulate the mesh deformation:

$$W = \int_{\Omega} \frac{\mu}{2}(I_1 - 3) + \frac{\kappa}{2}(J - 1)^2 \, \mathrm{d}V \tag{4}$$

where $\mu$ and $\kappa$ are artificial shear and bulk moduli, respectively, and the invariants $J$ and $I_1$ are given by

$$J = \det\left(\nabla_{X^x}\right)$$
$$I_1 = J^{-\frac{2}{3}} \mathrm{tr}\left(\nabla_{X^x}\right)^{\mathrm{T}} \nabla_{X^x} \tag{5}$$

The fluid and solid sub-domains of the cell can be coupled via one-way or two-way method. Coupling is one-way if the changes in one physics affect the other and in return the second physics does not influence the first one. However, if the influenced physics also affects the first one, then this is called a two-way coupling. The problem becomes more complicated if cell motion is induced by fluid flow, and at the same time, the fluid flow pattern is influenced by cell motion. Variation in the pressure and viscous forces on the particle surface leads to a rise in the forces acting on RBC and can be computed in two ways:

1) Integration of fluid stresses along the RBC contour, *i.e.*, integration of forces acting on the vicinity of the cell.

2) Integration of all the forces and stresses along the boundaries of the computational domain, *i.e.*, forces on the body in terms of the momentum flux far from the body, which does not require knowledge of the flow in the vicinity of the body.

A constant and uniform velocity at the inlet was applied to the RBC. At the outlet, zero pressure was applied, and a no-slip wall condition was set up along the inner walls.

## 3. Material Properties

The cell was assumed to have hyperelastic properties, which is defined as a material that can completely recover its original shape. The model used in the present work involves an elastic cell with a membrane structure filled with cytoplasm having plasma density and viscosity equal to 6e−3 Pa-s. The literature values for Young's modulus of RBC vary from 26 kPa to 54 kPa; the value of 30 kPa was chosen for use in this study as an average value [12]. The value of Poisson's ratio used during the simulation was 0.49.

The blood vessel was assumed to be rigid, and plasma was assumed as homogeneous. Incompressible Newtonian fluid was set to have viscosity of $\mu$ = 1.2e−3 Pa-s and density of $\rho$ = 1057 kg/m$^3$ (assuming the presence of alcohol at 1 ml/kg). The Arbitrary Lagrange-Euler (ALE) mesh was used to simulate the moving part of the fluid due to large deformation of the cell. The cytoplasm was considered to have the same density as plasma, and with viscosity of $\mu$ = 6e−3 Pa-s [13].

## 4. Numerical Implementation

In solving the stated problem, a commercial code for finite element analysis known as COMSOL Multiphysics (V 4.2a) was utilized. A 2-D model with dynamic meshes was created, and a fully coupled solver was used by coupling a solid and a fluid domain together via a direct two-way coupling formulation. Thus, fluid forces deformed the solid domain, and in turn, deformed solid affected the fluid domain. **Figure 1** shows the model used for simulating the motion and deformation of a single RBC.

First, a transient analysis of a single RBC in the microcapillary vessel filled with fluid was carried out by using direct computation. The movement of the cell was restricted only to a small displacement before the mesh deformed. Many inverted coordinates were present, and mesh quality became too low to ensure accurate computation. A prescribed mesh displacement methodology was used to overcome this problem. For this purpose, the cell was placed into a rectangular frame that was given a displacement, and at the same time, fluid structure interaction between the RBC and fluid was calculated inside the rectangular frame. In this case, the mesh was not skewed, and an acceptable quality remained until the end of the simulation. The forces acting on the cell were obtained by integrating the entire surface. Deformation plots were obtained using the post-processing tools. **Figure 2** shows the schematic of the model with triangular meshes. **Figure 2(a)** illustrates the initial condition of a cell in the mesh, and **Figure 2(b)** represents the deformed moving cell and mesh.

## 5. Results and Discussion

The 2-D model of a single moving RBC through a microvessel under normal and abnormal velocities and viscosities induced by alcohol was created using ALE method for finite element analysis. Simulations with different velocity and viscosity magnitudes were performed to distinguish varied RBC behaviors during its motion through the microcapillary.

The simulation results showed that microcirculation can be improved by increasing flow velocity. With velocity of $v$ = 0.001 m/s, RBC moved along the length of the channel with a deformed bio-concave so-called parachute shape, as shown in **Figure 3(a)**. With a velocity of $v$ = 0.0001 m/s, RBC started rotating with a partially deformed bio-concave shape, and after the longitudinal axis of the RBC nearly paralleled the flow direction, it began oscillating about this position, as shown in **Figure 3(b)**. Reduced blood flow velocity results in poor microcirculation, which leads to oxygen and nutrient deficiency in tissues and organs.

In implementing the effects of alcohol-induced reduced viscosity on the behavior of RBC, the viscosity of plasma was set to 5e−3 Pa-s, and for cytoplasm, it was 6e−3 Pa-s. For the same value of velocity and with different viscosities of plasma and cytoplasm, the deformability of RBCs increases [14]. **Figure 4** illustrates that by changing the cytoplasm-to-plasma viscosity ratio, RBC behavior changes. **Figure 4(a)** shows the total displacement of the moving cell, as well as the extent of cell deformation while moving through the vessel in case of velocity of 0.001 or 0.0001 m/s, respectively. The figures confirm the increased deformation of RBC compared with that in **Figure 3(a)**. **Figure 4(b)** also presents increased cell deformation, and RBC shows more stable motion through the capillary. However, a rare part of RBC in **Figure 4(a)** and **Figure 4(b)** are smaller compared with those in **Figure 3(a)** and **Figure 3(b)**.

**Figure 5** and **Figure 6** represent the shear rate plot at different time steps depending on the inlet velocity and viscosity of the plasma flowing through the microcapillary. For the same flow conditions, an increase in viscosity of the plasma solution decreased the shear rate by half, becoming constant after 0.03 s. On the other hand, an increase in velocity increased the shear rate up to 0.03 s for the same viscosity. These two results verify the relation between velocity and viscosity.

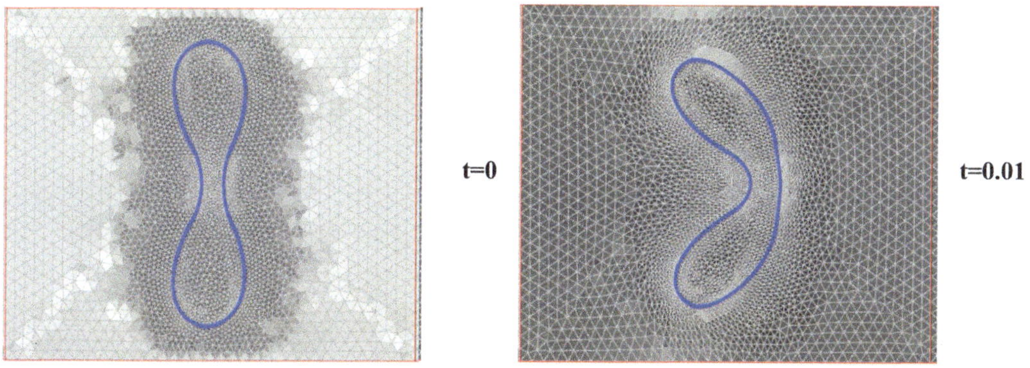


(a) (b)

**Figure 2.** Schematic of the moving mesh used for calculation of a single cell motion through microvessel. (a) Initial condition of a RBC; (b) Condition of a RBC and mesh deformation with t = 0.01 under velocity $v$ = 0.001 m/s.

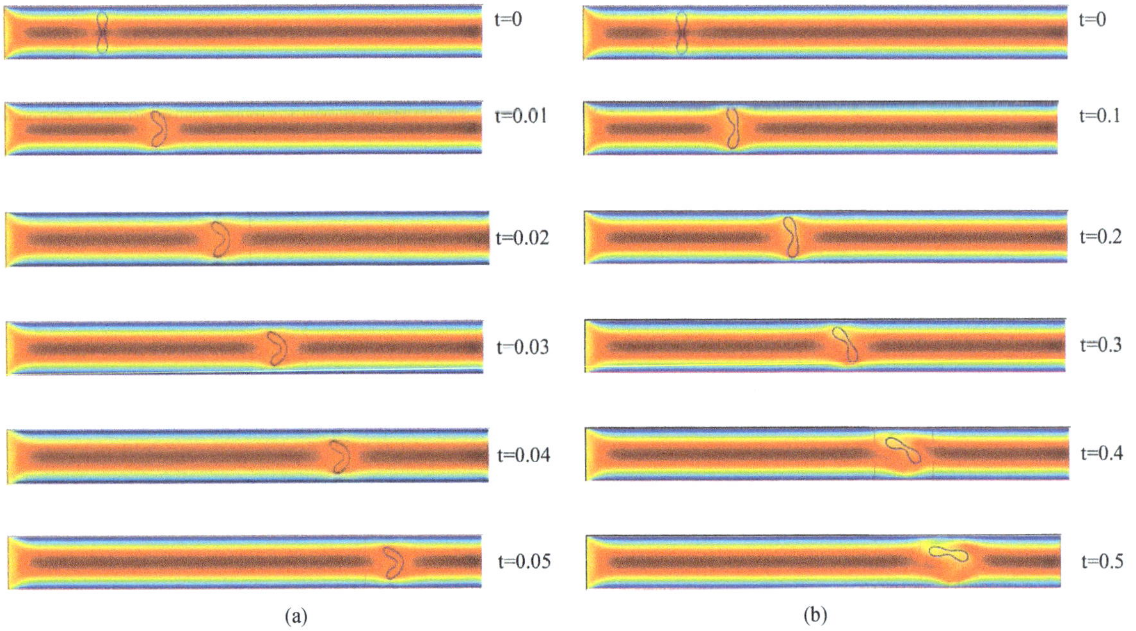


**Figure 3.** RBC motion in microcapillary for plasma viscosity $\mu$ = 1.2e−3 kg/m·s at (a) Inlet velocity $v$ = 1e−3 m/s; (b) Inlet velocity $v$ = 1e−4 m/s.

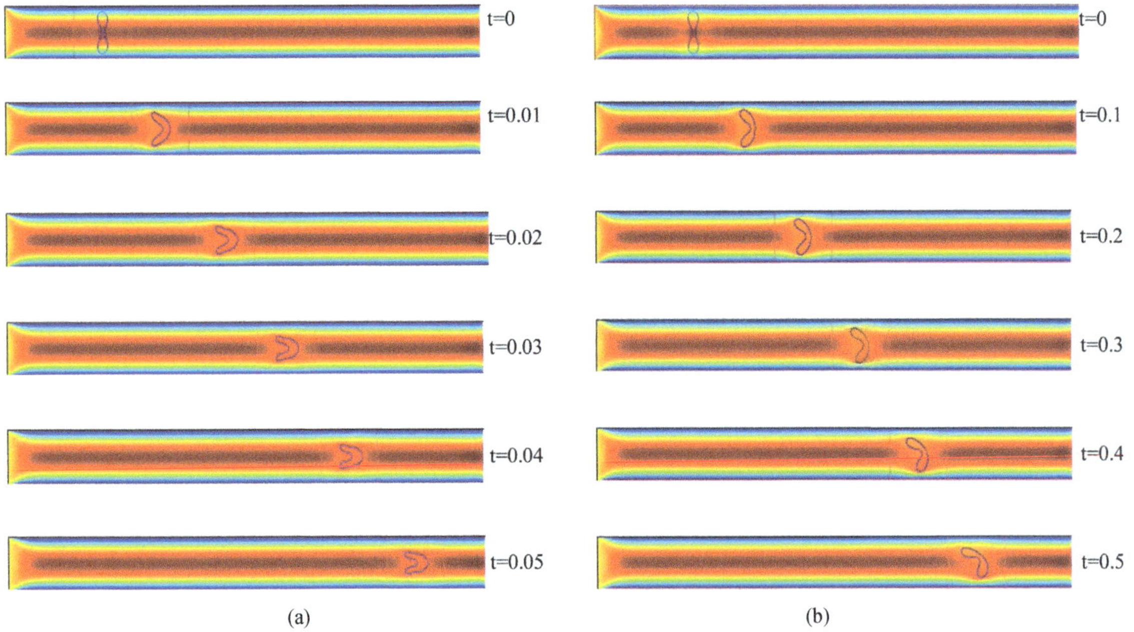


**Figure 4.** RBC motion in microcapillary for plasma viscosity $\mu$ = 5e−3 kg/m·s at (a) Inlet velocity $v$ = 1e−3 m/s; (b) Inlet velocity $v$ = 1e−4 m/s.

**Figure** 7 illustrates the variation of the ratio of transverse diameter over axial diameter with respect to time for different cytoplasm viscosity and plasma velocity values. Alcohol influences blood velocity. The physiological and pharmacological actions of ethanol are influenced by molecules penetrating into the membrane, which leads to changes in shape, function, and interactions of the membrane proteins [15]. The proposed model could shed light on the issue of how the different factors influence RBC behavior in a microvessel. In this approach, the RBC was represented by a 2-D viscoelastic element filled with liquid, namely, cytoplasm. The 2-D model allowed significant reduction of the calculation time and simplification of modeling. In the present model, the

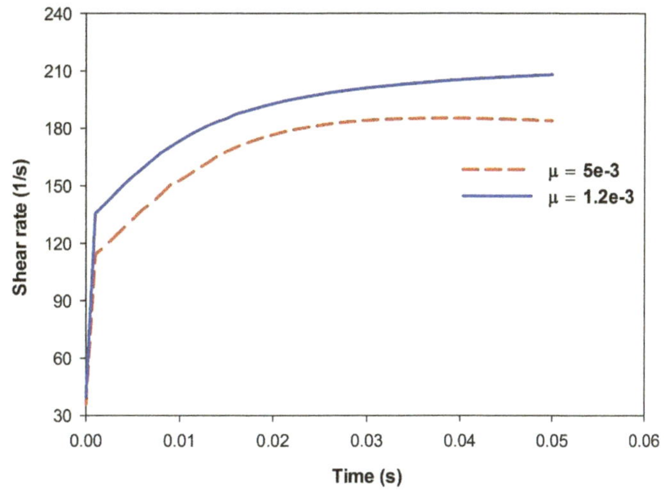


**Figure 5.** Graphical plot for plasma shear rate for different viscosities at $v$ = 1e−4 m/s.

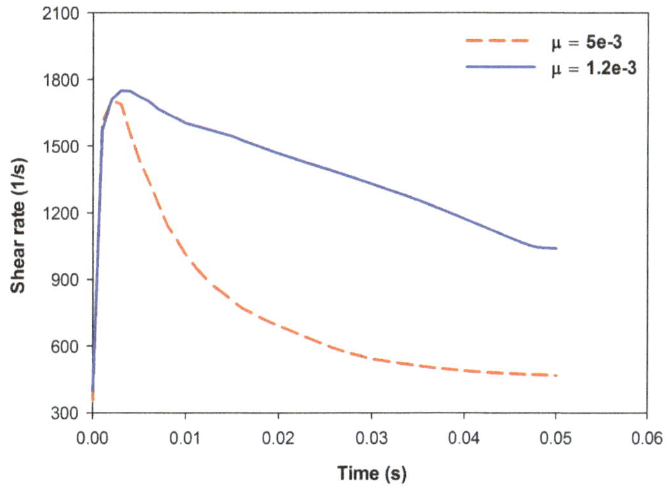


**Figure 6.** Graphical plot for plasma shear rate for different viscosities at $v$ = 1e−3 m/s.

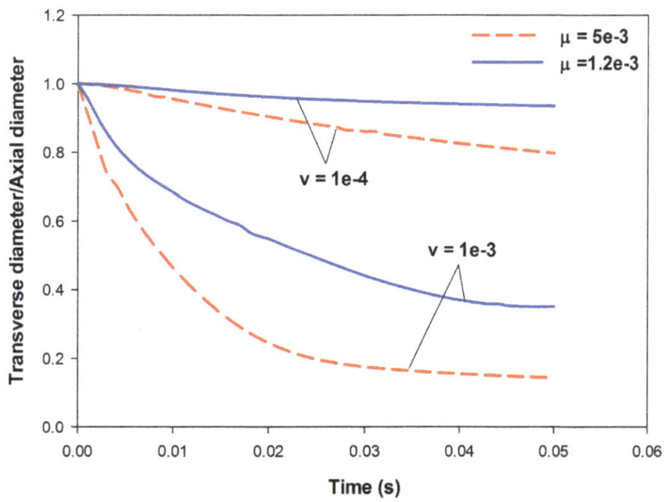


**Figure 7.** Variation of the ratio of transverse diameter over axial diameter with respect to time.

moving mesh technique with physics-controlled mesh was utilized to calculate the mechanical behavior of the single RBC moving through a microcapillary. The model aided the investigation on the relationship among cell deformability and various other parameters, such as plasma viscosity, blood flow velocity, and material properties.

The model successfully represented the phenomenon of fluid structure interaction of RBC floating in plasma through a microcapillary vessel. It can be useful for simulation of fully coupled problems, such as fluid and cell interaction, as it takes into account many parameters during calculation, which is helpful in understanding the blood flow mechanism from a micro-scale point of view. Although the model is simplified, compared with developed approaches and methods, it can be extended and applied in simulating more complex and realistic problems.

## 6. Conclusions

A 2-D model has been developed to predict shapes and behavior of a single RBC moving in a microcapillary filled with plasma alcohol solution under different velocities and elasticity parameters. This simulation provides access to a large panel of data for better understanding of the RBC, particularly involving alcohol-influenced behavior at the micro level. Moreover, it helps to explain the consequences of pathologies related to alcohol consumption. Damaged RBCs do not clot properly because the blood becomes too thin. In terms of diameter, microvessels are about one billion times larger than RBC, which make them very important for blood circulation. As such, the rheological properties and behavior of a single RBC have great influence on blood flow. Therefore, problems related to RBC behavior, such as decreased flow velocity in the microcapillary, increased viscosity, and others, will lead to severe conditions for the entire human body.

The numerical simulations are made by utilizing a time-dependent coupling: ALE equations, Navier-Stokes equations, and structural mechanics model of RBCs with non-zero thickness membrane. The obtained results for plasma flow and biconcave RBC motion reveal that the proposed model successfully simulates RBC deformation during capillary flow, and shows RBC dependence on plasma velocity and viscosity. The deformed, so-called parachute shape of RBC has been observed in experiments. Further work on this problem will allow investigating and establishing the behavior of RBC that moves through a stenotic structure in alcohol plasma solution.

## Acknowledgements

This work was supported by the Korea Research Foundation (NRF) of Korea, a grant funded by the Korean Government (MEST) (No. 2012R1A2A2A01046099), and a grant from the Priority Research Centers Program through NRF funded by MEST (No. 2010-0020089).

## References

[1] Jeong J.H., Sugii Y., Minamiyama, M. and Okamoto, K. (2006) Measurement of RBC Deformation and Velocity in Capillaries *in Vivo*. *Microvascular Research*, **71**, 212-217. http://dx.doi.org/10.1016/j.mvr.2006.02.006

[2] Secomb, T.W., Styp-Recowska, B. and Pries, A.R. (2007) Two-Dimensional Simulation of Red Blood Cell Deformation and Lateral Migration in Microvessels. *Annals of Biomedical Engineering*, **35**, 755-765. http://dx.doi.org/10.1007/s10439-007-9275-0

[3] Tsubota, K., Wada, S. and Yamaguchi, T. (2006) Particle Method for Computer Simulation of Red Blood Cell Motion in Blood Flow. *Computer Methods and Programs in Biomedicine*, **83**, 139-146. http://dx.doi.org/10.1016/j.cmpb.2006.06.005

[4] Fischer, T.M. (2004) Shape Memory of Human Red Blood Cells. *Biophysical Journal*, **86**, 3304-3313. http://dx.doi.org/10.1016/S0006-3495(04)74378-7

[5] Secomb, T.W. (1987) Flow-Dependent Rheological Properties of Blood in Capillaries. *Microvascular Research*, **34**, 46-58. http://dx.doi.org/10.1016/0026-2862(87)90078-1

[6] Zhang Z.G. and Zhang, X.W. (2011) Mechanical Behavior of the Erythrocyte in Microvessel Stenosis. *Science China Life Sciences*, **54**, 450-458. http://dx.doi.org/10.1007/s11427-011-4152-3

[7] Gdovinova Z., Vesela, D., Eiben, E., Kafka, J. and Shahpesandy, H. (1996) Transcranial Doppler Sonography in Patients with Syndrome of Alcohol Dependence. *European Journal of Neurology*, **3**, 134.

[8] Gdovinova, Z. (2001) Blood Flow Velocity in the Middle Cerebral Artery in Heavy Alcohol Drinkers. *Alcohol and*

*Alcoholism*, **36**, 346-348. http://dx.doi.org/10.1093/alcalc/36.4.346

[9] Hillbom, M., Kaste, M. Tarssanen, L. and Johnson, R. (1983) Effect of Ethanol on Blood Viscosity and Erythrocyte Flexibility in Healthy Men. *European Journal of Clinical Investigation*, **13**, 45-48. http://dx.doi.org/10.1111/j.1365-2362.1983.tb00063.x

[10] Beauge, F., Niel, E., Perrotin, R., Thepot, V., Boynard, M. and Nalpas, B. (1994) Red Blood Cell Deformability and Alchol Dependence in Humans. *Alcohol and Alcoholism*, **29**, 59-63.

[11] Sonmez, M., Ince, H.Y., Meiselman, H.J. and Baskurt, O.K. (2012) The Effect of Alcohols on Red Blood Cell Mechanical Properties Depends on Their Molecular Size. *ISB and ISCH* 2012.

[12] Rotsch, C., Jacobson, K. and Radmacher, M. (1999) Dimensional and Mechanical Dynamics of Active and Stable Edges in Motile Fibroblasts Investigated by Using Atomic Force Microscopy. *Proceedings of the National Academy of Sciences of the United States*, **3**, 921-926. http://dx.doi.org/10.1073/pnas.96.3.921

[13] Ramanujan, S. and Pozdrikidis, C. (1998) Deformation of Liquid Capsules Enclosed by Elastic Membranes in Simple Shear Flow: Large Deformations and the Effect of Capsule Viscosity. *Journal of Fluid Mechanics*, **361**, 117-143. http://dx.doi.org/10.1017/S0022112098008714

[14] Gdovinova, Z. (2006) Cerebral Blood Flow Velocity and Erythrocyte Deformability in Heavy Alcohol Drinkers at the Acute Stage and Two Weeks after Withdrawal. *Drug and Alcohol Dependence*, **81**, 207-2013. http://dx.doi.org/10.1016/j.drugalcdep.2005.07.006

[15] Tadahiro, O. and Kanako, S. (2000) Ethanol Improves Decreased Filterability of Human Red Blood Cells through Modulation of Intracellular Signaling Pathways. *Alcoholism: Clinical and Experimental Research*, **24**, 352-356. http://dx.doi.org/10.1111/j.1530-0277.2000.tb04621.x

# 22

# The Large Scale Instability in Rotating Fluid with Small Scale Force

**Michael Kopp[1], Anatoly Tur[2], Vladimir Yanovsky[1]**

[1]National Academy of Science Ukraine, Institute for Single Crystals, Kharkiv University, Kharkov, Ukraine
[2]Institut de Recherche en Astrophysique et Planétologie, CNRS, Université de Toulouse [UPS], Toulouse, France
Email: anatoly.tour@irap.omp.eu

## Abstract

**In this paper, we find a new large scale instability in rotating flow forced turbulence. The turbulence is generated by a small scale external force at low Reynolds number. The theory is built on the rigorous asymptotic method of multi-scale development. The nonlinear equations for the instability are obtained at the third order of the perturbation theory. In this article, we explain the nonlinear stage of the instability and the generation vortex kinks.**

## Keywords

**Large Scale Vortex Instability, Coriolis Force, Multi-Scale Development, Small Scale Turbulence, Vortex Kinks**

## 1. Introduction

It is well known, that the rotating effects play an important role in many practical and theoretical applications for fluid mechanics [1] and are especially important for geophysics and astrophysics [2] [3] when one has to deal with rotating objects such as Earth, Jupiter and Sun. Rotating fluids could generate different waves and vortex motions, for example, gyroscopic waves, Rossby waves, internal waves, located vortices and coherent vortex structures [4]-[7]. Among the vortex structures, the most interesting are the large scale ones, since they carry out the efficient transport of energy and impulse. The structures which have characteristic scale much more than the scale of turbulence or of the external force which generates this turbulence, are understood as large scale ones. At present, we can state that there are a lot of instabilities which generate the large scale vortex structures (see for example [8]-[14]), in particular, in rotating fluid with the non-homogeneous turbulence [15]. In this work, we find the new large scale instability in rotating fluid, under impact of small external force which keeps up turbulent fluctuations. The nonlinear large scale helical vortex structures of Beltrami type or localized kinks with internal helical structure appear as a result of the development of this instability in rotating fluid. We can consider that external small scale force substitutes the action of small scale turbulence. It is supposed that external force is in plane ($X$, $Y$), which is perpendicular to the rotation axis, for example, axis $Z$ is directed along the vector of angular velocity of rotation $\mathbf{\Omega}$. Helical 2D field of velocity $W_x$, $W_y$ turns around axis $Z$, when $Z$ changes in the kink which links the hyperbolic point and the stable focus (**Figure 1**). Moreover, this field does some turns in the kink, which links instable and stable focuses (**Figure 2**). The found instability belongs to the class of instabilities called hydrodynamic $\alpha$-effect. For these instabilities, the positive feedback between velocity components of $W_x$, $W_y$ is typical.

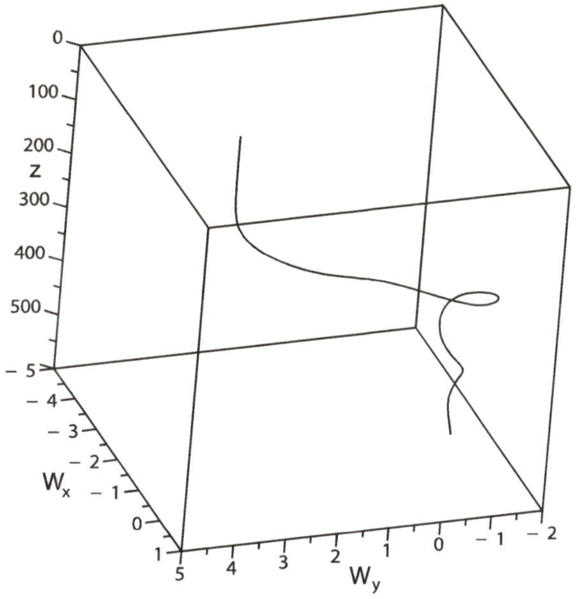


**Figure 1.** The kink which connects the hyperbolic point with stable knot with $D$ = 1, $C_1$ = 0.04, $C_2$ = 0.04. When approaching the stable knot one can see rotations of velocity field.

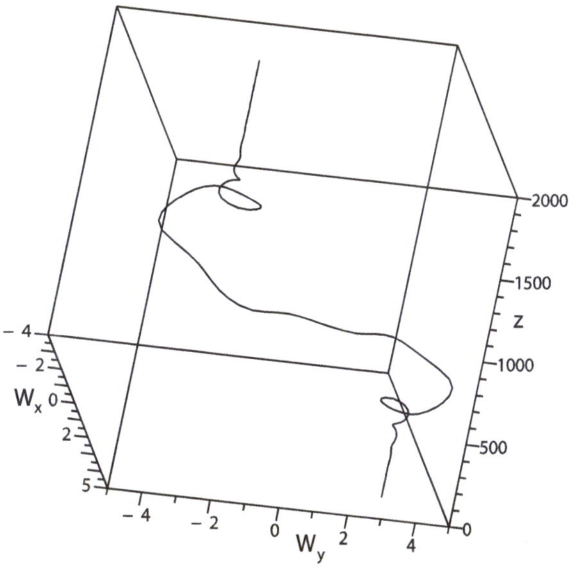


**Figure 2.** The kink which connects the instable and stable focuses with $D$ = 1, $C_1$ = 0.04, $C_2$ = 0.04. One can see the internal helical structure of the kink.

$$\partial_T W_x - \Delta W_x - \alpha_y \frac{\partial}{\partial z} W_y = 0,$$

$$\partial_T W_y - \Delta W_y + \alpha_x \frac{\partial}{\partial z} W_x = 0,$$

and leads to the instability. The $\alpha$-effect is taking its origins from magnetic hydrodynamics, where it engenders the increase of large scale magnetic fields (see for example [16]). It was generalized later for ordinary hydrodynamics. For the time being some examples of hydrodynamics $\alpha$-effect [8]-[14] are already known. From this point of view, in this work we found a new example of $\alpha$-effect. The theory of this instability is developed rigourously using the method of asymptotic multi-scale development, similar to what was done by Frisch, She and Sulem for the theory of the AKA effect [13]. This method allows finding the equations for large scale perturbations as secular equations of asymptotical theory in order to calculate the Reynolds stress tensor and to find the instability. The small parameter of asymptotical development is the Reynolds number $R$, $R \ll 1$. Our paper is organised as follows: in Section 2 we formulate the problem and the main equations in rotating system of coordinates; in Section 3 we examine the principal scheme of the multi-scale development and we give the secular equations. In Section 4 we calculate the velocity field of zero approximation. In Section 5 we describe the calculations of the Reynolds stress and find the large scale instability. In Section 6 we discuss the saturation of the instability and find non linear stationary vortex structures. The results obtained are discussed in the conclusions given in Section 7.

## 2. The Main Equations and Formulation of the Problem

Let us examine the equations of motion for non-compressible rotating fluid with external force $\boldsymbol{F}_0$ in rotating coordinates system:

$$\frac{\partial \boldsymbol{V}}{\partial t} + (\boldsymbol{V}\nabla)\boldsymbol{V} + 2\boldsymbol{\Omega}\times\boldsymbol{V} = -\frac{1}{\rho_0}\nabla P + \nu\Delta\boldsymbol{V} + \boldsymbol{F}_0, \tag{1}$$

$$div\boldsymbol{V} = 0. \tag{2}$$

The external force $\boldsymbol{F}_0$ is divergence-free. Here $\boldsymbol{\Omega}$-angular velocity of fluid rotation, $\nu$-viscosity, $\rho_0$-constant fluid density. Let us design characteristic amplitude of force $f_0$, and its characteristic space and time scale $\lambda_0$ and $t_0$ respectively.

Then $\boldsymbol{F}_0 = f_0\boldsymbol{F}_0\left(\frac{\boldsymbol{x}}{\lambda_0}, \frac{t}{t_0}\right)$. We will design the characteristic amplitude of velocity, generated by external force as $v_0$. We choose the dimensionless variables $(t, \boldsymbol{x}, \boldsymbol{V})$:

$$\boldsymbol{x} \to \frac{\boldsymbol{x}}{\lambda_0}, \quad t \to \frac{t}{t_0}, \quad \boldsymbol{V} \to \frac{\boldsymbol{V}}{v_0}, \quad \boldsymbol{F}_0 \to \frac{\boldsymbol{F}_0}{f_0}, \quad P \to \frac{P}{\rho_0 P_0},$$

$$t_0 = \frac{\lambda_0^2}{\nu}, \quad P_0 = \frac{v_0\nu}{\lambda_0}, \quad f_0 = \frac{v_0\nu}{\lambda_0^2}, \quad v_0 = \frac{f_0\lambda_0^2}{\nu}.$$

Then, in dimensionless variables the Equation (1) takes forme:

$$\frac{\partial \boldsymbol{V}}{\partial t} + R(\boldsymbol{V}\cdot\nabla)\boldsymbol{V} + \boldsymbol{D}\times\boldsymbol{V} = -\nabla P + \Delta\boldsymbol{V} + \boldsymbol{F}_0, \tag{3}$$

$R = \frac{\lambda_0 v_0}{\nu}$, $|D| = \sqrt{Ta}$ where $R$ and $Ta = \frac{4\Omega^2\lambda_0^4}{\nu^2}$ are respectively the Reynolds number and the Taylor number on scale $\lambda_0$. Further we will consider the Reynolds number as small $R \ll 1$ and will construct on this small parameter the asymptitical development. Concerning the parameter $D$, we do not choose any range of values for the moment. Let us examine the following formulation of the problem. We consider the external force as being small scale and of high frequency. This force leads to small scale fluctuations in velocity. After averaging, these quickly oscillating fluctuations vanish. Nevertheless, due to small nonlinear interactions in some orders of perturbation theory, nonzero terms can occur after averaging. This means that they are not oscillatory, that is to say, they are large scale. From a formal point of view, these terms are secular, *i.e.*, they create the conditions for the solvability of a large scale asymptotic development. So the purpose of this paper is to find and study the solvability equations, *i.e.*, the equations for large scale perturbations. Let us denote the small scale variables by $x_0=(\boldsymbol{x}_0,t_0)$, and the large scale ones by $X=(\boldsymbol{X},T)$. The small scale partial derivative operation $\frac{\partial}{\partial x_0^i}$, $\frac{\partial}{\partial t_0}$, and the large scale ones $\frac{\partial}{\partial \boldsymbol{X}}$, $\frac{\partial}{\partial T}$ are written, respectively, as $\partial_i$, $\partial_t$, $\nabla_i$ and $\partial_T$. To construct a multi-scale asymptotic development we follow the method which is proposed in [16].

## 3. The Multi-Scale Asymptotic Development

Let us search for the solution to Equations (2) and (3) in the following form:

$$V(\boldsymbol{x},t)=\frac{1}{R}\boldsymbol{W}_{-1}(X)+\boldsymbol{v}_0(x_0)+R\boldsymbol{v}_1+R^2\boldsymbol{v}_2+R^3\boldsymbol{v}_3+\cdots, \tag{4}$$

$$T(\boldsymbol{x},t)=\frac{1}{R}T_{-1}(X)+T_0(x_0)+RT_1+R^2T_2+R^3T_3+\cdots, \tag{5}$$

$$P(\boldsymbol{x},t)=\frac{1}{R^3}P_{-3}(X)+\frac{1}{R^2}P_{-2}(X)+\frac{1}{R}P_{-1}(X)+P_0(x_0)+R\left(P_1+\overline{P}_1(X)\right)+R^2P_2+R^3P_3+\cdots. \tag{6}$$

Let us introduce the following equalities: $\boldsymbol{X}=R^2\boldsymbol{x}_0$ and $T=R^4t_0$ which lead to the expression for the space and time derivatives:

$$\frac{\partial}{\partial x^i}=\partial_i+R^2\nabla_i, \tag{7}$$

$$\frac{\partial}{\partial t}=\partial_t+R^4\partial_T, \tag{8}$$

$$\frac{\partial^2}{\partial x^j\partial x^j}=\partial_{jj}+2R^2\partial_j\nabla_j+R^4\partial_{jj}. \tag{9}$$

Using indicial notation, the system of equation can be written as

$$\begin{aligned}&\left(\partial_t+R^4\partial_T\right)V^i+R\left(\partial_j+R^2\nabla_j\right)\left(V^iV^j\right)+D^j\varepsilon_{ijk}V^k\\&=\left(\partial_j+R^2\nabla_j\right)P+\left(\partial_{jj}+2R^2\partial_j\nabla_j+R^4\nabla_{jj}\right)V^i+F_0^i,\end{aligned} \tag{10}$$

$$\partial_tT-\partial_{jj}T=-V^z-R\partial_j\left(V^iT\right), \tag{11}$$

$$\left(\partial_i+R^2\nabla_i\right)V^i=0. \tag{12}$$

Substituting these expressions into the initial Equations (2) and (3) and then gathering together the terms of the same order, we obtain the equations of the multi-scale asymptotic development and write down the obtained equations up to order $R^3$ inclusive. In the order $R^{-3}$ there is only the equation

$$\partial_iP_{-3}=0\Rightarrow P_{-3}=P_{-3}(X). \tag{13}$$

In order $R^{-2}$ we have the equation

$$\partial_iP_{-2}=0\Rightarrow P_{-2}=P_{-2}(X). \tag{14}$$

In order $R^{-1}$ we get a system of equations:

$$\partial_tW_{-1}^i-\partial_{jj}W_{-1}^i+D^j\varepsilon_{ijk}W_{-1}^k=-\left(\partial_iP_{-1}+\nabla_iP_{-3}\right)-\partial_jW_{-1}^iW_{-1}^j, \tag{15}$$

$$\partial_iW_{-1}^i=0.$$

The system of Equations (17) and (18) gives the secular terms

$$-\nabla_iP_{-3}=D^j\varepsilon_{ijk}W_{-1}^k, \tag{16}$$

which corresponds to a geostrophic equilibrum equation.

In zero order $R^0$, we have the following system of equations:

$$\partial_t v_0^i - \partial_{jj} v_0^i + \partial_j \left( W_{-1}^i v_0^j + v_0^i W_{-1}^j \right) + D^j \varepsilon_{ijk} v_0^k = -\left( \partial_i P_0 + \nabla_i P_{-2} \right) + F_0^i, \tag{17}$$

$$\partial_i v_0^i = 0.$$

These equations give one secular equation:

$$\nabla P_{-2} = 0 \Rightarrow P_{-2} = \text{Const.} \tag{18}$$

Let us consider the equations of the first approximation $R$:

$$\partial_t v_1^i - \partial_{jj} v_1^i + D^j \varepsilon_{ijk} v_1^k + \partial_j \left( W_{-1}^i v_1^j + v_1^i W_{-1}^j + v_0^i v_0^j \right) = -\nabla_j \left( W_{-1}^i W_{-1}^j \right) - \left( \partial_i P_1 + \nabla_i P_{-1} \right), \tag{19}$$

$$\partial_i V_1^i + \nabla_i W_{-1}^i = 0. \tag{20}$$

From this system of equations there follows the secular equations:

$$\nabla_i W_{-1}^i = 0, \tag{21}$$

$$\nabla_j \left( W_{-1}^i W_{-1}^j \right) = -\nabla_i P_{-1}. \tag{22}$$

The secular Equations (27) and (29) are satisfied by choosing the following geometry for the velocity field (Beltrami field):

$$\boldsymbol{W} = \left( W_{-1}^x (Z), W_{-1}^y (Z), 0 \right); \tag{23}$$

$$\nabla P_{-1} = 0 \Rightarrow P_{-1} = \text{Const.}$$

In the second order $R^2$, we obtain the equations

$$\begin{aligned} &\partial_t v_2^i - \partial_{jj} v_2^i - 2\partial_j \nabla_j v_0^i + \partial_j \left( W_{-1}^i v_2^j + v_2^i W_{-1}^j + v_0^i v_1^j + v_1^i v_0^j \right) + D^j \varepsilon_{ijk} v_2^k \\ &= -\nabla_j \left( W_{-1}^i v_0^j + v_0^i W_{-1}^j \right) - \left( \partial_i P_2 + \nabla_i P_0 \right), \end{aligned} \tag{24}$$

$$\partial_i v_2 + \nabla_i v_0 = 0. \tag{25}$$

It is easy to see that there are no secular terms in this order.

Let us come now to the most important order $R^3$. In this order we obtain the equations

$$\begin{aligned} &\partial_t v_3^i + \partial_T W_{-1}^i - \left( \partial_{jj} v_3^i + 2\partial_j \nabla_j v_1^i + \nabla_{jj} W_{-1}^i \right) + \nabla_j \left( W_{-1}^i v_1^j + v_1^i W_{-1}^j + v_0^i v_0^j \right) \\ &+ \partial_j \left( W_{-1}^i v_3^j + v_3^i W_{-1}^j + v_0^i v_2^j + v_2^i v_0^j + v_1^i v_1^j \right) + D^j \varepsilon_{ijk} v_3^k = -\left( \partial_i P_3 + \nabla_i \overline{P}_1 \right), \end{aligned} \tag{26}$$

$$\partial_i v_3 + \nabla_i v_1 = 0.$$

From this we get the main secular equation:

$$\partial_T W_{-1}^i - \Delta W_{-1}^i + \nabla_k \overline{\left( v_0^k v_0^i \right)} = -\nabla_i \overline{P}_1. \tag{27}$$

There is also an equation to find the pressure $P_{-3}$:

$$-\nabla_i P_{-3} = D^j \varepsilon_{ijk} W_{-1}^k. \tag{28}$$

## 4. The Velocity Field in Zero Approximation

It is clear that the most important is Equation (36). In order to obtain these equations in closed form, we need to calculate the Reynolds stresses $\nabla_k \left( \overline{v_0^k v_0^i} \right)$. First of all we have to calculate the fields of zero approximation $v_0^k$. From the asymptotic development in zero order we have

$$\partial_t v_0^i - \partial_{jj} v_0^i + W_{-1}^k \partial_k v_0^i + D^j \varepsilon_{ijk} v_0^k = -\partial_i P_0 + F_0^i. \tag{29}$$

Let us introduce the operator $\hat{D}_0$:

$$\hat{D}_0 \equiv \partial_t - \partial_{jj} + W^k \partial_k. \tag{30}$$

Using $\hat{D}_0$, were write Equations (29):

$$\hat{D}_0 v_0^i + D^j \varepsilon_{ijk} v_0^k = -\partial_i P_0 + F_0^i. \tag{31}$$

Pressure $P_0$ can be found from condition $divV = 0$.

$$P_0 = \frac{[\boldsymbol{D} \times \partial]_i v_0^i}{\partial^2}. \tag{32}$$

Let us introduce designations for operatores:

$$\hat{P}_{ij} = \partial_j \frac{[\boldsymbol{D} \times \partial]_i}{\partial^2} \tag{33}$$

and for velocities: $v_0^x = u_0$, $v_0^y = v_0$, $v_0^z = w_0$. Then excluding pressure from (31), we obtain the system of equations to find the velocity field of zero approximation:

$$\begin{gathered}
\left(\widehat{D_0} + \widehat{P_{xx}}\right) u_0 + \left(\widehat{P_{yx}} - D_z\right) v_0 + \left(\widehat{P_{zx}} + D_y\right) w_0 = F_0^x, \\
\left(\widehat{P_{xy}} + D_z\right) u_0 + \left(\widehat{D_0} + \widehat{P_{yy}}\right) v_0 + \left(\widehat{P_{zy}} - D_x\right) w_0 = F_0^y, \\
\left(\widehat{P_{xz}} - D_y\right) u_0 + \left(\widehat{P_{yz}} + D_x\right) v_0 + \left(\widehat{D_0} + \widehat{P_{zz}}\right) w_0 = F_0^z.
\end{gathered} \tag{34}$$

For simplicity, we choose the systeme of coordinates so that the axis $Z$ coincides with the direction of angular velocity of rotation $\boldsymbol{\Omega}$. Then $D_x = 0$, $D_y = 0$, $D_z = D$ In order to solve this system of equations we have to set the force in the explicit form.Let us choose now the external force in the rotating system of coordinates in the following form:

$$\begin{gathered}
F_0^z = 0, \boldsymbol{F}_{0\perp} = f_0 \left(\boldsymbol{i}\mathrm{Cos}\varphi_2 + \boldsymbol{j}\mathrm{Cos}\varphi_1\right); \varphi_1 = \boldsymbol{k}_1 \boldsymbol{x} - \omega_0 t, \varphi_2 = \boldsymbol{k}_2 \boldsymbol{x} - \omega_0 t, \\
\boldsymbol{k}_1 = k_0 (1,0,1), \boldsymbol{k}_2 = k_0 (0,1,1).
\end{gathered}$$

It is obvious that divergence of this force us equal to zero. Thus, external force is given in plane $(x, y)$, orthogonal to rotation axis.

The solution for equations system (34) can be found easily in accordance with Cramer's Rule:

$$u_0 = \frac{\Delta_1}{\Delta}, v_0 = \frac{\Delta_2}{\Delta}, w_0 = \frac{\Delta_3}{\Delta}. \tag{35}$$

Here $\Delta$ is the determinant of the system (34):

$$\Delta = \begin{vmatrix} \widehat{D_0} + \widehat{P_{xx}} & \widehat{P_{yx}} - D & \widehat{P_{zx}} \\ \widehat{P_{xy}} + D & \widehat{D_0} + \widehat{P_{yy}} & \widehat{P_{zy}} \\ \widehat{P_{xz}} & \widehat{P_{yz}} & \widehat{D_0} + \widehat{P_{zz}} \end{vmatrix}, \tag{36}$$

$$\Delta_1 = \begin{vmatrix} F_0^x & \widehat{P_{yx}} - D & \widehat{P_{zx}} \\ F_0^y & \widehat{D_0} + \widehat{P_{yy}} & \widehat{P_{zy}} \\ 0 & \widehat{P_{yz}} & \widehat{D_0} + \widehat{P_{zz}} \end{vmatrix}, \tag{37}$$

$$\Delta_2 = \begin{vmatrix} \widehat{D_0} + \widehat{P_{xx}} & F_0^x & \widehat{P_{zx}} \\ \widehat{P_{xy}} + D & F_0^y & \widehat{P_{zy}} \\ \widehat{P_{xz}} & 0 & \widehat{D_0} + \widehat{P_{zz}} \end{vmatrix}, \tag{38}$$

$$\Delta_3 = \begin{vmatrix} \widehat{D_0}+\widehat{P_{xx}} & \widehat{P_{yx}}-D & F_0^x \\ \widehat{P_{xy}}+D & \widehat{D_0}+\widehat{P_{yy}} & F_0^y \\ \widehat{P_{xz}} & \widehat{P_{yz}} & 0 \end{vmatrix}. \tag{39}$$

After writing down the determinants in the explicit form, we obtain:

$$\begin{aligned} u_0 &= \frac{1}{\Delta}\left[\left(\widehat{D_0}+\widehat{P_{yy}}\right)\left(\widehat{D_0}+\widehat{P_{zz}}\right)-\left(\widehat{P_{yz}}\right)\left(\widehat{P_{zy}}\right)\right]F_0^x \\ &+\frac{1}{\Delta}\left[\left(\widehat{P_{zx}}\right)\left(\widehat{P_{yz}}\right)-\left(\widehat{P_{yx}}-D\right)\left(\widehat{D_0}+\widehat{P_{zz}}\right)\right]F_0^y, \end{aligned} \tag{40}$$

$$\begin{aligned} v_0 &= \frac{1}{\Delta}\left[\left(\widehat{P_{xz}}\right)\left(\widehat{P_{zy}}\right)-\left(\widehat{P_{xy}}+D\right)\left(\widehat{D_0}+\widehat{P_{zz}}\right)\right]F_0^x \\ &+\frac{1}{\Delta}\left[\left(\widehat{D_0}+\widehat{P_{xx}}\right)\left(\widehat{D_0}+\widehat{P_{zz}}\right)-\left(\widehat{P_{xz}}\right)\left(\widehat{P_{zx}}\right)\right]F_0^y, \end{aligned} \tag{41}$$

$$\begin{aligned} w_0 &= \frac{1}{\Delta}\left[\left(\widehat{P_{xy}}+D\right)\left(\widehat{P_{yz}}\right)-\left(\widehat{P_{xz}}\right)\left(\widehat{D_0}+\widehat{P_{yy}}\right)\right]F_0^x \\ &+\frac{1}{\Delta}\left[\left(\widehat{P_{xz}}\right)\left(\widehat{P_{yx}}-D\right)-\left(\widehat{D_0}+\widehat{P_{xx}}\right)\left(\widehat{P_{yz}}\right)\right]F_0^y. \end{aligned} \tag{42}$$

$$\begin{aligned} \Delta &= \left(\widehat{D_0}+\widehat{P_{xx}}\right)\left[\left(\widehat{D_0}+\widehat{P_{yy}}\right)\left(\widehat{D_0}+\widehat{P_{zz}}\right)-\left(\widehat{P_{yz}}\right)\left(\widehat{P_{zy}}\right)\right] \\ &-\left(\widehat{P_{yx}}-D\right)\left[\left(\widehat{P_{xy}}+D\right)\left(\widehat{D_0}+\widehat{P_{zz}}\right)-\left(\widehat{P_{xz}}\right)\left(\widehat{P_{zy}}\right)\right] \\ &+\left(\widehat{P_{zx}}\right)\left[\left(\widehat{P_{xy}}+D\right)\left(\widehat{P_{yz}}\right)-\left(\widehat{D_0}+\widehat{P_{yy}}\right)\left(\widehat{P_{xz}}\right)\right]. \end{aligned} \tag{43}$$

In order to calculate the expressions (40)-(43) we present the external force in complex form:

$$F_0^x = \frac{f_0}{2}\left(e^{i\varphi_2}+e^{-i\varphi_2}\right), \quad F_0^y = \frac{f_0}{2}\left(e^{i\varphi_1}+e^{-i\varphi_1}\right). \tag{44}$$

Then all operators in formulae (40)-(42) act from the left on their eigenfunctions. In particular:

$$\begin{aligned} &\widehat{D_0}e^{i\varphi_2} = e^{i\varphi_2}\widehat{D_0}\left(\boldsymbol{k}_2,-\omega_0\right), \ \widehat{D_0}e^{i\varphi_1} = e^{i\varphi_1}\widehat{D_0}\left(\boldsymbol{k}_1,-\omega_0\right), \\ &\Delta e^{i\varphi_2} = e^{i\varphi_2}\Delta\left(\boldsymbol{k}_2,-\omega_0\right), \ \Delta e^{i\varphi_1} = e^{i\varphi_1}\Delta\left(\boldsymbol{k}_1,-\omega_0\right). \end{aligned} \tag{45}$$

To simplify the formulae, let us choose $k_0 = 1, \ \omega_0 = 1$.
We will designate

$$\widehat{D_0}\left(\boldsymbol{k}_2,-\omega_0\right) = 2+i\left(w_y-1\right) = A_y, \ \widehat{D_0}\left(\boldsymbol{k}_1,-\omega_0\right) = 2+i\left(w_x-1\right) = A_x. \tag{46}$$

Before doing further calculations, we have to note that some components of tensors $\hat{P}_{ij}\left(\boldsymbol{k}_1\right)$ and $\hat{P}_{ij}\left(\boldsymbol{k}_2\right)$ vanish. Let us write the non-zero components only:

$$\hat{P}_{yx}\left(\boldsymbol{k}_1\right) = \frac{1}{2}D, \ \hat{P}_{xz}\left(\boldsymbol{k}_2\right) = -\frac{1}{2}D, \ \hat{P}_{xy}\left(\boldsymbol{k}_2\right) = -\frac{1}{2}D, \ \hat{P}_{yz}\left(\boldsymbol{k}_1\right) = \frac{1}{2}D. \tag{47}$$

Taking into account the formulae (45)-(47), we can find the determinant:

$$\Delta\left(\boldsymbol{k}_1\right) = A_x^3 + \frac{1}{2}D^2A_x, \ \Delta\left(\boldsymbol{k}_2\right) = A_y^3 + \frac{1}{2}D^2A_y. \tag{48}$$

In a similar way we find velocity field of zero approximation:

$$u_0 = f_0\frac{A_y}{2}\frac{e^{i\varphi_2}}{A_y^2+\frac{1}{2}D^2} + f_0\frac{D}{4}\frac{e^{i\varphi_1}}{A_x^2+\frac{1}{2}D^2} + C.C., \tag{49}$$

$$v_0 = -f_0 \frac{D}{4} \frac{e^{i\varphi_2}}{A_y^2 + \frac{1}{2}D^2} + f_0 \frac{A_x}{2} \frac{e^{i\varphi_1}}{A_x^2 + \frac{1}{2}D^2} + C.C., \tag{50}$$

$$w_0 = f_0 \frac{D}{4} \frac{e^{i\varphi_2}}{A_y^2 + \frac{1}{2}D^2} - f_0 \frac{D}{4} \frac{e^{i\varphi_1}}{A_x^2 + \frac{1}{2}D^2} + C.C.. \tag{51}$$

## 5. Reynolds Stress and Large Scale Instability

To close the Equations (27) we have to calculate the Reynolds stresses $\overline{w_0 u_0}$ and $\overline{w_0 v_0}$.

These terms are easily calculated with help of formulae (49)-(51). As a result we obtain:

$$\begin{aligned} \overline{w_0 u_0} &= \frac{f_0^2}{2} \frac{D}{\left|A_y^2 + \frac{1}{2}D^2\right|^2} - \frac{f_0^2}{8} \frac{D^2}{\left|A_x^2 + \frac{1}{2}D^2\right|^2}, \\ \overline{w_0 v_0} &= -\frac{f_0^2}{8} \frac{D^2}{\left|A_y^2 + \frac{1}{2}D^2\right|^2} - \frac{f_0^2}{2} \frac{D}{\left|A_x^2 + \frac{1}{2}D^2\right|^2}. \end{aligned} \tag{52}$$

Now Equations (27) are closed and take form:

$$\begin{aligned} &\partial_T W_x - \Delta W_x + \frac{\partial}{\partial z} \overline{w_0 u_0} = 0, \\ &\partial_T W_y - \Delta W_y - \frac{\partial}{\partial z} \overline{w_0 v_0} = 0. \end{aligned} \tag{53}$$

We calculate the modules and write the Reynolds stresses (52) in the explicit form:

$$\begin{aligned} \overline{w_0 u_0} &= \frac{f_0^2}{2} \frac{D}{16\left(w_y - 1\right)^2 + \left[4 + \frac{1}{2}D^2 - \left(w_y - 1\right)^2\right]^2} - \frac{f_0^2}{8} \frac{D^2}{16\left(w_x - 1\right)^2 + \left[4 + \frac{1}{2}D^2 - \left(w_x - 1\right)^2\right]^2}, \\ \overline{w_0 v_0} &= -\frac{f_0^2}{8} \frac{D^2}{16\left(w_y - 1\right)^2 + \left[4 + \frac{1}{2}D^2 - \left(w_y - 1\right)^2\right]^2} - \frac{f_0^2}{2} \frac{D}{16\left(w_x - 1\right)^2 + \left[4 + \frac{1}{2}D^2 - \left(w_x - 1\right)^2\right]^2}. \end{aligned} \tag{54}$$

With small $W_x$, $W_y$ Reynolds stresses (52) can be expanded in a series in the small parameters $W_x$, $W_y$. Taking into account the formula:

$$\frac{1}{\left|A^2{}_{x,y} + \frac{1}{2}D^2\right|^2} = \text{Const.} - \frac{32\left(D^2 - 10\right)}{\left[\left(D^2 + 6\right)^2 + 64\right]^2} w_{x,y} + \cdots$$

We obtain the linearized Equations (53):

$$\begin{aligned} &\frac{\partial}{\partial T} W_x - \frac{\partial^2}{\partial z^2} W_x - \frac{\alpha f_0^2 D}{2} \frac{\partial}{\partial z} W_y + \frac{\alpha f_0^2 D^2}{8} \frac{\partial}{\partial z} W_x = 0, \\ &\frac{\partial}{\partial T} W_y - \frac{\partial^2}{\partial z^2} W_y + \frac{\alpha f_0^2 D^2}{8} \frac{\partial}{\partial z} W_y + \frac{\alpha f_0^2 D}{2} \frac{\partial}{\partial z} W_x = 0. \\ &\alpha = \frac{32\left(10 - D^2\right)}{\left[\left(D^2 + 6\right)^2 + 64\right]^2}. \end{aligned} \tag{55}$$

We will search for the solution of linear system (55) in the form:

$$W_x, W_y \sim \exp(\gamma T + ikZ). \tag{56}$$

We substitute (56) in Equation (55) and obtain the dispersion equation:

$$\gamma = -ik\frac{\alpha f_0^2 D^2}{8} \pm k\frac{\alpha f_0^2 D}{2} - k^2. \tag{57}$$

The dispersion Equation (57) shows that equation system (55) has instable oscillatory solutions with oscillatory frequency $\omega = k\frac{\alpha f_0^2 D^2}{8}$ and instability growth rate $\gamma = k\frac{\alpha f_0^2 D}{2} - k^2$. The instability is large scale because the instable term dominates over dissipation on large scales: $\frac{\alpha f_0^2 D}{2} \succ k$. The maximum growth rate of instability is equal to $\gamma_{\max} = \frac{\alpha^2 f_0^4 D^2}{16}$, and is achieved on the wave vector $k_{\max} = \frac{\alpha f_0^2 D}{4}$. As a result of the development of instability the large scale helical circular polarized vortices of Beltrami type are generated in the system.

## 6. Saturation of Instability and Nonlinear Vortex Structures

It is clear that with increasing of amplitude nonlinear terms decrease and instability becomes saturated. Consequently stationary nonlinear vortex structures are formed. To find these structures let us choose for Equations (54) $\frac{\partial}{\partial T} = 0$ and integrate equations one time over $Z$. We obtain the system of equations:

$$\begin{aligned} \frac{\mathrm{d}}{\mathrm{d}Z} W_x &= \overline{w_0 u_0} + C_1, \\ \frac{\mathrm{d}}{\mathrm{d}Z} W_y &= \overline{w_0 v_0} + C_2. \end{aligned} \tag{58}$$

From Equations (58) follows:

$$\frac{\mathrm{d}w_x}{\mathrm{d}w_y} = \frac{\overline{w_0 u_0} + C_1}{\overline{w_0 v_0} + C_2}, \tag{59}$$

After integrating the system of Equations (59) we obtain:

$$\int \overline{w_0 v_0}\,\mathrm{d}w_x + C_2 w_x = \int \overline{w_0 u_0}\,\mathrm{d}w_y + C_1 w_y. \tag{60}$$

Integrals in expression (60) are calculated in elementary functions (see [17]), which give the expression for first integral of motion $J$ of Equations (59):

$$\begin{aligned} J = {} & \frac{D^2}{8}\frac{w_x}{\left[4+\frac{1}{2}D^2-(w_y-1)^2\right]^2+16(w_y-1)^2} + \frac{D}{2^{5/2}(8+D^2)}\ln\frac{(w_x-1)^2+(w_x-1)D\sqrt{2}+4+\frac{1}{2}D^2}{(w_x-1)^2-(w_x-1)D\sqrt{2}+4+\frac{1}{2}D^2} \\ & + \frac{D}{8(8+D^2)}\operatorname{arctg}\frac{(w_x-1)^2-4-\frac{1}{2}D^2}{4(w_x-1)} - \frac{D^2}{8}\frac{w_y}{\left[4+\frac{1}{2}D^2-(w_x-1)^2\right]^2+16(w_x-1)^2} \\ & + \frac{D}{2^{5/2}(8+D^2)}\ln\frac{(w_y-1)^2+(w_y-1)D\sqrt{2}+4+\frac{1}{2}D^2}{(w_y-1)^2-(w_y-1)D\sqrt{2}+4+\frac{1}{2}D^2} + \frac{D}{8(8+D^2)}\operatorname{arctg}\frac{(w_y-1)^2-4-\frac{1}{2}D^2}{4(w_y-1)} + C_1 w_y + C_2 w_x. \end{aligned}$$

Equations (58) can be easily calculated numerically using standard tools. In particular, this allows to construct phase portrait of the dynamical system (58) (**Figure 3**) and to get the most interesting solutions which link singular points on phase plane. See for example **Figure 1**, where the hyperbolic singular point is connected with

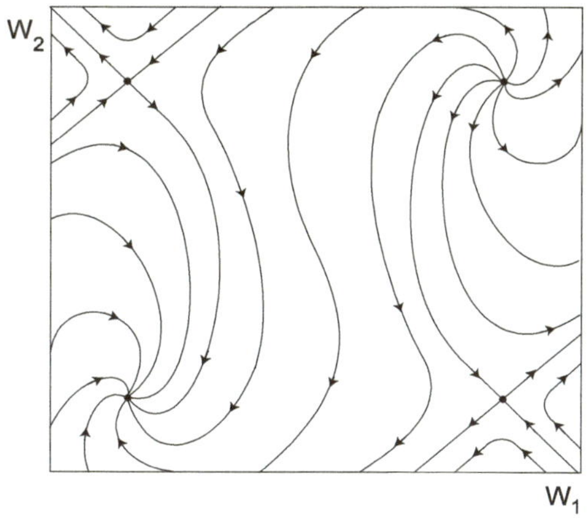


**Figure 3.** Phase portrait of the dynamical system (58), with $D = 1$, $C_1 = -0.03$, $C_2 = 0.03$. One can see two hyperbolic singular points and stable and instable knots.

the stable knot and **Figure 2**, where the solution connects instable and stable focuses. All these solutions correspond to the large scale localized vortex structures of kink type with rotation, generated by the instability which has been found in this work.

## 7. Conclusions and Discussion of the Results

In this work we find the new large scale instability in rotating fluid. It is supposed that the small scale vortex external force in rotating coordinates system acts on fluid which maintains the small velocity field fluctuations (small scale turbulence with small Reynolds number $R$, $R \ll 1$). For the real applications this Reynolds number should be calculated with help of the turbulent viscosity. The asymptotic development of motion equations by small Reynolds number allows obtaining motion equations for the large scale. These equations are of the hydrodynamic $\alpha$-effect type, in which velocity components $W_x$, $W_y$ are connected by the positive feedback. This may result in the appearance of the large scale vortex instability. The large scale vortices of Beltrami type are formed due to this instability in rotating fluid with small scale exterior force. With further increase of amplitude, the instability stabilizes and passes to stationary mode. In this mode the nonlinear stationary vortex structures form. Different vortex kinks belong to the most interesting structures. These kinks link stationary points of dynamical system (58). Kinks which link the hyperbolic point with the stable knot rotate around the stable knot as shown on **Figure 1**. In the kink which links instable and stable focuses, vector field turns around two singular points, see **Figure 2**.

Let us note that unlike previous works about hydrodynamic $\alpha$-effect in rotating fluid, the use of the asymptotic development allows constructing naturally the nonlinear theory and studying the stationary nonlinear vortex kinks.

## References

[1] Grinspen, H.P. (1990) The Theory of Rotating Fluids. Breukelen Press, Brookline.

[2] Roberts, P.H. and Soward, A.M. (1978) Rotating Fluids in Geophysics. Academic Press, London.

[3] Clarke, C. and Carswell, B. (2007) Principles of Astrophysical Fluid Dynamics. Cambridge University Press, Cambridge. http://dx.doi.org/10.1017/CBO9780511813450

[4] Vallis, G.K. (2010) Atmospheric and Oceanic Fluid Dynamics. Cambridge University Press, Cambridge.

[5] Abramowicz, M.A., Lanza, A., Spigel, E.A. and Szuszkiewicz, E. (1992) Vortices on Accretion Disks. *Nature*, **356**, 41-43. http://dx.doi.org/10.1038/356041a0

[6] Brandt, P.N., Scharmer, G.B., Ferguson, S., Shine, R.A., Tarbell, T.D. and Title, A.M. (1988) Vortex Flow in the Solar Photosphere. *Nature*, **335**, 238. http://dx.doi.org/10.1038/335238a0

[7] Dritschel, G. and Legras, B. (1993) Modeling Oceanic and Atmospheric Vortices. *Physics Today*, **46**, 44.

http://dx.doi.org/10.1063/1.881375

[8] Moiseev, S.S., Sagdeev, R.Z., Tur, A.V., Khomenko, G.A. and Yanovsky, V.V. (1983) A Theory of Large-Scale Structure Origination in Hydrodynamic Turbulence. *Soviet Physics—JETP*, **58**, 1149.

[9] Moiseev, S.S., Rutkevich, P.B., Tur, A.V. and Yanovsky, V.V. (1988) Vortex Dynamos in a Helical Turbulent Convection. *Soviet Physics—JETP*, **67**, 294.

[10] Lupyan, E.A., Mazurov, A.A., Rutkevich, P.B. and Tur, A.V. (1992) Generation of Large-Scale Vortices through the Action of Spiral Turbulence of a Convective Nature. *Soviet Physics—JETP*, **75**, 833.

[11] Khomenko, G.A., Moiseev, S.S. and Tur, A.V. (1991) The Hydrodynamic Alpha-Effect in a Compressible Fluid. *Journal of Fluid Mechanics*, **225**, 355. http://dx.doi.org/10.1017/S0022112091002082

[12] Levina, G.V., Moiseev, S.S. and Rutkevich, P.B. (2000) Hydrodynamic Alpha-Effect in a Convective System. *Advances in Fluid Mechanics*, **25**, 111.

[13] Frisch, U., She, Z.S. and Sulem, P.L. (1987) Large-Scale Flow Driven by the Anisotropic Kinetic Alpha Effect. *Physica D*, **28**, 382. http://dx.doi.org/10.1016/0167-2789(87)90026-1

[14] Tur, A.V. and Yanovsky, V.V. (2013) Non Linear Vortex Structures in Stratified Fluid Driven by Small-Scale Helical Force. *Open Journal of Fluid Dynamics*, **3**, 64. http://dx.doi.org/10.4236/ojfd.2013.32009

[15] Kitchatinov, L.L., Rudiger, G. and Khomenko, G. (1994) Large-Scale Vortices in Rotating Stratified Disks. *Astronomy & Astrophysics*, **287**, 320.

[16] Moffat, H.K. (1978) Magnetic Field Generation in Electrically Conducting Fluids. Cambridge University Press, Cambridge.

[17] Gradshteyn, I.S. and Ryzhik, I.M. (2000) Table of Integrals, Series, and Prodacts. Academic Press, London.

# 23

# The Tangential Force Distribution on Inner Cylinder of Power Law Fluid Flowing in Eccentric Annuli with the Inner Cylinder Reciprocating Axially

**Junzhe Ma, Yu Wei, Jianan Zhang, Hongjian Liu, Xianjun Meng**

Enhanced Oil and Gas Recovery Key Laboratory of Ministry of Education, Northeast Petroleum University, Daqing, China

Email: 147974485@qq.com

## Abstract

**Based on the governing equations of the inner cyinder of the unsteady flow of the power law fluid in eccentric annuli with the inner cylinder reciprocating axially in bipolar coordinate system, the calculation formulae of tangential force were established, and the relevant numerical calculation method was given. Taking the aqueous solution of partially hydrolyzed polyacrylamides (HPAM) for examples, the tangential forces were calculated by using the formulae and numerical calculation method mentioned above; the curves of the tangential force on the wall of the inner cylinder of HPAM aqueous solution were plotted; and the effects on the tangential force of the flow behavior index of the power law fluid, the stroke and the stroke frequency of the inner cylinder were analyzed.**

## Keywords

**Tangential Force, Power Law Fluid, Eccentric Annuli, Reciprocate Axially**

## 1. Introduction

At present in our country, there is few theoretical research on the tangential force on the wall of the inner cylinder of the power law fluid flowing in eccentric annuli with the inner cylinder reciprocating axially. This force for sucker rod eccentric wear research has an important influence. Therefore, research on the tangential force has a practical significance. Based on the governing equations of the inner cyinder of the unsteady flow of the power law fluid in eccentric annuli with the inner cylinder reciprocating axially in the bipolar coordinate system [1], the calculation formulae of tangential force are established; the relevant numerical calculation method is given; and taking the aqueous solution of partially hydrolyzed HPAM for examples, the tangential forces are calculated and analyzed in this article.

## 2. Mathematical Models

### 2.1. Assumed Condition

Power law fluid with the inner cylinder reciprocating axially, $w(t)$ is axial velocity, $T$ is the motion cycle, $R_o$ is the radius of the inner cylinder of annuli, $R_i$ is the radius of the inner cylinder, $e$ is the eccentricity of the eccentric annulus.

### 2.2. Governing Equations

The governing equations of the unsteady flow of the power law fluid in eccentric annuli with the inner cylinder reciprocating axially in the bipolar coordinate system by Haiqing Cui, Tao Gao [1]-[3].

$$\frac{\partial}{\partial \xi}\left[\eta(I_2)\frac{\partial w}{\partial \xi}\right]+\frac{\partial}{\partial \zeta}\left[\eta(I_2)\frac{\partial w}{\partial \zeta}\right]=\frac{C^2}{c^2+s^2}\left(\rho\frac{\partial w}{\partial t}-P\right) \quad (1a)$$

### 2.3. Initial Conditions and Boundary Conditions

By assumption flow conditions can be obtained the boundary conditions of dynamic flow equation.

$$w(\xi,\zeta,0)=w_p(\xi,\zeta) \quad (1b)$$

$$w(\xi_i,\zeta,t)=W\sin(2\pi ft) \quad (1c)$$

$$w(\xi_o,\zeta,t)=0 \quad (1d)$$

where $\eta(I_2)$ is the viscosity function of the power law fluid; $I_2$ is the second invariant of the component of the first order Rivlin-Ericksen tensor; $\xi$, $\zeta$ are the dipolar coordinates, and $\xi<0$, $\zeta\geq 0$; $\rho$ is the density of the power law fluid; $t$ is the time; $p$ is the pressure gradient; $w_p(\xi,\zeta)$ is the velocity distribution of the power law fluid in eccentric annuli with the motionless inner cylinder; $W$ is the amplitude of the velocity of the inner cylinder reciprocating axially; $f$ is the stroke frequency of the inner cylinder.

### 2.4. Tangential Force on the Wall of the Inner Cylinder

From the formulae of shear stress on the wall of the inner cylinder

$$\tau_{\xi z}=\frac{\mathrm{ch}\xi-\cos\zeta}{C}\eta(I_2)\left(\frac{\partial w}{\partial \xi}\right) \quad (2)$$

$$\tau_{\zeta z}=\frac{\mathrm{ch}\xi-\cos\zeta}{C}\eta(I_2)\left(\frac{\partial w}{\partial \zeta}\right) \quad (3)$$

$$\tau=\sqrt{\tau_{\xi z}^2+\tau_{\xi z}^2} \quad (4)$$

the calculation formulae of tangential force are given as follows

$$F_z=2R_i\int_0^{\pi}\left(\tau(\xi_i,\zeta,t)\cos\theta\right)\mathrm{d}\theta \quad (5)$$

which can be expressed as

$$F_z=2R_i\int_0^{\pi}f_z(\zeta)\mathrm{d}\zeta \quad (6)$$

where

$$f_z(\zeta)=\frac{\operatorname{sh}\xi_i}{(\cos\zeta-\operatorname{ch}\xi_i)}\sqrt{\left(\frac{\operatorname{ch}\xi_i-\cos\zeta}{C}\eta(I_2)\left(\frac{\partial w}{\partial \xi}\right)\right)^2+\left(\frac{\operatorname{ch}\xi_i-\cos\zeta}{C}\eta(I_2)\left(\frac{\partial w}{\partial \zeta}\right)\right)^2}. \quad (7)$$

## 3. Calculation Method

From the density of the power law fluid $\rho$, the flow behavior index of the power law fluid $n$, the consistency coefficient of the power law fluid $k$, the radius of the inner cylinder of annuli $R_o$, the radius of the inner cylinder $R_i$, the eccentricity of the eccentric annulus $e$, the stroke $S$, the stroke frequency of the inner cylinder $f$ and the pressure gradient $P$, the distribution of the tangential force $F_z$ can be calculated: firstly, through the equations (1a), (1b), (1c) and (1d), the velocity distribution of the power law fluid at anytime by using the finite difference method is calculated; then, through the formula (6), the tangential force by using the numerical integration method is calculated.

## 4. Calculation Examples

Take HPAM aqueous solution as an example, the radius of the inner cylinder of annulus $R_o = 2.960 \times 10^{-2}$ m, the radius of the inner cylinder $R_i = 0.885 \times 10^{-2}$ m, the pressure gradient of the aqueous solution $P = 61.061$ Pa/m and by using the calculation method mentioned above, the tangential forces on the wall of the inner cylinder of the HPAM aqueous solution are calculated and the relevant distribution curves are plotted as **Figure 1** to **Figure 4** (the minus of the tangential force indicates that the inner cylinder moves upwards).

Under different flow behavior index $n$, the curves of the tangential force to the cycle number $N$ are shown as **Figure 1**, where $N = t/T$. It shows that, with the eccentricity, the stroke and the stroke frequency of the inner cylinder being given as constants, as the flow behavior index $n$ increases, the peak values (or valley values) of the tangential force $F_z$ change a lot.

Under different eccentricities $e$, the curves of the tangential force to the cycle number are shown as **Figure 2**. It shows that, with the flow behavior index $n$, the stroke $S$ and the stroke frequency of the inner cylinder $f$ being given as constants, as the eccentricity $e$ increases, the peak values (or valley values) of the tangential force $F_z$ change a lot.

Under different strokes $S$, the curves of the tangential force to the cycle number are shown as **Figure 3**. It shows that, with the flow behavior index $n$, the eccentricity $e$ and the stroke frequency of the inner cylinder $f$ being given as constants, as the strokes $S$ increases, the peak values (or valley values) of the tangential force $F_z$ change a lot.

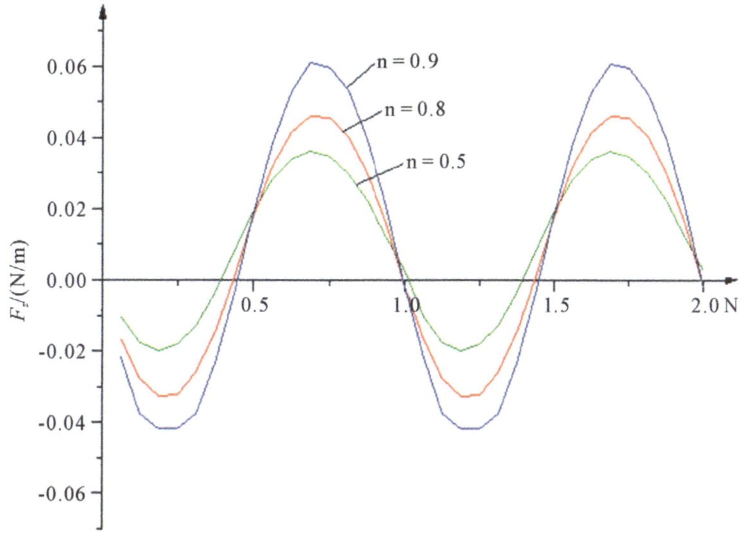


**Figure 1.** The curves of the distribution of the tangential force of the HPAM aqueous solution under different flow behavior index ($\rho$ = 998 Kg/m$^3$, $k$ = 10.886 × 10$^{-2}$ Pa·s$^n$, $e$ = 0.965 × 10$^{-2}$ m, $S$ = 1.0 m, $f$ = 0.167 Hz).

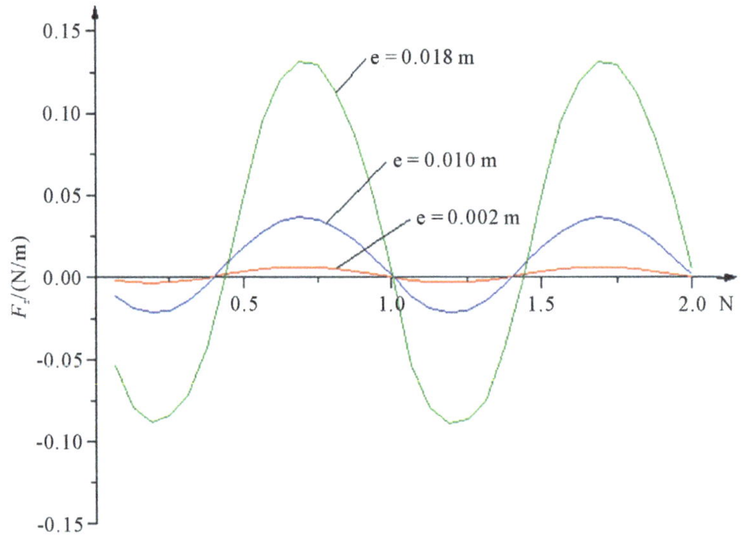


**Figure 2.** The distribution of the tangential force of the HPAM aqueous solution of quality concentration 0.1% under different eccentricities ($\rho$ = 998 Kg/m$^3$, $n$ = 0.556, $k$ = 10.886 × 10$^{-2}$ Pa·s$^n$, $S$ = 1.0 m, $f$ = 0.167 Hz).

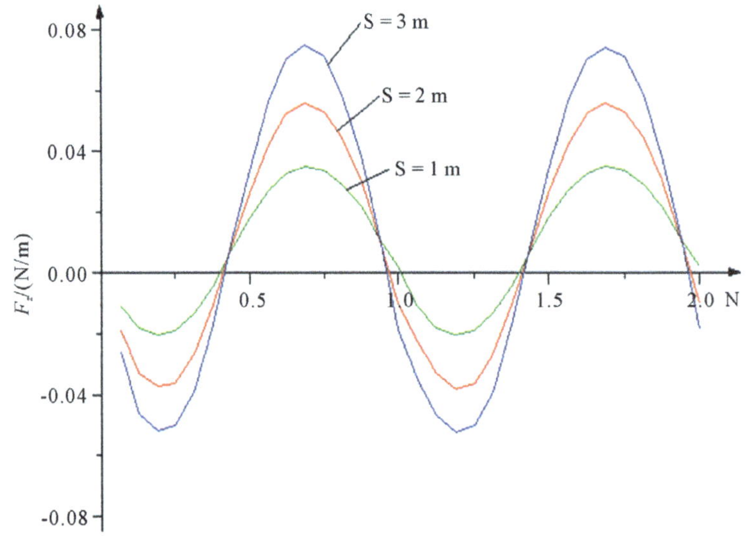


**Figure 3.** The distribution of the tangential force of the HPAM aqueous solution of quality concentration 0.1% under different strokes ($\rho$ = 998 Kg/m$^3$, $n$ = 0.556, $k$ = 10.886 × 10$^{-2}$ Pa·s$^n$, $e$ = 0.965 × 10$^{-2}$ m, $f$ = 0.167 Hz).

Under different frequencies $f$, the curves of the tangential force distribution to the cycle number are shown as **Figure 4**. It shows that, with the flow behavior index $n$, the eccentricity $e$ and the stroke $S$ being given as constants, as the stroke frequency $f$ increases, the peak values (or valley values) of the tangential force $F_z$ change a lot.

## 5. Summary

1) Based on the governing equations of the inner cyinder of the unsteady flow of the power law fluid in eccentric annuli with the inner cylinder reciprocating axially in the bipolar coordinate system, the calculation formulae of tangential force on the wall of the inner cylinder of the power law fluid flowing in eccentric annuli with the inner cylinder reciprocating axially in the bipolar coordinate system were established, and the relevant numerical calculation method was given.

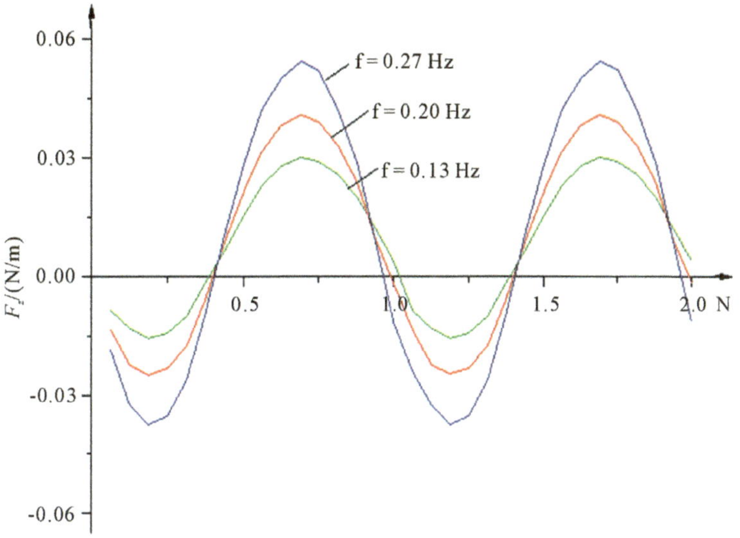


**Figure 4.** The distribution curves of tangential force of the HPAM aqueous solution of quality concentration 0.1% under different frequency ($\rho$ = 998 Kg/m$^3$, $n$ = 0.556, $k$ = 10.886 × 10$^{-2}$ Pa·s$^n$, $e$ = 0.965 × 10$^{-2}$ m, $S$ = 1.0 m).

2) Taking the HPAM aqueous solution for examples, by using the calculation formulae and method mentioned above, the tangential forces of the HPAM aqueous solution on the wall of the inner cylinder were calculated and the tangential force curves were plotted.

3) The effects on the tangential force of the flow behavior index of the power law fluid, the eccentricity of the eccentric annulus, the stroke and the stroke frequency of the inner cylinder were analyzed; the influence of the eccentricity was more obvious.

## References

[1] Cui, H.Q., Sun, Z. and Gao, T. (2003) Velocity Distribution of Unsteady Flow of Non-Newtonian Fluid in Eccentric Annuli with the Inner Cylinder Reciprocating Axially. *Journal of Hydrodynamics*, **18**, 711-715.

[2] Yang, Y.J., Cui, H.Q. and Gao, T. (2004) Flow Rate Distribution of the Unsteady Flow of Power Law Fluid in Eccentric Annuli with Inner Cylinder Reciprocating Axially. *Journal of Daqing Petroleum Institute*, **28**, 17-19.

[3] Meng, X.J., Liu, H.J., Zhang, H.Z., Gao, T. and Cui, H.Q. (2013) The Normal Force Distribution on the Wall of the Inner Cylinder of the Power Law Fluid Flowing in the Eccentric Annuli with the Inner Cylinder Reciprocating Axially. *Applied Mechanics and Materials*, **487**, 527-531. http://www.scientific.net/AMM.487.527

# Permissions

All chapters in this book were first published in OJFD, by Scientific Research Publishing; hereby published with permission under the Creative Commons Attribution License or equivalent. Every chapter published in this book has been scrutinized by our experts. Their significance has been extensively debated. The topics covered herein carry significant findings which will fuel the growth of the discipline. They may even be implemented as practical applications or may be referred to as a beginning point for another development.

The contributors of this book come from diverse backgrounds, making this book a truly international effort. This book will bring forth new frontiers with its revolutionizing research information and detailed analysis of the nascent developments around the world.

We would like to thank all the contributing authors for lending their expertise to make the book truly unique. They have played a crucial role in the development of this book. Without their invaluable contributions this book wouldn't have been possible. They have made vital efforts to compile up to date information on the varied aspects of this subject to make this book a valuable addition to the collection of many professionals and students.

This book was conceptualized with the vision of imparting up-to-date information and advanced data in this field. To ensure the same, a matchless editorial board was set up. Every individual on the board went through rigorous rounds of assessment to prove their worth. After which they invested a large part of their time researching and compiling the most relevant data for our readers.

The editorial board has been involved in producing this book since its inception. They have spent rigorous hours researching and exploring the diverse topics which have resulted in the successful publishing of this book. They have passed on their knowledge of decades through this book. To expedite this challenging task, the publisher supported the team at every step. A small team of assistant editors was also appointed to further simplify the editing procedure and attain best results for the readers.

Apart from the editorial board, the designing team has also invested a significant amount of their time in understanding the subject and creating the most relevant covers. They scrutinized every image to scout for the most suitable representation of the subject and create an appropriate cover for the book.

The publishing team has been an ardent support to the editorial, designing and production team. Their endless efforts to recruit the best for this project, has resulted in the accomplishment of this book. They are a veteran in the field of academics and their pool of knowledge is as vast as their experience in printing. Their expertise and guidance has proved useful at every step. Their uncompromising quality standards have made this book an exceptional effort. Their encouragement from time to time has been an inspiration for everyone.

The publisher and the editorial board hope that this book will prove to be a valuable piece of knowledge for researchers, students, practitioners and scholars across the globe.

# List of Contributors

**Khaled S. Al-Mashrafi and Ibrahim A. Eltayeb**
Department of Mathematics and Statistics, College of Science, Sultan Qaboos University, Muscat, Oman

**Ebrahim Ahmadloo**
Young Researchers and Elite Club, Darab Branch, Islamic Azad University, Darab, Iran

**Najmeh Sobhanifar**
Young Researchers and Elite Club, Shoushtar Branch, Islamic Azad University, Shoushtar, Iran

**Fatemeh Sadat Hosseini**
Young Researchers and Elite Club, Shiraz Branch, Islamic Azad University, Shiraz, Iran

**Arun Kumar Ghosh**
Jadavpur University, Kolkata, India

**Sanjib Kumar Datta**
Department of Mathematics, University of Kalyani, Kalyani, India

**Pulakesh Sen**
Department of Mathematics, D. N. C. College, Aurangabad, India

**Xin Chen, Quan Kuang, Yang Li, Songbo Wang, Yunling Ye and Shesheng Zhang**
Wuhan University of Technology, Wuhan, China

**Khaled S. Al-Mashrafi**
Section of Applied Sciences, Department of Human Resources Development, General Directorate of Education in Eastern Region, Ministry of Education, Sur, Oman

**Ibrahim A. Eltayeb**
Department of Mathematics and Statistics, College of Science, Sultan Qaboos University, Muscat, Oman

**Heyuan Wang**
College of Sciences, Liaoning University of technology, Jinzhou, China

**Yan Gao**
Liaoning Petro-chemical Vocational Technology College, Jinzhou, China

**João Victor Nunes de Sousa, Wanderley Ferreira de Amorim Junior and Antonio Gilson Barbosa de Lima**
Department of Mechanical Engineering, Center of Sciences and Technology, Federal University of Campina Grande, Campina Grande, Brazil

**Antônio Roberto Lins de Macêdo**
ARMTEC Technology in Robotics, Fortaleza, Brazil

**Javed I. Siddique**
Department of Mathematics, Pennsylvania State University, York Campus, York, USA

**Forrest A. Landis**
Department of Chemistry, Pennsylvania State University, York Campus, York, USA

**Muhammad R. Mohyuddin**
Department of Mathematics, FAST University, Islamabad, Pakistan
NCBA & E, Gujrat, Pakistan

**Lili Liu, Shuren Yang, Di Xu and Yue Cui**
Petroleum Engineering Department, Northeast Petroleum University, Daqing, China

**Lihui Wang**
Daqing Petroleum Equipment Group, Daqing, China

**Liang Zhao**
Daqing Oilfield Tianyu Engineering Design Co., Ltd., Daqing, China

**Alessandro Cubeddu and Cornelia Rauh**
Institute of Fluid Mechanics, Technical Faculty, University of Erlangen-Nuremberg, Erlangen, Germany
Department of Food Biotechnology and Food Process Engineering, Faculty III – Process Science and Engineering, Berlin Institute of Technology, Berlin, Germany

**Antonio Delgado**
Institute of Fluid Mechanics, Technical Faculty, University of Erlangen-Nuremberg, Erlangen, Germany

**Michael Kopp and Vladimir Yanovsky**
Institute for Single Crystals, National Academy of Sciences of Ukraine, Kharkov, Ukraine
V.N. Karazin Kharkiv National University 4 Svobody Sq., Kharkov, Ukraine

**Anatoly Tur**
Université de Toulouse [UPS], CNRS, Institut de Recherche en Astrophysique et Planétologie, Toulouse Cedex, France

**Shinya Okuhara**
Support Center for Practical Education, National Institute of Technology, Matsue College, Matsue, Japan

**Manabu Takao and Akiyasu Takami**
Department of Mechanical Engineering, National Institute of Technology, Matsue College, Matsue, Japan

**Hideki Sato**
Advanced Engineering Faculty, National Institute of Technology, Matsue College, Matsue, Japan

**Toshiaki Setoguchi**
Institute of Ocean Energy, Saga University, Saga, Japan

**Swetha Ravi**
Department of Mathematics, Gudlavalleru Engineering College, Gudlavalleru, India

**Jagdish Prakash**
Department of Mathematics, University of Botswana, Gaborone, Botswana

**Viswanatha Reddy Gottam and Vijaya Kumar Varma Sibyala**
Department of Mathematics, S. V. University, Tirupati, India

**Pradip Kumar Gaur and Abhay Kumar Jha**
Department of Mathematics, JECRC University, Jaipur, India

**Aniruddha Mitra and Manoranjan Khan**
Dept. of Instrumentation Science & Centre for Plasma Studies, Jadavpur University, Kolkata, India

**Labakanta Mandal**
Dept. of Physics, Budge Budge Institute of Technology, Kolkata, India

**Rajkumar Roychoudhury**
Advanced Centre for Nonlinear & Complex Phenomena, Kolkata, India

**Hadi Taheri, Ferdinand P. Schmidt and Martin Gabi**
Institute of Fluid Machinery, Department of Mechanical Engineering, Karlsruhe Institute of Technology, Karlsruhe, Germany

**Edvaldo Angelo**
Grupo de Simulação Numérica (GSN), Universidade Presbiteriana Mackenzie, São Paulo, Brazil

**Gabriel Angelo**
Department of Mechanical Engineering, Centro Universitário da Fundação Educacional Inaciana Padre Sabóia de Medeiros (FEI), São Bernardo do Campo, Brazil
Instituto de Pesquisas Energéticas e Nucleares, IPEN CNEN/SP, Universidade de São Paulo, São Paulo, Brazil

**Pedro Henrique Di Giovanni Santos and Delvonei Alves de Andrade**
Instituto de Pesquisas Energéticas e Nucleares, IPEN CNEN/SP, Universidade de São Paulo, São Paulo, Brazil

**Isaac Lare Animasaun**
Department of Mathematical Sciences, Federal University of Technology, Akure, Nigeria

**Sudarshan Ghonge, D. Nagendra Prasad, Swarnim Narayan, Hains Francis and Souri Banerjee**
Department of Physics, Birla Institute Technology and Science-Pilani, Hyderabad Campus, Hyderabad, India

**Astha Sethi**
Department of Physics and Frederick Seitz Materials Research Laboratory, University of Illinois, Urbana-Champaign, IL, USA

**Subimal Deb**
Department of Physics, GIT, GITAM University, Visakhapatnam, India

**Shin-Ichi Morita and Yasutaka Hayamizu**
Department of Mechanical Engineering, National Institute of Technology, Yonago College, Yonago, Japan

**Takanobu Yamada**
Department of Mechanical Engineering, Kitami Institute of Technology, Kitami, Japan

**Akihiko Horibe and Naoto Haruki**
Graduate School of Natural Science and Technology, Okayama University, Okayama, Japan

**Toshiaki Setoguchi**
Department of Advanced Technology Fusion, Saga University, Saga, Japan

**Kazuma Adachi**
Graduate School of Machine Creative Engineering, Nagaoka University of Technology, Nagaoka, Japan

**Aleksey Ni and Cheol Woo Park**
School of Mechanical Engineering, Kyungpook National University, Daegu, Korea

**Taqi Ahmad Cheema**
Department of Mechanical Engineering, GIK Institute of Engineering Sciences and Technology, Topi, Pakistan

**Michael Kopp and Vladimir Yanovsky**
National Academy of Science Ukraine, Institute for Single Crystals, Kharkiv University, Kharkov, Ukraine

**Anatoly Tur**
Institut de Recherche en Astrophysique et Planétologie, CNRS, Université de Toulouse [UPS], Toulouse, France

**Junzhe Ma, Yu Wei, Jianan Zhang, Hongjian Liu and Xianjun Meng**
Enhanced Oil and Gas Recovery Key Laboratory of Ministry of Education, Northeast Petroleum University, Daqing, China